WINTER MOUNTAIN LEADER MANUAL

U.S. MARINE CORPS

Mountain Warfare Training Center

Fredonia Books
Amsterdam, The Netherlands

Winter Mountain Leader Manual

by
United States Marine Corps
Mountain Warfare Training Center

ISBN: 1-4101-0885-6

Copyright © 2005 by Fredonia Books

Reprinted from the 2003 edition

Fredonia Books
Amsterdam, The Netherlands
http://www.fredoniabooks.com

All rights reserved, including the right to reproduce
this book, or portions thereof, in any form.

WINTER UNIT OPERATIONS (WMO)
WINTER MOUNTAIN LEADERS (WML)

INSTRUCTOR MANUAL
WINTER 2002-2003

TABLE OF CONTENTS

CHAPTER	CLASS TITLE
1	MOUNTAIN SAFETY (WINTER)
2	COLD WEATHER CLOTHING AND PERSONAL EQUIPMENT
3	COLD WEATHER MOUNTAIN LEADERSHIP CHALLENGES
4	MOUNTAIN HEALTH AWARENESS
5	WINTER WARFIGHTING LOAD REQUIREMENTS
6	MILITARY SKI EQUIPMENT
7	MOUNTAIN WEATHER
8	MOUNTAIN CASUALTY EVACUATIONS (WINTER)
9	AVALANCHE
10	SNOW STABILITY EVALUATION
11	AVALANCHE SEARCH ORGANIZATION
12	AVALANCHE TRANSCEIVERS
13	MILITARY SNOWSHOES
14	MILITARY SKI MOVEMENT
15	SKIJORING
16	MARINE CORPS COLD WEATHER INFANTRY KIT
17	TEN MAN TENT & YUKON STOVE
18	SPACE HEATER ARCTIC
19	BIVOUAC ROUTINE
20	SLED MOVEMENT
21	LIGHT AND NOISE DISCIPLINE IN A WINTER ENVIRONMENT
22	CAMOUFLAGE, COVER, AND CONCEALMENT
23	DEFENSIVE POSITIONS AND FIELD FORTIFICATIONS
24	ROUTE PLANNING IN COLD WEATHER OPERATIONS
25	COLD WEATHER PATROLLING
26	ICE RECONNAISSANCE AND BREACHING
27	COLD WEATHER NAVIGATION
28	EFFECTS OF COLD WEATHER ON INFANTRY WEAPONS AND OPTICS
29	COMMUNICATIONS CONSIDERATIONS IN A COLD WEATHER ENVIRONMENT
30	WINTER TRACKING
31	MOUNTAIN LOGISTICAL CONSIDERATIONS
32	COLD WEATHER AND MOUNTAIN HELO OPERATIONS
33	COLD WEATHER CONSIDERATIONS FOR NBC DEFENSE
34	FIRE SUPPORT IN A COLD WEATHER ENVIRONMENT

35	COLD WEATHER AND MOUNTAIN OPERATIONAL PLANNING
36	PLANNING COLD WEATHER OFFENSIVE OPERATIONS
37	PLANNING COLD WEATHER DEFENSIVE OPERATIONS
38	COLD WEATHER CONSIDERATIONS FOR THE 12 PATROL STEPS
39	REQUIREMENTS FOR SURVIVAL
40	EXPEDIENT SHELTERS AND FIRES
41	SURVIVAL NAVIGATION
42	SURVIVAL SIGNALLING AND RECOVERY
43	WATER PROCUREMENT
44	EXPEDIENT SNOWSHOES
45	CROSSING OBSTACLES IN WINTER
46	BIBLIOGRAPHY OF REFERENCES

UNITED STATES MARINE CORPS
Mountain Warfare Training Center
Bridgeport, California 93517-5001

WML
WMO
04/02/02

LESSON PLAN

MOUNTAIN SAFETY

INTRODUCTION (5 Min)

1. **GAIN ATTENTION**. The key to mountain safety in a mountainous environment is proper prior planning. Adhering to certain basic principles and predetermined actions will allow an individual or unit to efficiently perform their duties with minimum discomfort and maximum safety.

2. **PURPOSE**. The purpose of this period of instruction is to familiarize the student with the twelve mountain safety considerations and the acronym used to remember them. This lesson relates to all training conducted in a mountainous environment.

3. **INTRODUCE LEARNING OBJECTIVES**

    TERMINAL LEARNING OBJECTIVE.

    In a mountainous environment, execute preventive measures for mountain injuries, in accordance with the references.

    ENABLING LEARNING OBJECTIVE.

    a. Without the aid of references and given the acronym "BE SAFE MARINE", list in writing the 12 principles of mountain safety, in accordance with the references.

    b. In a mountainous environment, and given the necessary equipment, apply mountain safety to prevent injury and ensure mission readiness.

4. **METHOD/MEDIA**. The material in this lesson will be presented by lecture and demonstration. You will practice what you learn during upcoming field training exercises. Those of you who have IRF's please fill them out at the conclusion of this period of instruction.

5. **EVALUATION**. You will be tested later in the course by written and performance evaluations on this period of instruction.

TRANSITION: Having discussed our purposes, let's now look at the planning and preparation of a military operation.

BODY (50 Min)

1. (40 Min) **PLANNING AND PREPARATION**. As in any military operation, planning and preparation constitute the keys to success. The following principles will help the leader conduct a safe and efficient operation in any type of mountainous environment. We find this principle in the acronym "BE SAFE MARINE". Remember the key, think about what each letter means and apply this key in any type of environment.

　　B - Be aware of the group's ability.

　　E - Evaluate terrain and weather constantly.

　　S - Stay as a group.

　　A - Appreciate time requirements

　　F - Find shelter during storms, if required.

　　E - Eat plenty and drink lots of liquids.

　　M - Maintain proper clothing and equipment.

　　A - Ask locals about conditions.

　　R - Remember to keep calm and think.

　　I - Insist on emergency rations and kits.

　　N - Never forget accident procedures.

　　E - Energy is saved when warm and dry.

　　a. Be Aware of the Group's Ability. It is essential that the leader evaluates the individual abilities of his men and uses this as the basis for his planning. His evaluation should include the following:

(1) <u>Physical Conditioning</u>. Physical Fitness is the foundation for all strenuous activities of mountaineering. Leaders must be aware of their unit's state of fitness and take in account for the changes in altitude, climate, and amount of time for acclimatization.

(2) <u>Mental Attitude</u>. Units need to be positive, realistic, and honest with themselves. There needs to be a equal balance here. A "can do" attitude may turn into dangerous overconfidence if it isn't tempered with a realistic appraisal of the situation and ourselves.

(3) <u>Technical Skills</u>. The ability to conduct a vertical assault, construct rope installations, maneuver over snow covered terrain, conduct avalanche search and rescue operations, etc. The more a unit applies these skills increases their ability to operate in a mountainous environment effectively.

(4) <u>Individual skills</u>. At this point, you must choose who is most proficient at the individual skills that will be required for your mission, navigation techniques, security, call for fire, track plans, bivouac site selection, skijoring, etc.

b. <u>Evaluate Terrain and Weather Constantly</u>

(1) <u>Terrain</u>. During the planning stages of a mission, the leader must absorb as much information as possible on the surrounding terrain and key terrain features involved in his area of operation. Considerations to any obstacles must be clearly planned for. Will you need such things as ropes, crampons, climbing gear, skins, etc.

(a) Stress careful movement in particularly dangerous areas, such as loose rock and avalanche prone slopes.

(b) Always know your position. Knowing where you are on your planned route is important.

(2) <u>Weather</u>. Mountain weather can be severe and variable. Drastic weather changes can occur in the space of a few hours with the onset of violent storms, reduced visibility, and extreme changes. In addition to obtaining current weather data, the leader must plan for the unexpected "worst case". During an operation he must diagnose weather signs continually to be able to foresee possible weather changes.

(a) Constantly evaluate the conditions. Under certain conditions it may be advisable to reevaluate your capabilities. Pushing ahead with a closed mind could spell disaster for the mission and the unit.

(b) When in a lightning storm, turn off all radios, stage radios and weapons away from personnel. Have personnel separate in a preferably low-lying area, or around tall natural objects, however personnel should not come into direct contact with trees.

1. To calculate the approximate distance in miles from a flash of lightning, count in seconds the time from when you see the flash to when you hear the thunder, and divide by five.

c. <u>Stay as a Group</u>. Individuals acting on their own are at a great disadvantage in this environment.

   (1) Give adequate rest halts based upon the terrain and elevation, physical condition of the unit, amount of combat load and mission requirements.

   (2) Remember to use the buddy system in your group.

   (3) Maintain a steady pace that will allow accomplishment of the mission as all members of the unit reach the objective area.

d. <u>Appreciate Time Requirements</u>. Efficient use of available time is vital. The leader must make an accurate estimate of the time required for his operation based on terrain, weather, unit size, abilities, and the enemy situation. This estimate should also take into account the possibility of unexpected emergencies such as injuries or unplanned bivouacs due to severe conditions.

   (1) Time-Distance Formula (TDF). This formula is designed to be a guideline and should not be considered as the exact amount of time required for your movement. The TDF is made for acclimated troops on foot in the summertime and/or on skis with skins or snowshoes in the wintertime. The TDF will vary based on unit size, physical conditioning, experience, load carried, angle of slope, snow conditions, surface conditions, etc. A set of TDFs are used for planning as follows:

      (a) Mountain Leaders/Patrols with combat load and high level of mountain/winter experience - 3kph + 1 hour for every 300 meters ascent + 1 hour for every 800 meters descent.

      (b) Assault Climbers/Scout Skiers/Patrols with combat load – 2kph + 1 hour for every 300 meters ascent + 1 hour for every 800 meters descent.

      (c) Company/Bn movements with combat load – 1kph + 1 hour for every 300 meters ascent + 1 hour for every 800 meters descent.

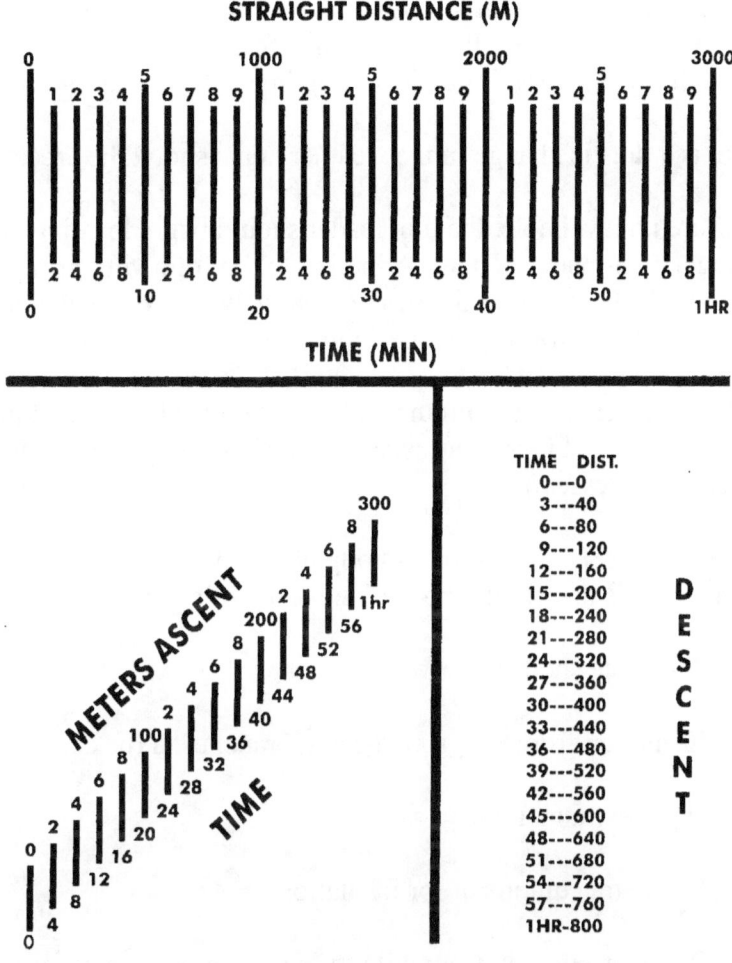

3KPH TDF

(2) Route Cards. Route cards are not to be used in place of an overlay, but as a tool to be used in route planning. They force a detailed map study. The route card is also an effective tool to use during storms/reduced visibility for dead reckoning navigation to the objective or for back azimuth escape. Detailed instruction on using route cards is in COLD WEATHER NAVIGATION CONSIDERATIONS.

(3) Route Planning. As in any military operation, route planning and execution are of vital importance. Prior to departure, the unit commander must submit a route card and/or route overlay to his higher headquarters and keep a duplicate copy for himself. This preplanned route should be followed as closely as possible, taking into account changes based on the tactical situation.

TRANSITION: Having planned our route and estimated the time it will take to move, let's examine our actions during movement.

e. <u>Find shelter during storms, if required</u>. Under certain conditions, inclement weather can provide tactical advantages to the thinking unit commander, but by the same token it can reduce the efficiency of a unit to nil if an incorrect evaluation of the situation is made. Being lost will not directly kill an individual. Starvation takes time, but hypothermia can manifest itself in a matter of hours resulting in death.

(1) If there is a drastic change in the weather, tents should be erected immediately.

(2) If tents are not available, the unit leader should begin locating natural shelter or began building a man-made shelter. Adhering to the following principles will give an individual the best chance to spend a relatively safe bivouac with the prospect of continued effort toward mission accomplishment.

    a. Make shelter. The requirements for expedient shelters and the building procedures will be covered in another chapter. The basic requirement for protection from the elements is essential.

    b. Keep warm. The retention of body heat is of vital importance; any action in which body heat is lost should be avoided. The following points should be considered:

        1. Adequate shelter

        2. Insulation from the ground using branches, a rucksack, etc.

        3. Wear extra clothing.

        4. Use extra equipment for insulation.

        5. Produce external heat while trying to conserve fuel for future use.

    c. Keep dry. Being wet causes the loss of body heat 32 times faster than when dry. Adequate protection from the elements is of prime importance to prevent the onset of hypothermia.

f. <u>Eat Properly and Drink Plenty of Fluids</u>

(1) <u>Food</u>. The human body can be compared to a furnace, which runs on food to produce energy (warmth). By planning the consumption of food to suit the specific situation, adequate nutrition and extra warmth can be supplied.

(2) <u>Water</u>. The intake of adequate amounts of water will maintain the body in proper working order. Danger from dehydration is as high in mountain regions as in hot dry areas. Loss of liquids is easily seen and felt in hot climates; whereas in the mountains, the loss of body fluids is much less noticeable. High water intake, at least 4 to 6 quarts per day when in bivouac, 6 to 8 quarts per day when active, will help to prevent dehydration.

g. <u>Maintain Proper Clothing and Equipment</u>

   (1) Clothing will be covered in detail in the chapter COLD WEATHER CLOTHING AND PERSONAL EQUIPMENT.

   (2) Equipment. In the mountains a man should never be separated from his gear. Here are some basic and essential items that should be considered during your planning stage.

   (a) Assault load

   1. There should be at least one combat load per squad for safety.

   (b) Combat load

   (c) Existence load.

   (d) Map and compass. Every individual in a leadership position and his assistant should carry a map and compass. The maps should be weatherproofed and extra maps should be distributed throughout the unit.

   (e) Repair kit. This kit should include those items necessary to do emergency repairs on your equipment.

   (f) Survival Kit. Always carried on your person. The contents of a survival kit will be covered in another chapter.

h. <u>Ask Locals About Conditions</u>. An often-overlooked source of information is the indigenous population of an area. Local weather patterns, rock slide/avalanche areas, watering points, and normal routes can all be obtained by careful questioning. The unit leader must obtain current information of the actual conditions along his intended route.

i. <u>Remember to Keep Calm and Think</u>

   (1) Having recognized that an emergency situation exists, the following principles should be followed:

   (a) Keep calm and do not panic. At this point you must make every effort to conserve body heat and energy.

   (b) Think. When an individual is cold, tired, hungry, or frightened he must force himself to organize his thoughts into a logical sequence.

   (c) The group must try to help itself by either finding the way back to safety or by preparing shelters and procuring food.

(d) Above all else, the group must act as a tight-knit unit. In emergency situations, individual dissension can cause a total loss of control and unit strength.

(2) If the decision is reached that the group should seek its way back to safety, several possibilities exist. In most situations, the safest approach will be to retrace the route to the last known point and continue from there. The other course of action is to get a group consensus on the present location and send out a small search party to locate a known point. This party must ensure that they mark their trail adequately to return to the group. If all attempts at finding a way back to known terrain fail, a definite emergency situation exists and actions discussed later in this section must be instituted.

j. <u>Insist on Emergency Rations and Kits</u>. Just like what was covered in the WINTER WARFIGHTING LOAD REQUIREMENTS chapter, emergency rations and a survival kit should always be carried.

k. <u>Never Forget Accident Procedures</u>

(1) Causes of accidents. The general procedures used to handle accidents differ little in this environment, but several distinct points should be kept in mind. The most frequent causes of accidents are as follows:

(a) Overestimation of physical and technical abilities.

(b) Carelessness.

(c) General lack of observation of one's surroundings.

(d) Lack of knowledge and experience by leaders.

(e) The failure to act as a group.

(f) Underestimation of time requirements to move through mountainous terrain and underestimation of the terrain itself.

(2) Preventive measures. The only truly effective preventive measures for the above lie in the education and experience of leaders at all levels. Too often, leaders sit by watching, rather than participating, during training and as a result have no concept of the requirements involved in the mountainous environment. Only by active involvement can a leader gain the knowledge and experience needed to effectively lead in this environment.

(3) General procedures for handling an accident. These require only a good dose of common sense as outlined below.

(a) Perform basic first aid.

(b) Protect the patient from the elements to include insulation on top and bottom.

(c) Evacuate if necessary.

(d) Send for help if required.

(e) If possible, never send a man for help alone.

(f) Send the following information regarding the accident:

1. Time of accident.

2. Nature and location of accident.

3. Number injured.

4. Best approach route to accident scene.

(4) If one man of a two-man team is injured, the injured man must be given all available aid prior to going for help. If the injured man is unconscious, he should be placed in all available clothing and sleeping gear and anchored if on steep terrain. A note explaining the circumstances, and reassuring him, should be left in a conspicuous spot. This note must also contain the following information:

(a) When you expect to return.

(b) Where you went.

(c) What you did before you left (medication, etc.).

(5) International distress signal.

(a) Six short blasts in 1 minute from person requesting help.

(b) The return signal is three blasts in 1 minute from the respondent.

(6) Other methods of signaling:

(a) Red pyrotechnics.

(b) SOS, (...---...)

(c) "Mayday" by voice communications.

l. <u>Energy is Saved When Warm and Dry</u>. With the previous 11 principles in mind this one should fall right into place. Save your heat and energy by following these steps:

(1) Dress properly.

(2) Eat properly.

(3) Drink properly.

(4) Ensure shelter meets criteria.

(5) Produce external heat (fires, stove, extra clothing, etc.) to save body heat and energy for future use.

(6) Avoid getting wet, this increases body heat loss.

TRANSITION: By applying these 12 principles, you increase your chance for survival.

PRACTICE (CONC)

    a. Students will apply mountain safety principles throughout the duration of the course.

PROVIDE HELP (CONC)

    a. The instructors will assist the students when necessary.

OPPORTUNITY FOR QUESTIONS (3 Min)

1. QUESTIONS FROM THE CLASS

2. QUESTIONS TO THE CLASS

    Q. What is the "B" in Be Safe Marine?

    A. Be aware of the group's ability.

    Q. What is the "F" in Be Safe Marine?

    A. Find shelter before storms, it required.

SUMMARY (2 Min)

    a. Safety in the mountains doesn't come naturally, it must be practiced or the results can be devastating. Constant observation and common sense are the keys to success and safety.

    b. Those of you with IRF's please fill them out at this time; we will take a short break before the next class.

UNITED STATES MARINE CORPS
Mountain Warfare Training Center
Bridgeport, California 93517-5001

WML
WMO
08/15/02

LESSON PLAN

COLD WEATHER CLOTHING AND PERSONAL EQUIPMENT

INTRODUCTION (5 Min)

1. **GAIN ATTENTION**. This environment could easily kill the unprepared, ill-equipped Marine. Even the most up-to-date equipment, if used improperly, would only prolong the end. This class will teach you the proper use and care of your gear so that you can fight effectively in a cold environment. As Marines, we are going to look for the easiest way to anything, or get through any situation as comfortably as possible. In the mountains, that would mean walking away with all of your fingers and toes, and sign up for the next ski tour when you get back south. If all the points of this class are taken to heart, you will not only be able to do this, but also defeat your most dangerous and ever-present enemy-"the cold".

2. **PURPOSE**. The purpose of this period of instruction is to familiarize the students with the cold weather clothing system and gear used in the Marine Corps. Also, we will discuss cold weather dressing theory, layers, fabrics, and the individual components of the ECWCS. This lesson relates to Warfighting Load Requirements.

3. **INTRODUCE LEARNING OBJECTIVES**

   a. TERMINAL LEARNING OBJECTIVE. Given cold weather clothing in a cold weather and/or mountainous environment, wear and maintain cold weather clothing, in accordance with the references.

   b. ENABLING LEARNING OBJECTIVES

   (1) Without the aid of references, state in writing the types of cold weather environments, in accordance with the references.

   (2) Without the aid of references, state in writing the meaning of each letter of the acronym "COLD", in accordance with the references.

(3) Without the aid of references, state in writing the three principles of design in the cold weather clothing system, in accordance with the references.

4. **METHOD/MEDIA**. The material in this lesson will be presented by lecture and demonstration. You will practice what you have learned during upcoming field training exercises. Those of you with IRF's please fill them out at the conclusion of this period of instruction.

5. **EVALUATION**. You will be tested later in the course by written and performance evaluations.

TRANSITION: Now that we know what we want to accomplish, let's discuss the most basic factor, cold weather itself.

BODY (50 Min)

1. (5 Min) **COLD WEATHER ENVIRONMENTS**. Cold weather can be characterized as: wet cold, dry cold, intense cold, and extreme cold. Weather often will move across 2 of these adjacent boundaries, requiring a unit to be prepared for the next lower category due to weather changes.

    a. Wet Cold. Wet cold conditions occur when temperatures are near freezing and variations in the day and night temperatures cause alternate freezing and thawing. The temperature ranges from 20 to 40 degrees F, not including wind chill. These conditions are often accompanied by wet snow (you can make a snow ball) and rain causing the ground to become mushy and muddy. With these conditions, Marines require clothing that consists of a waterproof, wind resistant outer layer, and an inner layer with sufficient insulation to provide protection down to 20 degrees F.

    b. Dry Cold. Moderate cold conditions occur when average temperatures are lower than 20°F, the ground is usually frozen, and the snow is dry (you can not make a snow ball). Temperatures range from –5 to 20 degrees F for Moderate cold, not including wind chill. Insulating layers must protect to –5 degrees F, these layers must be protected by a water resistant, wind resistant outer layer.

    c. Intense Cold. Intense cold temperatures range from –5 to –25 degrees F, not including wind chill. Substantial insulating layers are required. All tasks and movement is severely slowed down. Extreme care must be taken to avoid environmental casualties. This is the operational limit for offensive operations and the milspec design limits of weapons and equipment in the US inventory.

    d. Extreme Cold. Extreme cold temperatures are –26 degrees F and below, not including wind chill. Defensive operations only, with a frequent (15 – 30 minute) rotation of personnel manning defensive positions into warming shelters. Expect substantial weapons and equipment failures.

TRANSITION: Now that we know what we are dealing with, let's discuss some accepted theories and facts to help combat the cold.

2. (5 Min) **PRINCIPLES OF USE**. To help us remember come of the basic principles of wearing our clothing; we use the acronym "COLD".

    a. C - Keep Clothing CLEAN. Clothing keeps you warm by trapping warm air against your body and in the pores of the clothing itself. However, if these pores are filled with dirt, sweat, or other grime, it will not be able to do its job as efficiently.

    b. O - Avoid OVERHEATING. Allowing just enough clothes and body activity to keep you cool, and the environment to cool you down, will keep your clothes from getting sweaty and dirty, and therefore more effective. Overheating can also cause you problems, not with just your clothes. Several cold weather injuries can be caused by overheating such as dehydration, heat exhaustion, and hypothermia.

    c. L - LOOSE and LAYERED

        (1) Loose Clothing. You want to keep your clothes loose for comfort. If clothing is too tight:

            (a) It may act somewhat like a tourniquet, causing blood in your extremities, i.e., arms, legs, fingers, toes, etc., to pool there, not allowing it to get back into your body core and rewarm, thereby causing that limb to get cold.

            (b) Little or no air can be trapped between your body and clothes. It's this warm air that keeps you warm, not the clothes.

        (2) Layering. This is another important principle for staying warm in the cold. The more layers, the more warm air that is going to be trapped. Strangely enough, several thin layers working together will work better than one thick layer working alone.

    d. D - Keep Clothing DRY. This means not only from the outside, such as putting on rain gear during sleet or when walking through wet snow, but also from the inside, such as taking a layer off when you start sweating. Once your clothes are wet, the water or sweat starts to evaporate, drawing off warmth with it.

TRANSITION: Let's now take a look at the principles of design.

3. (5 Min) **PRINCIPLES OF DESIGN**. The principles of designing the military cold weather clothing systems are the Vapor Transmission Layer, Insulating Layer, and the Protective Layer.

    a. Vapor Transmission Layer. Better called a "sweat transfer layer'", this is a hydrophobic layer that does not absorb water (your perspiration) and draws it away from your skin so you will remain dry and warm. Significant progress has been made with synthetics, such as polypropylene, which draws water away from the body, but stays dry.

   b. <u>Insulating Layer</u>. This can be one layer or several layers, which hold the warm air around your body. Preferably, it is light-weight, very compressible, and fast-drying.

   c. <u>Protective Layer</u>. This protects the insulating layer(s) front getting wet or dirty. It should be made of a wind proof/water proof substance.

NOTE: These are the three main layers to consider in the military clothing system. There may be times when one or more layers are not used, or when the insulating layer may be several layers thick.

<u>TRANSITION</u>: Next, we'll discuss the Extended Cold Weather Clothing Systems.

4. (35 Min) **<u>THE EXTENDED COLD WEATHER CLOTHING SYSTEM (ECWCS)</u>**.

   a. <u>General</u>. There are three clothing systems currently in use in the Marine Corps; the current Extended Cold Weather Clothing System (ECWCS), the old M-1950 Cold Wet/Cold Dry seven layer system and the Specialty Clothing System developed to fit the specific needs of aviation and maintenance personnel (reference NAVAIR Manual 1316.7 Chapter 5). During this period of instruction, we will only discuss the ECWC System since the M-1950 seven-layer system is being phased out.

      (1) <u>Extended Cold Weather Clothing System (ECWCS)</u>. The ECWCS was developed to provide a light weight, compressible, fast-drying clothing system that is better suited to the modern cold weather battlefield. This system uses modern synthetic materials, which uses moisture management principles to transfer perspiration away from the skin so that the user will remain warm and dry. ECWCS is a layered insulating system adjustable to personal preference, metabolism, and prevailing weather conditions. It is designed to maintain adequate environmental protection between +40°F and -25°F. The Extreme Cold Weather Boots protect down to -50°F.

   b. <u>ECWCS Clothing Items</u>. Described below are the component items of the ECWC System, which combines many new items with a few items that have remained from the older M-1950 clothing system.

      (1) <u>Vapor Transmission Layer</u>. There are both light and heavy weight long underwear sets, each consists of a long sleeve undershirt and drawers.

         (a) <u>Description of Item</u>. The heavy weight type is tan polypropylene. The undershirt is a turtleneck that has a center front zipper, which extends to the middle of the chest area. The polypropylene drawers have elastic cuffs and waist. The light weight type is green capilene and of similar design as the heavier version.

(b) <u>Concept of Use</u>. The underwear layer next to the skin acts as a vapor transmission layer. This draws moisture away from the skin while transferring it to the outer layers of the clothing system. **Do not wear cotton undershirts and underwear when using this clothing**. The wearing of issued cotton undergarments will negate the wicking action of the polypropylene by causing moisture to be trapped against the skin.

> The heavy, tan polypro is for static work use.
> The light-weight, green capilene is for movement and physically active work.

(2) <u>Insulating Layer</u>. 300 wt fleece jacket and bibs have replaced the fiberpile (buffalo) ones. Fleece is 100% polyester that is lighter, less bulky, and more comfortable than fiberpile. There is also a 100 wt anorak style green fleece shirt as general issue, which is a great light weight insulating layer.

(a) <u>Description of Item</u>. The 300 wt jacket is black, has reinforced shoulders, upper back, upper chests and elbow patches, a convertible turtleneck collar, front zipper, elastic shock cord waist, Velcro fastened wrist straps, two mesh pit type pockets inside the chest area, and two hand-warmer pockets with zippers. The 300 wt bib is black, has adjustable elastic suspenders with quick release buckles located in the front, front zipper, and full-length zippers at the outside seams. The 100 wt shirt is green, has a front zipper to mid-chest, turtleneck, and a hand-warmer tube pocket.

(b) <u>Concept of Use</u>. The polyester fleece layer serves as the primary insulating layer. Recommended to be worn by Marines only when stationary or during periods of intense/extreme cold.

(3) <u>Protective Layer</u>. The Gore-Tex camouflage parka and trouser make up the ECWS protective layer.

(a) <u>Description of Item</u>. The parka has an integral hood, (the newest generation of parkas have a pocket in the collar for stowage of the hood), two inside breast pockets which can be accessed without unzipping the parka, two large cargo pockets, upper arm pockets, and a two-way, full-front zipper to provide full face protection, leaving only the eyes uncovered. There is an elastic drawcord at the hem and waist, Velcro wrist tabs, underarm ventilation with zippers and a rank tab at center chest. The trousers have seat and knee reinforced patches, pass through pockets, and inserts in the seams of the leg openings to allow easy donning and doffing without removing the boots.

(b) <u>Concept of Use</u>. The parka and trousers serve as the windproof and waterproof outer protective layer. The polytetrafluoroethylene (PTFE) laminate in the garment has the property to block water while allowing perspiration (water vapor) to be expelled.

(4) Snow Camouflage Parka and Trousers. The snow camouflage parka and trousers (overwhites) are standard carry over items from the M-1950 issue.

    (a) Description of Item. The hooded white parka has drawstrings for adjustment at the waist, pass through side and hip pockets, knee pleats and drawstrings at the ankles of the trouser. There are old cotton and new nylon overwhites in the system. The nylon overwhites do not hold water/freeze solid, are much lighter and more compact than the cotton. Cotton overwhites should not be issued for combat operations.

    (b) Concept of Use. The overwhite parka and trousers are used as a camouflage outer layer in snow covered terrain and is not a substitute for an outer garment. They are worn over all other clothing, if terrain dictates their use.

(5) Balaclava. The head wear in the ECWC System.

    (a) Description of Item. The hood consists of wool, knitted cap, which covers the entire neck and face with holes for the eyes and nose. It is a pullover ski mask style that comes in either green or black.

    (b) Concept of Use. The cap is intended to provide protection in cold weather to the neck, head, and face.

(6) Hand Wear. The CTAP cold weather issue hand wear items are: Two pair of Contact Gloves, Gore-Tex Gloves w/liners, Gore-Tex Mittens w/liners.

    (a) Description of Item. The Contact gloves are green Nomex with goat skin palms. The Gore-Tex gloves are black nylon with a rubber palm and an insulating liner that can be removed to speed drying. The Gore-Tex mittens are black nylon with rubber palms, trigger finger, and an insulating liner that can be removed to speed drying.

    (b) Concept of Use. Contact gloves are used to handle cold metal objects, such as weapons or hand tools. They are NOT heavy duty work gloves. They can be used as a light weight liner for either the Gore-Tex glove or mitten SHELLS. Contact gloves are NOT to be worn inside either the glove or mitten when the liners are in them (this could lead to frostbite). The Gore-Tex gloves can be used alone, or with their liners, or with the contact gloves. The Gore-Tex mittens can be used alone, or with the contact gloves, or with their liners. They combine as one large modular system with redundancy in case one becomes wet or unserviceable.

c. Accessory Items. The items in this section are considered part of the ECWCS issue.

(1) Suspenders. The suspenders (Suspenders, Trousers M-1950) are a carry over item and are used with the field trousers, Gore-Tex trousers, and overwhite trousers.

(a) <u>Description of Item</u>. The olive drab suspender straps are scissor-back style (cross over in the back). The suspenders have two slide buckles and two hooks, which attach to the trousers.

(b) <u>Concept of Use</u>. The suspenders are to be used instead of a belt to allow body heat to rise to through the torso and they help in holding multiple trousers up at the same time.

d. <u>Care, Use, and Maintenance of ECWCS</u>

(1) <u>General</u>. The individual Marine is responsible for keeping his ECWCS items in good serviceable condition. The ECWCS will protect him only if he takes care of it and wears it properly. Check the label to see if the size is correct. This is extremely important in order to achieve maximum user satisfaction using the layering principle.

(2) <u>Donning and Doffing Procedures</u>. The ECWCS is an insulating system consisting of the following five primary layers (including the overwhites, when necessary) and accessories:

(a) <u>Layer 1</u> – Polypropylene OR Capilene long underwear tops and bottoms. This is the next-to-skin layer.

(b) <u>Layer 2</u> – Fleece shirt, jacket, and bibs.

(c) <u>Layer 3</u> – Camouflage Utility Uniform.

(d) <u>Layer 4</u> – Gore-Tex camouflage parka and trousers.

(e) <u>Layer 5</u> - Snow camouflage parka and trousers (overwhites).

(f) <u>Accessories</u> - Handwear, footwear and headwear.

(3) <u>Layering</u>. Unit leaders dictate the outer camouflage layer. Inner layers are adjusted according to preference, metabolism, and weather conditions in order to avoid overheating when on the move or cold weather injuries when stopped. The polypropylene and capilene underwear have the ability to draw moisture away from the skin and transfer it to the outer layers of the system. Always wear one of these vapor transmission layers if wearing an insulating layer. Doff insulating layers during movement to keep them dry, don the dry insulating layer upon halting. Add or take off insulating layers as needed to avoid sweating and/or chills. Overwhites are not a protective outer garment, they are worn only as a camouflage outer layer.

(4) <u>Inspection</u>. Examine the ECWCS items regularly for tears, punctures, bubbling (delaminating), or damage to the material. Punctures on the outer layer will produce leaks and eventually ruin the material if not properly maintained. Repairs should be made as soon as possible.

(5) <u>Attaching Rank Insignias</u>. Attach rank insignia on the parka to the rank tab, which is provided at the center of the chest. Either the pin on or sewn on rank insignia may be used. Be careful not to puncture or snag the outer layer of the jacket itself when attaching rank as punctures will produce leaks.

(6) <u>Cleaning</u>. Clean ECWCS clothing items regularly when in use. Dirty clothes wear out quicker because dirt cuts textile fibers, mats down insulating fibers so its colder, gets saturated due to inhibiting of water repellency, and retains moisture on the inside from perspiration due to inhibited breathability (vapor transfer). Prior to laundering and drying, make sure all the drawcords are tied together, all zippers are zipped, and all snaps and hooks are fastened. Securing these items will result in a better laundered garment.

    (a) Vapor transmission and insulating layers: When laundering, use delicate or gentle fabric wash cycle or by hand, using cold water (up to 85°F/29°C) and cold water laundry detergent. Rinse in clean cold water. <u>DO NOT USE BLEACH OR STARCH</u>. Tumble dry at the lowest fabric cycle, delicate/gentle, do not exceed 90°F/32°C. Remove immediately at the end of drying. <u>AVOID OVER DRYING</u>. To drip dry, remove water and place on a rustproof hanger. <u>DO NOT PRESS</u>.

    (b) Gore-Tex protective layer: Machine wash warm with <u>powdered detergent</u> on regular cycle. <u>Do not use bleach, fabric softener, liquid detergents, or starch</u>. Tumble dry on <u>high heat</u> to help restore water repellency of DWR finish. Steaming Gore-Tex with an iron ½ inch above the garment will also help restore water repellency, but <u>do not press</u>. When the water repellent finish is completely worn off (water will not bead after washing and drying), there are various brands of water repellent finishes on the market that are approved by the manufacturer for Gore-Tex, such as Rain-X. These products are not durable like the one put on during manufacturing, so repeat applications of these products will be necessary periodically, until you can survey the garment.

e. <u>Footwear</u>. USMC footwear consists of: Vapor Barrier boots, Ski/March boots, Intermediate Cold-Wet boots. These are all focused on cold weather, not mountainous terrain. A mountaineering/cold weather boot is in development to address this issue.

(1) <u>Vapor Barrier (VB) Boot</u>. These boots use an inner and an outer boot made of rubber and filled with either wool fleece, felt, or closed cell foam (neoprene) insulation. The rubber acts to stop the movement of moisture from the feet. Heat is transferred quickly by the moisture in the air. By trapping the moisture, the boots trap heat. The boots also act to keep the moisture out. Whenever possible, the VB boots should be removed in order to air dry feet throughout the day. Change sox frequently, switching between 2 pair in order to dry out the sox.

VB BOOTS

(a) <u>VB Boots</u>. There are two types of VB boots:

  1. <u>Boots, Cold Weather (Type 1, Black)</u>. These boots are worn in the cold wet environment and will protect the feet down to -20°F.

  2. <u>Boots, Extreme Cold Weather (Type 2, White)</u>. These boots are worn in the cold dry environment and will protect the feet down to -50°F.

Note: The valve on the side of the boot should always be closed to keep moisture out, except during air transport in order to equalize pressure in the insulating cells.

(2) <u>Ski March Boot</u>. These boots are all leather and can either be single or double in construction. It has a box-toe and a grooved heel that is designed to work with our standard issue ski and the NATO 120 binding. The boot has a full leather tongue and collar with Thinsulate and Evapor insulation.

  (a) <u>Felt Liners</u>. The felt liners are designed to add insulation, absorb moisture that would otherwise be absorbed by the boot itself, and to form to the foot for a more comfortable fit. The boot is provided with two liners to allow a wear/dry rotation. Sleeping with one liner next to your body will both dry it and warm it prior to putting the boot on. Take the liners out of the boots at night to speed drying of the boots and liners.

  (b) <u>Sizing</u>. Proper sizing of the boot is perhaps the most important thing. If done incorrectly, the Marine may suffer from blisters or frostbite. Sizing must be done with the felt liner in the boots and wearing the complete sock system. The boot should fit snugly in the heel area to avoid blisters, but not so tight that it cuts off circulation. The toes should have some movement, but avoid side slippage, which not only causes blisters but also reduces control while skiing.

(c) <u>Waterproofing</u>. There are many products on the market that will work, but MWTC we uses Snow Seal. Instructions for use are on the package.

(d) <u>Care of the Ski March Boot</u>. Caring for your ski/march boot is much the same as any other leather boot. Dry your boot whenever possible, being careful not to use open flames or any method that will dry the boot to quickly. Strive to keep the boot dry to prevent freezing. Using foot powder to absorb excess moisture is okay as long as it is kept to a minimum, in order to avoid a paste. DO NOT LEAVE FULL GAITORS ON THE BOOTS TO DRY, pull the toe of the gaitor off the toe box to avoid boot curl.

(3) <u>Intermediate Cold/Wet Boot (ICW)</u>. These are 10" high leather boots with Gore-Tex and Thinsulate linings. ICW boots have insoles that should be removed at night to dry in the sleeping bag. They are designed to be used in cold wet environments down to 20 degrees F. However, because they are Gore-Tex lined, they take a long time to dry if immersed. Size the boot while wearing the winter sock system (usually one size larger than the combat boot size).

(3) <u>Accessories Mountain/Cold Weather Boots</u>.

   (a) <u>Gaiters</u>. These leggings are worn in conjunction with the boots to provide protection from getting snow or debris into the boot itself. There are two types, three-quarter and full gaiters.

   1. <u>Three-Quarter Gaiters</u>. These gaiters can be worn with any boot; the VB boots, standard combat boots, ICW boots, the ski march boots, etc. It is a nylon Gore Tex legging with a front Velcro separation seam for donning and doffing. It also has an adjustable bottom strap, which is placed under the boot instep, and an adjustable top draw cord.

   2. <u>Full Gaiters</u>. These are also known as "Super" gaiters. It is a nylon Gore Tex legging that is fully insulated. NOT FOR USE WITH VB BOOTS. To place on the boot, feed the toe of the boot through the front hole on the bottom of the gaiter. Slide the heel of the boot through the rear hole in the rubber bottom. Pull the toe of the gaiter over the toe of the boot, ensuring that the rubber seal is snug against the welt of the boot. If the rubber seal will not stay in place along the toe of the boot, try a cord or strap (<u>do not secure permanently or boot will curl</u>).

   (b) <u>Overboots</u>. Overboots are worn over a leather boot, such as ski/march boots or combat boots. The overboots are fully insulated, waterproof, and have a hard sole for walking. They are for static work, not prolonged movement.

   (c) <u>Care of the Gaiters/Overboots</u>. Caring for the gaiters and the overboots is essentially the same. Keep as clean and dry as possible. Open the gaiter occasionally while wearing to allow condensation to evaporate. If the rubber parts start to dry out, coat them with a silicone spray.

(4) <u>Components of the Sock System</u>. The sock system currently is a three sock-layered system. All the layers can be worn together or just parts, personal preference.

    (a) The first layer is a thin, light, vapor transmission layer made of Transpor fiber. It should fit snugly. It is designed to wick moisture away from the foot and prevent blisters by reducing friction.

    (b) The intermediate layer of the system is a vapor barrier sock. This layer is only worn in extreme cold temperatures. Working on the same principle as the Vapor Barrier (VB) boot, it traps all of the heat created from the feet. The problem is that it also traps all of the moisture. This sock should be worn between the vapor transmission and the insulating layers. This keeps the foot warm and protects the insulating layer from perspiration. Never wear the VB sock over the insulating sock, as it will cause it to become saturated and lose its heat retention properties. Remember the "D" in COLD. Also, be careful of using the VB sock when it is warm as it can cause blisters due to excessive sweating.

    (c) The outside layer is a hook stitch pile fiber made of 50 % wool and 50% polypropylene. This combination provides the warmth needed for prolonged movements and still allows the moisture to pass through the sock.

    (d) <u>Use of the Sock System</u>. Three pairs of the vapor transmission and insulating socks are issued. This enables the wearer to continually rotate the socks, allowing the other pair to be dried in whatever method is available. Body heat works well. Only 1 pair of VB socks are issued.

    (e) <u>Care of the Sock System</u>. Care of the sock system falls into two areas, field care and garrison care.

        1. <u>Field care</u>. It is important to keep the socks as dry and clean as possible to prevent them from losing their specific properties. Crinkle and shake your socks to keep them as free as possible from dirt and body oils, which render them less effective. This is in keeping with the "C" in COLD.

        2. <u>Garrison care</u>. Wash the polypropylene inner socks in the same manner as the polypropylene long underwear. The wool socks should be washed in cold water with a mild detergent.

g. <u>Cold Weather Personal Equipment</u>. Personal equipment for use in cold weather environments is especially designed to provide protection and be as lightweight as possible.

  (1) <u>Sleeping System</u>. The sleeping system consists of two sleeping bags, a vapor-permeable bivy cover, a compression sack, and sleeping mat.

(a) The modular sleeping bag system:

1. <u>Sleeping Bag, Green Modular, Light Weight</u>. For temperatures down to +40°F, uses polyester batting for insulation and weighs 2.5 lbs.

2. <u>Sleeping Bag, Black Modular, Intermediate Cold</u>. For temperatures down to 0°F, uses polyester batting for insulation and weighs 5.5 lbs.

3. <u>Vapor-Permeable Bivy Cover</u>. This shell is made out of gore-tex and allows body vapor to escape while helping to protect the bag from condensation inside a tent, dripping water in a snow shelter, or from snowfall and wind when without a shelter. It also provides an additional 10°F for warmth when combined with either or both bags and weighs 2.25 lbs.

NOTE: The green modular sleeping bag with bivy cover is rated to +30°F. The intermediate cold sleeping bag with bivy cover is rated to -10°F. When both sleeping bags are combined with the bivy cover, they are rated to -30°F. Again these temperature ranges are dependent upon how many layers of the ECWCS are worn within the sleeping bag and it is possible to extend the range to -50°F with additional clothing.

(b) <u>Compression Sack</u>. This is used to store the sleeping bags when not in use. The four straps are designed to compress the bags so that it will take up less space in the pack.

(c) <u>Sleeping Mat</u>. It provides excellent insulation from ground cold and can be used for sitting, sentries, when consolidating following assaults, and in ambush positions when personnel must lie prone for long periods of time.

h. <u>Load Carrying Equipment</u>. Marines are now issued the LCS-88 pack for use in cold weather/mountainous environments. This pack has an internal frame, fully adjustable suspension, a map flap, side external ski tunnel pockets, an internal divider for the zipper opened sleeping bag compartment, a radio pocket, and numerous attachment points for ALICE equipment. Packing of this pack is covered in the Winter Warfighting Load Requirements chapter.

i. <u>Miscellaneous Gear</u>

(1) <u>Sunglasses</u>. Snow blindness is a very real problem in snow covered terrain. Snow blindness is painful, requires bed rest to treat, and a snow blind Marine is a liability to the unit. Always use the issue sunglasses or sunglasses with side shields designed to filter out ultraviolet rays. Even on overcast days, the possibility exists for snow blindness to develop. Leaders must ensure that Marines have their sunglasses and require their wear.

(2) <u>Canteens</u>. The plastic canteen will freeze very quickly in cold weather. If used, it must be carried in the interior of the clothing or deep in the pack wrapped in spare clothing.

The two-quart, collapsible canteen is very useful in cold weather operations, but it too must be carried next to the body.

j. <u>Tricks of the Trade</u>. The final subject we will talk about is some of the things that can be done to take care of yourself, and your gear.

   (1) <u>Ideas for Taking Care of Yourself</u>

   (a) <u>Cold Fingers</u>. When using gloves or even mittens, your fingers tend to get cold. Simply pull your fingers out of their compartments, keeping them in the gloves, and make a fist. Your warm palm surrounding the fingers will warm them up. You can also swing your arms in big circles to force warm blood down into your fingers.

   (b) <u>Sleeping Bag</u>. Along with the sleeping mat, branches or MRE sleeves will help insulate the bottom. For the top, any rubberized gear will hold the heat in the bag. However, in the morning, the surface condensation should be quickly dried off. A balaclava should be worn while sleeping. Also ensure that the flap is pulled over the zipper to prevent it from freezing. Wear the vapor transmission layer, at the minimum, to keep body oils from being absorbed into the sleeping bag. Put all clothing for the morning inside the bag (except Gore-Tex) if not being worn, so it will be warm when you wake up. The Gore-Tex can be placed under the sleeping bag/on the mat or between the bivy cover and sleeping on top to dry and act as insulation.

   (c) <u>Food Before Sleeping</u>. You heat your sleeping bag, it doesn't heat you. Therefore, just before going to sleep, eat something high in carbohydrates like noodles, cereal or crackers to give you fuel to burn during the night. A hot liquid will also do the job. Having a topped off hot thermos is nice if you wake up with a chill.

   (2) <u>Ideas for Your Clothes</u>

   (a) <u>Waterproofing</u>. As clothes get old and worn, they lose their water repellency. The water repellency can be restored on the ECWCS as discussed earlier. Everything in the pack needs to be in waterproof bags if it can absorb water.

   (b) <u>Storing Clothes</u>. Although you don't want to sleep with your clothes on, you also don't want to leave them out to get cold. Therefore, you can put them inside your sleeping bag along the sides. Boots can be placed inside a waterproof bag.

   (c) <u>Canteens</u>. Keep it 2/3 full to prevent freezing, and wrap it in your pack, or keep it next to your body. When going to sleep, they can be filled with hot water, and placed inside the bag. This will not only keep them from freezing, but will also warm the bag.

   (d) <u>Drying</u>. For drying small items like socks or gloves, wear them next to your body during the day, and rotate as needed. Larger items can be dried in the tent. If your clothes are wet, hang them in the tent to dry. Small, damp items such as gloves and socks can be dried out by bringing them into your sleeping bag at night.

TRANSITION: These are just a few of the tricks that have been learned through experience. To find out what works best for you, you should trial and error with different techniques. Are there any questions at this time?

PRACTICE (CONC)

    a. Students will practice what was taught in upcoming field evolutions.

PROVIDE HELP (CONC)

    a. Instructors will assist the students when necessary.

OPPORTUNITY FOR QUESTIONS (3 Min)

1. QUESTIONS FROM THE CLASS

2. QUESTIONS TO THE CLASS

    Q. What are the types of cold?

    A. (1) Wet cold.
       (2) Moderate cold.
       (3) Intense cold.
       (4) Extreme cold.

    Q. What does the acronym "COLD" stand for?

    A. C - Keep clothing CLEAN.

       O - Avoid OVERHEATING.

       L - Wear clothing LOOSE and in LAYERS.

       D - Keep clothing DRY.

    Q. What are the three principles of design?

    A. (1) Vapor transmission.
       (2) Insulating.
       (3) Protective.

SUMMARY (2 Min)

a. This period of instruction has discussed cold weather clothing systems, paying particular attention to the ECWCS. Also discussed were the principles of design, different layers, fabrics, care and usage of systems, gear important to a cold weather environment, and some hints to make this environment a little more bearable.

b. Those of you with IRF's please fill them out at this time and turn them in to the instructor. We will now take a short break.

UNITED STATES MARINE CORPS
Mountain Warfare Training Center
Bridgeport, California 93517-5001

SML
SMO
09/17/02

LESSON PLAN

COLD WEATHER AND MOUNTAIN LEADERSHIP PROBLEMS FOR THE SMALL UNIT LEADER

INTRODUCTION (3 Min)

1. GAIN ATTENTION. Leadership is a vital aspect of military operations in all environments, particularly in the extreme conditions that we normally encounter in mountainous operations.

2. OVERVIEW. The purpose of this period of instruction is to emphasize the vital role of leadership in the conduct of successful operations and to promote among leaders at all levels an understanding of the problems common to units operating in a mountainous or cold weather environment. This lesson relates to all operations in the mountains or winter.

3. **INTRODUCE LEARNING OBJECTIVES**

    a. **TERMINAL LEARNING OBJECTIVE** (SML/WML, SMO/WMO) While assigned to a leadership billet in a field exercise in a mountainous and/or cold weather environment, lead Marines in accordance with the references.

4. METHOD / MEDIA. The material in this lesson will be presented by lecture.

5. EVALUATION. You will be tested by performance evaluation during upcoming field exercises.

TRANSITION: Your Marines will find themselves up against many new and challenging problems-but none that a properly trained Marine cannot overcome.

BODY (25 Min)

1. **SUPPORTING PUBLICATIONS.** In addition to your manuals, MCRP 3-35.1a Small Unit Leaders Guide to Cold Weather Operations and MCRP 3-35.2a Small Unit Leaders Guide to Mountain Operations, both in draft form, are great sources of knowledge for the small unit leader. These manuals may be downloaded from the doctrine division's page on the USMC home page.

    a. Further more, the Mountain Warfare Training Center has authored two Marine Corps Warfighting Publications, one on MAGTF Cold Weather Operations (MCWP 3-35.1), and one on Mountain Operations (MCWP 3-35.2).

    b. Also authored by MWTC were three X-Files (Cliff Assault, Mule Packing, and Water Procurement), the Military Nordic Ski Manual, and the Assault Climbers Guide.

    c. The aforementioned manuals and publications discuss specific techniques and procedures for operating in a cold weather / mountainous environment. Your small unit leaders should become very familiar with these publications.

TRANSITION: All leadership traits and principles are of importance to leaders assigned to units operating in high altitude areas. The extreme conditions so often found are the cause of some peculiar problems for leaders.

2. **UNDERSTANDING THE COLD WEATHER ENVIRONMENT**

    d. The Russo-Finnish War: Conducted during the winter of 1939-1940 for 105 days. The Soviets outnumbered the Finnish 40-1 in personnel, 30-1 in aircraft and 100-1 in tanks and artillery. The soviets suffered somewhere between 200,000, according to Field Marshal Mannerheim, and 2 million, according to Secretary Khrushchev, casualties. The Finnish's ability to operate freely in the AO against the road bound Soviet Army was the main reason why the Soviet 7 day war lasted 105 days. The use of motti tactics, the destruction of Soviet Combat Service Support, and lack of Soviet winter training are cited keys to Finnish success. Inevitably the Finns capitulated when their economy was no longer able to support the war effort.

    e. U.S. Army WWII: The U.S. Army suffered 84,000 cold weather injuries during World War II. 90% of these injuries were in the infantry. On average, each case required 87 days in the hospital.

    f. German Army: During Operation Barbarosa (the invasion of the Soviet Union) the German Army suffered 100,000 cold weather injuries. 15,000 required a limb to be amputated.

g. Korean War: MWTC was established as a result of environmental injuries sustained by USMC forces during the winter drive to the Yalu River (Chosin Campaign). Cold weather injuries accounted for 10 percent of all injuries during the Korean War, most occurring during the winter of 1950-51.

h. Falklands War: After 25 days of combat, the Argentine Army surrendered. Without getting into the feats of the British Royal Marines and Paras, 75% British forces suffered from immersion foot or other cold weather injuries. Considering all factors, if the campaign were extended, this could have posed a serious problem to the British.

TRANSITION: Good leadership at any time can make the difference between life and death for your subordinates, but this is especially true in climates and locales that can be so hostile, such as the mountains.

3. **UNDERSTANDING THE MOUNTAIN ENVIRONMENT**.

   i. Italian Campaign WWII: Considered to be the soft under belly of Europe, Italy proved to be a meat grinder for General Mark Clark's 5th Army, a drain on material, and frustration for Allied command. After the Anzio landing, the allied forces quickly became bogged down with advances measured in yards. It was not until the 10th Mountain Division arrived on scene did the allies begin to make significant advances and break into the Po valley. Mules were used extensively by allied forces to transport gear up the mountainous terrain.

   j. Russian Afghanistan War: Soviet Doctrine was a major factor in the Soviet defeat in Afghanistan. With their infantry tied to their vehicles, Assault and Offensive Air Support not coordinated with maneuver elements and focus on controlling the cities, Soviet forces were easy targets for the Mujahedin who controlled rural Afghanistan. The Soviets were unable to effectively hunt this guerrilla in his terrain.

   k. Chechnya: Although in the last two years, the Russians appear to be winning the war against the Islamic Extremists in Chechnya, the initial campaign resembled the debacle in Afghanistan. Notably, the Russian inability to sustain their forces logistically, resulted in numerous environmental injuries as well as the rampant spread of disease. Typhus, dysentery and hepatitis posed serious problems to Russian operations during the initial campaign. Cold weather clothing and equipment was also a contributing factor to cold weather injuries as most of their clothing was made of cotton.

TRANSITION: Understanding the mountain and its environment will aid the leader in making good decisions. There are keys to good leadership that you should know.

4. **KEYS TO GOOD LEADERSHIP**. There are four keys to good leadership that will be discussed in this chapter.

a. **Pre-environmental training.** Most casualties sustained in mountainous or cold weather environments are due to Marines and Sailors not being physically and mentally prepared for the environment. Based off your units TEEP, you should begin your work-up as soon as possible. As you develop your unit's training package, you need to set goals for what you expect the individual and unit to be able to accomplish prior to deployment. Additionally, you will need to keep in mind what assets you will have to support you prior to and once deployment has occurred. Pre-environmental training can be accomplished by your unit Mountain Leaders or by an MTT from the Mountain Warfare Training Center. At a minimum, pre-environmental training should encompass the following:

   1. <u>Physical Fitness</u>: This should entail moving heavy loads over medium distances at a quick pace. Loads should consist of a Combat Load, individual and crew-served weapons initially. Continue to increase weight and speed as the physical fitness program develops. Also, circuit courses combining cardio-vascular training as well as iso-tonic exercises serves to prepare individuals.

   2. <u>Individual Training</u>. Individual training should initially focus on the use of clothing and equipment for this environment, then on cross training of weapon systems. Most cold weather injuries result from the improper wearing of cold weather clothing.

   3. <u>Unit Training</u>. Establish your SOPs while you are still in garrison and refine them once you are in the environment; it will save you time and frustration in the environment. TDG's are great tools, as well as historical studies, for preparing subordinates prior to them getting in the environment. Finally, get use to working without communications and key leaders. This will better prepare your subordinates to assume higher's mission.

b. **Prepare for increases casualties**. A 500 man infantry Battalion normally suffers 15 – 30 injuries during summer operations and 30 – 45 injuries during the winter while training at Bridgeport. These injuries are normally exaggerated twisted ankles and Marines trying to avoid training.

   1. Understanding that you will take increased casualties, as illustrated in the historical examples, you need to make sure that you are capable of still accomplishing the mission despite personnel shortfalls. As an example, it takes 3 Marines to transport and carry a 60 mm mortar system. If you lose one or two Marines from that section, you are only able to employ two systems, cutting your ID fires by 1/3 at the company level. You need to ensure your 0311s are just as capable at employing that weapon system.
   2. Another example, is the RTO the only Marine who knows how to operate the AN/PRC-119, if so, you are out of luck if you lose him. Can

you do it? After all, in addition to your 3 Squad Leaders, it is your primary weapon system as a Platoon Commander.

3. Also, have you trained your squad leaders to assume the platoon commanders job, if the answer is yes, have you tested them? Can your FO run your FIST Team? Finally, are you Marines capable of being the Marine on site to provide essential First Aid? After all, the buddy aid does come before corpsman aid.

4. Casualties will significantly hinder the ability to accomplish the mission. Depending on the CASEVAC plan, to move one casualty from the point of injury to the casualty collection point may take the Marine's entire squad. Leaders must ensure the CASEVAC plan is supportable at all levels.

c. **Understanding your capabilities and limitations**. In the estimate of the situation (METT-TSL) we talk about the enemy's capabilities and limitations, which is derived from his composition, disposition and strength, which leads us to assume what we think his most probable and most dangerous course of action will be. Understanding our capabilities and limitations is what most leaders fail to do in this environment. Time and time again, leaders at all levels fail to plan operations in the environment that are within their personnel and equipment's capabilities and limitations. The classic example of this is the platoon commander who is assaulting and consolidating on the objective alone, except for the RTO whom he managed to drag along by the handset, and wondering why his platoon is still a terrain feature away.

1. The fact is you cannot achieve success in this environment if you do not realistically plan your operation from crossing the line of departure to consolidation, re-supply and CASEVAC. You must account for everything.

2. Past performance documentation will indicate future capabilities: How many times do we have to re-invent the wheel? For example, every time we conduct a movement, we create a route card. What do we normally do with that route card? Throw it away when we are done. How many leaders have taken the time estimate that he uses during planning, compared that to actual performance, and created a time estimate formula that better reflects his unit's ability?Realistic training results in accurate assessment: In training, the more things you fairy dust, the more you detract from an accurate assessment of your unit. How many Battalions have ever required their Marines to conduct CASEVACS to the BAS? How can you ever determine if the CASEVAC plan would work if you do not test it? The sad thing is, most units will learn that it will fail when they are in a real world situation.

d. **Controlling the situation**. When Marines are tired or cold, the first thing they lose is individual discipline. This problem is exacerbated in this environment because Marines will always be tired or cold as it is the nature of the environment. With that, Marines get lazy and the first thing that goes is continuing actions and a tactical mindset.

   1. Individual and unit discipline: Marines are on patrol, as they start getting tired, fewer and fewer Marines, including the leaders, execute continuing actions (looking at the ground. kneel or prone), the laziness expands, re-entry of friendly lines is administrative, they get back to their tents and throw their packs on the ground and go to sleep gear maintenance is not conducted… Marine wakes up with immersion foot…

   2. Leadership presence: Leadership presence through inspections and supervision is required at all levels. Discipline at levels must be exercised. This includes not only self-discipline, but the enforcement of discipline on Marines for violations of orders and SOPs. This must be the norm from the beginning. Once you let one transgression slide, it will open the floodgates. Leaders who do not make the mark need to be dealt with swiftly and sternly. DO NOT BE AFRAID TO FIRE ANY LEADER WHO IS NOT MAKING THE MARK. HE WILL GET MARINES INJURED OR KILLED IN THIS ENVIROMENT!

   3. Keeping the Marines informed: This goes a long way to keeping Marines involved in the mission. The more a Marine is involved in the mission, the better he performs.

TRANSITION: As in any environment there are certain trends in the mountains that will affect all leaders and their subordinates.

5. **COMMON LEADERSHIP PROBLEMS**. There are five specific problems that you as a leader will face in this environment. Those problems are: cocooning, group hibernation, loss of personal contact and communication, time and space planning, and sustainment.

   a. <u>Cocooning</u>. Turning inward on yourself and being more concerned about your own comfort than the environment around you. It results in diminished situational awareness and having no tactical mindset. Basically, you are cold or tired and all you care about is getting some sleep or getting warm. We have all seen this at one time or another, except in this environment it is endemic. Marines are not use to operating in a cold weather environment. How do we deal with this?

      1. <u>Environmental training</u>. Teach the Marines how to properly wear the clothing and equipment that is issued to them. Are they wearing every piece of clothing they have and are still freezing? Well the problem is

probably that their clothing is not loose and layered. Or, are they wearing cotton utilities that are soaking wet instead of gore-tex?

    2. <u>Physical activity</u>. When Marines cocoon, they focus in on themselves. Give them a mission, which causes them to focus on something else. There is always something that needs to be accomplished; like improving the defensive positions. Or, rotate Marines out of security more frequently. Increase the frequency of security patrols. Basically, Marines cocoon when their leadership does not supervise. This is a continual problem that needs to be dealt with.<u>Time</u>. Ultimately, most Marines will learn to deal with the environment the longer they are in it. However, poor time management and making Marines stand around only contributes to the problem. Also, taking your Marines out of the field after four days accomplishes nothing but building a four-day mindset.

b. <u>Group hibernation</u>. Cocooning ultimately leads to group hibernation or collective cocooning. The training schedule will say Night Company Movement. During the day, the temperature was cold and toward evening it started snowing and it was rather windy. As the Marines are new to the environment, the environment obviously mentally affects them. The Company Commander, who is suffering the same as the Marines, wants desperately to go to his tent. He confers with the XO, 1st SGT and Company Gunny, they want to go to the tent as well. They all agree to cancel training for the "safety" of the Marines. That is collective cocooning.

    1. <u>Remedy.</u> As the Leader, you have to be the sounding board. If you know that training is not going to endanger the safety of the Marines, explain it to the Commander. If that does not work, bring it to higher's attention.

c. <u>Loss of personal contact and communication</u>. A by-product of cocooning and group hibernation is the loss of personal contact and communication. Both cocooning and group hibernation degrade the ability of the chain of command to operate. Because Marines find sanctuary in the tent or where ever they throw themselves down, they do not want to leave it, and that becomes their world. Squad leaders no longer show up to the Platoon Sergeant's meeting, instead he sends one of his fire team leaders with some excuse why he could not be there. Additionally, the leaders tour the areas less frequently and as a result don't talk to their Marines. They have no idea what their state of mind is or how they are doing physically. Word no longer gets passed and people have no idea what is going on. This results in an informational and emotional isolation. Marines become apathetic to the world around them. He no longer knows what the mission of his adjacent unit is, nor cares. This significantly degrades from the moral of the unit as well as their ability to function as a cohesion unit, in the maneuver sense.

    1. <u>Remedy</u>. Insist on the proper use of the chain of command, talk to the Marines at the lowest level, and supervise the Marines as much as

possible. Check on the Marines manning the M-240G at 0230 to see how he is doing.

d. <u>Time and space planning</u>. This is another area where leaders and units consistently fail in this environment. What people need to understand is that everything takes longer, from breaking down tents to moving from A to B. In Camp Pendleton it may take a platoon 10 minutes to break down and pack up their tents, but here it will take at least double that time.

1. <u>Mobility</u>. Mobility is affected by the terrain and altitude (flat or mountains, at sea level or 2000 meters) the weather (34 degrees and raining, sunny or 12 inches of snow) the physical condition of the Marines and their moral. All these factors are going to determine how well your Marines will move a distance over time.

2. <u>Planning</u>. Very rarely do units cross the LD on time for a morning movement. This is because the leaders do not adequately back-plan from LD to pull pole to reveille. As talked about during knowing your capabilities and limitations, realistic assessment of mobility capabilities is essential to mission accomplishment. Furthermore, you need to issue timely warning orders to ensure your subordinate leaders have enough time to accomplish their tasks as well.

e. <u>Sustaining your Marines</u>. When it comes down to it, you accomplished your mission, good job, but are you ready to continue the fight? We all know that we can push our Marines to the breaking point to accomplish the mission, but if we do that we are actually failing, especially in this environment?

1. <u>Realistic Mission Planning</u>. Terrain selection, which minimizes the impact on the Marines, the gear selection essential to mission accomplishment, and the selection of personnel that will not hinder the mission. With all that Marines locate close with and destroy the enemy. However, inadequate mission planning results in your Marines having to suffer the wraths of poor terrain selection, extra / needless weight on their backs and making up for the Marine who cannot carry himself. Your Marines are spent long before they reach the objective.

2. <u>Logistical Support</u>. You need to ensure that your plan is logistically supportable. You can come up with some great big blue arrow plan of attack, but if you cannot get chow and water to your Marines, it is worthless. Furthermore, if your Marines are spent on the approach march, what good are they in the attack?

3. <u>Conservation of energy is key</u>. What it comes down to is, once your Marines are spent, it takes time for them to recover. Re-hydration takes at least 6 hours minimum and it takes almost 24 hours for your body to

recover energy from food. The time it takes you to move a distance has to be at a pace sustainable by the Marines. Getting to the objective still ready to fight is key in this environment. Especially true in this environment is that the more starved your body is, the harder it will be for the immune system to fight off disease.

TRANSITION: The mountains are a very dangerous place to survive and fight. They present many dangers to overcome. Are there any questions?

PRACTICE (CONC)

    a. Students will apply mountain leadership principles throughout duration of the course.

PROVIDE HELP (CONC)

    a. The instructors will assist the students when necessary.

OPPORTUNITY FOR QUESTIONS (3 Min)

1. **QUESTIONS FROM THE CLASS**

2. **QUESTIONS TO THE CLASS**

    Q. What are 2 of the key points of good leadership?

    A. (1) Pre environmental training.
        (2) Prepare for increased casualties.

    Q. What are the five leadership problems peculiar to mountain operations?

    A. (1) Cocooning.
       (2) Group hibernation.
       (3) Loss of personal contact and communications.
       (4) Time and space planning.
       (5) Sustaining your marines.

SUMMARY (2 Min)

    a. This period of instruction has discussed keys to good leadership, including: pre-environmental training, prepare for increased casualties, understanding your capabilities and limitations, controlling the situation. Also discussed were common leadership problems: cocooning, group hibernation, loss of personal contact and communication, time and space planning, and sustaining your Marines.

b. Those of you with IRF's please fill them out at this time and turn them in. We will now take a short break.

UNITED STATES MARINE CORPS
Mountain Warfare Training Center
Bridgeport, California 93517-5001

WML
WMO
09/31/00

LESSON PLAN

MOUNTAIN HEALTH AWARENESS

INTRODUCTION (5 Min)

1. **GAIN ATTENTION**. A study of military history in mountain regions reveals that success and failure rates are measured in terms of the regard held for the environment itself. The man who recognizes and respects these forces of nature can do his job and even use these forces to his advantage. The man who disregards or underestimates these forces is doomed to failure, if not destruction. In the mountains, care of the body requires special emphasis. If men fail to eat properly, or do not get sufficient liquids, efficiency will suffer. Lowered efficiency increases the possibility of casualties, either by high altitude injury or enemy action resulting in failure to accomplish the mission.

2. **OVERVIEW**. The purpose of this period of instruction is to provide the student with the information necessary to help prevent, recognize, and treat various health problems that can arise in a mountainous environment in both winter and summer. This lesson relates to all operations conducted while in a mountainous environment.

INSTRUCTORS NOTE: Have students read learning objectives.

3. **INTRODUCE LEARNING OBJECTIVES**.

   a. TERMINAL LEARNING OBJECTIVE. In a mountainous environment, execute measures for preventing, recognizing and treating illnesses and injuries, in accordance with the references.

   b. ENABLING LEARNING OBJECTIVES (MLC)

   (1) Without the aid of references, list in writing the five ways a body loses heat, in accordance with the references.

(2) Without the aid of references, define in writing dehydration, in accordance with the references.

(3) Without the aid of references, list in writing the ways to prevent dehydration, in accordance with the references.

(4) Without the aid of references, define in writing heat exhaustion, in accordance with the references.

(5) Without the aid of references, define in writing Acute Mountain Sickness (AMS), in accordance with the references.

(6) Without the aid of references, define High Altitude Cerebral Edema (HACE), in accordance with the reference.

(7) Without the aid of references, define High Altitude Pulmonary Edema (HAPE), in accordance with the reference.

(8) Without the aid of references, define in writing hypothermia, in accordance with the references.

(9) Without the aid of references, define in writing frostbite, in accordance with the references.

(10) Without the aid of references, list in writing the major risk factors for frostbite, in accordance with the references.

(11) Without the aid of references, define in writing trench foot, in accordance with the references.

(12) Without the aid of references, define snow blindness, in accordance with the references.

(13) Without the aid of references, state in writing how to prevent CO_2 poisoning, in accordance with the references.

(14) Without the aid of references, list in writing the ways to dispose of waste, in accordance with the references.

c. ENABLING LEARNING OBJFCTIVES (SMO)

(1) With the aid of references, list orally the ways a body loses heat, in accordance with the references.

(2) With the aid of references, define orally dehydration, in accordance with the references.

(3) With the aid of references, list orally the causes of dehydration, in accordance with the references.

(4) With the aid of references, list orally the ways to prevent dehydration, in accordance

with the references.

   (5) With the aid of references, define orally heat exhaustion, in accordance with the references.

   (6) With the aid of references, define orally Acute Mountain Sickness (AMS), in accordance with the references.

   (7) With the aid of references, define orally High Altitude Cerebral Edema (HACE), in accordance with the reference.

   (8) With the aid of references, define orally define High Altitude Pulmonary Edema (HAPE), in accordance with the reference.

   (9) With the aid of references, define orally hypothermia, in accordance with the references.

   (10) With the aid of references, define orally frostbite, in accordance with the references.

   (11) With the aid of references, list orally the major risk factors for frostbite, in accordance with the references.

   (12) With the aid of references, define orally trench foot, in accordance with the references.

   (13) With the aid of references, define orally snow blindness, in accordance with the references.

   (14) With the aid of references, state orally how to prevent C02 poisoning, in accordance with the references.

   (15) With the aid of references, list orally the ways to dispose of waste, in accordance with the references.

4. **METHOD/MEDIA**. The material taught in this lesson will be presented by lecture and practical application method. You will practice what you have learned during upcoming field training exercises. Those of you with IRF's please fill them out at the end of this period of instruction.

5. **EVALUATION**.

   a. Mountain Leaders - You will be tested by a written exam.

   b. Unit Operations - You will have a verbal examination.

TRANSITION: Are there any questions over the purpose, learning objectives, how the class will be taught, or how you will be evaluated? Let's first talk about how the body is affected by heat.
BODY (115 Min)

1. (5 Min) **GENERAL**. The body's ability to adjust to a harsh environment is greatly controlled by its physical condition, which is influenced by fitness, nutrition, water intake and other factors. Acclimatization is also vital, but to acclimatize properly, a person must force himself to enter the outdoor environment and work in it, which requires a healthy attitude. The importance of a healthy attitude cannot be stressed enough.

    a. Physical Fitness: This is fundamental to successfully perform in a mountainous environment. The more fit an individual is, the easier it is for him to move, to carry heavy loads, and to adapt to the stresses of the mountains.

    b. Nutrition: An adequate number of calories must be consumed on a daily basis.

        (1) 1 MRE provides 1,200 – 1,300 calories and are slightly high in sodium and protein.

        (2) The RCW comes in six varieties and have nearly 4,500 calories.

    c. Water Intake: Obviously, a person cannot exist for long without adequate water intake. In the mountains, water intake becomes even more important than in the lowlands.

TRANSITION: Are there any questions about the body in general? If not, let us discuss how the body generates heat.

2. (5 Min) **HEAT GENERATION**. The human body can be compared to a constantly running furnace. To run efficiently, it must always maintain a temperature within a small range. The food we eat is our fuel. The amount of energy in the food we eat is measured in calories. Typically, about 25% of the food we take in is used by the body to rebuild itself, while the remaining 75% is used to produce heat to maintain that small temperature range (usually from 96°F to 99°F, with the average being 98.6°F). This generated heat is then distributed to various parts of the body. In the mountains at high altitude, especially in the winter, several other points need to be remembered:

    a. The Diet.

        (1) Caloric intake. At sea level, a Marine in the field typically requires about 2,800 – 3,600 calories per day. But at altitude, in the mountains and especially in the winter, the need for calories almost doubles to at least 4,500 – 6,500 calories per day. This is because one is much more physically active, and because of cold, calories needed for heat generation greatly increases.

        (2) Carbohydrate intake. Carbohydrates are simple foods like sugars, bread, rice, and pasta. In the mountains an increase in carbohydrate intake is recommended because at high altitude these tend to taste better and are more easily converted into heat energy by the body.

        (3) Hot meals and hot wets. These should be consumed whenever possible.

b. <u>The Body at rest produces heat at a specific rate</u>. With physical activity, heat production increases.

   (1) Moderate Exercise. This can increase heat production up to 5 - 6 times the normal rate, and can be tolerated for long periods of time.

   (2) Moderate Shivering. This can increase heat production by 20 times normal, but only for a few minutes. Shivering is not an efficient means of heat production as it quickly leads to exhaustion.

   (3) Intense Shivering. When muscle activity is at its maximum rate, as in intense shivering, heat production can increase by up to 50 times normal. However, exhaustion follows within 30 seconds.

TRANSITION: Are there any questions about how the body generates heat? If not, let us talk about the different ways a body loses heat.

3. (10 Min) **FIVE WAYS THE BODY LOSES HEAT**

   a. <u>Radiation:</u> is direct heat loss from the body to its surroundings. If the surroundings are colder than the body, the net result is heat loss. A nude man loses about 60% of his total body heat by radiation. Specifically, heat is lost in the form of infrared radiation. Infrared targeting devices work by detecting radiant heat loss.

   b. <u>Conduction:</u> is the direct transfer of heat from one object in contact with a colder object.

      (1) Most commonly conduction occurs when an individual sits or rests directly upon a cold object, such as snow, the ground, or a rock. Without an insulating layer between the Marine and the object (such as an isopor mat), one quickly begins to lose heat. This is why it's important to not sit or sleep directly on cold ground or snow without a mat or a pack acting as insulation.

      (2) Metals, like rocks, conduct heat very rapidly.

      (3) Water conducts heat away from the body 25 times faster than air.

   c. <u>Convection:</u> is heat loss to the atmosphere or liquid such as water in the following manner:

      (1) Air and water can both be thought of as "liquids" running over the surface of the body. Water or air, which is in contact with the body, attempts to absorb heat from the body until the body and air or water is both the same temperatures. However, if the air or water is continuously moving over the body, the temperatures can never equalize and the body keeps losing heat.

      (2) Most commonly one encounters convection through the wind-chill effect. Whether walking, skiing, or moving in open vehicles, wind must be taken into account to

determine the effective temperature experienced by the unprotected body. For example: At a 32°F air temperature, the effective temperature, or wind-chill, with a 25 mph wind is actually 3°F. At 5°F, a 25 mph wind will give a wind-chill of 37°F below zero (-37°F)!

   (a) However, if dressed properly (with the appropriate protective layers) wind-chill effects are minimized, except for areas of exposed skin. If Marines are cold because of wind-chill, it means that they are not dressed properly and that is a leadership failure.

d. <u>Evaporation</u>. Heat loss from evaporation occurs when water (sweat) on the surface of the skin is turned into water vapor. This process requires energy in the form of heat and this heat comes from the body.

   (1) This is the major method the body uses to cool itself down. This is why you sweat when you work hard or PT. One quart of sweat, which you can easily produce in an hour of hard PT, will take about 600 calories of heat away from the body when it evaporates.

e. <u>Respiration.</u> When you inhale, the air you breathe in is warmed by the body and saturated with water vapor. Then when you exhale, that heat is lost. That is why breath can be seen in cold air. Respiration is really a combination of convection (heat being transferred to moving air by the lungs) and evaporation, with both processes occurring inside the body.

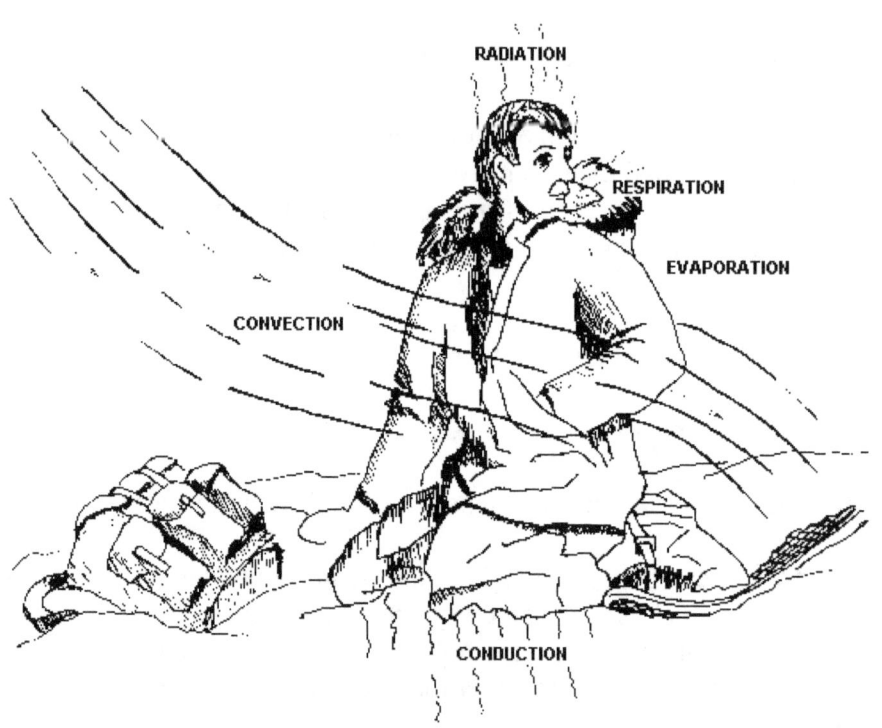

<u>5 WAYS THE BODY LOSES HEAT</u>

TRANSITION: Now that we know the five ways a body can lose heat, are there any questions? Let discuss how the body responds to heat.

4. (5 Min) **PHYSICAL RESPONSES TO HEAT.** When the body begins to create excess heat, it responds in several ways to rid itself of that heat.

   a. Initially, the blood vessels in the skin expand, or dilate. This dilation allows more blood to the surface where the heat can more easily be transferred to the surroundings.

   b. Soon afterwards, sweating begins. This contributes to heat loss through convection and evaporation.

TRANSITION: Now that we have discussed how heat can be lost, are there any questions? Now let's talk about how the body reacts to cold.

5. (5 Min) **PHYSICAL RESPONSES TO COLD.** Almost the opposite occurs as with heat.

   a. First, blood vessels at the skin surface close down, or constrict. This does two things:

      (1) Less blood goes near the surface of the body so that less heat is lost to the outside.

      (2) More blood goes to the "core" or the center of the body, to keep the brain, heart, lungs, liver, and kidneys warm. This means fingers and toes tend to get cold.

   b. If that is not enough to keep the body warm, the next step is shivering. Shivering is reflexive regular muscular contractions, this muscular activity causes heat production. As mentioned before, shivering can only last for a short time before exhaustion occurs. With shivering you will either warm up, as usually occurs, or continue to get colder and start to become hypothermic. Hypothermia will be discussed later.

   c. Another effect of more blood flow going to the body's inner core is that the kidneys are "fooled" into thinking that the body has more blood than it really does. The kidneys respond by producing more urine, and this can contribute to dehydration, which we will talk about next.

TRANSITION: Thus far, we have talked about the body when it is responding normally, are there any questions? Now let's look at things when they go wrong.

6. (5 Min) **DEHYDRATION**

   a. Definition: Dehydration is defined as a deficit of total body water.

   b. Causes. Dehydration is the most common illness seen, both in the winter and in the summer. Ultimately, the reason someone becomes dehydrated is:

      (1) Excessive loss.

         (a) Urination. Increased as a response to the cold and high altitude.

(b) Cold, dry air. In most mountainous areas, like the high Sierras, the air is often cold and always dry. Thus, inhaled air must be humidified and warmed by the body and this takes water.

(c) Strenuous activity. Marines in the mountains are always involved in strenuous activity, and this leads to large amounts of sweating, even in the winter.

(d) Coffee and tea. These are mild diuretics; that is, they stimulate the kidneys to produce excess urine.

(2) <u>Inadequate intake.</u>

(a) Thirst. Thirst is not a good indicator of your state of hydration, especially in a high altitude, mountainous environment. *If you are thirsty, it is too late, you are already dehydrated.*

(b) Water inaccessibility. It is possible, though not likely, that adequate amounts of water may not be available. This should only be a factor if you are in a survival situation.

c. <u>Symptoms of Dehydration</u>

(1) Headache.

(2) Nausea.

(3) Dizziness.

(4) Fainting.

(5) Constipation.

(6) Dry mouth

(7) Weakness

(8) Lethargy.

(9) Stomach cramps.

(10) Leg and arm cramps.

d. <u>Signs of Dehydration</u>

(1) Swollen tongue.

(2) Dark urine.

(3) Low blood pressure.

(4) Rapid heart rate, greater then 100 beats per minute.

e. Prevention of Dehydration

(1) Since it is impossible to limit the water that you lose (except by limiting your coffee and tea consumption), you must then ensure adequate intake.

(a) Drink a minimum of 6 - 8 quarts per day.

(b) Watch the color of your urine. Try to keep it crystal clear. The more yellow it gets, the more you need to drink.

(c) Do not rely on thirst as an indicator.

f. Field Treatment of Dehydration.

(1) Oral Fluids. Give at least 6 quarts per day.

(2) Intravenous (IV) fluids. Severe cases may require fluid by IV.

g. Leadership. A healthy Marine should *NEVER* become dehydrated. It is an entirely preventable condition. In the Israeli Army, when a soldier becomes dehydrated, his platoon commander and platoon sergeant are court-martialed. It is the responsibility of small unit leaders to ensure their men have adequate access to water and that they drink it. Failure to prevent dehydration in a healthy Marine is a failure in leadership.

TRANSITION: Are there any questions about dehydration? Let us now talk about heat cramps.

7. (5 Min) **HEAT CRAMPS**

a. Definition: Heat cramps are painful spasms of skeletal muscle as a result of excessive loss of body salt.

b. Causes. Sweat is composed of water and salt. When a Marine is involved in strenuous activity that leads to excessive sweating, and replaces the lost water but *not* the lost salt, a salt imbalance within the body may result. This salt imbalance may then lead to muscle cramps.

c. Symptoms of Heat Cramps.

(1) Muscle cramps in the arms, legs, or abdomen.

d. <u>Prevention of Heat Cramps.</u> Prevention is always better than treatment.

   (1) Avoid overheating by proper ventilation.

   (2) Eat correctly. There is no need to take salt tablets as long as proper diet is maintained. MRE's and RCW's contain more than enough salt.

e. <u>Field Management for Heat Cramps</u>

   (1) Have the victim stop moving. (Rest)

   (2) Gentle massage of the affected muscles may help relieve the spasm.

   (3) Stretch-out the muscle.

   (4) Ensure the victim is adequately hydrated. Replace the victim's salt by adding either 1 salt tablet or 1 tablespoonful of table salt (from an MRE accessory packet) to a quart of water. Have the victim sip the salted water over a period of a few hours.

<u>TRANSITION</u>: Now that we have discussed heat cramps, are there any questions? If not, let us talk about heat exhaustion.

8. (5 Min) **HEAT EXHAUSTION**

   a. <u>Definition</u>: Heat exhaustion occurs when body salt losses and dehydration from sweating are so severe that a person can no longer maintain adequate blood pressure.

   b. <u>Causes</u>. Heat exhaustion is really a severe form of dehydration combined with or as a result of strenuous physical activity. Blood is made up of mostly of water. When a large amount of water is lost in the form of sweat, the amount of blood volume in the body drops. When the blood volume drops low enough, in combination with tough physical exercise, heat exhaustion results.

   c. <u>Symptoms of Heat Exhaustion</u>.

   (1) Headache.

   (2) Nausea.

   (3) Dizziness.

   (4) Fatigue.
   (5) Fainting.

   d. <u>Prevention of Heat Exhaustion.</u>

(1) It is the same as for heat cramps. Dress properly with adequate ventilation to avoid overheating. Dress comfortably cool.

e. <u>Field Management for Heat Exhaustion</u>.

(1) Lay the victim down, with his feet higher than his head. Insulate the victim from the cold ground with an isopor mat.

(2) Ensure that he is well ventilated. Unzip his parka or take it off, until he feels cool. Make sure he doesn't get too cold.

(3) Fluids. If he is awake and not vomiting, he may be given fluids by mouth. Usually 3 quarts at a minimum are required.

<u>TRANSITION</u>: Heat exhaustion can be a serious ailment, but heat stroke is even more deadly if not treated in a timely manner.

9. (5 Min) **HEAT STROKE**

a. <u>Definition</u>: Also known as Sunstroke, it is defined as a failure of the body's cooling mechanisms that rid the body of excessive heat build up.

NOTE: Heat stroke is LETHAL in up to 40% of cases, while a majority of those who live may suffer permanent brain damage.

b. <u>Risk Factors for Heat Stroke</u>. Heat stroke occurs when the body is unable to rid itself of excess heat, such as when exercising in a hot, humid environment. Typically the air in a mountainous environment is cool and dry; nonetheless, heat stroke can and does occur in the mountains, even in the winter. The elevation of body temperature levels is usually greater then 103°F.

c. <u>Symptoms of Heat Stroke.</u> In the majority of cases, the onset of heat stroke is sudden, and the victim becomes delirious or comatose, before he begins to complain of symptoms. However, approximately 20% of victims will complain of:

(1) Headache.

(2) Nausea.

(3) Dizziness.

(4) Fatigue.

d. <u>Signs of Heat Stroke.</u>

(1) Usually, victims are delirious or comatose.

(2) Pupils maybe pinpoint.

(3) Flushed skin may or may not be present.

(4) Rectal temperatures of 103'F or greater.

(5) Sweating. It is often taught that in Heat Stroke, sweating is absent. THIS IS **UNTRUE!!!** Sweating often is present in sunstroke, so do not assume a victim does not have heat stroke simply because he is sweating. As with hypothermia, the only way to absolutely diagnose a victim as having heat stroke is with a rectal thermometer. Anybody with abnormal behavior (hallucinations, bizarre behavior, confusion, etc.) and a rectal temperature of 103°F or greater has heat stroke, until proven otherwise.

e. <u>Preventive Measures for Heat Stroke</u>. For the most part, the same principles apply as with Heat Exhaustion. That is, drink 6 - 8 quarts of water per day and keep as well ventilated as possible. When the temperature and humidity are high, however, physical activity must be reduced.

f. <u>Field Management for Heat Stroke.</u> Remember that heat stroke is a true life and death emergency. The longer the victim remains overheated, the more likely it's irreversible.

(1) Reduce heat immediately by dousing the body with large amounts of cool water or by applying wet, cool towels to the neck, the groin, chest, and armpits. If cold packs are available then use them.

(2) Maintain an open airway.

(3) Remove as much of the victim's clothing as possible.

(4) Give him nothing by mouth.

(5) When his rectal temperature has dropped below 102°F, you may discontinue cooling. Be sure to recheck the temperature every 5 minutes, if his temperature rises to 103°F or greater, begin re-cooling.

(6) Casevac immediately!

TRANSITION: Now that we have discussed problems with the body, are there any questions? Now let us talk about problems of altitude.

10. (5 Min) **ACUTE MOUNTAIN SICKNESS (AMS)**

a. <u>Definition</u>: AMS is a self-limited illness resulting from the rapid exposure of an un-acclimatized individual to high altitude.

b. <u>Risk Factors for AMS</u>. Anyone ascending rapidly from sea level to over 7,000 feet may develop AMS. Approximately 25% of individuals who ascend rapidly to 8,000 - 9,000 feet will develop AMS. Virtually all un-acclimatized persons who rapidly ascend to 11,000 - 12,000 feet will develop AMS. Factors which will increase your chance of developing AMS or make it worse are overexertion at altitude and dehydration. The cause of AMS, or altitude illness in general, is not well understood. However, it is known that the lower levels of oxygen's barometric pressure found at high altitude leads to a state of hypoxia, which means low levels of oxygen in the blood. The way in which the body responds to this hypoxia can lead to AMS or other altitude illness. Symptoms of AMS will usually occur 6 - 48 hours after reaching altitude.

c. <u>Symptoms of AMS</u>.

   (1) Headache, the most common symptom, may be severe.

   (2) Nausea.

   (3) Decreased appetite.

   (4) Difficulty sleeping; due to irregular breathing.

   (5) Weakness, loss of coordination.

   (6) Easily fatigued.

   (7) Dizziness.

   (8) Apathy.

d. <u>Signs of AMS</u>.

   (1) Rapid breathing or an irregular breathing pattern.

   (2) Rapid pulse.

   (3) Vomiting.

e. <u>Prevention of AMS</u>.

   (1) The best prevention of AMS is a staged ascent. For Marines going to altitude from sea level, the following ascent rates should be adhered to:

ALTITUDE	RATE OF ASCENT
<8,000 ft	Unlimited rest for 48 -hrs @ 8,000 ft, then proceed
8,000 - 10,000 ft	Rest for 24 hrs @ 10,000 ft, and proceed
10,000 - 14,000 ft	1,000 ft per day.

| >14,000 ft | 500-1,000 ft per day |

    (2) After 48 hours with no symptoms of altitude illness, proceed no higher than 10,000 ft. Remain at 10,000 for 24 hrs and then if no symptoms of altitude illness are present, proceed no faster than 1,000 feet per day.

    (3) Above 14,000 ft, ascend no faster than 500-1000 feet per day. If no symptoms occur after 48 hrs at a given altitude, it is safe to assume you can ascend, but remember, there are no steadfast rules or guarantees.

    (4) Certain drugs, can be used to treat or even prevent AMS. These drugs are used ONLY under the direction of a medical department personnel. These drugs include diamox.

  f. Treatment of AMS

    (1) Light duty.

    (2) Fluids - ensure adequate fluid intake. (AMS is a fluid retention condition, be careful not to over hydrate)

    (3) Drugs for AMS:

      (a) Tylenol or aspirin can be given for the headache.

      (b) Diamox as prescribed.

    (4) **DESCEND**. Most cases of AMS should resolve with 2 - 3 days of the above measures. However, if this does not occur or if the symptoms are severe or worsening, then a descent of 1,000 - 3,000 feet should greatly improve the condition of the victim. He can re-ascend after several days.

TRANSITION: Now that we discussed AMS, let us talk about the affects of the brain due to lack of oxygen in a high altitude environment.

11. (5 Min) **HIGH ALTITUDE CEREBRAL EDEMA (HACE)**

  a. Definition: HACE is a high altitude illness, which is characterized by swelling of the brain.

  b. Symptoms/Signs of HACE

    (1) Usually the symptoms of AMS are also present. In fact, the two may look exactly alike. However, because HACE is lethal and AMS is not, you must be able to distinguish the two. Testing the victim's balance easily does this. Simply have him walk heel to toe (just like a field sobriety test). A HACE victim will have difficulty executing this maneuver, where an AMS victim will not.

(2) Other symptoms may include:

   (a) Bizarre behavior.

   (b) Hallucinations.

   (c) Confusion.

   (d) Excessive fatigue.

   (e) In severe cases--coma.

c. Prevention of HACE. The preventive measures for HACE are the same as for AMS.

d. Treatment of HACE. Immediate descent is mandatory!

   (1) Drugs prescribed by medical personnel include decadron (a steroid), diamox, and oxygen if available.

   (2) A device called a Gamow Bag. This is a man-portable (14 lbs.) hyperbaric chamber. A HACE (or HAPE) victim is placed in the bag and zipped up. Using a foot pedal operated pump; the pressure in the bag is increased. This simulates a decrease in altitude. Altitude "decreases" of up to 6,000 feet may be achieved. However, the use of this bag should only be reserved for emergencies when rapid descent is delayed. This is very labor intensive and only a temporary measure.

   (3) Remember: ***HACE KILLS*** if not treated.

TRANSITION: The lungs can also be effect at high altitude, it is called high altitude pulmonary edema.

12. (5 Min) HIGH ALTITUDE PULMONARY EDEMA (HAPE)

a. Definition: HAPE is a high altitude illness, which is characterized by filling of the lungs with fluid.

b. Risk Factors for HAPE. They are the same as for AMS and HACE, except HAPE is rarely seen below 10,000 feet.

c. Symptoms of HAPE
   (1) The two key symptoms to look for are:

   (a) A cough that persists even at rest. Cough will be initially dry and the progress over several hours to days to produce a pink frothy sputum.

(b) Severe shortness of breath, which also persists even at rest.

(2) Other symptoms of AMS are also usually present and include:

(a) Disorientation.

(b) Fainting.

d. Signs of HAPE

(1) Cool, clammy skin.

(2) Rapid breathing.

(3) Rapid, weak pulse.

(4) Blue lips.

(5) Undue fatigue.

e. Prevention of HAPE. The same as for AMS, slow graded ascent.

f. Treatment of HAPE

(1) The best treatment is descent-rapid descent, to as low as possible, preferably to sea level.

(2) The victim should be given 100% oxygen by mask if available.

(3) Medical can prescribe a drug called nifedipine. Diamox may also be given.

(4) A device called a Gamow Bag. This is a man-portable (14 lb.) hyperbaric chamber. A HAPE (or HACE) victim is placed in the bag and zipped up. Using a foot pedal operated pump; the pressure in the bag is increased. This simulates a decrease in altitude. Altitude "decreases" of up to 6,000 feet may be achieved. However, the use of this bag should only be reserved for emergencies when rapid descent is delayed. This is very labor intensive and only a temporary measure.

(5) Remember: **HAPE KILLS** if not treated.

TRANSITION: These are the high altitude injuries that you are going to see. Hopefully you will not fall victim to them during training or otherwise. Another aspect of the high altitude is the much colder temperatures; therefore, we need to be aware of the potential for cold weather injuries. Let's talk about Hypothermia.

13. (5 Min) **HYPOTHERMIA**

   a. Definition. Hypothermia is defined as the state when the body's core temperature falls to 95°F or less.

   b. A core temperature is the temperature at the center of the body. Taking an oral or armpit (auxiliary) temperature is not an accurate way to determine body core temperature. The best way to measure core temperature in the field is to take the temperature rectally. This must be done with a special low range rectal thermometer, which should be carried by all officers, SNCO's and corpsmen. These thermometers are available through the federal stock system. (FSN 6515-00-139-4593)

   c. Commonplace Misconceptions.

   (1) Exposure. At times you may hear that an individual has "exposure". While the term is usually used in reference to hypothermia, it is without real meaning and should not be used to describe hypothermia.
   (2) Extreme cold. It is a common belief that extreme cold is needed for hypothermia to occur. In fact, most cases occur when the temperature is between 30°F and 50°F. This temperature range is quite common in the Fall, Winter, and Spring months.

   d. Causes of Hypothermia. The ways in which the body generates and loses heat has been discussed earlier. Quite simply, hypothermia occurs when heat loss from the body exceeds the body's ability to produce heat. Contributing factors include:

   (1) Ambient temperature. Outside air temperature.

   (2) Wind chill. This only affects improperly clothed individuals.

   (3) Wet clothing.

   (4) Cold water immersion.

   (5) Improper clothing.

   (6) Exhaustion.

   (7) Alcohol intoxication, nicotine and drugs such as barbiturates and tranquilizers.

   (8) Injuries. Those causing immobility or major bleeding, major burn and head trauma.

   e. Signs and symptoms of Hypothermia

   (1) The number one sign to look for is altered mental status; that is, the brain is literally getting cold. These signs might include confusion, slurred speech, strange behavior, irritability, impaired judgment, hallucinations, or fatigue.

(2) As hypothermia worsens, victims will lose consciousness and eventually slip into a coma.

(3) Shivering. Remember that shivering is a major way the body tries to warm itself early on, as it first begins to get cold. Shivering stops for 2 reasons:

   (a) The body has warmed back up to a normal temperature range.

   (b) The body has continued to cool. Below 95°F shivering begins to decrease and by 90°F it ceases completely.

   (c) Obviously, continued cooling is bad. So if a Marine with whom you are working, who was shivering, stops shivering, you must determine if that is because he has warmed up or continued to cool.

(4) A victim with severe hypothermia may actually appear to be quite dead, without breathing or a pulse. However, people who have been found this way have been successfully "brought back to life" with no permanent damage. So remember, *you are not dead until you are warm and dead*.

f. <u>Prevention of Hypothermia</u>

(1) Obviously, prevention is always better (and much easier) than treatment.

(2) Cold weather clothing must be properly warm and cared for.

   (a) Keep your clothing as dry as possible.

   (b) If your feet are cold, wear a hat. Up to 80% of the body's heat can escape from the head.

(3) Avoid dehydration. Drink 6 - 8 quarts per day.

(4) Eat adequately. At least 4,500 calories per day.

(5) Avoid fatigue and exhaustion. A Marine in a state of physical exhaustion is at increased risk for hypothermia.

(6) Increase levels of activity as the temperature drops. Do not remain stationary when the temperature is very low. If the tactical situation does not permit moving about, perform isometric exercises of successive muscles.

(7) Use the buddy system to check each other for signs/symptoms of hypothermia.

g. Treatment of Hypothermia.

   (1) Make the diagnosis.

   (2) Prevent further heat loss.

   (a) Remove the victim from the environment where he became hypothermia, that is, bring him into the BAS, a tent, a snow cave, etc.

   (b) As soon as possible, remove the victim's cold, wet clothes.

   (3) Insulate the victim.

   (a) First, wrap the victim in a vapor barrier liner (VBL). A VBL will prevent heat loss as a result of evaporation and slow down heat loss from convection. The easiest way to do this in the field is by wrapping the victim in plastic trash bags. (Be sure not to cover the face.)

   (b) Next, place the victim in a sleeping bag.

   (4) Re-warm the victim

   (a) The easiest way to do this in the field is to zip two sleeping bags together. Place the victim in the zipped up bags with 2 stripped volunteers. While this may not agree with Marines, it could save the victims life.

   (b) In addition to the two stripped volunteer's place warmed materials on either side of the victim's neck, armpits, and his groin. Items such as warmed rocks, bags of warm water, or heat packs can be used. Be advised the warmed materials should not be hot, and the stripped volunteers should be in contact with the items as well. A hypothermia victim may not be able to tell if his skin is burning, but the volunteers will.

   (5) Evacuate the victim. A casualty evacuation may not be possible due to the tactical situation, weather, or other factors. However, the sooner a victim can be evacuated, the better. ***Severe hypothermia is a medical emergency.***

   (6) Other Points to Remember.

   (a) Fluids. If the victim is mildly hypothermic, he may be given hot wets. Otherwise give him nothing by mouth.

   (b) Avoid, if possible, excessive movement of the victim, as his heart may stop beating if it is jarred.

(c) Major Wounds. Apply first aid to major wounds first, before attempting to re-warm the victim. Re-warming a victim who has bled to death does little good.

(d) Never give alcohol to hypothermia victims.

(e) Even after you have started re-warming a victim, he must be constantly monitored. Don't forget about him.

TRANSITION: Not only can cold temperatures cause casualties through hypothermia, but cold can also lead to frostbite.

14. (5 Min) **FROSTBITE**

   a. Definition: Frostbite is the actual freezing of tissues.

   b. Risk Factors of Frostbite. The high-risk areas are fingers, toes, nose, cheeks, and ears.

     (1) Three Major Risk Factors. There are many factors that cause frostbite, of which three stand out as contributing to the majority of injuries:

       (a) Improper clothing or improper care of clothing. This is a major factor in frostbite.

         1. Wearing gloves when mittens should be worn.

         2. Failure to dry gloves or liners after they have become wet.

         3. Wet clothing of any kind.

         4. Improper footwear, such as wearing summer combat boots when VB or ski/march boots should be worn.

         5. Improper care of footwear. Failing to remove boots at night, sleeping with boots on, or failing to dry boots when they become wet.

         6. Wearing boots which are too tight.

         7. The proper use of cold weather clothing, as well as its proper maintenance in the field, is dependent on small unit leadership. Small unit leaders must ensure that their men are adequately clothed, as well as the clothing being adequately maintained.

       (b) Dehydration. This is another major contributing factor in frostbite. Marines who are well hydrated are much better equipped to fight off frostbite.

       (c) Poor diet or starvation. This is another major contributing factor in frostbite. Remember that the body can be thought of as a furnace, and that the fuel is food.

When food intake is low, there is less fuel to feed the furnace and the risk for frostbite goes up.

(2) Other factors that contribute to frostbite.

(a) Outside temperature. Obviously, the colder it is, the greater the risks.

(b) Snow or Ground temperature. The snow temperature can be 30°F - 40°F colder than the air temperature.

(c) Wind chill. As mentioned previously, wind-chill should not be a factor with properly dressed Marines.

(d) Cold metals. Never touch very cold metals with bare flesh. Use contact gloves.

(e) Petroleum products. Fuels and oils freeze at a much lower temperature then water Spilling cold fuel (such as white gas, gasoline, etc.) on bare skin can cause immediate, severe frostbite.

(f) Exhaustion. The body's natural defense mechanisms in general are lowered when you are exhausted.

(g) Hypothermia.

(h) Race/Place of Birth. African Americans and those from the south are at increased risk for frostbite.

(i) Other factors include prolonged immobility (as when sitting in an ambush position), wounds with blood loss, previous cold injury, and tobacco use.

c. Signs and Symptoms of Frostbite.

(1) Signs of Frostbite.

(a) The skin may appear red, white, yellow, gray, blue, frosty, or even normal.

(b) The skin may feel woody or firm.

(c) The joints may be stiff or immobile.

(d) The affected part may feel like a block of wood or even ice.
(e) Pulses may or may not be present.

(2) Symptoms of Frostbite. A victim may complain of:

(a) Tingling.

(b) Burning.

(c) Aching cold.

(d) Sharp pain.

(e) Increased warmth.

(f) Decreased sensation.

(g) No sensation at all. The victim may describe the affected part as clumsy, lifeless, bulky or club like.

d. <u>Classification of Frostbite.</u> Like burns, frostbite has been divided into 1st, 2nd, 3rd, and 4th degrees. But it is much easier to divide it up into Frosting, Superficial Frostbite, and Deep Frostbite.

(1) Frosting is something we all have experienced at one time or another. It is when some part of the body (toes, fingers, or nose usually) becomes painfully cold, but does not freeze. It is a harmless condition and the affected part returns to normal with re-warming.

(2) Superficial Frostbite. This is when the skin freezes, but not the tissue beneath (such as muscle, nerves, and bone).

(a) Skin appears red, gray, or even blue, and has a waxy feel to it.

(b) Pulses will be present, but decreased.

(c) The sensation of pain and light touch may be absent, but deeper sensations such as pressure will be intact.

(d) The joints will be mobile, but stiff.

(e) Movement of the part by the victim will be possible, although it may be difficult.

(3) Deep Frostbite.

(a) Initially, the skin may appear the same as above.

(b) Pulses will not be present.

(c) The skin will feel woody, firm or even rock hard.

(d) Tissues below will feel doughy or hard.

(e) All sensation will be absent.

(f) Skin will not move easily or not at all.

(g) Joints will be stiff or immobile.

(h) Movement of the affected part will be minimal or absent.

(4) It is often difficult to say exactly how severe a case of frostbite is until several weeks have passed. Therefore, it is wise to assume the worst.

NOTE: Frostbite may be present in different degrees in the same affected part, for example: a frostbitten hand may have deep at the fingers, superficial at the palm and frosting at the wrist.

e. <u>Prevention of Frostbite.</u>

(1) Frostbite is an entirely preventable injury. Obviously, there is little one can do about the weather, but Marines can ensure that the other risk factors that can lead to frostbite are minimized. The best way to prevent frostbite is to prevent the three major risk factors: Improper clothing or improper care of clothing, dehydration, and starvation.

(a) Dress in layers. Keep comfortably cool. If you begin to become uncomfortable, add layers. If your hands or feet become uncomfortable, do not ignore them - you may have to add more layers, or you may want to change socks or gloves.

(b) Keep clothes dry. This is vitally important. If your boots, socks, or gloves get wet, then dry them. This may mean you have to change socks up to 4 - 5 times a day (especially with Vapor Barrier boots). If your gloves or liners are wet, warm and dry them. Do not continue to wear wet clothing.

(c) Dress properly. This may seem obvious, but Marines have gotten frostbite because they did not. If the wind is blowing, wear the correct protective layer. Always have a balaclava or watch cap available, and if it's cold - wear it. If your fingers are getting cold in gloves, wear mittens.

(d) Avoid Dehydration. When you become dehydrated, the amount of blood available to warm your fingers and toes goes down, greatly increasing your risk of frostbite.

(e) Avoid Starvation. Remember - Food is Fuel - and the body uses that fuel to make heat. When you are low on fuel, you will be low on heat.

(2) U.S. Marines are the toughest people in the world. However, Mother Nature is tougher. If you notice your fingers or toes are getting cold even after you have tried to warm them, do not ignore it. Let your leaders know. Ignoring the problem will not make it go

away, it will only get worse.

(3) Small Unit Leaders must ensure that preventive measures are taken. Like dehydration, frostbite results from *failure of leadership.*

f. Field Management for Frostbite. Only frosting should be treated in the field, all others should be evacuated immediately. If you don't know, assume the worst and evacuate.

(1) Treatment of Frosting. This is easily done in the field using the 15 minute rule. Frosting will revert to normal after using this technique of body heat re-warming. Hold the affected area, as described below, skin to skin for 15 minutes. If the affected area does not return to normal, assume a frostbite injury has occurred and report it to your seniors.

(a) Re-warm face, nose, ears with hands.

(b) Re-warm hands in armpits, groin or belly.

(c) Re-warm feet with mountain buddies armpits or belly.

(d) ***DO NOT RUB ANY COLD INJURY WITH SNOW-EVER!***

(e) Do not massage the affected part.

(f) Do not re-warm with stove or fire, a burn injury may result.

(g) Loosen constricting clothing.

(h) Avoid tobacco products.

(2) Treatment of Superficial or Deep frostbite. Any frostbite injury, regardless of severity, is treated the same – evacuate the casualty and re-warming in the rear. Unless the tactical situation prohibits evacuation or you are in a survival situation, *no consideration should be given to re-warming frostbite in the field.* The reason is something-called freeze – thaw – re-freeze injury.

(a) Freeze – Thaw – Re-freeze injury occurs when a frostbitten extremity is thawed out, then before it can heal (which takes weeks and maybe months) it freezes again. This has devastating effects and greatly worsens the initial injury.

(b) In an extreme emergency it is better to walk out on a frostbitten foot than to warm it up and then have it freeze again.

(c) Treat frozen extremities as fractures - carefully pad and splint.

(d) Treat frozen feet as litter cases.

(e) Prevent further freezing injury.

(f) Do not forget about hypothermia. Keep the victim warm and dry.

(g) Once in the rear, a frostbitten extremity is re-warmed in a water bath, with the temperature strictly maintained at 101°F - 108°F.

TRANSITION: Now that we discussed frostbite, let us talk about immersion foot.

15. (5 Min) **TRENCHFOOT / IMMERSION FOOT**

   a. Definition. This is a cold - wet injury to the feet or hands from prolonged (generally 7 - 10 hours) exposure to water at temperatures above freezing.

   b. Causes of Trenchfoot/Immersion Foot. The major risk factors are wet, cold and immobility.

   c. Signs and Symptoms of Trenchfoot/Immersion Foot.

   (1) The major symptom will be pain. Trench foot is an extremely painful injury.

   (2) Trench foot and frostbite are often very difficult to tell apart just from looking at it. Often they may both be present at the same time. Signs include:

   (a) Red and purple mottled skin.

   (b) Patches of white skin.

   (c) Very wrinkled skin.

   (d) Severe cases may leave gangrene and blisters.

   (e) Swelling.

   (f) Lowered or even absent pulse.

   (3) Trench foot is classified from mild to severe.

   d. Prevention of Trench foot/Immersion Foot is aimed simply at preventing cold, wet and immobile feet (or hands).

   (1) Keep feet warm and dry.

   (2) Change socks at least once a day. Let your feet dry briefly during the change, and wipe

out the inside of the boot. Sock changes may be required more often.

(3) Exercise. Constant exercising of the feet whenever the body is otherwise immobile will help the blood flow.

e. Treatment of Trench foot/Immersion Foot.

(1) All cases of trench foot must be evacuated. It cannot be treated effectively in the field.

(2) While awaiting evacuation:

(a) The feet should be dried, warmed, and elevated.

(b) The pain is often severe, even though the injury may appear mild; it may require medication such as morphine.

(3) In the rear, the healing of trench foot usually takes at least two months, and may take almost a year. Severe cases may require amputation. *Trench foot is not to be taken lightly.*

TRANSITION: We have just discussed the definition, signs and symptoms, prevention, and treatment of trench foot. Are there any questions? The next injury, which is easy to acquire at high altitude, is snow blindness.

16. (5 Min) **SNOW BLINDNESS**

a. Definition. Sunburn of the cornea.

b. Causes of Snow Blindness. There are two reasons Marines in a winter mountainous environment are at increased risk for snow blindness.

(1) High altitude. Less ultraviolet (UTV) rays are filtered out, UV rays are what cause snow blindness (as well as sunburn). So at altitude, more UV rays are available to cause damage.

(2) Snow. The white color of snow reflects much more LTV rays off of the ground and back into your face.

c. Signs and Symptoms of Snow Blindness.

(1) Painful eyes.

(2) Hot, sticky, or gritty sensation in the eyes, like sand in the eyes.

(3) Blurred vision.

(4) Headache, may be severe.

(5) Excessive tearing.

(6) Eye muscle spasm.

(7) Bloodshot eyes.

d. Prevention of Snow Blindness. Prevention is very simple. Always wear sunglasses, with UV protection. If sunglasses are not available, then field expedient sunglasses can be made from a strip of cardboard with horizontal slits, and charcoal can be applied under the eyes to cut down on reflection of the sun off the snow.

e. Treatment of Snow Blindness.

(1) Evacuation, when possible.

(2) Patch the eyes to prevent any more light reaching them.

(3) Wet compresses, if it is not too cold, may help relieve some of the discomfort.

(4) Healing normally takes two days for mild cases or up to a week for more severe cases.

TRANSITION: Are there any questions about snow blindness? If not, let's discuss carbon monoxide poisoning.

17. (5 Min) **CARBON MONOXIDE POISONING**

a. Definition. Carbon Monoxide (CO) is a heavy, odorless, colorless, tasteless gas resulting from incomplete combustion of fossil fuels. CO kills through asphyxia even in the presence of adequate oxygen, because oxygen-transporting hemoglobin has a 210 times greater affinity for CO than for oxygen. What this means is that CO replaces and takes the place of the oxygen in the body causing Carbon Monoxide poisoning.

b. Signs/Symptoms. The signs and symptoms depend on the amount of CO the victim has inhaled. In mild cases, the victim may have only dizziness, headache, and confusion; severe cases can cause a deep coma. Sudden respiratory arrest may occur. The classic sign of CO poisoning is cherry-red lip color, but this is usually a very late and severe sign, actually the skin is normally found to be pale or blue.

(1) CO poisoning should be suspected whenever a person in a poorly ventilated area suddenly collapses. Recognizing this condition may be difficult when all members of the party are affected.

c. Treatment. The first step is to immediately remove the victim from the contaminated area.

(1) Victims with mild CO poisoning who have not lost consciousness need fresh air and light duty for a minimum of four hours. If oxygen is available administer it. More severely affected victims may require rescue breathing.

(2) Fortunately, the lungs excrete CO within a few hours.

d. <u>Prevention</u>. Prevention is the key. Ensure that there is adequate ventilation when running vehicle engines, operating stoves in closed spaces (tents), or when cooking over open flames.

TRANSITION: Let us now discuss personal hygiene and ways to take care of our body through field expedient measures.

18. (5 Min) **<u>PERSONAL HYGIENE.</u>** The five most important areas of personal hygiene are:

a. <u>Body</u>

(1) The body should be washed frequently in order to minimize the chances of small cuts and scratches developing into full blown infections and as a defense against parasitic infections.

(2) A daily bath or shower consisting of soap and hot water is ideal. However, when this is not possible you should:

(a) Give yourself a sponge bath using soap and water, making sure particular attention is given to body creases i.e., armpits, groin area, face, ears, and hands.

(b) If water is in extremely short supply, you should take an air bath. To do this:

1. Remove all clothing and hang it up to air.

2. Expose the body for two hours to sunlight, which is ideal, but the effects will basically be the same if done indoors or during an overcast day. **BE CAREFUL NOT TO SUNBURN.**

b. <u>Hair</u>

(1) Should be cleaned frequently.

(2) Should be inspected at least once a week for parasites.

c. <u>Fingernails</u>

(1) Should be trimmed to prevent accidentally scratching yourself.

(2) Should be kept clean to prevent harborage areas for bacteria.

d. <u>Feet</u>. They are your primary source of transportation.

(1) The feet should be inspected frequently for:

(a) Blisters.

(b) Infections. Bacterial and fungal.

(2) They should be kept dry by:

(a) Frequent sock changes (one to three times daily) in conjunction with:

1. Foot powders.

2. Antiperspirants.

e. <u>Oral Hygiene.</u> The mouth and teeth should be cleaned at least daily to prevent tooth decay and gum disease.

(1) Ideally, cleaning should be done with:

(a) Toothbrush.

(b) Toothpaste.

(c) Dental floss.

(2) If these items are not available, the following methods can be used:

(a) Make a chew stick from a clean twig about 8" long and about finger width. Chew one end until it becomes frayed and brush-like, and then brush the interior and exterior surfaces thoroughly.

(b) The gums should be stimulated at least once a day by rubbing them vigorously with a clean finger.

(c) Field expedient dental floss can be made from the inner strands of a 10 inch section of paracord.

<u>TRANSITION</u>: Now that we have discussed personal hygiene, are there any questions? Since we have in mind our own personal hygiene, let's consider another item which can effect our health.

19. (5 Min) **WATER PURIFICATION**. Water purification simply consists of removing or destroying enough impurities to make water safe to drink. Giardia cysts are an ever-present danger in clear mountain water, even though this water appears clean and safe to drink.

   a. Boiling is the oldest way of water disinfection. Recent studies have shown that the old recommendation of boiling water for 10 minutes and adding 1 minute of boiling for each 1000 feet in elevation is not necessary and wasteful of limited fuel supplies. The studies found that the thermal death point of microorganisms is reached in shorter time at higher altitudes, while lower temperatures are effective with a longer contact period. Therefore the minimum critical temperature needed to render water safe to drink is well below the boiling point at elevation. With these findings in mind it can now be safely said that once water is brought to a boil, and allowed to cool it is disinfected and safe to drink. For an extra margin of safety the Wilderness Medical Society recommends boiling water for 11minute no-matter what altitude you're at to render water safe to drink.

TRANSITION: Are there any questions over water purification? Now we will discuss what the human animal produces in the area of waste products and how to dispose of them.

20. (5 min) **WASTE DISPOSAL**. Waste should be disposed of by burning, burying, or hauling it away.

   a. The importance of waste disposal cannot be overemphasized.

      (1) It serves to:

         (a) Eliminate harborage areas for rodents and vermin.

         (b) Preclude an attractant for rodents and vermin.

         (c) Prevent a source of pathogenic contamination.

      (2) Two basic types:

         (a) Organic wastes

            (1) Human waste - burn or haul away to a designated waste pit area.

            (2) Urine - Use only assigned, marked areas away from food and water sources.

            (3) Edible garbage - Burn. Do not leave exposed for animals, vermin, or the enemy.

         (b) Non-organic wastes

            (1) Papers -burn.

            (2) Metals - haul away or bury.

(3) Liquids – burn or bury.

TRANSITION: Improper disposal of waste is hazardous in more ways than one. Are there any questions at this time?

PRACTICE (CONC)

a. Students will practice what was taught in upcoming field evolutions.

PROVIDE HELP (CONC)

a. The instructors will assist the students when necessary.

OPPORTUNITY FOR QUESTIONS (3 Min)

1. QUESTIONS FROM THE CLASS

2. QUESTIONS TO THE CLASS

    Q. What are the five ways a body loses heat?

    A. (1) Radiation
       (2) Conduction
       (3) Convection
       (4) Evaporation
       (5) Respiration

    Q. What are the two causes of dehydration?

    A. (1) Inadequate intake
       (2) Excessive loss

    Q. What is the best treatment for HAPE and HACE?

    A. Descent.

SUMMARY (2 Min)

a. During this period of instruction we have covered the way a body loses heat, dehydration, heat injuries, high altitude sickness, hypothermia, frostbite, trench foot, snow blindness, carbon monoxide poisoning, and personal hygiene.

b. Those with IRF's please fill them out at this time. We will now take a short break.

UNITED STATES MARINE CORPS
Mountain Warfare Training Center
Bridgeport, California 93517-5001

WML
WMO
02/11/02

LESSON PLAN

WINTER WARFIGHTING LOAD REQUIREMENTS

INTRODUCTION   (5 Min)

1. **GAIN ATTENTION**. Marines operating in a mountainous environment will be carrying allot of extra gear just to survive the elements and negotiate the difficult terrain. This equipment, plus the gear they need to wage war, will be carried on their backs. This period of instruction will cover the different types of loads that Marines will be required to carry and also provide some general guidelines on how to properly carry it.

2. **PURPOSE**. The purpose of this period of instruction is to introduce the student to the different levels of gear requirements in accordance MCRP 3-35.1A, Small Units Leader's Guide to Cold Weather Operations. This lesson relates to all ski training and tactical evolutions.

3. **INTRODUCE LEARNING OBJECTIVES**

   a. TERMINAL LEARNING OBJECTIVES. In a cold weather mountainous environment, pack for ski borne/snowshoe movement, in accordance with the reference.

   b. ENABLING LEARNING OBJECTIVES

   (1) Without the aid of references, list in writing the seven pocket items required to be carried by each individual, in accordance with the references.

   (2) Without the aid of references, list in writing the eight required items for the Assault Load, in accordance with the references.

   (3) Without the aid of references, list in writing the six required items that are added to the Assault Load to make the Combat Load, in accordance with the references.

4. **METHOD / MEDIA**. The material in this lesson will be presented by lecture and demonstration. You will practice what you have learned during upcoming field training exercises. Those of you with IRF's please fill them out at the end of the lesson and turn them in.

5. **EVALUATION**. You will be tested later in this course by written and performance evaluations on this period of instruction.

TRANSITION: The first thing we need to talk about is the uniform requirements and the necessary gear worn on your person and carried in your pockets.

BODY (35 Min)

1. (5 Min) **BASIC UNIFORM REQUIREMENTS**.

    a. The following list of uniform items comprise the basic cold weather uniform. This list may vary depending upon the severity of the weather, the activity level of the Marine, and the individual metabolism of the Marine. However, the outer camouflage layer should be dictated by the unit leader.

    (1) ECWCS Parka and Trousers

    (2) C/W Trousers w/Suspenders

    (3) C/W Hat or Balacalava

    (4) Polypropylene or Capilene long underwear

    (5) VB Boots or Ski March Boots w/appropriate sock system and gaiters

    (6) Gloves with Inserts and Trigger-finger Mittens

    (7) Overwhites (parka, trouser, pack cover, and over mittens)

    (8) Helmet w/Camo cover (white)

    b. As part of the basic cold weather uniform, each man should be required to have in his possession at all times some required pocket items. These seven items should be carried in the pockets of your ECWCS uniform:

    (1) Pocket knife

    (2) Whistle

    (3) Pressure Bandage

    (4) Chapstick and sunscreen

(5) Sunglasses

(6) Survival Kit and rations

    (a) Signaling items

    (b) Food items

    (c) Water procuring items

    (d) First aid items

    (e) Fire starting items

    (f) Shelter items

(7) Notebook with pen/pencil

c. Some additional items that should be carried in your pockets at all times are:

(1) Contact gloves

(2) Avalanche cord (10 meters)

(3) Flashlight w/ tactical lens and spare batteries

(4) Chemlights or route marking equipment

(5) High energy, lightweight snacks

TRANSITION: Next, we will discuss the items that make up the basic equipment required to survive and accomplish our mission.

2. (5 Min) **ASSAULT LOAD**

   a. The Assault Load (Level 1) is equipment in addition to the basic cold weather uniform requirements, and is carried in the load bearing vest (LBV), butt pack and the pack system. This is the equipment carried for short duration missions such as security patrols or during the final assault phase. It is carried at all times when you are away from your bivouac site.

      (1) An extra insulating layer (Fleece jacket, wooly pully, etc.)

      (2) Protective layer (ECWCS parka and trousers ,if not worn)

      (3) LBV with 2 quarts of water and first aid kit

(4) Rations for the time away from your bivouac site

(5) Extra socks and gloves

(6) Cold weather hat or balaclava

(7) Isopor mat (strapped to assault pack or carried on the ski pole)

(8) Over-the-snow mobility (skis, poles, wax kit and/or snowshoes)

(9) Mission essential gear as required:

    (a) T/O weapon w/accessories (sling, magazines, cleaning gear, bayonet/K-bar, and basic allowance of ammunition)

    (b) *Extra ammunition, demolitions, and pyrotechniques

    (c) *Optical gear (binoculars, night vision devices, etc.)

    (d) *Communications equipment (field phones, spare batteries, etc.)

    (e) *Navigational equipment

NOTE: Mission essential gear items indicated with an * are spread-loaded throughout the unit as the mission dictates. Also, it may be required for designated personnel (such as RTO's) to carry the assault load in the large Vector pack vice in the small assault pack.

TRANSITION: The next level of equipment is the Combat Load.

3. (5 Min) **COMBAT LOAD**. The Combat Load (Level 2) is the equipment carried for longer duration missions such as movements to contact. It is carried in the large Vector pack and consists of essential gear required in the event of an unplanned bivouac and the gear required to conduct medevacs. The following items are in addition to the items already being carried in the Assault Load:

    a. Sleeping bag (w/ bivy bag if issued) inside a WP bag

    b. Snow shovel (to dig expedient shelters, fighting positions, or rescue avalanche victims)

    c. Individual/squad stove

    d. Fuel bottle w/ fuel

    e. Thermos

    f. Poncho (for expedient shelters or medevac purposes)

NOTE: If the gear list dictates that each man carries the Assault Load, then 1 man per squad will also bring the Combat Load items. These items may be spread-loaded throughout the squad to prevent over-burdening 1 man with the extra weight. If all personnel are carrying the Combat Load, then items b, c, d, e, and f are 1 each per 2 men.

TRANSITION: The final individual load is the Existence Load.

4. (5 Min) **EXISTENCE LOAD**. The Existence Load (Level 3) is any extra gear that is required that can be brought up to the forward combat elements once the situation allows. Ideally, each fire team packs their excess gear in one seabag, and it comes forward on the log train. It includes, but is not limited to:

   a. Extra insulating layers

   b. Extra socks

   c. Extra glove and mitten liners

   d. Toiletries

   e. Repair kit

TRANSITION: Now that we know about the different loads we are required to carry, let's now cover some general guidelines that come in handy.

5. **PACKING THE PACK**. Because most Marines are familiar with how to pack a pack, these are general guidelines only.

   a. Keep your pack as compact as possible. The bulk of the weight should be next to your body, low and centered. This will assist you in maintaining your balance when skiing or snowshoeing.

   b. Your shoulder and waist straps should be properly adjusted. They should be snug yet allow freedom of movement to prevent cutting off circulation.

   c. Limit the amount of gear and miscellaneous equipment you have strapped to the outside of your pack. This will reduce the amount of snagging when moving through dense vegetation and should prevent lost gear that may save your life.

   d. Ensure you properly waterproof any gear you do not want to get wet. Rain or snow can quickly soak through your pack and make your last pair of dry socks wet.

   e. Use stuff sacks to keep small or related gear neatly stowed where you can readily locate it, even in the dark.

f. When not wearing your protective layer or insulating layers, keep it handy at the top of the pack. When taking a break during a movement, you will be able to quickly don a layer to prevent getting chilled.

g. Keep your stove and fuel bottle in the outside pockets of your pack. They may leak and soak your equipment with fuel if stored inside your pack.

h. Keep your ski wax kit readily available in the outside pockets, this allows easy access to your wax during short halts.

i. Keep some high energy snacks handy in the exterior pockets of your pack. This will allow you to refuel your system on short breaks during movements without the need to dump your gear.

j. When not wearing your overwhites, keep them handy under the map flap of your pack, so you can quickly change your camouflage as the terrain requires.

k. Snowshoes should be secured to your pack with straps, paracord, bungy cord, etc. with the tails pointing up and the shovel wrapping in under the bottom of the pack.

l. Skis should be carried on the pack by sliding the tails down through the external side pockets. The tips are then strapped together with a toe strap. The tails are slid into the small pockets on the waist band. Do not allow the tails to hang too low, as this will irritate your legs and interfere with walking.

TRANSITION: Properly carrying the needed equipment in a cold weather environment call make the difference between an exhausted Marine and a combat-ready Marine.

PRACTICE (CONC)

a. Students will carry the different fighting loads throughout the upcoming field evolutions.

PROVIDE HELP (CONC)

a. The instructors will assist the students when necessary.

OPPORTUNITY FOR QUESTIONS (3 Min)

1. **QUESTIONS FROM THE CLASS**

2. **QUESTIONS TO THE CLASS**

    Q. What are the 7 required pocket items required to be carried by all individuals?

    A. (1) Pocket Knife
       (2) Whistle

(3) Pressure Bandage
(4) Notebook w/ pen /pencil
(5) Chapstick and Sunscreen
(6) Sunglasses
(7) Survival kit and Rations

Q. What are the 6 items that are added to the Assault Load to make the Combat Load?

A. (1) Sleeping Bag   (4) Fuel bottle w/ fuel
   (2) lsopor mat   (5) Thermos
   (3) Individual/squad stove (6) Poncho

SUMMARY (2 Min)

a. During this period of instruction we have covered the basic fighting load requirements for cold weather operations, the 3 fighting loads, the group stores, and some general guidelines for packing your packs.

b. Those of you with IRF's please fill them out at this time. We will now take a short break.

UNITED SATES MARINE CORPS
Mountain Warfare Training Center
Bridgeport, California 93517-5001

WML
WMO
09/31/00

LESSON PLAN

MILITARY SKI EQUIPMENT

INTRODUCTION (5 Min)

1. **GAIN ATTENTION**. Marines must be able to move over the snow to be successful in winter warfare. Snowshoes, augmented by helicopters, provide one of the means. Skis provide another means and once mastered, are much faster and more efficient than snowshoes.

2. **PURPOSE**. The purpose of this period of instruction is to introduce the student to military ski equipment, its nomenclature, and how to care for it. This lesson relates to ski movement.

3. **INTRODUCE LEARNING OBJECTIVES**

    a. TERMINAL LEARNING OBJECTIVE. Given a pair of military skis with poles, maintain military ski equipment, in accordance with the references.

    b. ENABLING LEARNING OBJECTIVES

    (1) Without the aid of references, given a diagram of military skis and ski poles, identify in writing the nomenclature of military skis and poles, in accordance with the references.

    (2) Given a pair of military skis with bindings, adjust the ski bindings, in accordance with the references.

    (3) Given a pair of military skis and ski waxes in prevailing snow conditions, wax skis to provide both grip and glide, in accordance with the references.

4. **METHOD / MEDIA**. The material in this lesson will be presented by lecture and demonstration. You will practice what you have learned during upcoming field training exercises. Those of you with IRF's please fill them out at the end of this period of instruction.

5. **EVALUATION**. You will be tested later in the course by written and performance evaluations.

TRANSITION: The first topic we'll discuss will be the different types of skis.

BODY (65 Min)

1. (5 Min) **TYPES OF SKIS**. To understand the design of skis, we should first look at what their differences and uses are. For simplicity we can break skiing into three categories.

    a. Nordic Skiing (Cross Country). There are several different types of cross country skis. They are as follows:

    (1) Racing. This type of ski is the narrowest of the cross country skis. The racer wants lightweight skis that do not drag in the tracks (The racer always skis in a prepared track so support is not a concern). There are two types of racing skis; the classic and the marathon skating ski.

    (2) Light Touring. This type of ski is narrow to medium in size. The light tour skier wants skis suited to racing but they must be wide enough to support his weight in the powder and for off track skiing.

    (3) Touring. This is a medium to wide ski. The tour skier wants support when breaking trail and skiing the powder in the back country where a wider ski is a must.

    (4) Keep in mind that all these skies mentioned have no metal edges, therefore they are not desirable for descending steep slopes.

    b. Alpine Skiing (Downhill). This is the type of skiing done at ski resorts. The alpine ski makes skiing very easy for the normal skier. It is designed to go down a slope with minimum effort, but it is nearly impossible to move on gentle rolling terrain on this gear.

    c. Mountaineering Skiing. This is the widest of the ski types. The mountain skier wants skis for support and turning. A wider ski is designed to turn more easily and the increased width supports better in the backcountry. The mountaineering ski is a cross between Nordic and Alpine type skis with a metal edge and bindings that allow free movement of the heel.

TRANSITION: Let's now look at the nomenclature of the military skis.

2. (10 Min) **MILITARY SKIS**. The current military skis in the system is the Asnes double cambered skis. These skis vary in length from 180cm to 210cm and have a hole in the tips for towing. The size of the skis issued to individuals will be dependant upon their weight. Heavier Marines should receive longer skis to assist in flotation.

a. Nomenclature of the military ski.

   (1) <u>Tip</u>. The obvious forward point of the ski.

   (2) <u>Shovel</u>. The curvature at the front of the ski that helps push aside the snow as it moves.

   (3) <u>Tail</u>. This is the rear of the ski. It has a notch in the center for attaching climbing skins.

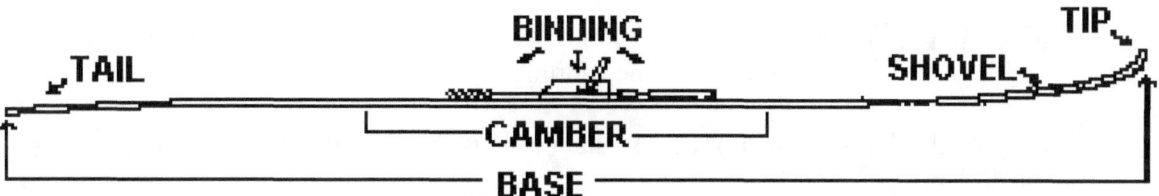

   (4) <u>Base</u>. This is the bottom of the ski which is made from a synthetic material called P-Tex.

   (5) <u>Camber</u>. If you put a pair of skis together, you will notice that there is a bow in the center portion of the skis, what ski manufactures call a "camber". It is often referred to as the wax pocket. When you put weight on the ski you will notice that the bow will flatten out. The amount of weight needed to flatten the bow depends on the skier's weight and ability.

      (a) <u>Single camber</u>. The camber in the ski is soft so that more of the ski is in contact with the snow. This makes steering easier and provides more control.

      (b) <u>Double camber</u>. This is a term that is used to describe the "stiffness" in the camber of the ski. A double camber ski will require more force to flatten out which causes steering to be more difficult, but the stiffer ski will have a better gliding and grip wax pocket.

   (6) <u>Binding</u>. The NATO 120 binding is an all-metal binding consisting of a cable clamp, toe plate and a cable. This will be discussed in further detail in the next section.

b. Other features

   (1) <u>Tracking Groove</u>. If you look at the bottom of almost every cross country ski, you will see a u-shaped or l-shaped cut called a "tracking groove" going from just below the tip down to the tail. It is designed to help the ski run over the snow in a straight line. Without the groove, the ski will tend to wobble or move from side to side.

   (2) <u>Kick Zone</u>. This area is located within the camber and is where the wax is applied. The kick zone will vary between skiers due to his weight, ability and the terrain being skied. As a general guideline, the zone begins 6 inches before the binding and extends to 6 inches beyond the binding.

(3) <u>Glide Zone</u>. This is the area that remains in contact with the snow surface. The glide zone runs from the ends of the kick zone to the respective ends of the ski.

(4) <u>Metal Edge</u>. This is an alpine ski feature that is essential for mountaineering skis. The edge can be offset or flush with the running surfaces of the ski.

(5) <u>Side Wall</u>. The side cover of the ski that protects the core of the ski from warping due to water damage.

(6) <u>Sidecut (Waist)</u>. The difference in width measurements from tip to tail is referred to as the ski's sidecut. The sidecut makes the ski easier to turn when pressure is applied on the ski at an angle to the snow surface.

(7) <u>Flex</u>. There are three things involving flex:

(a) <u>Tip flex</u>. Soft tips follow the terrain by easily flowing over bumps, dips and irregularities in the snow. If the tip is too soft, the ski tends to wander and become difficult to control in turns. Moderate tip flex is more desirable for back country touring and mountain skiing, thus providing better flotation in powder and adequate control when turning.

(b) <u>Tail flex</u>. Tail flex is similar to tip flex in its response to snow and turning. If it is too soft, the ski may wash out or not hold an edge while turning.

(c) <u>Torsional flexibility</u>. This refers to the twisting action from side to side that a ski goes through while in a turn or track. A good touring or mountain ski has a torsionally stiffer tip which gives the ski more holding power and better edge control when turning.

TRANSITION: Let's now take a look at our ski bindings.

3. (10 Min) **SKI BINDING**. The NATO 120 Binding is a versatile binding because it can be fitted to a variety of boots. The Vapor Barrier (VB) boot and the 75mm box-toed leather ski march boot being the most commonly used in the Marine Corps.

   a. Nomenclature of the NATO 120 Binding

      (1) <u>Toe plate</u>. This consists of a wing nut fastener, locking lever, and two adjustable toe plates designed for proper emplacement of the toe of a boot toe.

(2) <u>Cable clamp</u>. This is located in the front of the binding and is designed to tension the cable around the boot. The cable clamp also has a retractable nut, which allows for two full sizes of adjustment of a cable to a boot.

(3) <u>Cable</u>. This is a plastic coated cable with a coil spring portion that fits behind the back of the heel. The cables come in four sizes with a different colored band representing the size rating of that cable. The color and corresponding boot size are as follows:

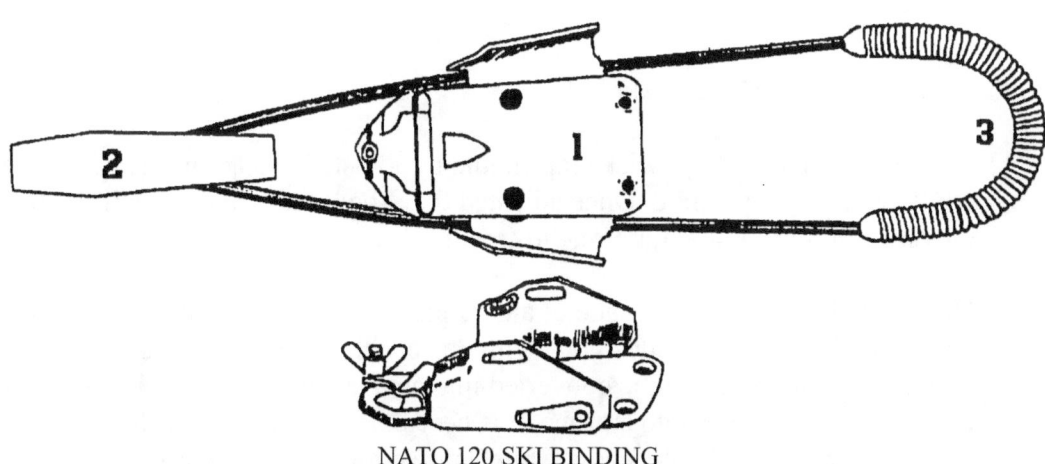

NATO 120 SKI BINDING

CABLE SIZING CHART		
BAND COLOR	**SIZE**	**BOOT SIZE**
Blue band	XLG	12 to 14
Black band	LG	10 to 12
Green and yellow band	MED	8 to 10
Red band	SM	6 to 8

NOTE: Blue band cables need to be ordered separately.

b. Adjustment of the NATO 120 Binding

(1) <u>Toe plate</u>. Due to the different size boots used with this binding, the adjustment of the toe plate is necessary. Start by loosening the wing nut and lifting the locking lever up. Place the toe of the boot behind the wing nut ensuring that the heel of the boot is centered on the ski. Align the toe plates against the welt of the boot and push the locking lever back down into its original position. Tighten the wing nut until no movement can be felt from the toe plate. It may be necessary to mark which ski is left or right since each toe plate will be adjusted differently.

(2) <u>Cable clamp</u>. Ensure that the proper tension is attained and secure the clamp down. Minor adjustments to the tension can be done by unscrewing or tightening the nut. Once the cable is clamped down, there should be no lateral movement of the boot.

(3) <u>Cable</u>. No adjustment is needed for the cable. The cable clamp will compensate for the proper adjustment. Attention to the correct cable length for the boot size is important.

<u>TRANSITION</u>: The next topic we will discuss is the military ski poles.

4. (5 Min) **<u>MILITARY SKI POLES</u>**. The ski poles aids the skier in movement, balance and timing.

   a. Nomenclature of the Military Ski Poles.

      (1) <u>Wrist straps</u>. The leather wrist strap should be adjusted to support the wrist for pushing while cross country skiing. Once adjusted they should not be cut as different types of gloves or mittens will require readjustment.

      (2) <u>Handgrip</u>. The handgrip is made of a hard plastic and is where the wrist straps are attached. Some poles are designed with detachable color-coded handgrips, which indicate that the poles can be converted into an avalanche probe pole. In order to transform the ski poles into a probe, it is necessary to have a red colored top handgrip, representing a "male" pole, and a white colored top handgrip representing a "female" pole.

         (a) To convert the ski poles to an avalanche probe pole, remove the color-coded handgrips by unscrewing them from the poles.

         (b) Screw the male end into the female end.

         (c) Remove a basket from one end of the pole.

      (3) <u>Shaft</u>. The shaft is made from one piece of hollow aircraft aluminum.

      (4) <u>Basket</u>. The basket is located near the end of the pole. This basket allows the pole to remain above the surface of the snow during pole plants.

      (5) <u>Point</u>. This is located at the end of the pole, also known as a Ferrule. This is what penetrates the snow surface during pole plants.

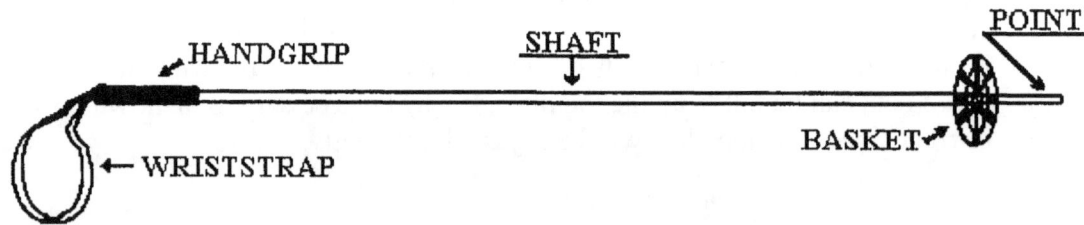

MILITARY SKI POLE

b. <u>Other considerations</u>

(1) <u>Length</u>. The military ski pole comes in three lengths, 130cm, 137cm and 147cm. To properly size a pole to an individual, place the tip on the deck, the handgrip should fit snugly under the individual's armpit.

(2) <u>Weight</u>. Each pole weighs approximately one pound.

<u>TRANSITION</u>: Now that we know about the nomenclature of the military skis/ski poles, we are ready to learn about the care and maintenance of the military ski.

5. (10 Min) **<u>CARE AND MAINTENANCE OF THE MILITARY SKI</u>**. Serviceability checks and proper maintenance is a regular routine for ski equipment.

a. <u>Serviceability Checks</u>.

(1) <u>Breakage</u>. Skis are vulnerable to load stress, particularly the tips and tails.

(2) <u>Delamination</u>. This occurs when the plastic coating separates from the ski causing water damage to the inner core. Skis should be frequently checked for nicks and gouges of the coating.

(3) <u>Ski base</u>. The ski base should be uniformly flat and smooth. Check for possible gouges and cuts. These may hamper the glide of the ski and create problems in waxing. Any gouges or cuts should be filled to prevent an unstable ski.

(4) <u>Ski edges</u>. If the ski edge is separated from the ski, turning will become a problem. Generally, if large pieces of the edges are missing, mobility will become difficult, especially when turning or edging. The ski should be replaced.

(5) <u>Bindings</u>. Check all metal parts for stress fractures and missing parts. Ensure that the cable is not missing large sections of the plastic coating or that the coil spring is not over stretched. This can possible damage your boots.

(6) <u>Detuning</u>. If the skis are new, the metal edges are very sharp. Detuning the tips and tails approximately six inches on both sides will help prevent you from "catching an edge".

b. <u>Care of the Ski</u>

(1) <u>Heat</u>. Don't place the ski next to direct heat, because the bottom of the ski could easily melt. Anytime heat is placed near a ski, it should only be for a few seconds. Caution should be taken not to attach skis too close to the exhaust pipes on tracked vehicles.

(2) <u>Snow/ice</u>. Remove snow and ice from the skis before bivouacing. Icing may occur making the ski difficult to wax later on.

(3) <u>Waxes</u>. Remove all waxes before bivouacing, or you may be skiing with yesterday's wax on today's snow conditions.

(4) <u>Staging skis</u>. Never place the tails of the skis straight into the snow. This can cause damage to the tails by striking a possible hidden object i.e. rocks, tree stumps, etc. Skis should be staged during breaks by directing the base of the ski toward the sun using the poles for support. This will keep the grip wax fresh and applicable.

c. <u>Maintenance of the Ski</u>.

(1) <u>Base preparation</u>. Base glide wax is used to protect the ski's synthetic base and to maximize the forward speed of the ski. The simplest method for waxing is to rub the wax directly on the base. However; hot waxing allows the wax to penetrate deeper and last longer. Here's how it is done:

(a) Ensure that the base is clean and dry.

(b) If the base is damaged due to gouges, a P-TEX candle can be melted into the problem areas. Ensure that the excess is smoothed evenly across the base. Repeat the process if the patch shrinks below the surrounding base surface.

(c) With a hot wax iron, apply the based glide wax by dripping it over the base of the ski.

(d) Run the wax iron over the ski to reheat the wax and to spread it evenly along the base of the ski. Always keep the iron moving to avoid melting the synthetic base.

(e) Remove excess wax with a scraper to include the tracking groove.

(2) <u>Metal edge tuning</u>. When the ski metal edge is dulled or pitted, turning the ski can be difficult. Therefore, the edges should be sharpened as needed, except in the shovel and tail area. These two areas should be divided (detuned) for maximum turning efficiency.

(a) To file the ski edges, hold the file parallel and lengthwise against the side of the edge. Your thumb should be on top, and your fingers curled under, acting as a guide along the ski itself.

(b) File from the tip to tail, while being careful to keep the file at a 90 degree angle to the base; avoid excessive filing. To smooth out the operation of the ski, the first 6 inches at the tip, and the last 3 inches at the tail, should be slightly dulled.

TRANSITION: Now lets discuss an important aspect of military skiing - waxing.

6. (10 Min) **SKI WAXING**

a. <u>Factors Affecting Wax</u>. Wax is applied to the base of the skis to prevent slipping, influence momentum, and help maintain glide. The goal of waxing is to find the proper wax combination for optimum grip and glide, without sacrificing either. As the snow conditions and temperature change, the wax required will also change.

b. <u>Ski Waxes</u>

(1) Each wax has its range of ideal snow condition. The type of snow (wet or dry) and the temperature will play a role in what wax you choose and how you will apply it. The wax chart below provides examples of some waxes commonly used.

WAX CHART

WAX	**SNOW TEMPERATURE (°F)**	**USAGE**
Polar	-22° to 5°	Very cold snow condition wax. Frequently used in the polar regions of Europe and North America.
Green Special	5° to 12°	Cold snow condition wax and ideal for those who want a long kick zone.
Green	9° to 19°	Cold snow wax.
Blue Special	16° to 23°	Cold snow wax with a wide range.
Blue	18° to 27°	Used on moderate cold snow.
Blue Extra	19° to 32°	Demands new snow and low air humidity (below 45-50%).
Violet Special	30° to 32°	When Blue Extra becomes somewhat slippery, a thin layer of Violet Special probably is the right wax.
Violet	32°	To be used when the snow is in a stage going from cold to wet, around freezing (32°).
Violet Extra	32° to 34°	Used when the snow is moist.
Red Special	30° to 36°	Wax for moist, new snow, but might go into some colder snow and ball up the snow in your kick zone.
Red	32° to 38°	Wax for moist to wet new snow, when making snowballs is very easy.
Red Extra	34° to 38°	For wet, new snow and glazed tracks. Must be applied in an even layer to prevent balling.

KLISTER CHART

KLISTER	**SNOW TEMPERATURE (°F)**	**USAGE**

Green	-13° to 27°	To be used as a first layer as a binder for other klisters or hard waxes in very abrasive conditions. Used alone for icy conditions at very cold temperatures. As a first layer, it should be heated into the base of ski.
Blue	5° to 32°	For frozen, icy conditions. Can also be used as a base for wet-snow.
Violet	25° to 37°	Use for snow on both sides of freezing, but also when conditions are changeable and mixed with fine grained snow.
Red	32° to 41°	Used on changing, coarse grained snow on both sides of freezing. Best on the warm side of freezing when the snow is wet.
Orange	36° to 52°	Used when the snow has a high water content such as slush and the air temperature is well above freezing.

    (2) Two wax system. Instead of carrying around all the above mentioned waxes, we can keep it simple using a two wax system. This is what training units will be issued at MWTC.

TWO WAX SYSTEM

WAX	**SNOW TEMPERATURE (°F)**	**USAGE**
Blue	5° to 32°	A wide range wax for normal cold snow.
Red	32° and warmer	Ideal for conditions around freezing and slightly warmer.

  c. <u>Wax and it's effects on snow</u>. When wax is spread on the ski's base, it provides a cushion for thousands of tiny snowflakes to stick into. For a brief moment, when all the skier's weight is on one ski, a multitude of snow crystals are embedded in the wax, holding the ski firm while the skier pushes (or kicks) off. As the ski begins to slide forward, the pressure release and friction of the sliding ski releases this bond. A thin, microscopic layer of water (as a result of friction between the ski's base and the snow) causes the ski to glide until it stops and downward pressure is again applied.

  d. <u>Classifications of Snow</u>

    (1) Fresh fine-grain snow. This usually occurs after a snowstorm when temperatures are cold. Each flake has a very distinctive feature, like the fingers of an outstretched hand.

    (2) Coarse-grain snow. This develops one to three days after a snowstorm. This snow undergoes a transformation in which the crystals begin to lose their shape. Evaporation and compression cause a rounding of the fresh snow's sharp points, like taking an outstretched hand and starting to form a fist.

(3) Granular snow. This type of snow develops when the temperature rises above freezing and the snow melts and refreezes. It is similar in shape to a closed fist, with no relationship to the original snowflake.

(4) Wet snow. This type of snow is usually found during the spring. It may also occur at other times, particularly in regions of moderate climate. This type of snow can be made into a heavy, solid snowball. In extreme conditions, wet snow will become slushy, and contain a large amount of water.

(5) Dry snow. This type of snow is generally associated with winter at its height, but it can occur in late autumn, as well as in the spring, when abnormally low temperatures occur. This snow is light and fluffy. It cannot be compressed into a hardball unless the snow is made moist by holding it in the hand. At extremely low temperatures, such as those found in the far northern regions, this snow is like sand, and has very poor sliding qualities.

e. <u>Proper Selection and Application of Ski Waxes</u>. Grip waxes are formulated to provide optimum gliding and gripping characteristics for various types of snow and temperatures. Each type is labeled with appropriate instructions on its intended use, i.e., wet, moist or dry snow conditions, or temperature. Since the type of wax varies between manufacturers, no particular type of wax can be prescribed for each classification of snow; however, the instruction on each container specifies the weather conditions and the type of snow where performance of the wax is best. To provide a proper grip, varying amounts, combinations, and methods of application of different waxes may be used.

(1) <u>Tips for apply grip wax</u>:

(a) Wax in the temperature being skied.

(b) Ski should be dry when waxing.

(c) The wax should be corked in from tip to tail.

(d) Grip waxes may be crayoned, then corked in.

(e) When in doubt start with a harder wax.

(f) Don't apply a harder on top of a softer wax.

(g) Several thin layers work better than one thick layer.

(h) During movement carry wax in an inside pocket to keep it warm.

(i) Do not put newly waxed skis on the snow, until the wax has cooled to air temperature. If the lead group has a specific wax on for the conditions of starting out, the trailing group might want to use a warmer or colder wax due to different track conditions left by the lead group.

   f. <u>Test</u>. Before you move out on a ski march or any other type of movement, test the waxing job prior to movement and re-wax if necessary. It will normally take several hundred feet of skiing for the wax to function properly.

   g. <u>Wax Kit</u>. The wax kit should include:

   (1) Two grip waxes (The two wax system)

   (2) Cork

   (3) Scrapper

TRANSITION: Now let us finally talk about climbing skins.

7. (5 Min) **CLIMBING SKINS**. Climbing skins were originally made from sealskin. Modern day skins are made from synthetic fur called "mohair". These mohair strips attach to the bottom of the skis, which allow the skier to slide forward, but not back. Used for ascending moderate to steep terrain or pulling heavy loads over long distances.

   (1) Nomenclature of the climbing skin.

   (a) <u>Skin</u>. This is two sided. The adhesive side is placed against the base of the ski. The mohair side is the portion which comes in contact with the snow.

   (b) <u>Heel Clamp</u>. This secures the skin to the tail of the ski.
   (c) <u>Toe clamp</u>. This secures the skin to the tip of the ski. It is normally equipped with a rubber tensioning devise.

CLIMBING SKIN

   (2) Fitting. The M-buckle located on the toe clamp is held in place by inserting the actual skin through the buckle and folding in back of itself. To adjust skins, simply pull apart the folded adhesive apart and move the clamp forward or back. When the correct length is attained, fold skin back on itself.

(3) Maintenance and storage. To maintain the adhesive side after each use, the skins must be air-dried. To store when not in use, find the midpoint and fold the two adhesive sides back on themselves, then store in carrying bag.

TRANSITION: Skis are our most effective and reliable means of transportation in snow covered terrain. Are there any questions?

PRACTICE (CONC)

    a. Students will practice what was taught in upcoming field evolutions.

PROVIDE HELP (CONC)

    a. The instructors will assist the students when necessary.

OPPORTUNITY FOR QUESTIONS (3 Min)

1. **QUESTIONS FROM THE CLASS**

2. **QUESTIONS TO THE CLASS**

    Q. What are the six parts of the military ski?

    A. (1) Tip
       (2) Shovel
       (3) Base
       (4) Tail
       (5) Binding
       (6) Camber

    Q. What are the three items carried in a ski wax kit?

    A. (1) Grip Wax

       (2) Scraper

       (3) Cork

SUMMARY (2 Min)

    a. This period of instruction has covered the types of skis, those specific to military applications, their nomenclature, and their care.

    b. Those of you with IRF's please fill them out at this time and turn them in to the instructor. We will now take a short break.

UNITED STATES MARINE CORPS
Mountain Warfare Training Center
Bridgeport, California 93517-5001

WML
WMO
06/20/02

LESSON PLAN

MOUNTAIN WEATHER

INTRODUCTION (5 Min)

1. **GAIN ATTENTION**. Normally, as Marines in a temperate climate we think of bad weather as possibly a tactical advantage, giving us concealment in order to move undetected. But in a cold weather mountainous environment, bad weather could be devastating if not properly prepared. For you as a Mountain Leader, it is crucial that you be able to understand the fundamentals of meteorology and determine what to expect from incoming weather patterns.

2. **OVERVIEW.** The purpose of this period of instruction is to introduce you to weather patterns, the invisible aspects of weather, and some visible clues to help you forecast the weather. This lesson relates to all operations in the mountains.

INSTRUCTORS NOTE: Have students read learning objectives.

3. **INTRODUCE LEARNING OBJECTIVES**

   a. TERMINAL LEARNING OBJECTIVE. In a mountainous environment and 24 hours prior to weather conditions occurring, forecast weather, in accordance with the references.

   b. ENABLING LEARNING OBJECTIVES. (MLC)

   (1) Without the aid of references and given a list of air masses, describe in writing the types of air masses, in accordance with the references.

   (2) Without the aid of references, list in writing the ways that air could be lifted and cooled beyond its saturation point, in accordance with the references.

(3) Without the aid of references, list in writing the types of clouds, in accordance with the references.

(4) Without the aid of references, describe in writing the types of clouds, in accordance with the references.

(5) Without the aid of references, list in writing the cloud progression of each front, in accordance with the references.

c. ENABLING LEARNING OBJECTIVES. (SMO)

(1) With the aid of references and given a list of air masses, list orally the types of air masses, in accordance with the references.

(2) With the aid of references, list orally the ways that air could be lifted and cooled beyond its saturation point, in accordance with the references.

(3) With the aid of references, list orally the types of clouds, in accordance with the references.

(4) With the aid of references, describe orally the types of clouds, in accordance with the references.

(5) With the aid of references, list orally the cloud progression of each front, in accordance with the references.

4. **METHOD/MEDIA**. The material in this lesson will be presented by lecture with the use of computer visual aid and/or turn chart. You will practice what you learned during upcoming field training exercises. Those of you with IRF's please fill them out at the end of this period of instruction.

5. **EVALUATION**.

   a. MLC - You will be tested by a written exam.

   b. SMO - You will be tested by a performance tests later in the course.

TRANSITION: Are there any questions over the purpose, learning objectives, how you will be taught, or how you will be evaluated? Let's start by looking at some of the tactical or backcountry considerations and general info you need to be aware of.

BODY (80 Min)

1. (5 Min) **THE WEATHER IN GENERAL**

    a. How well you are able to see the enemy, terrain and Marines around you is greatly affected by the type of weather you might have to deal with. Visibility affects your ability to navigate over the terrain as well as being able to identify the enemy upon contact.

    b. Route selection over snow covered mountainous terrain, is affected by the avalanche potential the weather brings with it.

    c. While you are traveling or sitting in your Patrol base ask these questions concerning the weather. What is the weather currently? What was the weather recently and when did it change last? When is the forecasted next change coming? Having access to this information can help you better prepare before going to the field.

    d. The earth is surrounded by the atmosphere, which is divided, into several layers. The world's weather systems are in the troposphere, the lower of these layers. This layer reaches as high as 40,000 feet.

    e. Dust and clouds in the atmosphere absorb or bounce back much of the energy that the sun beams down upon the earth. Less than one half of the sun's energy actually warms the earth's surface and lower atmosphere.

    f. Warmed air, combined with the spinning (rotation) of the earth, produces winds that spread heat and moisture more evenly around the world. This is very important because the sun heats the Equator much more than the poles and without winds to help restore the balance, much of the earth would be impossible to live on. Where the air cools, you can get clouds, rain, snow, hail, fog, frost, etc.

    g. The weather that you find in any place depends on many things. How hot the air is, how moist the air is, how it is being moved by the wind, and especially, is it being lifted or not?

TRANSITION: Are there any questions over the general information? Now let's take a look at how wind effects our weather.

2. (10 Min) **WINDS**. The uneven heating of the air by the sun and rotation of the earth causes winds. Much of the world's weather depends on a system of winds that blow in a set direction. This pattern depends on the different amounts of sun (heat) that the different regions get and also on the rotation of the earth.

    a. Above hot surfaces, air expands (air molecules spread out), and move to colder areas where it cools and becomes denser, and sinks to the earth's surface. This forms a circulation of air from the poles along the surface or the earth to the Equator, where it rises and moves towards the poles again.

b. Once the rotation of the earth is added to this, the pattern of the circulation becomes confusing.

c. Because of the heating and cooling, along with the rotation of the earth, we have these surfaces winds. All winds are named from the direction they originated from:

(1) Polar Easterlies. These are winds from the polar region moving from the east. This is air that has cooled and settled at the poles.

(2) Prevailing Westerlies. These winds originate from approximately 30 degrees North Latitude from the west. This is an area where prematurely cooled air, due to the earth's rotation, has settled back to the surface.

(3) Northeast Trade winds. These are winds that originate from approximately 30 degrees North from the Northeast. Also prematurely cooled air.

d. Here are some other types of winds that are peculiar to mountain environments but don't necessarily affect the weather:

(1) Anabatic wind. These are winds that blow up mountain valleys to replace warm rising air and are usually light winds.

(2) Katabatic wind. These are winds that blow down mountain valley slopes caused by the cooling of air and are occasionally strong winds.

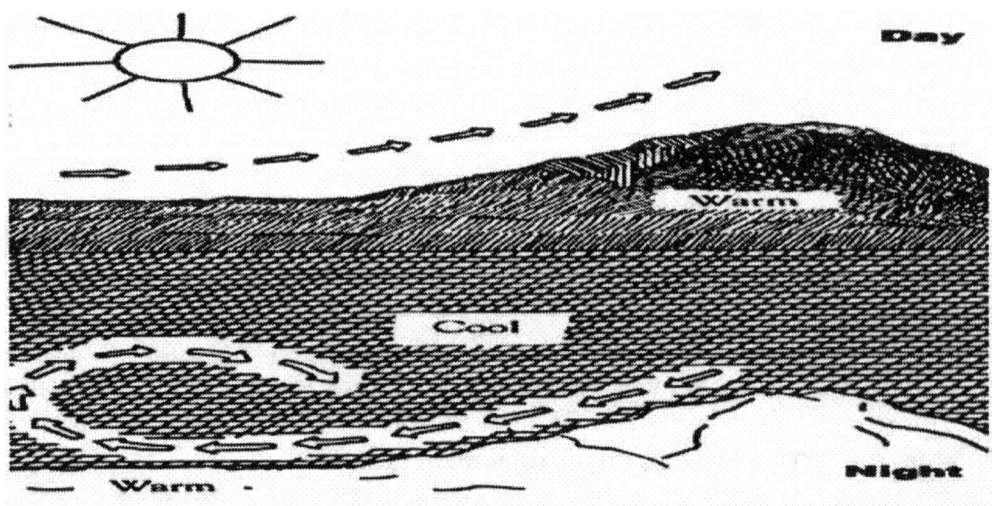

AIR FLOWS UP THE MOUNTAIN DURING THE DAY AND DOWN THE MOUNTAIN AT NIGHT.

e. Jet Stream. A jet stream can be defined as a long, meandering current of high speed winds near the tropopause (transition zone between the troposphere and the stratosphere) blowing from generally a westerly direction and often exceeding 250 miles per hour. The jet stream results from:

(1) Circulation of air around the poles and Equator.

(2) The direction of airflow above the mid latitudes.

(3) The actual path of the jet stream comes from the west, dipping down and picking up air masses from the tropical regions and going north and bringing down air masses from the polar regions.

NOTE: The average number of long waves in the jet stream is between three and five depending on the season. Temperature differences between polar and tropical regions influence this. The long waves influence day to week changes in the weather; there are also short waves that influence hourly changes in the weather.

f. Wind Speed. Determined by Sir Francis Beaufort in 1805. Combined with the air temperature will determine the wind chill index. The Wind chill is the actual temperature of the air on exposed skin.

THE BEAUFORT SCALE	WIND SPEED (MPH)
0-1	Smoke Rises Straight up; Calm
1-3	Smoke Drifts
4-7	Wind Felt on Face; Leaves Rustle
8-12	Leaves and Trigs Constantly Rustle; Wind Extends Small Flags
13-18	Dust and Small Paper raised; Small Branches Moved
19-24	Crested Wavelets form on Inland Waters; Small Trees Sway
25-31	Large Branches Move in Trees
32-38	Large Trees Sway; Must Lean to Walk
39-46	Twigs Broken from Trees; Difficult to Walk
47-54	Limbs Break from Trees; Extremely Difficult to Walk
55-63	Tree Limbs and Branches Break
64 on up	Widespread Damage with Trees Uprooted

WIND CHILL INDEX									
	TEMP	40	30	20	10	0	-10	-20	-30
WIND MPH	5	37	27	16	6	-5	-15	-26	-36
	10	28	16	2	-9	-22	-31	-45	-58
	15	22	11	-6	-18	-33	-45	-60	-70
	20	18	3	-9	-24	-40	-52	-68	-81
	25	16	0	-15	-29	-45	-58	-75	-89
	30	13	-2	-18	-33	-49	-63	-78	-94
	35	11	4	-20	-35	-52	-67	-83	-98
	40	10	-4	-22	-36	-54	-69	-87	-101

TRANSITION: Now that we have a better understanding of the winds, are there any questions? Let's discuss the air masses and their effects on weather.

3. (5 Min) **AIR MASSES**. As we know, all of these patterns move air. This air comes in parcels known as "air masses". These air masses can vary in size from as small as a town to as large as a country. These air masses are named from where they originate.

   a. Maritime. Over water.

   b. Continental. Overland.

   c. Polar. Above 60 degrees North.

   d. Tropical. Below 60 degrees North.

   e. Combining these give us the names and description of the four types of air masses:

   (1) Continental Polar (CP). Cold, dry air mass.

   (2) Maritime Polar (MP). Cold, wet air mass.

   (3) Continental Tropical CT). Dry, warm air mass.

   (4) Maritime Tropical (MT). Wet, warm air mass.

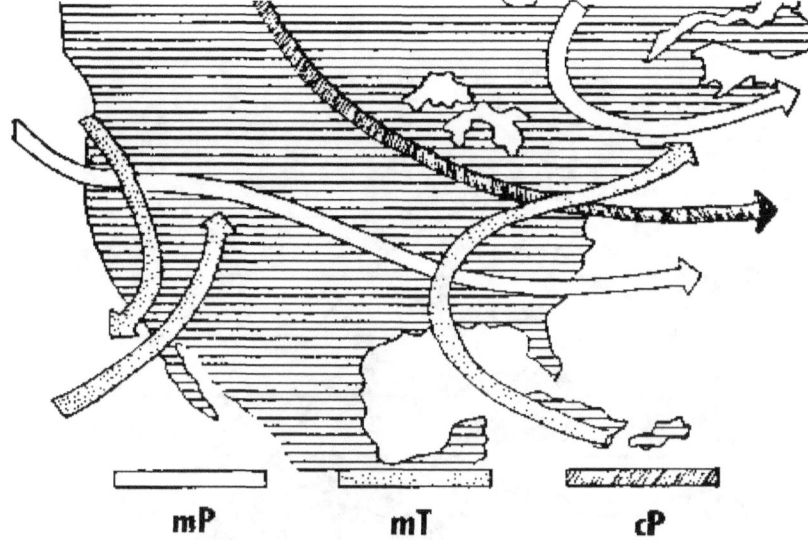

Movement of air masses: Maritime Polar (mP), Maritime Tropical (mT), and Continental Polar (cP).

    f. The thing to understand about air masses, they will not mix with another air mass of a different temperature and moisture content. When two different air masses collide, we have a front, which will be covered in more detail later in this period of instruction.

TRANSITION: Now that we have an understanding of meteorology, we can now look at the reason why we receive precipitation.

4. (10 Min) **HUMIDITY**. Humidity is the amount of moisture in the air. All air holds water vapor, although it is quite invisible.

    a. Air can hold only so much water vapors, but the warmer the air, the more moisture it can hold. When the air has all the water vapor that it can hold, the air is said to be saturated (100% relative humidity).

    b. If the air is then cooled, any excess water vapor condenses; that is, it's molecules join to build the water droplets we can see.

    c. The temperature at which this happens is called the "condensation point". The condensation point varies depending on the amount of water vapor and the temperature of the air.

    d. If the air contains a great deal of water vapor, condensation will form at a temperature of 68°F. But if the air is rather dry and does not hold much moisture, condensation may not form until the temperature drops to 32°F or even below freezing.

e. <u>Adiabatic Lapse Rate</u>. The adiabatic lapse rate is the rate that air will cool (-) on ascent and warm (+) on descent. The rate also varies depending on the moisture content of the air.

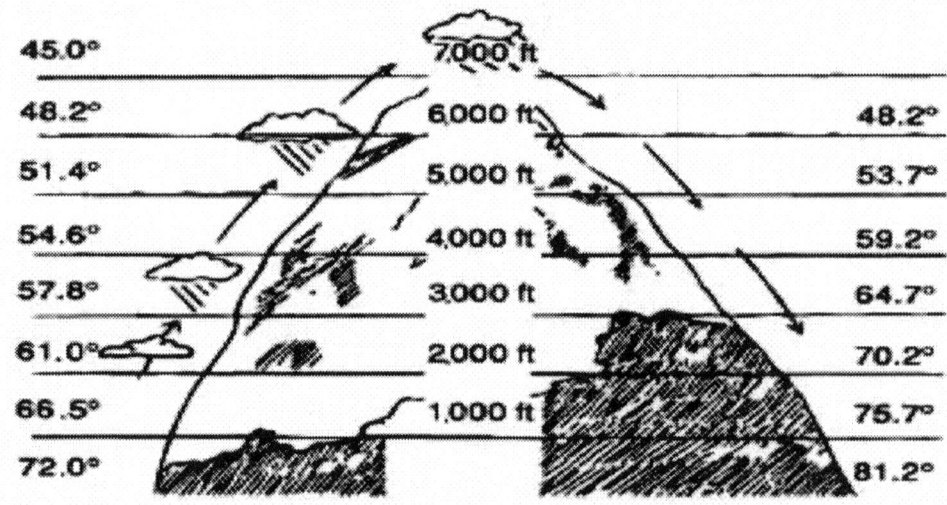

Figure 19 Add condensation/moisture to the mix and the air cools at a slower rate (approximately 3.2 degrees per 1,000 feet) and then warms at a slower rate too until the moisture evaporates, illustrating adiabatic cooling and warming.

   (1) Saturated Air = 3.2 degrees F per 1,000 feet.

   (2) Dry Air = 5.5 degrees F per 1,000 feet.

NOTE: For military planning purposes, 4 degrees F should be used.

TRANSITION: Are there any questions over humidity? Next, we'll talk about the effects of pressure on weather.

5. (10 Min) **PRESSURE**

   a. All of these factors are related to air pressure, which is the weight of the atmosphere at any given place. The lower the pressure, the more likely it is to rain and have strong winds.

   b. In order to understand this we can say that the air in our atmosphere acts very much like a liquid.

   c. Areas with a high level of this liquid would exert more pressure on the Earth and be called a "high pressure area".

d. Areas with a lower level would be called a "low pressure area".

e. In order to equalize the areas of high pressure it would have to push out to the areas of low pressure.

f. The characteristics of these two pressure areas are as follows:

   (1) <u>High pressure area</u>. Flows out to equalize pressure.

   (2) <u>Low pressure area</u>. Flows in to equalize pressure.

g. The air from the high-pressure area is basically just trying to gradually flow out to equalize its pressure with the surrounding air; while the low pressure is beginning to build vertically. Once the low has achieved equal pressure, it can't stop and continues to build vertically; causing turbulence, which results in bad weather.

NOTE: When looking on the weather map, you will notice that these differences in pressure resemble contour lines. These contour lines are called isobars and are translated to mean equal pressure area.

h. Isobars. Pressure is measured in millibars or another more common measurement is inches mercury.

i. Fitting enough, areas of high pressure are called ridges and areas of low pressure are called troughs or depressions.

j. As we go up in altitude, the pressure (or weight) of the atmosphere decreases.

EXAMPLE: At 18,000 feet in elevation it would be 500 millibars vice 1,013 millibars at sea level.

TRANSITION: Now that we have discussed pressure, are there any questions? Let's take a look at the affects of Lifting and Cooling.

6. (10 Min) **LIFTING/COOLING**. As we know, air can only hold so much moisture depending on its temperature. If this air is cooled beyond its saturation point, it must release this moisture in one form or another, i.e. rain, snow, fog, dew, etc. There are three ways that air can be lifted and cooled beyond its saturation point.

   a. <u>Orographic uplift</u>. This happens when an air mass is pushed up and over a mass of higher ground such as a mountain. Due to the adiabatic lapse rate, the air is cooled with altitude and if it reaches its saturation point we will receive precipitation.

OROGRAPHIC UPLIFT

b. <u>Convection effects</u>. This is normally a summer effect due to the sun's heat re-radiating off of the surface and causing the air currents to push straight up and lift air to a point of saturation.

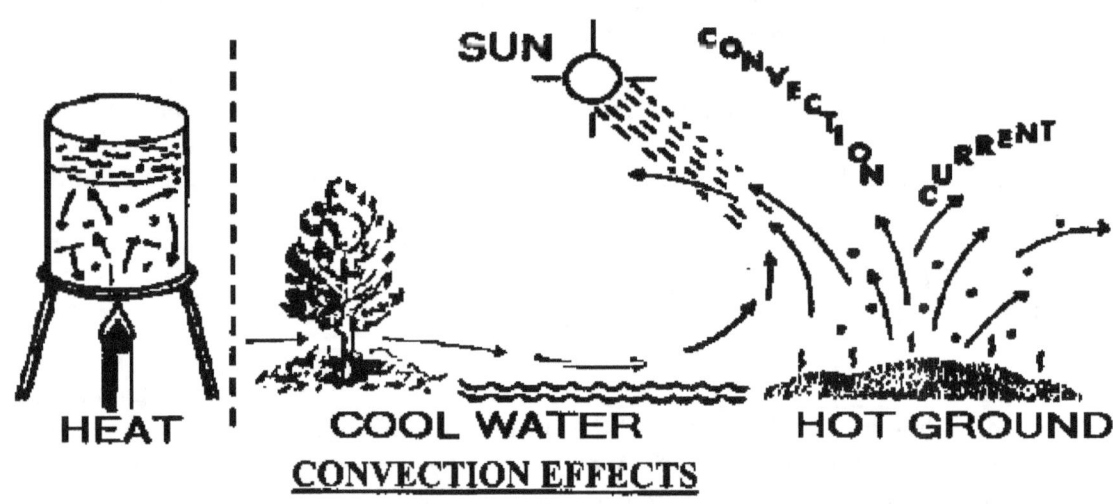

CONVECTION EFFECTS

c. <u>Frontal lifting.</u> As we know when two air masses of different moisture and temperature content collide, we have a front. Since the air masses will not mix, the warmer air is forced aloft, from there it is cooled and then reaches its saturation point. Frontal lifting is where we receive the majority of our precipitation. A combination of the different types of lifting is not uncommon.

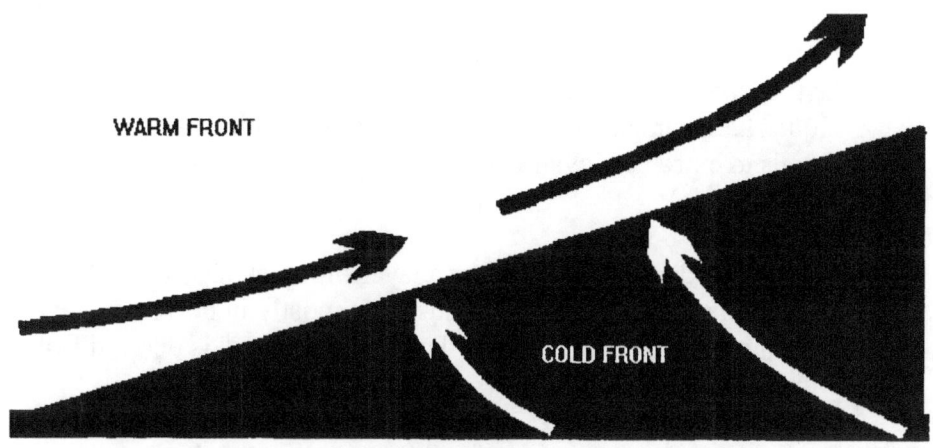

FRONTAL LIFTING

TRANSITION: Are there any questions over lifting and cooling of air masses? So far we've talked about are the invisible aspects of weather, now let's talk about the visible aspects of weather. The first one being clouds.

7. (15 Min) **CLOUDS**. Anytime air is lifted or cooled beyond its saturation point (100% relative humidity), clouds are formed. Clouds are one of our signposts to what is happening. Clouds can be described in many different ways, they can also be classified by height or appearance, or even by the amount of area covered, vertically or horizontally.

   a. Cirrus.
      (1) Cirrus clouds are formed of ice crystals at very high altitudes (usually 20,000 to 35,000 feet) and are thin, feathery type clouds.

      (2) Cirrus clouds can give you up to 24 hours warning of approaching bad weather, hundreds of miles in advance of a front (cold, warm, or occluded).

      (3) Cirrus clouds look thin, frail, feathery. There are sculpted types, such as "mare-tails" and "lenticulars" which show high winds in the upper atmosphere, dense cirrus layers, and/or scattered tufts which are a sign of fair weather.

   b. Cumulus.
      (1) Cumulus clouds are formed due to rising air currents and are prevalent in unstable air that favors vertical development.

      (2) Cumulus clouds are associated with short, heavy precipitation, strong winds, lightening, tornadoes, and hail.

(3) Cumulus clouds look piled or bunched up, like cotton balls. There are three different types to help us to forecast the weather: 1- Fair Weather Cumulus are scattered puffs of cotton in a blue sky, 2- Towering Cumulus are characterized by thick, vertical development like "cauliflower", 3- Cumulonimbus are heavy, dark, towering, and "anvil" shaped that produce precipitation. These clouds are characterized by violent updrafts, which carry the tops of the clouds to extreme elevations.

c. Stratus.
  (1) Stratus clouds are formed when a layer of moist air is cooled below its saturation point. Stratiform clouds form mostly in horizontal layers or sheets, resisting vertical development. The word "stratus" is derived from the Latin word "layer".

  (2) Stratus clouds are associated with long, light precipitation, such as drizzle or snow flurries.

  (3) Stratus clouds look uniform or flat, with a dull, gray appearance that resembles fog.

d. As previously stated, clouds are formed when air is lifted to a point where it cools to its saturation point. We also know that frontal lifting affects our fronts, which produce the largest portion of our precipitation.

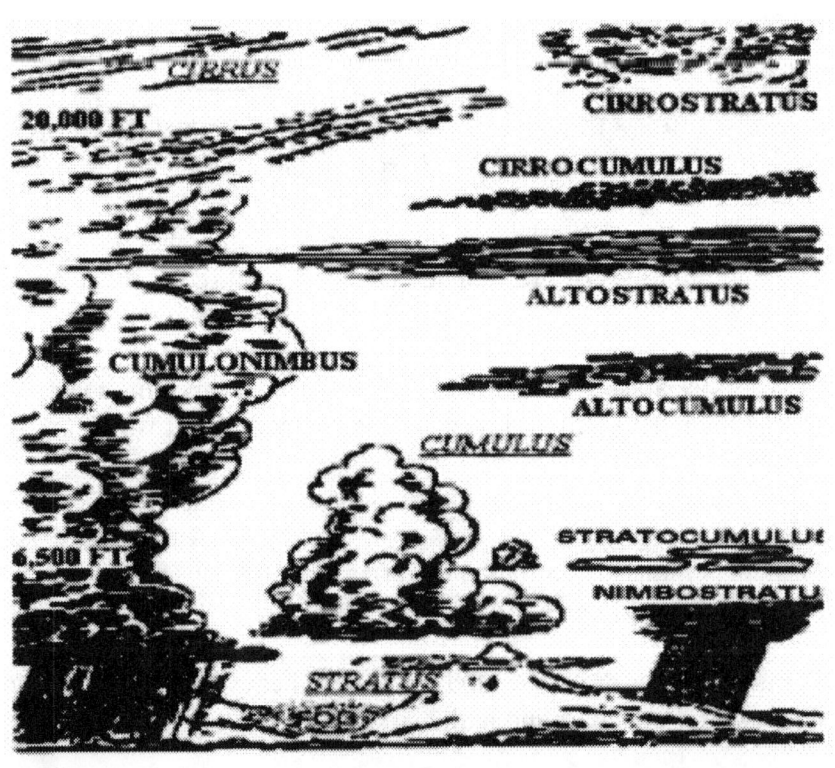

TRANSITION: Are there any questions over clouds? Let's look at the different types of fronts and the progression of clouds that accompany them.

8. (10 Min) **FRONTS**. Fronts often happen when two air masses of different moisture and temperature content interact. One of the ways we can identify this is happening is by the progression of the clouds.

   a. Warm Front. A warm front occurs when warm air moves into and over a slower (or stationary) cold air mass. Since warm air is less dense, it will rise naturally so that it will push the cooler air down and rise above it. The cloud you will see at this stage is cirrus. From the point where it actually starts rising, you will see stratus. As it continues to rise, this warm air is cooled by the cold air, and it receives moisture at the same time. As it builds in moisture, it darkens becoming "nimbostratus", which means rain from thunder clouds. At that point, some type of moisture will generally fall. In short, the cloud progression is cirrus to stratus to nimbostratus for a warm front.

   WARM FRONT

   b. Cold Front. A cold front occurs when a cold air mass (colder than the ground that it is traveling over) overtakes a warm air mass that is stationary or moving slowly. This cold air, being denser, will go underneath the warm air, pushing it higher. Of course, no one can see this "air", but they can see clouds and the clouds themselves can tell us what is happening. The cloud progression to look for is cirrus to cirrocumulus to cumulus to cumulonimbus for a cold front.

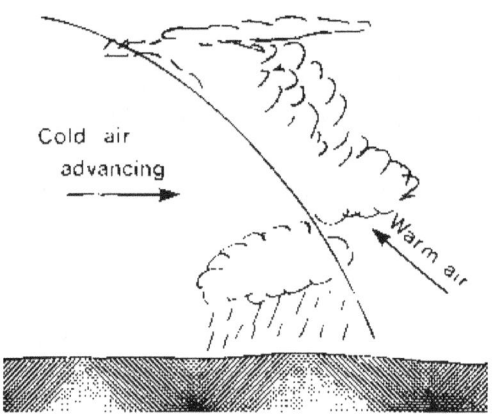

COLD FRONT

    c. Occluded Front. Cold fronts move faster than warm ones so that eventually a cold front overtakes a warm one and the warm air becomes progressively lifted from the surface. The zone of division between cold air ahead and cold air behind is called a "cold occlusion". If the air behind the front is warmer than ahead, it is a warm occlusion. Most land areas experience more occlusions than other types of fronts. In the progression of clouds leading to fronts, orographic uplift can play part in deceiving you of the actual type of front, i.e. progression of clouds leading to a warm front with orographic cumulus clouds added to these. The cloud progression for an occluded front is a combination of both progressions from a warm and cold front.

TRANSITION: Are there any questions over fronts? The clouds and their progression are our basis to our forecast. But there are other indicators to put in before making our decision.

9. (5 Min) **USING PRESSURE AS AN INDICATOR.** A very important factor of telling us what might happen is the pressure. As we know, low pressure or dropping pressure normally indicates deteriorating weather whereas high pressure usually gives us more good weather or clearing of bad weather. There are a couple of ways to monitor our pressure and are as follows:

    a. The Barometer. A barometer could be described as a pan of mercury with a tube leading out of the pan. Pressure from the atmosphere causes the mercury to rise in the tube.

        (1) The tube is marked in millibars and the station that's reading these millibars will know how much it should rise for that location. Once again, if it rises more than normal, it would be considered a high pressure reading.

    b. The Altimeter. Another means that is used to measure pressure is an altimeter, which is commonly used by mountaineers. It works like this:

INSTRUCTOR NOTE: Show an altimeter to the students.

> (1) As you rise in elevation the pressure becomes less, thus allowing the needle in the altimeter to rise. If the needle rises without you rising with it, there is less pressure in the atmosphere than before and thus, a low pressure area.

   c. <u>Contrail Lines</u>. A basic way of identifying a low pressure area is to note the contrail lines from jet aircraft. If they don't dissipate within two hours, that indicates a low pressure area in your area. This usually occurs about 24 hours prior to an oncoming front.

   d. <u>Lenticulars</u>. These are optical, lens-shaped cumulus clouds that have been sculpted by the winds. This indicates moisture in the air and high winds aloft. When preceding a cold front, winds and clouds will begin to lower.

<u>TRANSITION</u>: Are there any questions over using pressure as an indicator? Last of all, let's see how we can use some signs from nature to help us predict the weather.

10. (10 Min) **USING SIGNS FROM NATURE**. These signs will give you a general prediction of the incoming weather conditions. Try to utilize as many signs together as possible, which will improve your prediction. All of these signs have been tested with relative accuracy, but shouldn't be depended on 100%. But in any case you will be right more times than wrong in predicting the weather. From this we can gather as much information as needed and compile it along with our own experience of the area we are working in to help us form a prediction of incoming weather. The signs are as follows:

   a. A spider's habits are very good indicators of what weather conditions will be within the next few hours. When the day is to be fair and relatively windless, they will spin long filaments over which they scout persistently. When precipitation is imminent, they shorten and tighten their snares and drowse dully in their centers.

   b. Insects are especially annoying two to four hours before a storm.

   c. If bees are swarming, fair weather will continue for at least the next half day.

   d. Large game such as deer, elk, etc., will be feeding unusually heavy four to six hours before a storm.

   e. When the smoke from a campfire, after lifting a short distance with the heated air, beats downward, a storm is approaching. Steadily rising smoke indicates fair weather.

   f. "Red sky at night is a sailor's delight and red sky in the morning is a sailor's warning". This poem is correct in only some places of the world. When the sun rises in the morning and there is moisture present, the sky will be red. If the wind is moving west to east, that moisture has already past. This does not mean that it will not rain, it just means that the moisture making the sky red is already past. When the sun sets in the west, and there is moisture in the sky, the sunset will be red. If the winds are moving west to east, it means that the moisture in the west making the sky red will move east and possibly form as clouds later.

g. A gray, overcast evening sky indicates that moisture carrying dust particles in the atmosphere have become overloaded with water; this condition favors rain.

h. A gray morning sky indicates dry air above the haze caused by the collecting of moisture on the dust in the lower atmosphere; you can reasonably a fair day.

i. When the setting sun shows a green tint at the top as it sinks behind clear horizon, fair weather is probable for most of the next 24 hours.

j. A rainbow in the late afternoon indicates fair weather ahead. However, a rainbow in the morning is a sign of prolonged bad weather.

k. A corona is the circle that appears around the sun or the moon. When this circle grows larger and larger, it indicates that the drops of water in the atmosphere are evaporating and that the weather will probably be clear. When this circle shrinks by the hour, it indicates that the water drops in the atmosphere are becoming larger, forming into clouds, rain is almost sure to fall.

l. In fair weather, air currents flow down streams and hillsides in the early morning and start drifting back up towards sunset. Any reversal of these directions warns of a nearing storm.

m. When the breeze is such that the leaves show their undersides, a storm is likely on the way.

n. It is so quiet before a storm, that distant noises can be heard more clearly. This is due to the inactivity of wildlife a couple of hours before a storm.

o. When in the mountains, the sight of morning mist rising from ravines is a good sign of clear weather the rest of the day.

p. A heavy dew or frost in the morning is a sign of fair weather for the rest of the day. This is due to the moisture in the atmosphere settling on the ground in the form of precipitation such as rain snow etc.

TRANSITION: It is very important that we pay attention to all of the clues that are given to us so that we are not surprised by bad weather.

PRACTICE (CONC)

a. Students will forecast weather.

PROVIDE HELP (CONC)

a. The instructors will assist the students when necessary.

OPPORTUNITY FOR QUESTIONS (3 Min)

1. QUESTIONS FROM THE CLASS

2. QUESTIONS TO THE CLASS

   Q. What are the three types of clouds?

   A. (1) Cirrus
      (2) Stratus
      (3) Cumulus

   Q. What the three types of fronts?

   A. (1) Warm front
      (2) Cold front
      (3) Occluded front.

   Q. What are the three types of wind patterns?

   A. (1) Polar Easterlies
      (2) Prevailing Westerlies
      (3) Northeast Tradewinds

SUMMARY (2 Min)

a. Now that we have the basic knowledge to forecast the weather, you will find that you are not always right in your forecast. Though with time and experience in forecasting, your degree of accuracy will definitely improve.

b. Those of you with IRF's please fill them out and turn them into the instructor. We will now take a short break.

UNITED STATES MARINE CORPS
Mountain Warfare Training Center
Bridgeport, California 93517-5001

WML
WMO
06/24/02

LESSON PLAN

MOUNTAIN CASUALTY EVACUATIONS (WINTER)
LESSON DESIGNATOR:
TYPE: INFORMAL LECTURE DEMONSTRATION
TOTAL LESSON TIME: 50 MIN
REFERENCES: Mountaining the FREEDOM OF THE hills, 6th edition ,The Mountaineers, Seattle WA 1996
Rope Rescuce

INSTRUCTOR PREPARTION :
2 INSTRUCTORS, ONE TURN CHART,1 TEAM SLED,1LARGE SLED,CLASS ROOM INSIDE OR OUT SIDE ,BUILDING 44044,MWTC BRIDGEPORT,CA
SAFETY PRECAUTIONS: N/A

INTRODUCTION   (5 Min)

1. **GAIN ATTENTION**. "Looks like snow with strong gusty winds and possible whiteout conditions. We're going to have a rough time evacuating this casualty!" Adverse weather conditions are inherent to a mountainous cold weather environment. Under these conditions, even personnel with relatively minor injuries can become casualties which require evacuation.

2. **PURPOSE**. The purpose of this period of instruction is to familiarize the student with the problems inherent in evacuating casualties in a cold weather environment. Also, we will cover litters and movement considerations. This lesson relates to ski movement.

INSTRUCTOR NOTE
READ AND SHOW TLO'S AND ELO'S

TC2

3. **INTRODUCE LEARNING OBJECTIVES**

    a. <u>TERMINAL LEARNING OBJECTIVE</u>. In a cold weather/mountainous environment, evacuate casualties, in accordance with the references.

TC3

    b. <u>ENABLING LEARNING OBJECTIVES</u>.

(1) Without the aid of references, list in writing the eight general considerations for casevac procedures, in accordance with the references.

(2) Without the aid of references, list in writing the two types of standard litters used in a casevac, in accordance with the references.

TC4

(3) Given the necessary equipment in a cold weather/mountainous environment, perform methods of evacuating casualties, in accordance with the references.

4. **METHOD / MEDIA**. The material in this lesson will be presented by lecture and demonstration. You will practice what you have learned in upcoming field training exercises. Those of you with IRF's please fill them out at the end of this period of instruction.

5. **EVALUATION**. You will be tested later in the course by written and performance evaluations.

TRANSITION: In the event a casevac arises, there are several points that need to be considered in reference to trail selection and patient preparation.

BODY (50 Min)

TC5

1. (10 Min) **GENERAL CONSIDERATIONS**

TC6

    a. Apply essential first aid. Splints, pressure bandages, etc.

    b. Protect the patient from the elements. Provide the casualty with proper insulation and ensure that he is warm and dry.

TC7

    c. Avoid unnecessary handling of the patient.

    d. Select the easiest route. Send scouts ahead if possible, to break trail.

TC8

    e. Set up relay points and warming stations. If the route is long and arduous, set up relay points and warming stations with medical personnel at warming stations to:

        (1) Permit proper emergency treatment. Treat for shock, or other conditions that may arise.

(2) <u>Reevaluate the patient constantly</u>. If the patient develops increased signs of shock or other symptoms during the evacuation, he may be retained at an emergency station until stable.

TC9

f. <u>Normal litter teams must be augmented in arduous terrain.</u>

TC10

g. <u>Give litter teams specific goals to work towards</u>. This job is extremely tiring, both physical and mentally.

h. <u>Gear</u>. Ensure all of the patient's gear is kept with him throughout the evacuation.

<u>TRANSITION</u>: Let us now talk about the types of standard litters employed in cold weather casevacs.

TC11

2. (5 Min) **LITTERS**. There are two types of standard litters:

   a. <u>Team Sled</u>. This is the current fire team sled in the Marine Corps system.
   b. <u>Large Sled</u>. There are 2 large sleds in the system, the old Akhio and new Weapons/Casevac Variant of the Team Sled.

<u>TRANSITION</u>: Let's now learn how to prepare a casualty utilizing the litters.

3. (20 Min) **PREPARING A CASUALTY FOR MOVEMENT**

> **BE PREPARED TO DEMONSTRATE AND PRACTICE BOTH SLED TYPES DURING CASEVAC**

   a. <u>Team Sled</u>

   (1) <u>Insulating</u>. Place the casualty's isopor mat or extra clothing inside it. This will also help to pad the team sled.

   (2) <u>Sleeping bag</u>. Place the casualty's sleeping bag inside the team sled and place the casualty inside the sleeping bag.

   (3) <u>Securing the casualty</u>. Secure the straps across the chest, knees, and shins. You can also place skis under the victim to provide C-spine precautions and to help support the legs out of the sled.

   (4) <u>Movement</u>. When transporting, try to keep the patient's head uphill.

   b. <u>Large Sled</u>

(1) <u>Insulating</u>. Place the casualty's isopor mat or extra clothing inside it. This will also help to pad the sled.

(2) <u>Sleeping bag</u>. Place the casualty's sleeping bag inside the sled and place the casualty inside the sleeping bag.

(3) <u>Securing the casualty</u>. Secure the casualty with the internal straps and by closing the sled cover over the casualty (covering the face or not, will be dependent on weather conditions).

(4) <u>Movement</u>. Try to keep the patient's head uphill at all times. See Sled Movement outline for specific techniques.

<u>TRANSITION</u>: Let us now talk about evacuations conducted by ground transportation.

TC12

4. (3 Min) **EVACUATION BY GROUND VEHICLES**. Vehicle requirements will vary depending on the type of operation and the terrain. Wheeled vehicles are usually limited to maintained roads. Chains are frequently required, even with four-wheel drive. Amtracks and LAR vehicles are just as limited. Over-the-snow vehicles are best, i.e., snowmobiles, BV-206s, etc.

TC13

<u>TRANSITION</u>: Now that we talked about ground transportation, let us talk about air transportation.

5. (2 Min) **EVACUATION BY AIR**. Medical evacuation by air is ideal because it is quick and is frequently easier on the patient. There are NO ABSOLUTE CONTRADICTIONS TO EVACUATION BY AIR, however there may be restrictions placed on the aircraft.

<u>TRANSITION</u>: When the time comes to move a casualty, it is not necessary to stick solely with the choices presented here, but you should remember to apply the eight considerations every time and your imagination within these constraints will get the casualty safely to aid. Are there any questions at this time?

PRACTICE (CONC)

a. Students will practice what was taught in upcoming field evolutions.

PROVIDE HELP (CONC)

a. The instructors will assist the students when necessary.

OPPORTUNITY FOR QUESTIONS (3 Min)

1. QUESTIONS FROM THE CLASS.

2. QUESTIONS TO THE CLASS.

   Q. What are the eight general considerations for casevacing a casualty?

   A. (1) Apply essential first aid.
      (2) Protect the patient from the elements.
      (3) Avoid unnecessary handling of the patient.
      (4) Select the easiest route.
      (5) Set up relay points and warming stations.
      (6) Normal litter teams must be augmented in arduous terrain.
      (7) Give litter teams specific goals to work towards.
      (8) Keep all victims gear with him.

   Q. What are the two standard litters for evacuation?

   A. (1) Team Sled.
      (2) Large Sled.

SUMMARY (2 Min)

   a. This period of instruction has briefly described the general and specific characteristics of evacuating casualties from a cold weather environment.

   b. Those of you with IRF's please fill them out at this time and turn them in to the instructor. We will now take a short break. Take a 10 minute break.

UNITED STATES MARINE CORPS
Mountain Warfare Training Center
Bridgeport, California 93517-5001

WMO
WML
07/01/02

LESSON PLAN

AVALANCHE

INTRODUCTION (5 Min)

1. **GAIN ATTENTION**. Snow avalanches are powerful force of nature. A large one may transport not only ice and snow but also rock, soil and vegetation. Avalanches thus play a significant role carving and weathering the world's most spectacular peaks. Fortunately, most of this action occurs in places remote from civilization, but when man and avalanches meet the results can be terrifying. From 16 October of 1998 to 22 May 1999 approximately 198 people were killed in the world by avalanches. That's an average of 27 per month. These people include all sorts from backcountry travelers to soldiers on the Kashmir glacier.

2. **PURPOSE**. The purpose of this period of instruction is to introduce the student to avalanches, their characteristics, dangers, and how to cross them. This lesson relates to Snow Stability Evaluation.

3. **INTRODUCE LEARNING OBJECTIVES**

    a. TERMINAL LEARNING OBJECTIVE. In a cold weather / mountainous environment and given a snow covered slope, determine avalanche hazards, in accordance with the references.

    b. ENABLING LEARNING OBJECTIVES

    (1) Without the aid of references, list in writing the types of avalanches, in accordance with the references.

    (2) Without the aid of references, describe in writing Rounding, in accordance with the references.

    (3) Without the aid of references, state in writing the relative stability of Rounding, in accordance with the references.

(4) Without the aid of references, describe in writing Faceting, in accordance with the references.

(5) Without the aid of references, state in writing the relative stability of Faceting, in accordance with the references.

(6) Without the aid of references, describe in writing Melt-Freeze Metamorphism, in accordance with the references.

(7) Without the aid of references, state in writing the relative stability of Melt-Freeze Metamorphism, in accordance with the references.

(8) Without the aid of references, describe how wind effects the slopes in snow covered mountainous terrain, in accordance with the references.

(9) Without the aid of references, list in writing the considerations for crossing an avalanche prone slope, in accordance with the references.

(10) Without the aid of references, state in writing the individual preparations to take when crossing a potential avalanche prone slope, in accordance with the references.

4. **METHOD / MEDIA**. The material in this lesson will be presented by lecture and demonstration. You will practice what you have learned during upcoming field training exercises. Those of you with IRF's please fill them out at the conclusion of this period of instruction.

5. **EVALUATION**. You will be tested later in the course by written and performance evaluation.

INSTRUCTOR NOTE: All material covered in this outline is in reference to Snow Sense, 4th Edition, 1994, Jill A. Fredston and Doug Fesler and the American Institute for Avalanche Research and Education curriculum.

TRANSITION: Let's first learn the definition of an avalanche.

BODY (90 Min)

1. (1 Min) **DEFINITION**. Avalanches are falling masses of snow that can contain rock, soil or ice, which will travel over terrain of least resistance.

TRANSITION: Now that we know the definition, let's talk about the avalanche path.

2. (4 Min) **AVALANCHE PATH**. The term avalanche path defines the area in which an avalanche runs. An avalanche is generally divided into three parts: Starting Zone, Track, and Runout Zone.

   (1) Starting Zone. This is where the unstable snow failed and began to move.

(2) Track. This is the slope or channel down, which snow moves at a fairly uniform speed.

(3) Runout Zone. This is where the snow slows, debris is deposited, and the avalanche stops.

AVALANCHE ZONES

TRANSITION: Now that we discussed the avalanche path, let's talk about the different types of avalanches.

3. **TYPES OF AVALANCHES**

   a. (10 Min) <u>Loose Snow Avalanches</u>. Loose snow slides, also called point releases, start with a small amount of cohesionless snow and typically pick up more snow as they descend. From a distance, they appear to start at a point and fan out into a triangle. They usually are small, involving only upper layers of snow, but they are capable of being quite large and destructive depending upon how much material they entrain. Factors of a loose snow avalanche are:

   (1) The stress of the moving snow in a loose snow slide can also trigger larger and deeper slab releases.

   (2) Loose snow releases occur most often on steep slopes of 35° or higher.

(a) During or shortly after a snowstorm.

(b) During warming events caused by rain, rising temperatures, or solar radiation

LOOSE SNOW AVALANCHE

b. <u>Slab Avalanche</u>. Slab avalanches occur when one or more layers of cohesive snow break away as a unit. As the slab travels down slope, it splits up into smaller blocks or clods. Factors of a slab avalanche are:

(1) Slab failure is commonly initiated when the bond between the slab and the bed surface fails, thus placing tremendous stress on the other boundary regions which, in turn, are unable to hold the slab in place.

(2) Slab thickness can range from less than an inch to 35 feet or more, and range in width from a few yards to well over a mile.

(3) Slab material is also highly variable. Slabs may be hard or soft, wet or dry.

(4) The speed of a slab avalanche can range from roughly 65 mph for a wet slab, on up to 150 mph for a dry slide.

(5) Most slab avalanches release on slopes with angles between 35° and 40°.

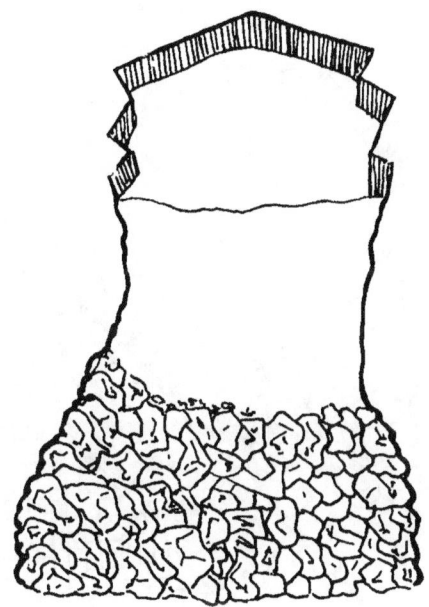

SLAB AVALANCHE

(6) Nomenclature of a Slab Avalanche

   (a) Crown. This is the breakaway wall of the top periphery of the slab. It is usually at a right angle to the bed surface. It is formed by tension fracture through the depth of the slab from bottom to top.

   (b) Bed Surface. This is the surface over which the slab slides. The bed surface can be the ground.

   (c) Flanks. These are the left and right sides of the slab.

   (d) Stauchwall. This is the lowest downslope fracture surface. It is usually overridden by the slab material and it consists of a diagonal shear fracture of wedge-like shape.

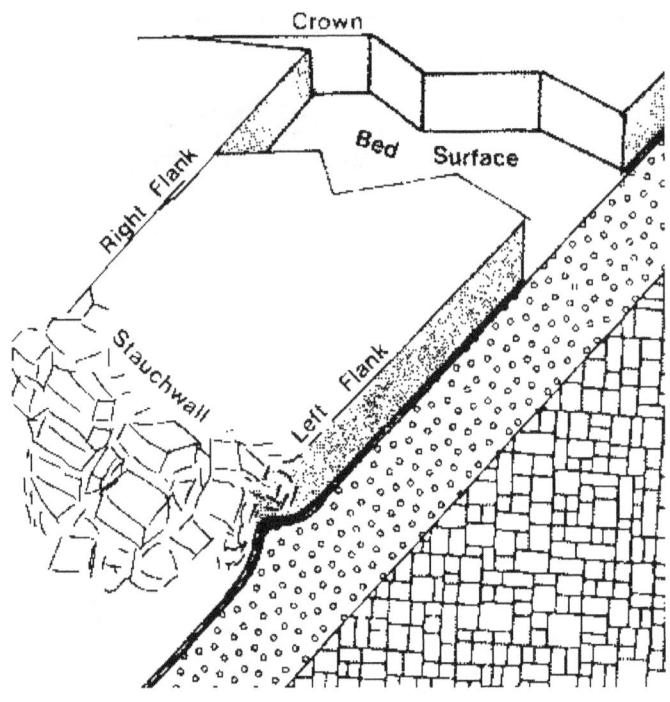

NOMENCLATURE OF SLAB AVALANCHE

TRANSITION: Now that we have talked about the different types of avalanches, let's now look at some different information to help us in predicting avalanches.

4. (5 Min) **AVALANCHE HAZARD EVALUATION PROCESS**

    a. The Data Triad. The interaction of three critical variables will help in determining whether or not an avalanche is possible. These three variables are:

        (1) The Terrain.

            (a) Terrain Analysis. Is the terrain capable of producing avalanches?

        (2) The Snow pack.

            (a) Stability Evaluation. Could the snow slide?

        (3) The Weather.

            (a) Avalanche Forecasting. Is the weather contributing to instability?

    b. To determine whether an avalanche hazard exists, we must add another variable… humans. Without the presence of humans, there is no hazard.

        (1) The Human Factor.

(a) Decision Making. What are your alternatives and their possible consequences?

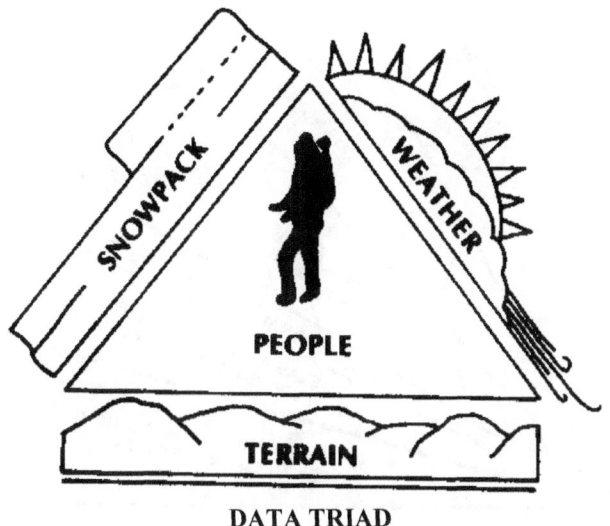

DATA TRIAD

TRANSITION: Let's begin the Avalanche Hazard Evaluation process by first discussing the terrain.

5. (15 Min) **TERRAIN**. Learning to recognize avalanche terrain is the critical starting point in the avalanche hazard evaluation process. Assuming that avalanches occur on only big slopes is a very common mistake. The following factors influence whether a given slope is capable of producing an avalanche and will help you recognize avalanche terrain:

   a. Slope Angle: Slope angle is the most important terrain variable on determining whether or not it is possible for a given slope to avalanche. The underlying concept is that as the slope angle increases, so does the stress exerted on all boundary regions of the slab.

      (1) Slope angles less then 25° will rarely slide due to lack of stress to the snowpack.

      (2) Slab avalanches in cold snow are possible between the slope angles of 25° and 60°.

      (3) Most slab avalanches release on slopes with starting zone angles between 35° and 40°.

      (4) Slope angles 60° or greater will continually sluff due to large amounts of stress to the snowpack.

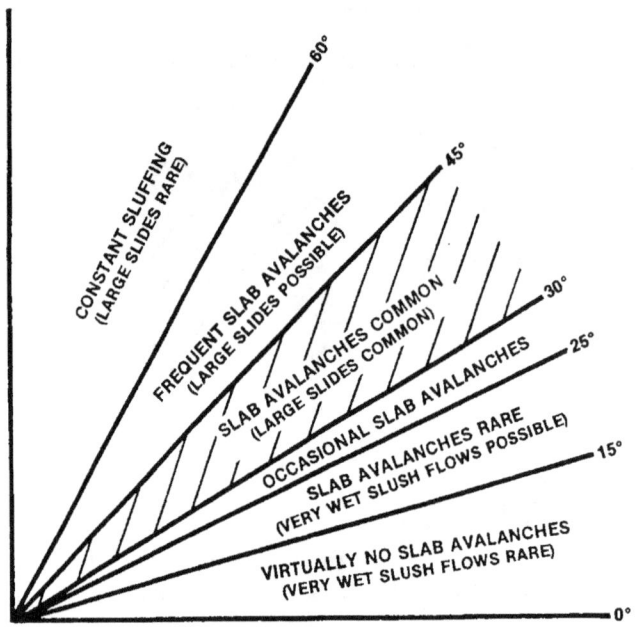

SLOPE ANGLE RELATIONSHIP TO AVALANCHES

NOTE: It is important to know that your motion or body weight can trigger an avalanche even if you are on a low angle slope or on the flats as long as this terrain is connected to a slope with a angle of roughly 25° and instability exists.

    b. <u>Slope Aspect (Orientation)</u>. Subtle changes in slope aspect can greatly affect snow stability.

        (1) Leeward. Deposition of wind-transported snow increases the stress on the snow pack and enhances slab formation.

        (2) Moderate warming by the sun can help strengthen and stabilize the snowpack.

        (3) Intense, direct sunlight has the opposite effect by weakening and lubricating the bonds between grains.

        (4) On shaded slopes, weak layers often persist or are more well-developed because of generally colder conditions and the absence of solar warming during much of the winter. Therefore, suspect instability on shadowed slopes.

NOTE: For operational planning purposes, north-facing and shaded slopes tend to be more dangerous during the mid-winter period. South-facing slopes tend to be most dangerous during spring thaw especially on a sunny day.

    c. <u>Terrain Roughness (Anchoring)</u>. Slopes with anchors are less likely to avalanche than open slopes.

(1) Boulders, trees, ledges will act as anchors and help hold the snow in place until they are buried.

(2) Smooth slopes (i.e. smooth granite, grassy) may only need 1 foot of snow to release.

(3) Anchors are commonly areas of stress concentration because the snow upslope of them is being held in place while the snow below or to the sides is being pulled downhill by gravity. For this reason, anchors can be starting points for initial failure to occur and fractures often run from tree to tree to rock.

TERRAIN ROUGHNESS

d. <u>Slope Shape</u>. Avalanches can happen on any snow-covered slope steep enough to slide.

(1) Convex Slopes. Slabs are most likely to fracture just below the bulge where stresses are greatest.

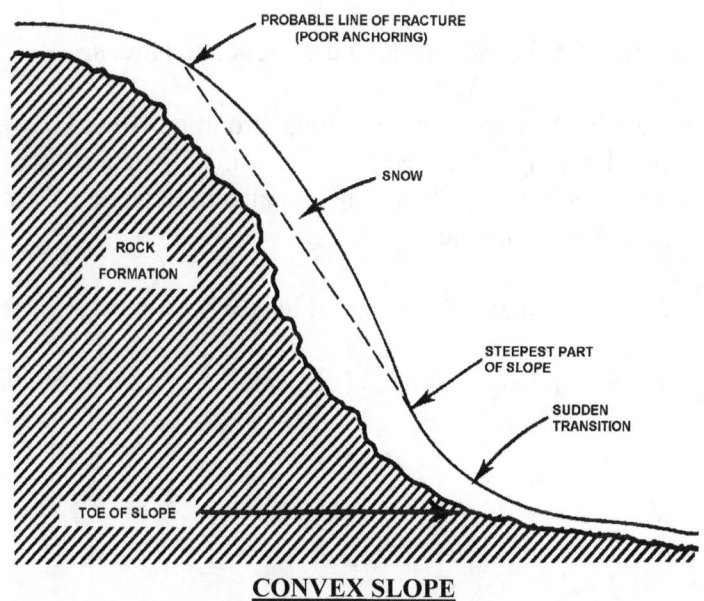

CONVEX SLOPE

9-9

(2) Planar Slopes. On these broad, smooth slopes, avalanches can happen anywhere. Slabs often fracture below cliff bands.

(3) Concave Slopes. These slopes provide a certain amount of support through compression at the base of the hollow, but they are still capable of avalanching, especially on large slopes.

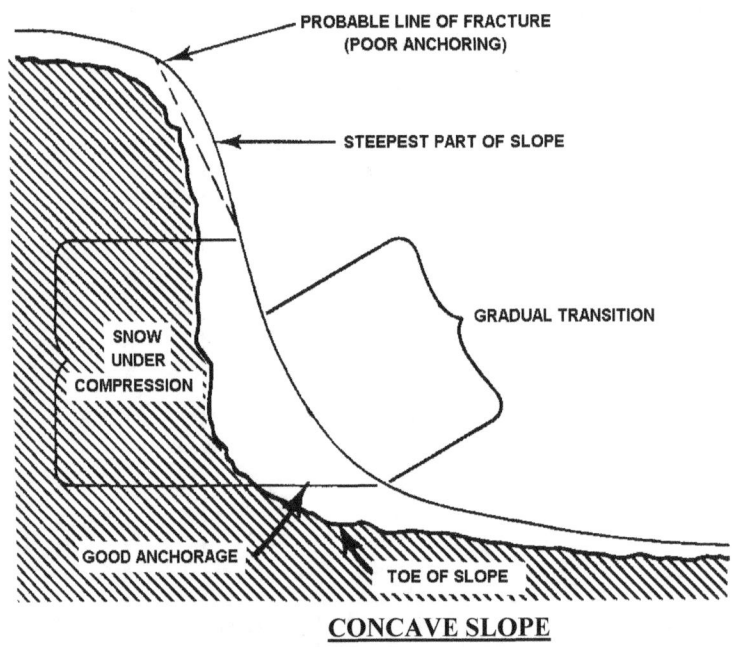

CONCAVE SLOPE

e. <u>Vegetation</u>. Vegetation can provide evidence of both the frequency and magnitude of past avalanche occurrences and thus indicate potential avalanche terrain as well as the capability of a given path. Vegetative indicators include:

(1) Swaths of open slope between forested or vegetated areas.

(2) Trees which are bent, broken, or uprooted, "broomed" trees (i.e., previously broken but with new growth tops), and vegetation which is polished or "flagged" (i.e., missing branches on the uphill side). Flagging can also indicate the flow height of the avalanches which have impacted the area.

(3) Presence of "disaster species" such as alders, willows, dwarf birch and cottonwoods.

(4) Marked difference in height of trees (i.e., smaller spruce in the path, larger on the edges).

VEGETATION INDICATOR OF AN AVALANCHE PRONE SLOPE

    f. <u>Elevation</u>. Temperature, wind and precipitation often vary significantly with elevation. Common differences are rain at lower elevations with snow at higher elevations or differences in precipitation amounts or wind speed with elevation. Never assume that conditions on a slope at a particular elevation reflect those of a slope at a different elevation.

    g. <u>Path History</u>. All avalanche paths have some sort of history, whether it be their magnitude or how often they slide. Before going into avalanche country, try and get as much information as possible.

<u>TRANSITION</u>. So far we've looked at some of the terrain variables, and how they effect an avalanche, now let's look at the snowpack and it's affects.

6. (15 Min) **SNOWPACK**. The snowpack accumulates layer by layer with each new snow or wind event. These layers are then subject to changes in texture and strength throughout the winter. The changes help determine snow strength by influencing how well individual snow grains are bonded to each other both within the layer and between layers. Many combinations of strong and weak layers can exist within the snowpack. The structure of the snowpack varies greatly depending upon the particular season, location, climate, slope aspect, inclination and shape.

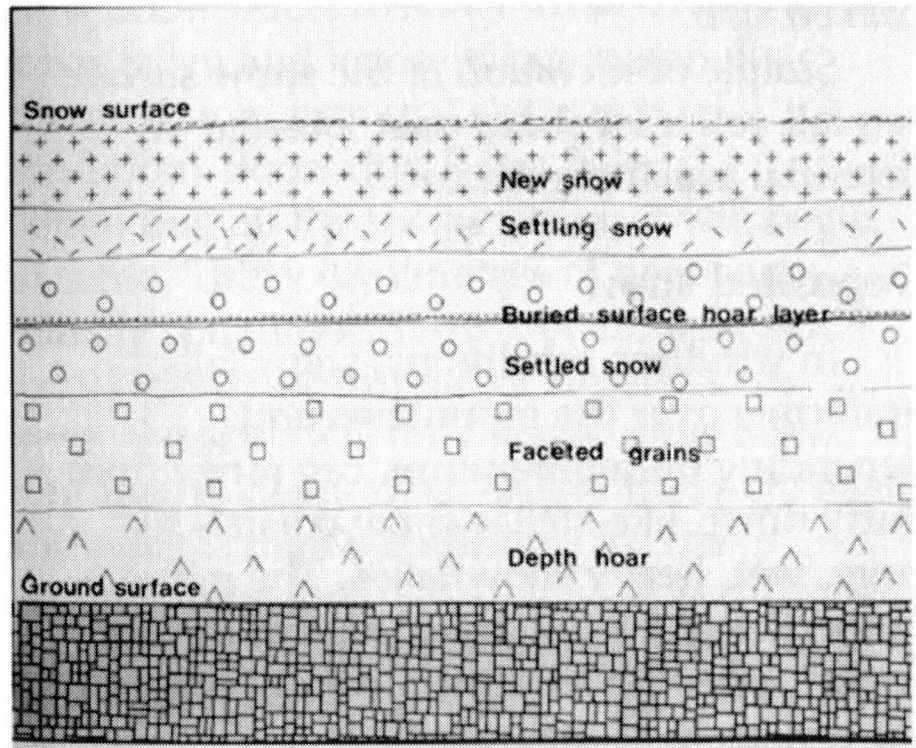

EXAMPLE SNOWPACK

a. Snow Metamorphism. This is the name given to describe the changes in structure that take place over time within the layers of the snowpack. There are several types of snow metamorphism. Each occurs under a different set of conditions and each affects the strength of the snowpack. As conditions change, the dominant type of metamorphism in a given layer may change. Also importantly, different types of metamorphism may be occurring in various layers of the snowpack at the same time. The following are snow metamorphic types encountered in a snowpack:

(1) Rounding. Rounded grains develop when temperatures in a layer or between layers are fairly uniform. There is no significant temperature change within the snowpack.

 (a) Individual grains become smaller and rounder. Bonds or necks between grains are developed. Thus, the equilibrium form process produces fine, rounded, well-bonded grains and the result is that relatively strong layers form within a snowpack.

 (b) As the equilibrium form process advances in stages, the smaller, rounder, and better bonded the grains will be.

 (c) Favorable conditions or habitat for the development of rounded grains are cloudy, mild weather or a thick snowpack.

9-12

(2) <u>Faceting</u>. Faceted grains develop when a significant temperature change exists within or between layers. In most areas, the temperature at the ground/snow interface is warmer than the air temperature. The larger the temperature differences, the quicker the vapor transfer and the process of change.

   (a) The shallower the snowpack, the greater the temperature change within the snowpack. Deep snowpacks tend to dampen this difference by adding many layers of installation between the relatively warm ground and cold air.

   (b) The trend of the kinetic growth form process is to produce large, angular grains which are poorly bonded and weak, especially in shear. This result is that weak layers may form within a snowpack.

   (c) Because faceted grains have as much the same consistency as sugar, they are sometimes referred to as "sugar snow". Advanced faceted grains are also known as depth hoar.

   (d) Faceted snow is often the layer that collapses and goes "whump". When subjected to significant loading or wind-transported snow it becomes very sensitive.

   (e) Favorable conditions for the development of faceted grains are cold weather and/or a thin snowpack.

(3) <u>Melt-freeze Metamorphism (MF)</u>. This type of metamorphism occurs during mid-winter thaws or in the spring, when meltwater or rain enters the snowpack and the snowpack temperature reaches 32°F.

   (a) The trend is toward the production of coarse, rounded grains and with repeated cycles of melting and freezing. These large grains are also known as "corn snow".

   (b) In the freeze phase, these grains are well-bonded and strong creating a stable snowpack. But resulting ice crust can make a good potential bed surface for slabs formed on top of them.

   (c) In the melt phase, wet grains weaken rapidly and are lubricated by the presence of free water. This creates an unstable snowpack and is why timing during movement is critical in the spring near steep slopes that are being subjected to warming.

b. Other Weak Layers. There are some other important weak layers, like surface hoar and unmetamorphosed new snow.

   (1) Surface hoar is the wintertime equivalent of summertime dew and is formed at the snow surface during cold, clear weather. Surface hoar crystals are loose, feathery, and poorly bonded.

(a) Surface hoar is a potentially deadly weak layer once buried because it persists for a long period of time, can form a thin shear plane that is difficult to detect, and can produce long-running "zipper" fractures.

(2) Unmetamorphosed new snow is snow that may have fallen during cool or windless period of a storm and then had denser, heavier snow deposited on top of it.

(a) These rounded, iced pellets often roll downslope and collect in depressions or at the bottom of cliff bands, thus forming an area that may be more sensitive once the next load is deposited.

TRANSITION: We've taken a look at the snowpack metamorphism, let us now discuss how the weather affects avalanche activities.

7. (15 Min) **WEATHER**. Weather has been termed the "Architect of Avalanches". Most natural avalanches occur during or shortly after storms because the snowpack often cannot adjust to the vast amounts of new weight added in a short time. Weather affects the stability of the snowpack by altering the critical balance between strength and stress. The three main contributing factors are the precipitation, wind, and temperature.

   a. Precipitation. The significance of precipitation is that it increases the stress exerted upon a snowpack by adding weight.

   (1) Snow. New snow can provide a certain amount of strength to a snowpack, but can also cause rapid loading during storm.

   (2) Rain. Heavy rain weakens the snowpack by warming and eroding the bond between grains and slab layers.

   b. Wind. Since wind speed and direction help determine which slopes are being loaded this is a very important avalanche consideration.

   (1) Top loading. Wind accelerates on the windward side of terrain features and picks up loose snow, carries it over the crest and deposits it on the leeward side where the wind decelerates and deposits the snow.

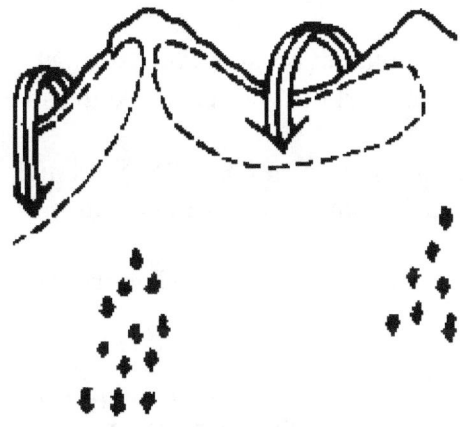

TOP LOADING

(2) Side loading. Also known as cross-loading, is sometimes more insidious because it can be harder to detect, especially in areas of gentle gullies.

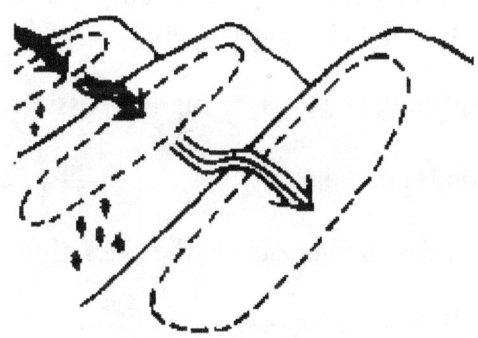

SIDE LOADING

c. <u>Temperature</u>. Changes in snow temperature can significantly affect snow stability. These changes are governed largely by ground and air temperatures, solar radiation and terrestrial radiation.

   (1) A warm snow pack will settle rapidly, becoming denser and stronger. This is associated with cloudy skies, since clouds trap warm air against the earth's surface.

   (2) Though gradual warming encourages strength and stabilization, intense warming weakens the bonds between grains and increasing the rate of the downslope deformation in affected layers.

   (3) On shaded slopes, the snowpack not only undergoes less (or slower) settlement due to warming but also the development of weak layers such as faceted snow or surface hoar.

This is because temperature gradients within the snowpack and at the snow surface can be more pronounced and persist for longer periods of time.

(4) Storms that start out cold and get progressively warmer are more likely to produce avalanches than those that start out warm and progressively become cooler.

(5) Any rapid, prolonged rise in temperature following long periods of cold weather could potentially lead to instability and should be noted as one of nature's signs.

TRANSITION: We have had a look at what happens when the weather plays a part. Let us now look at the human factor.

8. (15 Min) **HUMAN FACTOR**. It is possible to travel at times of high snow instability by choosing safe routes. Similarly, it is possible to get caught in an avalanche during periods of relatively low snow instability through poor route selection and stability evaluation. In other words, we create potential hazard by traveling in avalanche terrain. However; through careful route selection, preparation and decision-making, we can limit the amount of danger that is involved.

   a. Considerations for crossing an avalanche prone slope. Careful route selection can greatly reduce the chances of getting caught in an avalanche and in some areas, make it possible to travel during periods of high instability. The following route considerations should be taken before crossing a potential avalanche prone slope:

   (1) Determine starting zones and cross as high as possible.

   (2) Travel on high points and ridges.

   (3) When ascending or descending, stay to the sides of the start zone and track.

   (4) Avoid wind-loaded, leeward slopes.

   (5) Favor terrain with anchors, i.e. tree covered area, instead of open slopes.

   (6) Pick areas with flat, open run-outs so burial depth is decreased. Avoid areas that feed into gullies, crevasses and over cliffs.

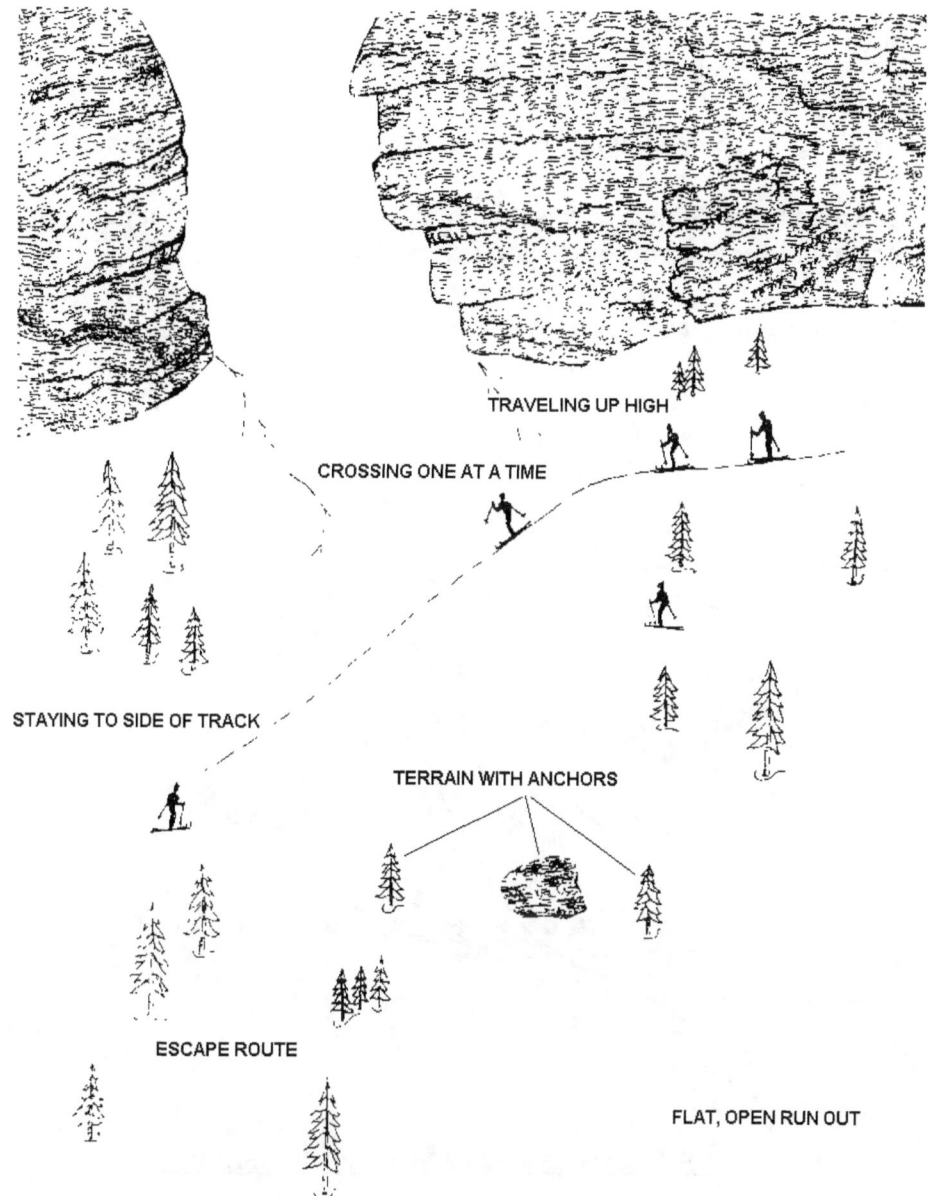

CROSSING AN AVALANCHE PRONE SLOPE

(7) Avoid V-shaped valleys and travel in wide U-shaped valleys. In V-shaped valleys, avalanches could run from either side and continue up the opposite side, so there may be little or no safe ground.

b. Preparations for Crossing Potential Avalanche Slopes. Occasionally, being reasonably sure that you will not fall, or that you can travel quickly, or knowing that all members of the patrol have rescue equipment and are proficient with it may allow you to travel across marginal areas. Before crossing a potential avalanche slope the following preparations should be taken:

(1) Loosen ski bindings and remove hands from ski pole straps.

(2) Loosen your pack and unbuckle the waist strap.

(3) Secure ECWCS hood tightly covering face and trail a 10m avalanche cord if available.

(4) Go around if possible. Travel straight downhill/uphill on foot rather than ski and look for possible escape routes, do not traverse back and forth.

(5) When traversing, cross as high as possible.

(6) Cross one at a time, belay if possible. If one Marine crosses it safely, that doesn't mean that it is safe passage for the rest.

CROSSING AN AVALANCHE PRONE SLOPE

c. Actions if Caught in an Avalanche.

(1) If your are caught in an avalanche, call out so the other members of your patrol know to watch you as you are carried down the slope, and then keep your mouth closed to prevent ingestion of snow.

(2) Discard pack, skis and poles although this is much easier said than done. This gear tends to drag you underneath the surface of the moving debris.

(3) Assess best line of escape.

(4) Delay your departure, i.e., let as much of the avalanche pass you as possible.
(5) Try and work to the side. There will be less force of the avalanche at the edge of the flow.

(6) Try to swim out using a swimming and rolling action to stay on the surface of the snow.

(7) As you feel the snow slow down, thrust your arm or hand or any part of your body above the snow surface so that it can be seen by others. You will probably be so disoriented that you won't know where the surface is so just guess and lunge.

(8) Before the snow comes to a rest, cup your arm or hand in front of your face to clear an air space. If possible, try to expand your chest during this time.

(9) If buried, stop fighting and relax to preserve oxygen. Remember, you are not supposed to panic!

d. Rescue Action. Use the Acronym STOP and GO. STOP means Stop, Think, Observe, Plan. GO means Go into action and Organize the rescuers. Do not panic! You are the victim's best chance of survival now.

(1) Watch the victim as he is carried down the slope. If the victim disappears under the moving snow, keep your eyes fixed on the mass of snow he was enveloped in, until it comes to a rest. The victim may be under the snow surface in that area.

(2) Also, using a ski pole, mark any position where he reappeared during his journey down the hill.

(3) Make a quick visual search of the area. Note any arms, legs, avalanche cord, and/or pieces of equipment that are sticking up and dig them out.

(4) At first, if nothing is apparent, make a quick surface search.

(5) If nothing is found, a more systematic search should be made from the bottom working up.

(6) If you again fail to find anything, your next step is to probe, which will be covered later.

(7) Stay on site and search. Almost all hope of a live rescue depends on you. Statistically, a victim has only about a 50% chance of survival if buried 30 minutes. The first 15 minutes are critical. Outside help cannot usually arrive fast enough.

TRANSITION: We have covered the definition of avalanches, the parts of the avalanche path, the types of avalanches, the avalanche evaluation process, terrain effects, snowpack, weather effects, and the human factor. Are there any questions?

PRACTICE (CONC)

a. Students will practice what was taught in upcoming field evolutions.

PROVIDE HELP (CONC)

    a. The instructors will assist the students as necessary.

OPPORTUNITY FOR QUESTIONS (3 Min)

1. <u>QUESTIONS FROM THE CLASS</u>

2. <u>QUESTIONS TO THE CLASS</u>

    Q. What are the three snow metamorphisms?

    A. (1) Rounding
       (2) Faceting
       (3) Melt-freeze

    Q. What should you do if you have to cross an avalanche prone slope?

    A. (1) Loosen ski bindings and remove hands from ski pole straps.
       (2) Loosen your pack leaving the downhill pack strap on your shoulder.
       (3) Secure ECWCS hood tightly covering face and trail a 15 foot avalanche cord if available.
       (4) Go straight downhill on foot rather than ski and look for possible escape routes.
       (5) Go straight down. Do not traverse.
       (6) If possible, cross as high as possible on concave slopes.
       (7) Cross one at a time. Just because one crosses it safely doesn't mean that it is safe passage for the rest. Belay everyone else across if possible.

SUMMARY (2 Min)

    a. Next to cold weather injuries, avalanches are the biggest threat to us in cold weather mountainous environment. Hopefully now, everyone has at least a basic knowledge of avalanches.

    b. Those of you with IRF's fill them out now and turn them into the instructor. We will now take a short break.

UNITED STATES MARINE CORPS
Mountain Warfare Training Center
Bridgeport, California 93517-5001

WML
WMO
07/01/02

LESSON PLAN

SNOW STABILITY EVALUATION

INTRODUCTION (5 Min)

1. **GAIN ATTENTION**. Snow stability evaluation, the process of determining if the snowpack is capable of avalanching, is just a fancy name for "hammering" on the snowpack to see how it will respond. It is an ongoing process which continues during every step of a climb and each turn of a ski, but it does not have to take a lot of time. This will help you identify what you know and don't know about what has been happening in the mountains. Always be willing to change your opinion based upon new information. Avalanche dangers can be avoided by applying common sense and performing a few simple tests.

2. **PURPOSE**. The purpose of this period of instruction is to introduce the student to snow stability evaluation and snow pit analysis. This lesson relates to Avalanche.

3. **INTRODUCE LEARNING OBJECTIVES**

   a. TERMINAL LEARNING OBJECTIVE. In a cold weather/mountainous environment and given a snow covered slope, determine avalanche hazards, in accordance with the references.

   b. ENABLING LEARNING OBJECTIVES

   (1) Without the aid of references, describe in writing the types of avalanche triggers, in accordance with the references.

   (2) Without the aid of references, describe in writing the signs of instability, in accordance with the reference.

   (3) Without the aid of references, describe in writing the signs of stability, in accordance with the references.

   (4) Without the aid of references, conduct a snow pit analysis, in accordance with the references.

   (5) Without the aid of references, conduct a rutschblock test on a slope, in accordance with the references.

(6) Given an Avalanche Decision Making Checklist in prevailing snow conditions, determine avalanche hazard, in accordance with the references.

4. **METHOD / MEDIA**. The material in this lesson will be presented by lecture and demonstration. You will practice what you have learned during upcoming field training exercises. Those of you with IRF's please fill them out at the conclusion of this period of instruction.

5. **EVALUATION**. You will be tested later in the course by written and performance evaluation.

INSTRUCTOR NOTE: All material covered in this outline is in reference to Snow Sense, 4th Edition, 1994, Jill A. Fredston and Doug Fesler and American Institute for Avalanche Research and Education.

TRANSITION: Let's begin by talking about avalanche triggers.

BODY (60 Min)

1. (5 Min) **AVALANCHE TRIGGERS**. There are two types of triggers, natural and artificial.

    a. Natural Triggers. These are not triggered directly by man or his equipment. A falling cornice, sluffing snow, stress change due to metamorphism in the snow pack, can all trigger avalanches.

    b. Artificial Triggers. These are triggered by man or his equipment. A skier passing, a mountaineer's weight, an explosive blast, a sonic boom, and the like commonly set off avalanches.

NOTE: An important fact is that artificial triggering leads to a far greater frequency of avalanches on a given path than if the path were left up to avalanche naturally.

TRANSITION: Now that we discussed avalanche triggers, let's talk about signs of instability and stability.

2. (10 Min) **SIGNS OF INSTABILITY AND STABILITY**. To prevent ourselves from being the trigger, we need to formulate an opinion about the stability of the snow early on. Nature can provide some clues for us to use in determining the sensitivity of the snowpack.

    a. Signs of Instability. Skilled avalanche hazard evaluation can be based upon a systematic decision-making process using signs of nature. You may experience these indicators by themselves or together. The following are signs of instability:

    (1) Recent avalanche activity on similar slopes and small avalanches under foot.

    (2) Booming, which is the audible collapse of the snow layers (normally a faceted layer collapsing).

    (3) Visible cracks shooting out from underfoot (severe tension in the snow pack).

    (4) Sluffing debris is evidence of avalanche activity occurring.

    (5) Sunballing, which is caused by rapid re-warming.

(6) Excessive snowfall, over 1 inch per hour for 24 hours or more.

(7) Heavy rain that warms and destroys the snow pack.

(8) Significant wind-loading causing leeward slopes to become overloaded.

(9) Long, cold, clear, calm period followed by heavy precipitation or wind-loading.

(10) Rapid temperature rise to near or above freezing after a long, cold period.

(11) Prolonged periods (more than 24 hours) of above freezing temperatures.

(12) Snow temperatures remaining at or below 25°F which slows down the settlement/strengthening process thus allowing unstable snow conditions to persist longer.

b. Signs of Stability. The following are signs of stability:

(1) Snow cones or settlement cones, which form around trees and other obstacles and indicate the snow around the object is settling.

(2) Creep and Glide. Creep is the internal deformation of the snow pack. Glide is slippage of the snow layer with respect to the ground. Evidence of these two properties on the snow pack is a ripple effect at the bottom of a slope. It is an indication that the snow is gaining equilibrium and strength through this type of settlement process.

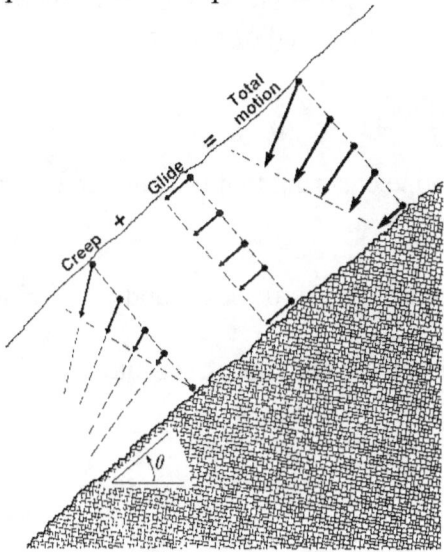

CREEP AND GLIDE

(3) Absence of wind during storms, which is indicated by snow accumulation in the trees.

(4) Snow temperatures remaining between 25°F and 32°F, which ordinarily settles the snow rapidly, becoming denser and stronger because of the effects of rounding metamorphism.

TRANSITION: Now that we talked about natural signs to look for, let's discuss expedient methods to determine slope angle.

3. (5 Min) **EXPEDIENT METHODS TO DETERMINE SLOPE ANGLE**. Slope angle is the most important variable determining whether or not it is possible for a given slope to avalanche. The following methods can be used to determine slope angle:

(a) Inclinometer Method. These devices are designed to get an approximate reading of a slope angle in degrees. They may be a part of a compass, or a separate piece of gear.

   (1) Place a ski pole on a slope ensuring that the pole is flushed with the surface.

   (2) Place the inclinometer on the ski pole.

   (3) The dial will indicate the angle of the slope.

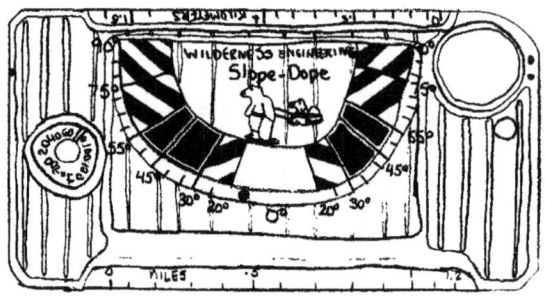

INCLINOMETER

(b) Protractor Method. The standard issued protractor needs to be constructed with a 12-inch string in its center with a weight placed on the end of the string.

   (1) Dig a small hole into a slope.

   (2) Place a ski pole on the slope ensuring that the pole is flushed with the surface and that it is aligned over the hole.

   (3) Place the protractor on the ski pole; 0° down and 90° facing down the slope.

   (4) Read the azimuth degree scale where the string intersects it. This will indicate the angle of the slope.

PROTRACTOR METHOD

TRANSITION: Now that we talked about expedient methods to determine slope angle, let's discuss field expedient stability tests.

4. (10 Min) **FIELD EXPEDIENT STABILITY TESTS**. Often, no single field test or observation will tell it all. You must piece together the story the snowpack is trying to tell by gathering up all available information. You will usually find that the various pieces of information back each other up and tell the same story. Field expedient stability tests are a good place to start gathering information, but keep in mind that they should be conducted on short slopes where no serious consequences would result.

   (a) Small Steep Hills. The tester can sometimes get very useful feedback from a slope that is only a few feet high by jumping from the top onto the slope. Remember to take note of how they respond.

   (b) Test Skiing. This is a stability test whereby a skier adds stress to the snow through his weight and/or by jumping and kicking. The tester can immediately observe the depth and type of weak layer that might have failed.

   (1) When traversing uphill on skis and have just turned a corner, jump just below your uphill ski track and see if you can get a "piece of the pie" to break into blocks. This indicates that the snow may be cohesive enough to propagate a fracture.

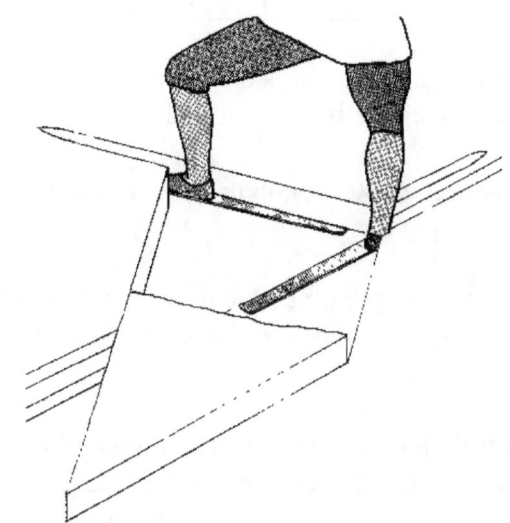

EXAMPLE OF A SKI TEST FOR STABILITY

   (c) Ski Pole Test. This test takes only seconds and should be done often as you travel.

   (1) Holding your ski pole at a right angle to the snow surface, gently push the pole into the snowpack.

   (2) Feel for the relative hardness and the thickness of the layers.

   (3) Be alert for well-consolidated layers that feel harder than underlying soft, weaker layers. This is a method to keep track of the depth and distribution of potential slabs.

   (4) If the basket of the ski pole interferes with probing, use the handle of the ski pole to probe instead.

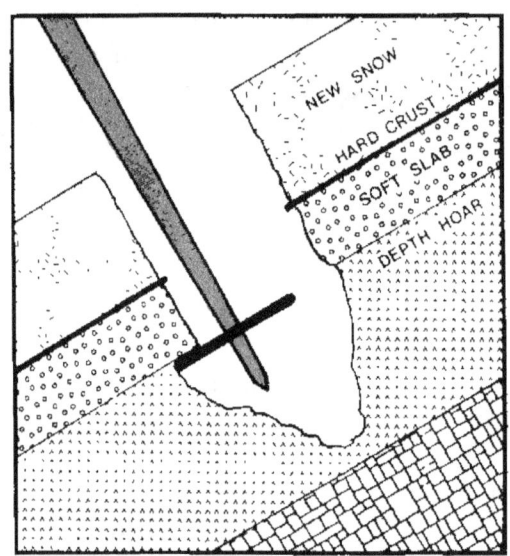

SKI POLE TEST

NOTE: One serious limitation of the ski pole test is that sometimes the weak layers are too thin to detect and it does not detect how well the layers are bonding together.

TRANSITION: Now that we have discussed expedient stability test methods, let us talk about different shear tests.

5. (15 Min) **SHEAR TESTS**. The principle objective of a shear test is to locate weak layers and interfaces. There are many different types of shear tests, but the two types we will discuss are the rutschblock and shovel/ski shear tests.

   (a) Rutschblock Test. The rutschblock test involves loading a block of snow by a person in several stages. The following is the procedure for constructing the rutschblock.

   (1) Construction
      a. Select a site on a slope with the same slope angle and aspect as the slope that you are concerned with. Personnel conducting the test may be belayed.

      b. Begin digging a pit approximately one ski length in width and at least 4 to 5 feet deep.

      c. From the ends of the pit, dig two narrow trenches uphill into the slope approximately the length of one ski pole. Ensure that the depths of the trenches are the same as the pit's depth.

      d. Being very careful not to disturb the area of the rutschblock, use a snow saw or a knotted length of cordage and cut the back of the wall. This will isolate the snow block.

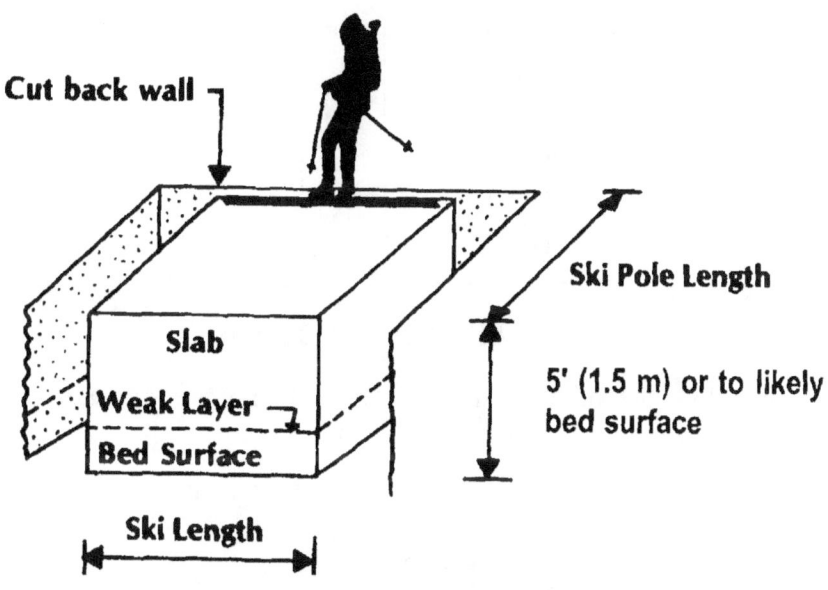

RUTSCHBLOCK TEST

(2) <u>Testing</u>. Carefully ski to the side of the test site and approach the upper cut of the rectangular block diagonally from above. Once your skis are perpendicular to the cut on the uphill side, gently move on to the block. The test is conducted in stages contained in the following chart. When the block fails the chart will give you an idea how stable the slope is.

STEP	REACTION	RESULT
One	Fails while excavating test site.	Extremely unstable
Two	Fails while approaching test site	Extremely unstable
Three	Fails while standing on shear block	Extremely unstable
Four	Fails while flexing your knees	Unstable
Five	Fails with one jump while wearing skis	Unstable
Six	Fails after repeated jumps with skis on	Relatively stable
Seven	Doesn't fail after repeated jumps with skis off	Stable or very stable

(b) <u>Shovel/Ski Shear Test</u>. This is a quick method to obtain information of a location where weak layers are suspected, without involving a lot of digging.

(1) <u>Construction</u>.

    a. Isolate a column in the uphill pit wall by cutting away the sides with a shovel or ski.

    b. The width of this column as well as the depth cut into the pit should be approximately 12 inches. Ensure that the column is both vertical and smooth.

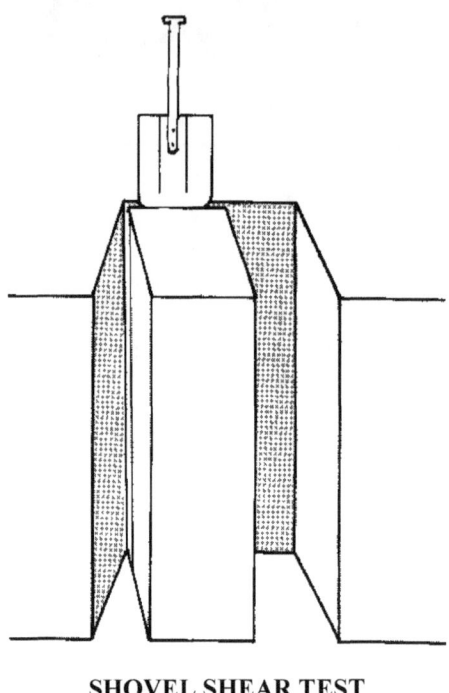

SHOVEL SHEAR TEST

(2) <u>Testing</u>. Insert a shovel/ski behind the column and exert steady pressure as you work your way down. Look for possible separation of the weak layers.

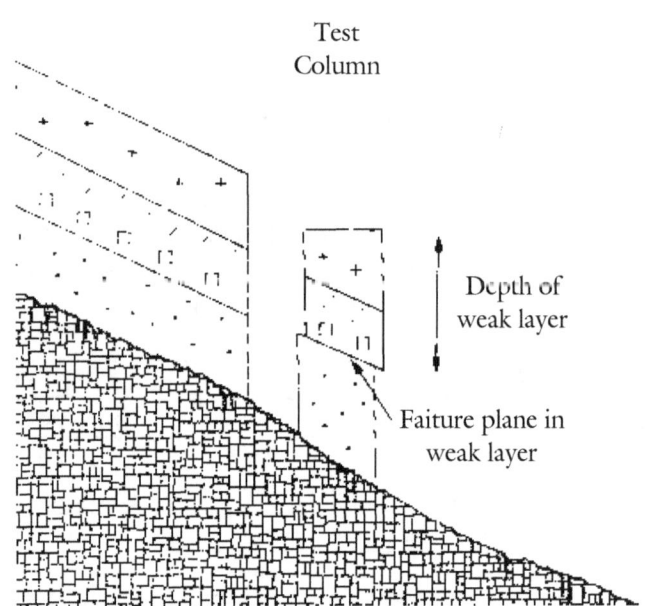

WEAK LAYER GIVING IN A SHOVEL SHEAR TEST

(c) <u>Shred Block Test (using snow shoes)</u>. The Shred Block test is like a Rutschblock, just with snow shoes instead of skis.

1. Construction. Dimensions of the shred block will be a column approximately 1.75 metres wide (across the fall line), 1.5 metres on each side (up the fall line), and somewhat deeper than the suspected failure layer (to a maximum of about 1.5 metres). This requires excavating a significant amount of snow. Dimensions will vary slightly depending on what method is used to cut the sides of the block.
2. Conduct the test same as the Rutschblock, the Marine just has snow shoes on instead of skis.

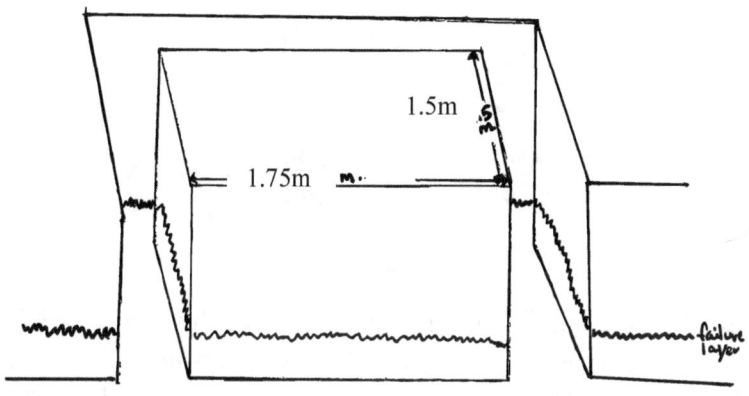

Shred Block (for snow shoes)

TRANSITION: These methods are an easy way to attain information quickly on weak layers. Now let's talk about a more extensive method to gain information on a snowpack as a whole, snowpit analysis.

6. (10 Min) **SNOWPIT ANALYSIS**. In this section we will discuss how to analyze the snowpack for instabilities by identifying weak layers. Snowpit analysis can be extremely complex, but we will deal only with the basic observations.

    a. Construction

        (1) Choose a location with the conditions similar to those you are trying to evaluate. They should at least be at a similar elevation, snow condition, slope angle and aspect as the slope you are concerned about.

        (2) Dig a pit 4 to 5 feet deep and wide enough to work in. Be careful not to disturb the snow surface surrounding the uphill portion of the pit.

        (3) With a shovel, smooth off the uphill pit wall and adjacent (side) wall. Ideally the adjacent wall should be shaded. These walls are where your tests will be conducted. It is important that they be smooth and vertical and that the snow above the uphill wall remains undisturbed.

    b. Identifying Layers

        (1) Stratigraphy Test. Using a whiskbroom, paint brush, hat or mitten, lightly brush the sidewall of the pit with uniform strokes parallel to the snow surface. This will quickly transform the wall from a plain white surface into a layered mosaic of snow history. The raised or ridged surfaces indicate the harder, stronger layers that may be possible slabs or sliding surfaces. The indented surfaces reveal softer, weaker layers.

        (2) Resistance Test. Insert a credit card, saw, or any straight edge into the top of the sidewall. Run the card down the wall, feeling the relative resistance of the layers and noting the boundaries of hard and soft layers. In helping to identify potential slab and weak layers, this test can help corroborate and expand upon the information gained from the stratigraphy test.

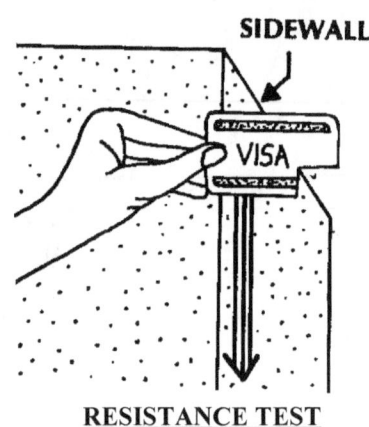

RESISTANCE TEST

(3) Hardness Test. Test the relative hardness of each layer by gently pushing your hand or fingers into the pit wall, applying approximately 10 lbs. of pressure. One layer might be so soft that you can easily push your whole fist into it while another's might require a knife to penetrate it. An example of a potential unstable slab configuration would be a cohesive 1-finger hard layer resting on top of a less cohesive fist hard layer.

HARDNESS TEST

7. (10 Min) **AVALANCHE DECISION MAKING CHECKLIST.** The checklist consists of 2 different checklist tools, the Decision Making Checklist (rates hazard low to extreme) and the Data Observation Checklist (go/no go levels). Using both checklists will give you the most accurate avalanche hazard assessment when moving your unit through a snow covered battlefield. The tactical situation will effect the determination of a go or no go hazard level. The Route Planning lesson plan covers ways to mitigate the avalanche threat for commanders. These checklist blanks can be laminated back-to-back for field use.

AVALANCHE DATA OBSERVATIONS

DATA CLASS	INFORMATION CATEGORY	OBSERVATIONS MADE	RED FLAG VALUES	YOUR OBSERVATION
Weather	Precipitation	Type	Rain/heavy wet snow	
		Intensity	>3cm (1 in.)/hour	
		Accumulation	>30 cm (12 in.)/12 hours	
	Wind	Speed	Strong enough to move snow	
		Direction	Moving snow onto/across terrain where you will travel	
		Duration	Long	
	Temperature	Current	$\geq 0°C/32°F$	
		Maximum/minimum	$\geq 0°C/32°F$	
		Trends	Rapid changes (especially from cold to warm, and through the freezing level)	
	Solar Radiation	Cloud cover	Allowing a lot of radiation to enter or intensifying radiation.	
		Intensity	Strong	
		Duration	Long	
Snowpack	Snow Cover	Height	<1.5m/5ft	
		Strength	Weak	
		Variability	High	
	Layers	Strength	Strong over weak	
		Temperature	Near/=0°C/32°F	
		Grain Characteristics	Large, loosely packed, angular	
	Bonding	Strength	Compression test ≤ 20 Rutschblock ≤ 4	
		Plane characteristics	Smooth, clean	
		Failure layer	Large, loosely packed, angular grains	
	Whumphing	Initiation	Natural/human trigger	
		Propagation	Far (>3m/10ft)	
		Extent	Widespread	
Avalanche Activity	When	Current	Observed	
		Recent	<24 hours (Maritime climate) <48 hours (Continental climate)	
		Past	If condition still exists	
	Where	Area	Widespread	
		Slope Angle	Over 30 degrees	
		Slope Aspect	Facing sun Leeward	
		Slope Shape	Concave	
		Terrain Traps	Traps exist where avalanches are running	
	What	Natural Triggers	All natural triggers	
		Human Triggers	All human triggers	
		Other Triggers	Remote triggers, arty, demo	
	How	Destructive Potential	$\geq$Class 2	
		Propagation	Wide fracture lines running far	
		Failure layer	Large, loosely packed, angular grains	

Current Danger Rating	Low (Green)	Moderate (Yellow)	Considerable (Orange)	High/Extreme (Red)
Danger Trend/Forecast	Improving/Steady	Rising Slowly	Rising Rapidly	

Continue? If yes, proceed to avalanche activity data:

	Low (Green)	Moderate (Yellow)	Considerable (Orange)	High/Extreme (Red)
When:	Past	Recent	Current	Current + Widespread + Same as you + Human/Natural + Large
Number:	None/Few	Many	Widespread	
Where:	Far away	In area	Same as you	
Triggers:	Large	Human	Human/Natural	
Characteristics:	Small	Medium/Slabs	Large	

Continue? If yes, proceed to snowpack data:

	Low (Green)	Moderate (Yellow)	Considerable (Orange)	High/Extreme (Red)
Average Depth:	> 2.0m	1.5 – 2.0m	< 1.5m	
Average Strength:	Strong	Moderate	Weak	
Variability From Average Depth/Strength:	Uniform	Somewhat variable	Highly variable	
Strong Over Weak Layering:	Little/None	Some	Pronounced	
Compression Tests/ Rutschblock Tests:	CT 30+ / RB 7	CT 20 - 30 / RB 5 - 6	CT 10 - 20 / RB 3 - 4	CT 0 - 10 / RB 1 - 2
Danger Signs (Cracking, Whumphing, etc.):	Few/None Heavy trigger Localized propagation	Isolated Mod. trigger Mod. propagation	Widespread Light trigger Wide propagation	

Continue? If yes, proceed to weather data:

	Low (Green)	Moderate (Yellow)	Considerable (Orange)	High/Extreme (Red)
Storm:	None	Snow 1 - 2 cm/hr. Winds move little snow in start zone. Cool and steady temps	Snow 2 - 3 cm/hr. Winds move some snow in start zone. Warm temps and/or Rapid temp rise.	Snow 3+ cm/hr + Winds move much snow in start zone + Very warm temps and/or Rapid temp rise.
Last No Go Storm Ended:	> 48 hours ago	36 - 48 hours ago	< 36 hours ago	
New Snow (12 hrs):	< 15 cm	15 - 30 cm	> 30 cm	
Blowing Snow:	None	Some recently	Much recently or currently	
Temperature: Solar Radiation:	Cold - Cool/ None – Little	Cool - Warm/ Some	≥0 and/or rapid rise/ Strong	

Continue? If yes, proceed to terrain assessment:

	Low (Green)	Moderate (Yellow)	Considerable (Orange)	High/Extreme (Red)
Incline:	< 25	25 - 35	> 35	
Wind Exposure/Aspect:	Windward	Some cross/lee	Much cross/lee	
Trigger Points:	None – Few	Some	Many	
Size/Traps:	Small/None - Few	Moderate/Some	Large/Many	

Other Pertinent Data:

	Go with normal caution. Consider human factors.	Consider Safer Options Go with increased caution Consider Human Factors	Consider Safest Options Travel not recommended on specific terrain or certain snowpacks. Consider Human Factors.	Travel Not Recommended Consider Human Factors

Discussion of decision (terrain/snowpack to avoid, human factors, etc.):

TRANSITION: There are many dangers in the mountains, none are as violent and frightening as avalanches. Every effort should be made to protect yourself and your Marines from these deadly forces of nature.

PRACTICE (CONC)

a. Students will practice what has been taught in upcoming field exercises.

PROVIDE HELP (CONC)

a. The instructors will assist the students when necessary.

OPPORTUNITY FOR QUESTIONS (3 Min)

1. QUESTIONS FROM THE CLASS

2. QUESTIONS TO THE CLASS

    Q. What are the types of avalanche triggers?

    A. (1) Natural
       (2) Artificial

    Q. How hard is a layer of snow when it is fist hard?

    A. Very soft

    Q. What should you do if you have to cross an avalanche prone slope?

    A. (1) Loosen ski bindings and remove hands from ski pole straps.
       (2) Loosen your pack leaving the downhill pack strap on your shoulder.
       (3) Secure ECWCS hood tightly covering face and trail an avalanche cord if available.
       (4) Go straight downhill on foot rather than ski and look for possible escape routes.
       (5) Go straight down. Do not traverse.
       (6) If possible, cross as high as possible on concave slopes.
       (7) Cross one at a time. Just because one crosses it safely doesn't mean that it is safe passage for the rest. Belay everyone else across if possible.

SUMMARY (2 Min)

a. During this period of instruction we have discussed avalanche triggers, signs of instability and stability, and different shear tests and how to construct them.

b. Those of you with IRF's fill them out at this time and turn them in to the instructor.

UNITED STATES MARINE CORPS
Mountain Warfare Training Center
Bridgeport, California 93517-5001

WML
WMO
05/22/02

LESSON PLAN

AVALANCHE SEARCH ORGANIZATION

INTRODUCTION (5 Min)

1. **GAIN ATTENTION**. Due to some unforeseen circumstance, or more likely through lack of observation to warning signs, your men may find themselves under several hundred pounds of snow. If you panic and search hither and yonder, you'll most likely miss something crucial and allow your men to perish. Searching for victims of avalanches can be done in several ways, by hasty search, probing, transceivers, dogs or radar.

2. **PURPOSE**. The purpose of this period of instruction is to familiarize the students with the basics of avalanche rescue to include probing techniques, search routines, and organization of a unit. This lesson relates to Avalanches, Snow Stability Evaluation and Avalanche Transceivers.

3. **INTRODUCE LEARNING OBJECTIVES**

    a. TERMINAL LEARNING OBJECTIVE. In cold weather/mountainous environment, demonstrate avalanche search techniques, in accordance with the references.

    b. ENABLING LEARNING OBJECTIVE. In a cold weather/mountainous environment and given an actual/simulated search victim, locate a victim using avalanche search and organization techniques, in accordance with the references.

4. **METHOD / MEDIA**. The material in this lesson will be presented by lecture. You will practice what you have learned during upcoming field training exercises. Those of you with IRF's please fill them out at the conclusion of this period of instruction.

5. **EVALUATION**. You will be tested later in the course by performance evaluation.

TRANSITION: Are there any questions concerning what we are going to cover during this period of instruction. If not, let's begin by discussing the hasty search.

BODY (55 Min)

1. (10 Min) **HASTY SEARCH**. A buried victim's life depends crucially on the action of survivors in the few minutes following an avalanche. Organized rescues are often slow getting to the scene due to the location and inaccessibility. Avalanche rescue is very similar to dealing with a drowning victim – "self-help is the key". When persons are caught in an avalanche, the survivors should react calmly and methodically. The following actions should be taken when witnessing an avalanche:

   a. Quickly look to make sure you are in a safe location.

   b. Immediately make note of the last seen point of a victim.

   c. Do a head count to determine who was caught in the avalanche. Also try to determine the last location of the missing victims when the avalanche occurred.

   d. Quickly assess the hazard of other possible avalanches. Post a guard (most likely the radio operator) at a safe spot. Make sure that all escape routes and warning signals are understood by all.

   e. Conduct a visual search of the deposit surface with concentration on the most likely burial areas. Look for parts of the victim and their equipment on the surface.

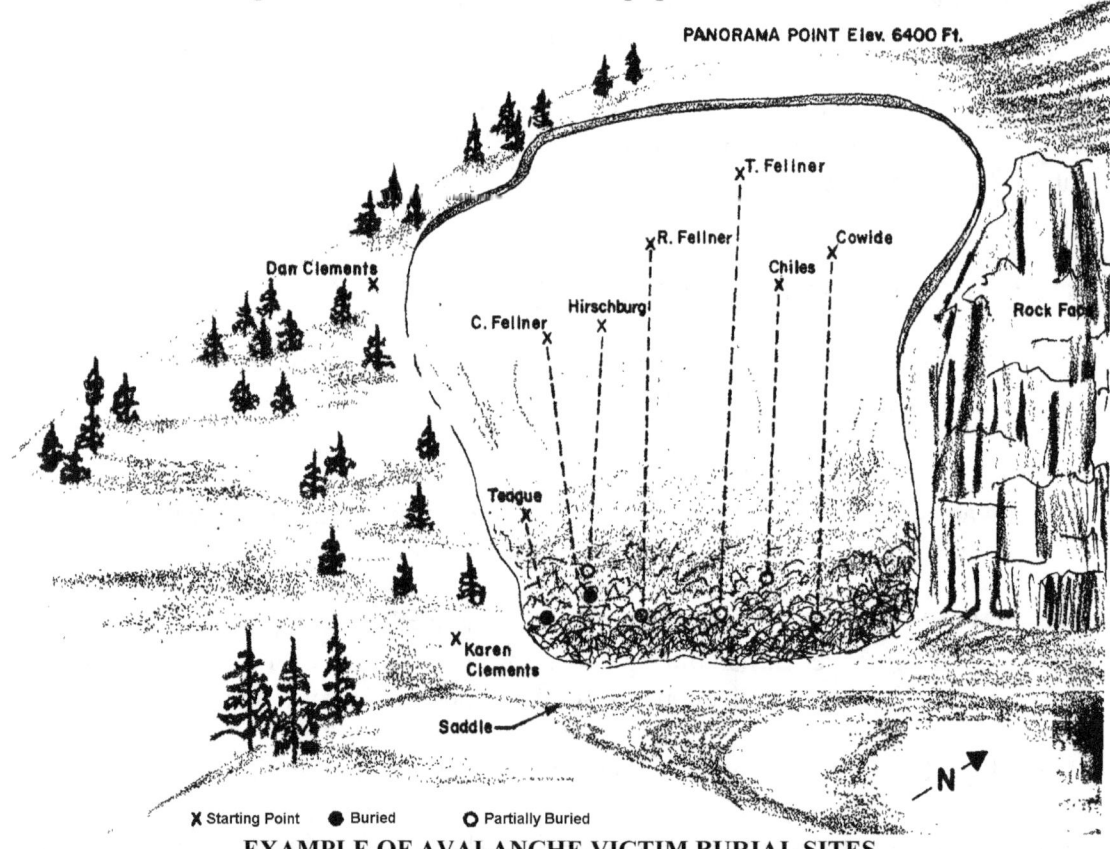

EXAMPLE OF AVALANCHE VICTIM BURIAL SITES

f. Draw a line from where the victim was caught, to the location of equipment and then of the victim's last seen point. The end of this line may point to the most likely burial site.

g. Have the radio operator alert the command of the situation to include location and number of victims involved. Request an avalanche search organization immediately.

h. Began searching the most likely burial areas. Simultaneously keep looking for signs of the victims and their equipment.

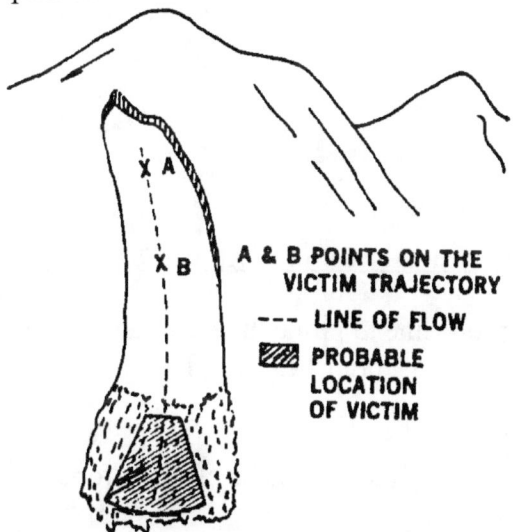

SUSPECTED LOCATION OF AVALANCHE VICTIM

i. Stop from time to time, call out to the victims and listen for voices of the buried persons.

j. Consider keeping packs on during the search so that shovels, probes and first-aid equipment are readily available.

k. Keep everyone involved in the search by probing and investigating most likely burial areas.

l. When victims are located, quickly dig them up and perform first aid as necessary.

m. If not all victims are located, mark the locations of clues found on the surface and spot probe around them.

n. Continue to spot probe most likely areas of burial to include areas behind trees, rocks, in depressions and in the run-out zone.

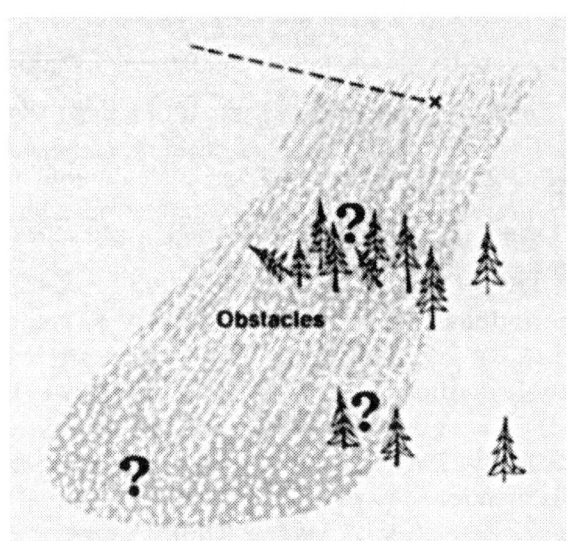

SUSPECTED SEARCH AREAS NEAR OBSTACLES AND RUN OUT ZONE

o. Organize a hasty probe line to probe the most likely burial areas and mark areas already probed with ski poles, skis, branches, etc. Techniques of probing will be discussed later in this chapter.

p. Try to keep the surface of the avalanche clean of food. This will prevent distraction of search dogs, if utilized.

q. Keep searching and probing until help arrives to take control over the search operation.

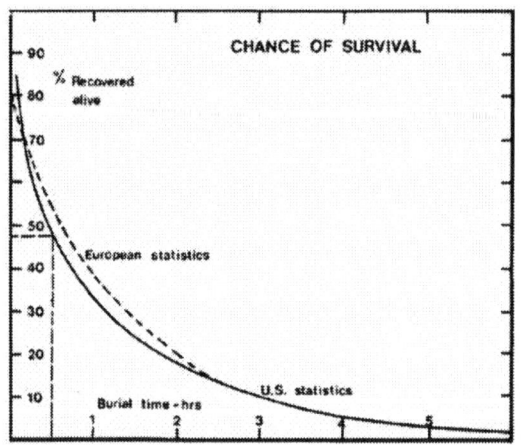

AMOUNT OF TIME BURIED AND SURVIVAL %

TRANSITION: Now that we have talked about the hasty search, let us discuss search organization.

2. (10 Min) **AVALANCHE SEARCH ORGANIZATION**. An organized rescue is conducted by a well trained and equipped unit. Normally this unit will be the unit of the victims. Once the commander is alerted that his Marines will be performing the search, he must collect his command element and proceed to the avalanche site.

a. Upon arriving at the site, the CO must make an estimate of the situation and make the following determinations, in order of priority:

   (1) Evaluate the accident site.

   (2) Post an avalanche guard and arrange for a warning signal, if not already established.

   (3) Designate escape routes.

   (4) Question witnesses and survivors at the scene about:

      (a) How the accident happened.

      (b) Persons buried.

      (c) Locations of the unit's members when the avalanche occurred.

      (d) Last seen point of the victims.

      (e) Search efforts conducted so far.

   (5) Provide care to the survivors.

   (6) Determine most likely burial areas.

   (7) Keep notes of actions and sketch a map of the avalanche site with location of clues.

   (8) Locate a safe location for the Command Post.

   (9) Locate a helicopter landing zone.

   (10) Consider the enemy situation.

b. Upon the company's arrival at the site, the CO must:

   (1) Have the platoons store their equipment at a safe area away from the avalanche site.

   (2) Inform platoon commanders about safety measures, the accident and action taken so far.

   (3) Delegate tasks to each platoon, i.e. hasty search teams, avalanche guards, probe lines etc.

   (4) Have platoon commanders organize the probe lines.

   (5) Consider equipment, food, and support that may be needed for a prolonged rescue.

(6) Keep the search organization focused.

TRANSITION: Are there any questions concerning search organization? If not, let us talk about considerations and procedures for assigned tasks.

3. (10 Min) **CONSIDERATIONS AND PROCEDURES FOR ASSIGNED TASKS**. As each platoon receives its task from the CO, there are certain considerations and procedures which should be adhered to for each task.

   a. Posting of avalanche sentries.

      (1) Above natural anchors and starting zones.

      (2) Sentries must prevent anyone from entering starting zone areas.

      (3) Sentries must be in a position to observe any adjacent starting zones and prevent anyone from entering this zone.

      (4) Sentries must be out of danger themselves.

      (5) Sentries must be equipped with a signal device that will warn everyone about possible avalanche threats.

      (6) The avalanche sentries may also serve as a security element for the search organization.

   b. Establishing the Command Post.

      (1) Set up warming tents for the searchers and victims, and prepare hot wets.

      (2) Set up an aid station.

      (3) Emergency medical sled(s) should be ready with sleeping bag, sled teams, and a corpsman.

      (4) Stamp out and mark the LZ.
      (5) The CP must be close enough to support the search yet far enough away from any existing avalanche hazards.

      (6) Ensure that all radios are all on the same net and that communication is established to the next higher command.

      (7) Provide guides to escort personnel from the road head to the accident site.

   c. Hasty Search Teams.

(1) Search gullies and ravines which could channelize a victim.

(2) Search uphill of catchment areas such as rock outcrops, trees, benches in the slope and fallen logs.

d. Probe Lines.

(1) Snowshoes should not be worn on the probe line as the debris of a hard slab avalanche will make snowshoe movement difficult.

(2) The ends of two adjacent probe lines must overlap by two men to insure that there is no gap between the lines.

(3) Since probes lines can easily get out of alignment and probers tend to take overly large steps, the Probe Line Leaders must ensure that the probe lines remain aligned and in order.

(4) Overlap the flanks and run-out zones by at least 20 feet. Victims have been recovered from these areas after being shoved there by the force of the avalanche, even though the snow surface remains undisturbed.

(5) All areas searched must be marked to avoid confusion.

e. Other considerations.

(1) Dogs and individuals with transceivers will search independently of the probe line. However, they should have probers and shovelers readily available to uncover any possible strikes.

NOTE: The tactical situation may dictate how long the search may be conducted.

TRANSITION: Now that we talked about considerations and procedures for assigned tasks, let's discuss the method of probing.

4. (10 Min) **PROBING**. There are two types of probes, the coarse probe and the fine probe. Each probe line will consist of two squads, and numerous markers and shovelers. The Platoon Sergeant will perform the duties as the Probe Line Leader. His job is to control the tempo of the line and ensures that the probers stay abreast.

a. Coarse Probe. The following steps will be taken for the coarse probe:

(1) With two squads on line, fingertip to fingertip, and at a designated area of search, each man will place the probe between his feet.

(a) The Probe Line Leader will be located behind each probe line.

(b) Also behind each probe line are the markers. Their mission is to place a mark where a strike has been indicated by a prober.

(c) Along with each marker is a team of shovelers whose job is to uncover the mark.

(2) The Probe Line Leader will give the command "DOWN PROBE", at that time the probes are then pushed down at a 15° angle out to the left. The probe is pushed through each layer of the snow, being careful not to impale a victim if a strike is made.

(3) The Probe Line Leader must always check the alignment, spacing and penetration of the probes.

(4) The next command given is "UP PROBE", and all of the probes will be withdrawn from the snow.

(5) The Probe Line Leader will give the command "DOWN PROBE", at that time the probes are then pushed straight down.

(6) The next command given is "UP PROBE", and all of the probes will be withdrawn from the snow.

(7) The Probe Line Leader will give the command "DOWN PROBE", at that time the probes are then pushed down at a 15° angle out to the right.

(8) The next command given is "UP PROBE", and all of the probes will be withdrawn from the snow.

(9) If a strike is made at any time, the prober signals to the marker to place a mark on the spot. The shovelers will then dig up this marked area. The line will never stop at a strike.

(10) At the command "STEP", each man takes a 30-inch step and the process repeats.

COARSE PROBE

b. Fine Probe. This is similar to a coarse probe with exception to the following:

(1) Probing is performed over the left, middle, and right foot

(2) A 15 inch step is taken rather than a 30 inch step.

(3) A fine probe is usually a body recovery and should only be started when all hope of a live recovery is exhausted.

(4) A fine probe search takes from four to five times longer than a coarse probe.

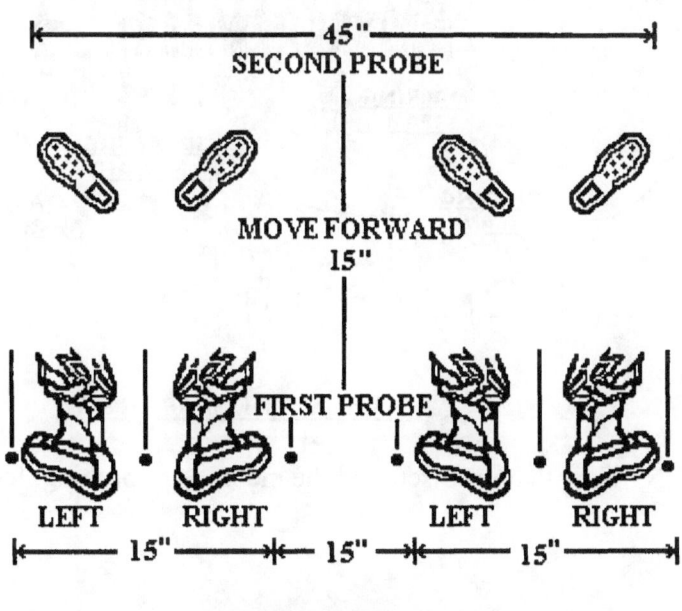

FINE PROBE

11-9

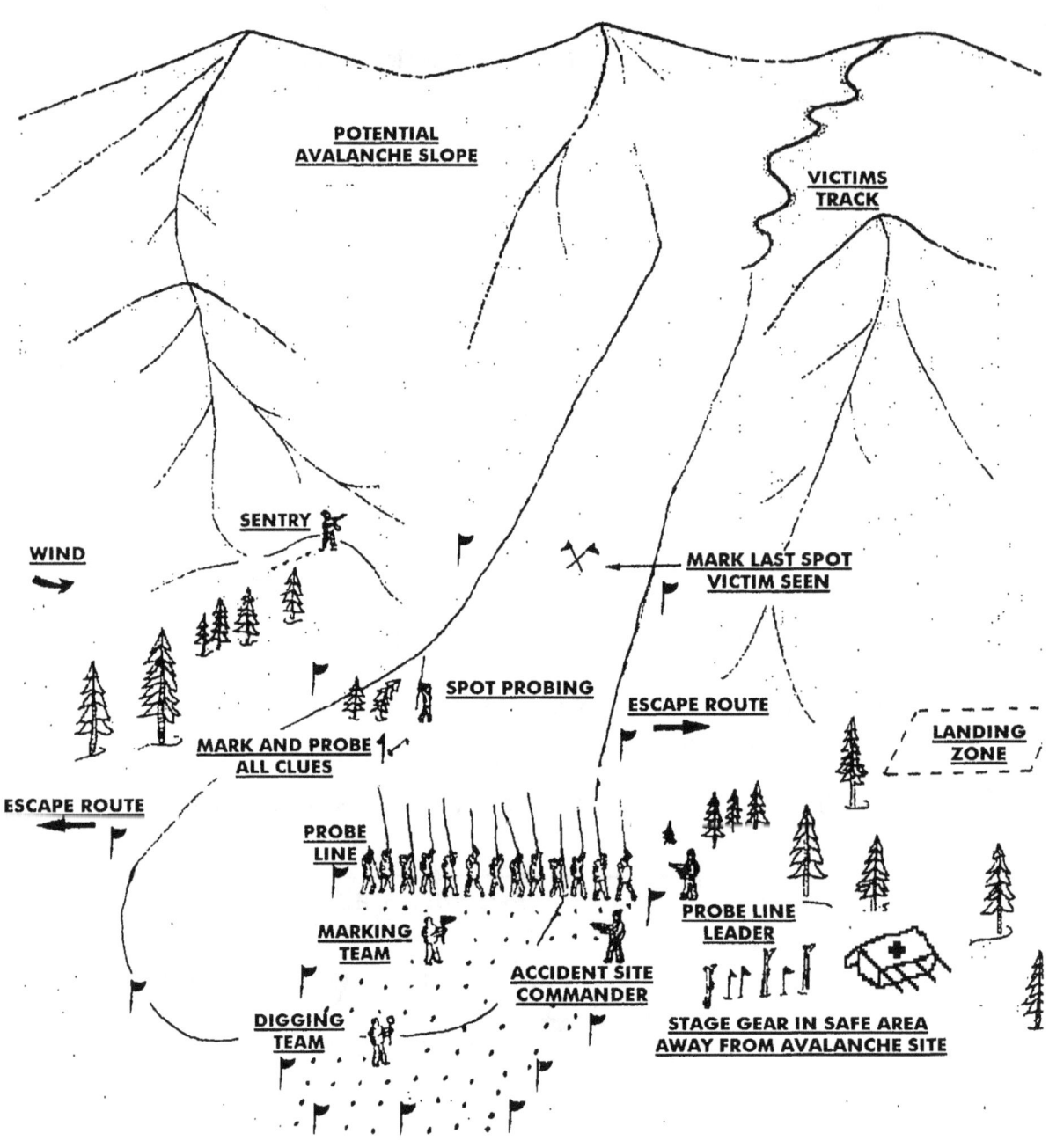

AVALANCHE SEARCH SITE

TRANSITION: Now that we discussed the methods of probing, let us now discuss specialized searches.

5. (5 Min) **SPECIALIZED SEARCHES**. These searches operate independently from other searchers.

   a. Dogs. A probe line takes approximately four hours, but a dog takes about 20 minutes at 15 feet in depth. Searchers will allow the dog handlers to perform their duty without interference.

   b. Transceivers. This technique will be discussed in further detail in Chapter 12.

   c. Sonar. Requires qualified operators as well as the obvious specialized equipment. Engineer units are normally outfitted with this equipment. If your unit has engineer support, than they should be assigned with this task.

TRANSITION: Remember, only prompt and decisive action can save an avalanche victim. Are there any questions on what we've just covered?

PRACTICE (1 Hr)

   a. Students will practice what was taught during upcoming field exercises.

PROVIDE HELP (CONC)

   a. The instructors will assist the students when necessary.

OPPORTUNITY FOR QUESTIONS (3 Min)

1. QUESTIONS FROM THE CLASS

2. QUESTIONS TO THE CLASS

   Q. What are the three types of probe techniques used?

   A. (1) Hasty
      (1) Coarse
      (2) Fine

   Q. What are the commands used to control the probe line?

   A. (1) Down Probe
      (2) Up Probe
      (3) Step

SUMMARY (2 Min)

a. During this period of instruction we have discussed the techniques used to conduct an avalanche search to included the types of searches, avalanche search organization, the duties and procedures for the search, and probing techniques.

b. Those of you with IRF's please fill them out at this time and turn them in to the instructor at this time. We will now take a short break.

UNITED STATES MARINE CORPS
Mountain Warfare Training Center
Bridgeport, California 93517-5001

WML
WMO
09/13/02

LESSON PLAN

AVALANCHE TRANSCEIVERS

INTRODUCTION (3 Min)

1. **GAIN ATTENTION**. Getting caught in an avalanche is a very dangerous endeavor. With all of the turbulence generated, it is easy to lose contact with where the victims come to rest. It can be very difficult to find a body in all the likely areas of burial. The search process can be expedited by dogs or sonar, but the simplest way to shift odds in your favor is by the use of transceivers.

2. **PURPOSE**. The purpose of this period of instruction is to introduce the student to avalanche transceivers, operational testing procedures and how to conduct a search using them. This lesson relates to Avalanche Search Organization.

3. **INTRODUCE LEARNING OBJECTIVES**

   a. TERMINAL LEARNING OBJECTIVE. In cold weather/mountainous environment, demonstrate avalanche transceiver search techniques, in accordance with the references.

   b. ENABLING LEARNING OBJECTIVE. In a cold weather/mountainous environment and given an actual/simulated victim, use avalanche transceivers, in accordance with the references.

4. **METHOD / MEDIA**. The material in this lesson will be presented by lecture. You will practice what you have learned during upcoming field training exercises. Those of you with IRF's please fill them out at the conclusion of this period of instruction.

5. **EVALUATION**. You will be tested later in the course by performance evaluation.

TRANSITION: Are there any questions concerning what we are going to cover during this period of instruction.

BODY (30 Min)

1. (2 Min) **AVALANCHE TRANSCEIVERS**. Transceivers (rescue beacons) are electronic devices that can transmit and receive radio signals.

    a. The 457-kHz radio frequency is now standard but numerous older units with a 2.275-kHz frequency and units that operate on both 457 and 2.275 kHz are still in use.

    b. Unit leaders must ensure that all members of the patrol carry transceivers that are compatible with each other.

TRANSITION: Now that we discussed the general information of a transceiver, let's talk about the nomenclature and functions.

2. (5 Min) **NOMENCLATURE AND FUNCTIONS**. The following list is the nomenclature and functions of a transceiver:

    a. On/off Plug. To turn the transceiver on, insert the cross plug into the designed socket. Lock into place by gently pushing the cross plug inward while turning 90°. To turn the transceiver off, reverse the procedures.

    b. Earphone Jack. The socket where an earphone can be attached.

    c. Battery Compartment. 2 AA (1.5V) premium quality alkaline. Do not use rechargeable batteries.

    d. Battery LED (Light Emitting Diode). Indicates the battery power.

    e. Transmit/Receive Switch with safety catch. Push in to transmit or pulled out to receive. Ensure safety catch is in place during the transit mode to prevent accidental employment of the receiving mode.

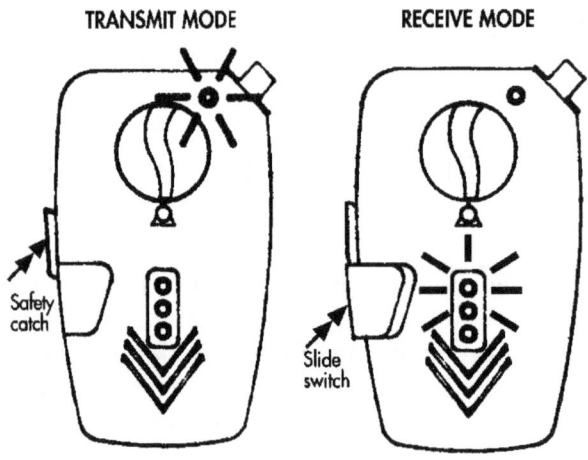

TRANSCEIVER IN TRANSMIT AND RECEIVE MODES

    f. Range Dial. Controls the volume of the signal.

g. Signal Strength Indicator LED. Indicates the strength of the signal being received. Accuracy to within 50 cm.

h. Search Direction Arrow. Indicates the direction the transceiver needs to be orientated during the search.

i. Casing. Watertight and shockproof plastic.

j. Straps. Nylon webbing adjustable straps with plastic buckle. Used to secure the transceiver around the body.

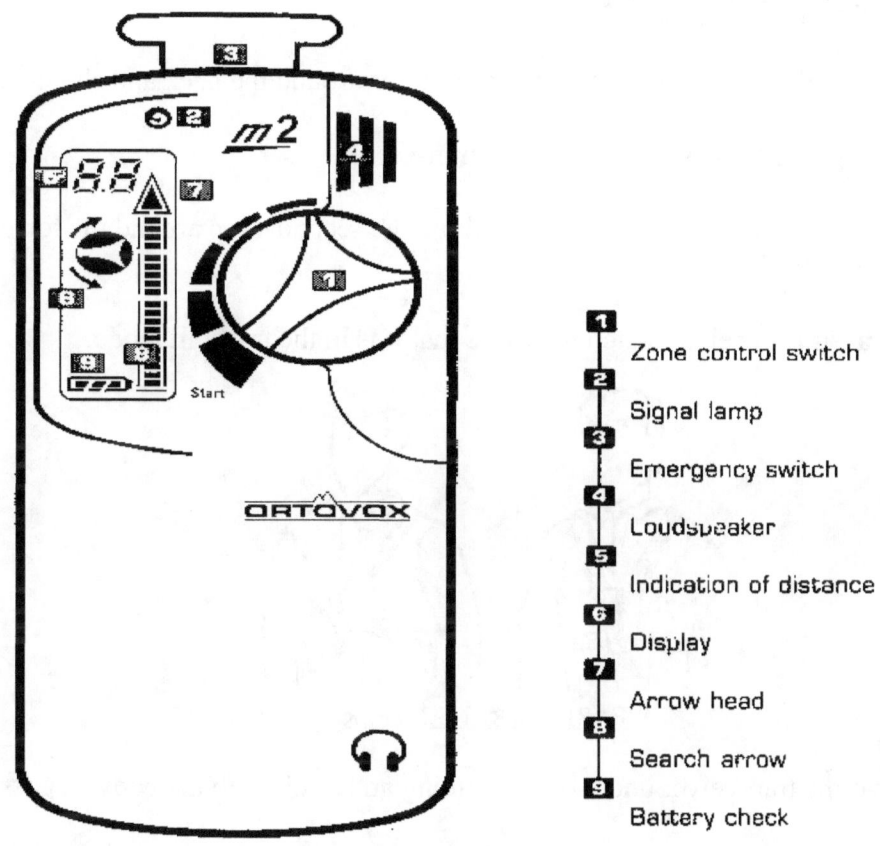

1. Zone control switch
2. Signal lamp
3. Emergency switch
4. Loudspeaker
5. Indication of distance
6. Display
7. Arrow head
8. Search arrow
9. Battery check

Orotovox M2

Avalanche transceiver with the fastest pulse rate of the new digital-analog generation. Adjustable elasticated body belt. Functional status and battery charge lights. Step wise volume switch. Earphone connection for searching with a group. Quartz stabilized 457kHkz frequency.

Technical data:
Housing: ergonomic, waterproof and secure against knocks
Size: 145x62x25 mm
Frequency: 457kHk

Range: up to 80m
Search strip length: up to 60m
Temperature tolerance: -30 to +50 C
Batteries: 2 x LR 1,5 V alkaline (no rechargeables) =2 "AA" batteries
Working life with battery :
Transmitting: approximately 300 hours
Receiving: approximately 40 hours
Weight: 230g

TRANSITION: Now that we have talked about the nomenclature and functions of the transceivers, let's discuss the wearing of it.

3. (5 Min) **WEARING THE TRANSCEIVER**.

   a. Place the strap with the cross plug around the neck or around the neck and shoulder.

   b. Insert cross plug into the case to activate the transceiver.

   c. Adjust the strap to a comfortable position and run the second strap around the torso and secure the plastic buckle.

   d. Check the Transmit/Receive Switch to ensure that it is in the transmit mode.

WEARING TRANSCEIVER

NOTE: Always wear the transceiver under outer clothing and as close to the body as possible.

TRANSITION: Now that we know how to wear a transceiver, let's discuss the operational test.

4. (5 Min) **OPERATIONAL TEST**. Before a patrol deploys from a safe area, the transceivers must be checked to ensure that all the transceivers are operable. The patrol leader should perform the following procedure:

   a. The leader will set his transceiver to the transmit mode while all other members set their transceivers to the receive mode.

   b. The leader will walk away from the patrol to a point where the patrol member's transceivers are no longer picking up the patrol leader's signal. This will ensure that all the patrol member's transceivers are receiving and that the patrol leader's transceiver is transmitting.

c. The leader will now set his transceiver to the receive mode while all the patrol members set their transceivers to the transmit mode.

d. One at a time the patrol members will file past the leader. The leader will ensure that each patrol member's transceiver is transmitting properly.

e. After each member has been checked, the leader will set his transceiver back into the transmit mode.

TRANSITION: Now that we have discussed the operational test prior to departure of friendly lines, let's talk about the bracketing method of search.

5. (10 Min) **BRACKETING METHOD OF SEARCH**. There are numerous methods of searching for victims. This method provides speed and accuracy when conducted properly.

   a. Carefully note and mark last seen point of the victim. If there is further danger of avalanches, post an avalanche guard and be prepared to switch back to transmit.

   b. Set all the transceivers to the receive mode. The leader must ensure that this is accomplished to prevent any misleading transmissions from other transceivers.

   c. The range dial will be turned to its highest range indicator and/or volume.

   d. Deploy line of searchers at a maximum of 20 meters apart to the last seen area and move down the slope.

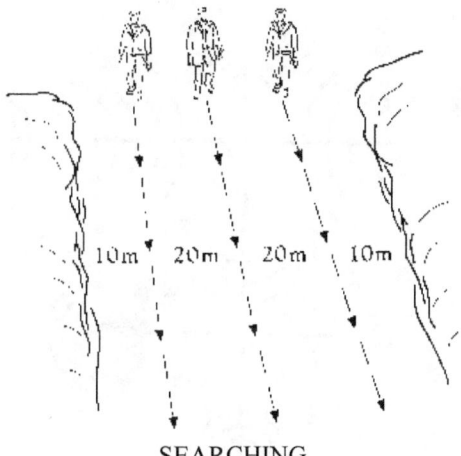

SEARCHING

   e. Slowly rotate the transceiver 120° from side to side to determine direction of strongest signal.

ROTATING THE TRANSCEIVER

f. Mark the spot where you receive the first signal and proceed in a straight line.

g. The signal will get stronger then weaker, mark the weakest point.

h. Find the middle of the two weak points, mark this spot, and turn the range dial/volume down so that the signal is low.

i. Proceed in a direction 90° to the first line. In one direction the signal will get weaker, in the opposite direction the signal will get stronger. Move in the direction of the stronger signal.

j. Proceed past the strong signal until it becomes weaker, mark this weak spot. Remember to turn down the range dial/ volume each time a new line of travel is taken.

k. Repeat this process until the exact location is found and begin probing/digging.

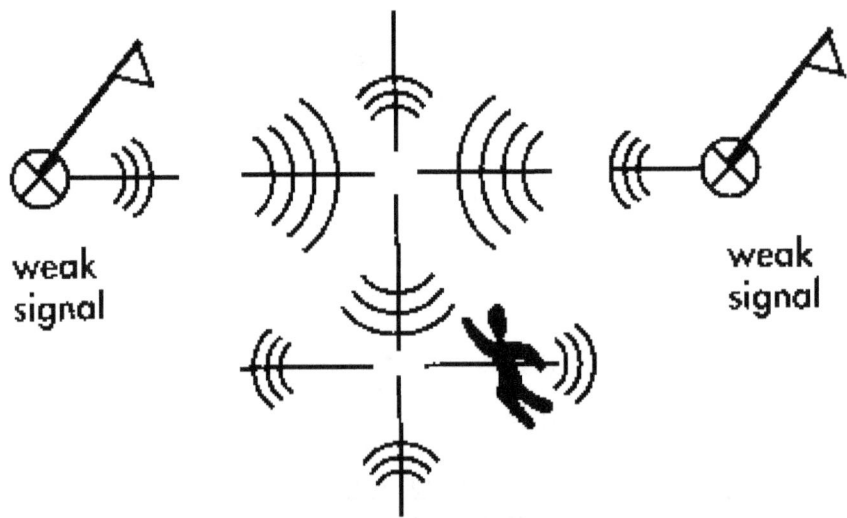

PINPOINTING LOCATION

NOTE: After the search and recovery is completed, ensure that all transceivers are set back to the transmit mode.

TRANSITION: Are there any questions concerning the bracket search method? Remember that these devices will increase the chances of a live recovery if employed properly.

PRACTICE (1 Hr)

    a. Students will practice what was taught during upcoming field exercises.

PROVIDE HELP (CONC)

    a. The instructors will assist the students when necessary.

OPPORTUNITY FOR QUESTIONS (2 Min)

1. QUESTIONS FROM THE CLASS

2. QUESTIONS TO THE CLASS

    Q. What is a transceiver?

    A. Transceivers are electronic devices that can transmit and receive radio signals.

    Q. What should the patrol leader do before deploying from a safe area?

    A. Perform an operational test.

SUMMARY (3 Min)

    a. During this period of instruction we have discussed avalanche transceivers to include the nomenclature and functions, wearing, operational checks and search method.

    b. Those of you with IRF's please fill them out at this time and turn them in to the instructor. We will now take a short break.

UNITED STATES MARINE CORPS
Mountain Warfare Training Center
Bridgeport, California 93517-5001

WML
WMO
09/05/02

LESSON PLAN

MILITARY SNOWSHOES

INTRODUCTION (5 Min)

1. **GAIN ATTENTION**. Much of what we do in the future in cold weather regions will be greatly influenced by our over the snow mobility. At some time in the future, it is almost certain that the welfare and even the survival of your men may depend on your knowledge of over the snow mobility. Because of this a sound knowledge of snowshoeing is important.

2. **PURPOSE**. The purpose of this period of instruction is to introduce the student to military snowshoes, including their nomenclature and their advantages and disadvantages. This lesson relates to mountain movement.

3. **INTRODUCE LEARNING OBJECTIVES**

   a. TERMINAL LEARNING OBJECTIVE. Given military snowshoes in snow covered mountainous terrain, execute snowshoe movement, in accordance with the references.

   b. ENABLING LEARNING OBJECTIVES

   (1) Without the aid of references and using skis in comparison, describe in writing the advantages of snowshoes, in accordance with the references.

   (2) Without the aid of references and using skis in comparison, describe in writing the disadvantages of snowshoes, in accordance with the references.

   (3) Given a diagram of a military snowshoe and without the aid of references, label the parts of a military snowshoe, in accordance with the references.

   (4) Given a pair of military snowshoes, adjust the military snowshoe bindings, in accordance with the references.

(5) Given a pair of military snowshoes in a snow covered mountainous environment, demonstrate snowshoe techniques, in accordance with the references.

4. **METHOD / MEDIA**. The material in this lesson will be presented by lecture and demonstration. You will practice what you have learned during upcoming field training exercises. Those of you with IRF's please fill them out at the conclusion of this period of instruction.

5. **EVALUATION**. You will be tested later in the course by written and performance evaluations.

TRANSITION: Let's discuss the advantages and disadvantages of snowshoeing.

BODY (50 Min)

1. (5 Min) **ADVANTAGES AND DISADVANTAGES OF SNOWSHOEING**

    a. Advantages of Snowshoes

        (1) Training. Little training time is required to gain a high degree of proficiency in their use.

        (2) Maintenance. Little maintenance is required.

        (3) Heavy loads. Carrying and pulling of heavy loads on gentle terrain is relatively easy.

        (4) Confined areas. Movement in confined areas and around equipment is relatively easy. Snowshoes are particularly useful for individuals working in confined areas, such as bivouac sites and supply dumps. Snowshoes are also helpful to drivers, gun crews, cooks and other support personnel.

    b. Disadvantages of Snowshoes

        (1) Rate of movement. The rate of movement for a unit is extremely slow, and inefficient in terms of energy expended.

        (2) Moderate to steep slopes. Movement on moderate to steep slopes is extremely difficult.

        (3) Thick or cut-off brush. Movement through thick or cut-off brush is difficult.

        (4) Fire and movement. Quick movement, as needed during fire and movement is difficult.

TRANSITION: Now let's examine the movement rates.

2. (3 Min) **SNOWSHOE MOVEMENT RATES** (Reduce by one-third for mountainous terrain).

MOVEMENT MODE	UNBROKEN TRAIL	BROKEN TRAIL
On foot with less than 1 foot of snow	1.5-3 KPH	2-3 KPH
On foot with more than 1 foot of snow	.5-1 KPH	2-3 KPH
Snowshoeing	1.5-3 KPH	3-4 KPH

TRANSITION: Let us look at the three types in the military supply system and some field expedient types.

3. (5 Min) **TYPES OF SNOWSHOES**. There are three basic types of snowshoes.

   a. Magnesium. The magnesium snowshoe is the lightest and most durable of the three types. The nylon binding used with this snowshoe is adaptable to all types of issued footwear. The magnesium snowshoe also has teeth under the sides which are intended to aid traction.

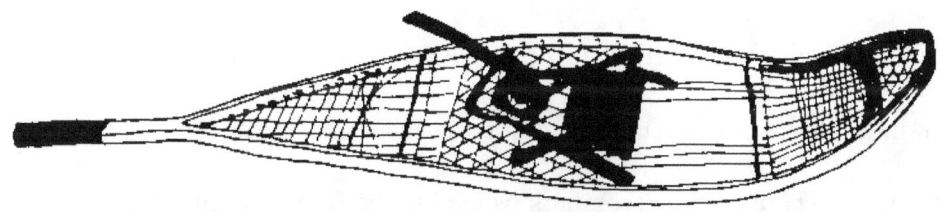

MAGNESIUM SNOWSHOE

   b. Assault. This type of snowshoe is short, wide and oval in shape, with no tail. It is best utilized when working near equipment and heavy weapons, due to it offering little floatation. This type is called a "bear paw" in civilian terminology.

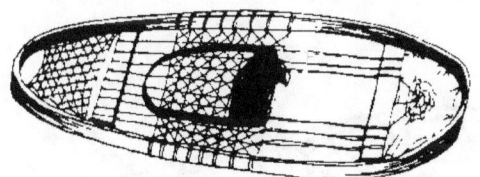

ASSAULT SNOWSHOE

   c. Improvised Snowshoes. Improvised snowshoes may be constructed by forming a frame from green, flexible branches, then weaving string, 550 cord, wire or branches to form a supporting surface. A satisfactory pair can also be made by tying branches from thick fir, or spruce trees, to one's feet. A very simple sasquatch binding can be made with a short length of cord or wire.

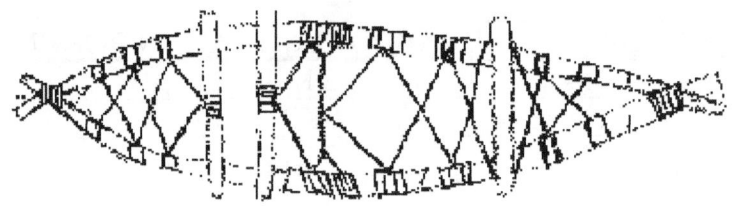

IMPROVISED SNOWSHOE

TRANSITION: Now that we have gone over the types of snowshoes, let's look at the nomenclature of the military snowshoe.

4. (10 Min) **NOMENCLATURE OF THE MILITARY SNOWSHOE**. The magnesium snowshoe consists of nine parts:

   a. Tip. This is the front portion of the snowshoe frame.

   b. Tail. This is the back portion of the snowshoe frame.

   c. Binding. This is constructed of a nylon material and fits the boot.

   d. Crossbars. There are two crossbars welded to the frame to reinforce it.

   e. Window. This is the opening in the snowshoe, which allows the toe of your boot to pivot through.

   f. Webbing. Made from galvanized aircraft cable covered with nylon.

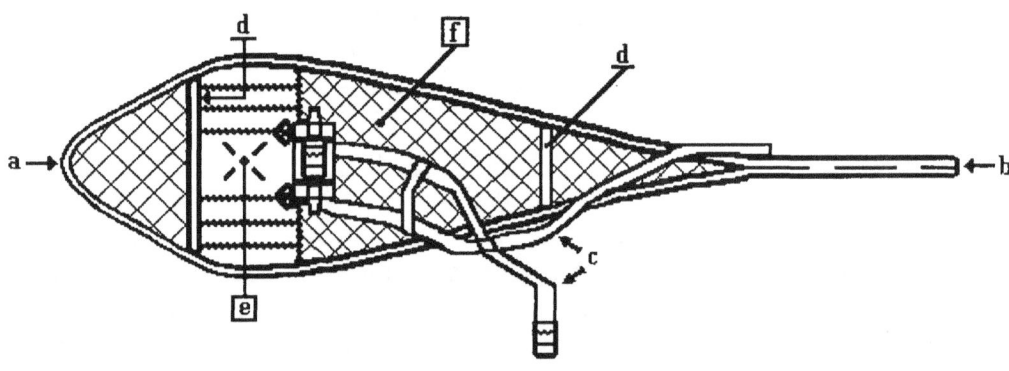

g. Shovel. Front of the snowshoe which has a shovel like resemblance.

h. Frame. Consists of a magnesium alloy.

i. Teeth. These are located on the underside of the frame and create more traction on the surface when worn.

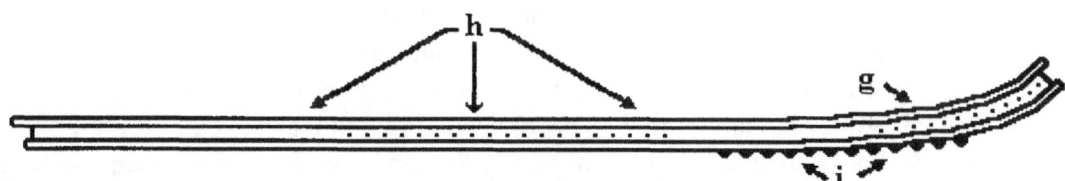

TRANSITION: Let's see how to properly care for and store any type of snowshoe.

5. (2 Min) **CARE AND STORAGE**

   a. Webbing and Bindings. Check load bearing webbing and bindings frequently.

   b. Magnesium Frames. Check frames of magnesium snowshoes for stress fractures.

TRANSITION: This is the proper way to care and store snowshoes. Now, how do we keep them on our feet? By proper binding adjustment.

6. (5 Min) **BINDING ADJUSTMENT**. Proper snowshoe binding adjustment will ensure that:

   a. Ball of Foot. The foot pivots freely about the ball of the foot, so that the toe of the foot moves through the window of the snowshoe.

   b. Heel of Foot. The heel of the foot is centered on the snowshoe.

PROPER SNOWSHOE BINDING ADJUSTMENT

   c. Fit. The binding fits snugly to provide adequate control, but not so tightly that circulation in the feet is impaired. A sloppy fitting snowshoe will make movement extremely difficult

13-5

d. <u>Toe strap</u>. May be used to secure the binding together, to help limit the number of blown bindings.

<u>TRANSITION</u>: We've just about covered everything except how to travel with snowshoes on our feet. It's all in using the proper technique.

7. (10 Min) **TECHNIQUE**

   a. <u>General</u>. There is little difference in snowshoeing compared to normal walking, except that the surface being walked on is inconsistent, and snowshoes are longer, wider, heavier, and consequently more awkward, than normal footwear. With standard military snowshoes, the stride is somewhat longer than in normal walking, but the shape of the snowshoe allows the snowshoer's stance to be a normal width, thereby reducing much strain and fatigue on his hips and legs. It should be stressed that the snowshoer should walk in a relaxed, and normal rolling toe manner, and should only lift the snowshoe high enough to clear the surface of the snow.

   b. <u>Turning</u>

      (1) <u>The kick turn</u>. This is normally the easiest way to change directions on level ground. One snowshoe is swung up to the front so that its tail is on the snow, then it is allowed to pivot toward the new direction. The other snowshoe is then brought around.

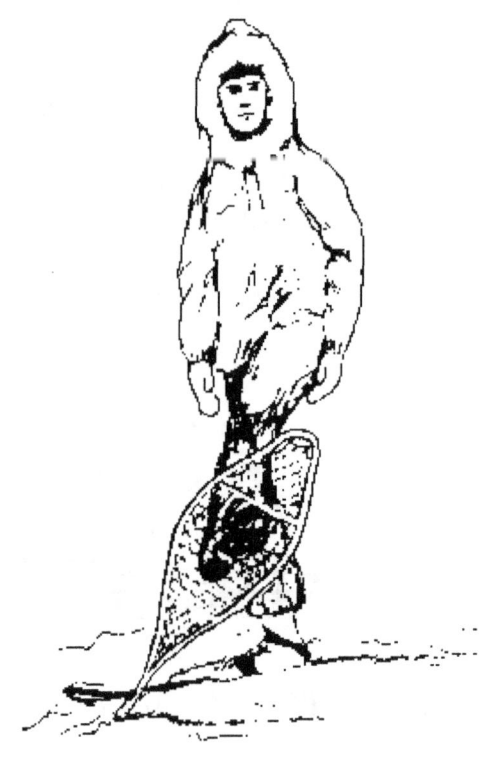

KICK TURN

(a) <u>On steep terrain</u>. It is important to remember to step off with the uphill foot, when changing direction. For example: If making a turn to the right, shift your weight to the left foot, face down the slope, and swing the right snowshoe around to point in the direction of the next switchback. Then stamp the right snowshoe into the snow. Make sure the tail is not on the left snowshoe. Now, gently shift your weight to your right foot and swing the left snowshoe around so it is parallel with your right snowshoe.

(b) <u>Each succeeding man</u>. When using the kick turn technique on steep terrain, try and stay well above your previous trail. This trail has undermined the snow on which you are now building the turn. As each succeeding man uses the turn, it will tend to slough off on the shoulders, and the men toward the end of the column will have a hard time getting around. This can be prevented if care is used by each man in placing his snowshoes precisely where those in front of him have placed theirs. If there is only one way around an obstacle, this can be very important.

(2) <u>The star turn</u>. This can also be used to change direction by simply executing a series of half facing movements.

(3) <u>Choosing a route</u>. When climbing, plan to use the gentlest places on a slope for turns. Look ahead, and pick the route and use the terrain to your advantage. Avoid the steep parts, and don't hesitate to make short switchbacks.

c. <u>Side Step</u>. This is used when the slope is at a critical angle.

d. <u>Herringbone</u>. This is used when the slope is at a gradual angle.

e. <u>Crossing Obstacles</u>. Here are a few simple rules to remember:

(1) Always step over obstacles. Do this to avoid damaging snowshoes and losing balance.

(2) Never bridge a gap. Never do this with your snowshoe so that the tip and tail are higher than the center.

(3) Shallow snow. In shallow snow, there is a danger of catching, and tearing the webbing on tree stumps, or snags, which are only slightly covered.

(4) Wet snow. This will frequently ball up under the feet, interfering with comfortable walking. This snow should be knocked off as soon as possible.

(5) Deep Snow. Breaking trail in deep snow uses a lot of energy. Frequent change of the lead man should be stressed.

(6) Water. Stepping into water with snowshoes can form ice, to which significant amounts of snow can cling, making the snowshoe very heavy.

TRANSITION: Proper training is required to get the most out of snowshoes. Are there any questions?

PRACTICE (CONC)

    a. Students will practice what was taught in upcoming field evolutions.

PROVIDE HELP (CONC)

    a. The instructors will assist the students when necessary.

OPPORTUNITY FOR QUESTIONS (3 Min)

1. QUESTIONS FROM THE CLASS

2. QUESTIONS TO THE CLASS

    Q. What are the four advantages of snowshoeing?

    A. (1) Little training time is required to gain a high degree of proficiency in their use.
        (2) Little maintenance is required.
        (3) Carrying and pulling of heavy loads on gentle terrain is relatively easy.
        (4) Movement in confined areas and around equipment is relatively easy.

    Q. What are the four disadvantage of snowshoeing?

    A. (1) The rate of movement for a unit is extremely slow and inefficient in terms of energy expanded.
        (2) Movement on moderate to steep slopes is extremely difficult.
        (3) Movement through thick or cut-off brush is difficult.
        (4) Quick movement, as needed during firing and movement, is difficult.

    Q. When crossing obstacles, do you step squarely on them?

    A. No.

SUMMARY (2 Min)

    a. We have covered a few basics of military snowshoeing, including nomenclature, advantages, and disadvantages.

    b. Those of you with IRF's please fill them out and turn them in to the instructor at this time. We will now take a short break.

UNITED STATES MARINE CORPS
Mountain Warfare Training Center
Bridgeport, California 93517-5001

WML
WMO
11/23/01

LESSON PLAN

MILITARY SKI MOVEMENT

INTRODUCTION (5 Min)

1. **GAIN ATTENTION.** The goal of skiing is to move Marines from the assembly area to the assault point. Marines must arrive ready to fight. Combat skiing is a combat multiplier. If done correctly, skiing will give Marines the edge in combat. Consider the ski as an over-the-snow support device. Combat skiing will enhance individual mobility and give Marine units the ability to move across snow-covered terrain, steep slopes, and through bush and tree covered terrain that cannot be crossed on foot or snowshoes. A force must have mobility at least equal to its enemy if it is to succeed. Snow covered terrain limits mobility, making cold weather operations more difficult than operations in other climates. For Marines to have the mobility to successfully maneuver, they must learn to ski effectively.

2. **PURPOSE.** The purpose of this period of instruction is to introduce the students to military ski movement skills, both flat ground and downhill, that are required to pass the military skier test. This lesson relates to military snowshoe movement.

3. **INTRODUCE LEARNING OBJECTIVES.**

   a. TERMINAL LEARNING OBJECTIVE. In a winter mountainous environment, given military ski equipment, combat load and variable terrain, execute over the snow mobility, in accordance with the references.

   b. ENABLING LEARNING OBJECTIVES.

   (1) Given military ski equipment and flat ground terrain, execute all Military Ski Instructor (MSI) flat ground ski skills, in accordance with the references.

   (2) Given military ski equipment and downhill terrain, execute all Military Ski Instructor (MSI) downhill ski skills, in accordance with the references.

(3) Given a group of novice skiers with military ski equipment, instruct basic flat ground or downhill skills, in accordance to the references.

(4) Given military ski equipment, combat load and variable terrain, execute a 5 to 10-kilometer movement, in a described time limit.

4. **METHOD/MEDIA.** The material in this lesson will be presented by lecture and demonstration. You will practice what you have learned during upcoming field training exercises. Those of you who have IRF's please fill them out at the conclusion of this period of instruction.

5. **EVALUATION.** You will be tested later in the course by performance evaluations on this period of instruction.

TRANSITION: Many potential threat areas are snow covered and special skills must be mastered in order for Marines to successfully operate in such areas.

BODY (30 Min)

1. (10 Min) GENERAL

    a. Marine ski training has evolved a great deal since its "modern" inception here at MWTC in the late seventies. At that time troops were trained using antiquated equipment with very little individual attention offered. Instructor/student ratios, typically 1 to 50 or greater, offered no opportunity for poor skiers to improve their skills. The instructors themselves had little or no formal training to prepare them as ski instructors. Consequently, units undergoing ski training often had high casualty rates and suffered from frustration and low unit morale.

    b. Marine units deploying to Norway were often victims of encirclement due to their inability to move over the deep snow, whereas Allied units like the British, Dutch, and Italians could move much more quickly and effectively on skis. As Marines gained experience through successive Norway deployments, commanders began to recognize the need for increased ski training. Initially, we relied on the expertise of our British and Norwegian Allies in over snow mobility through the use of exchange billets and winter training courses, which emphasized individual skills. Graduates of these courses brought back ideas and training methods that helped shape the current programs.

    c. Another valuable source of input to the military ski program has been the civilian community in the form of the Professional Ski Instructors of America, (PSIA). MWTC established a working liaison with this organization in the fall of 1985. PSIA provides MWTC with state of the art methods of ski instruction and techniques that were being used daily with great success at ski schools across the nation. The Marine Corps ski program is broken down into four levels: Basic Military Skier, Scout Skier (SS), Military Skier Instructor (MSI) and Military Ski Examiner (MSE).

(1) The Basic Military Skier Program is designed to introduce Marines going through Battalion training to ski equipment and basic flat and downhill ski techniques. This is familiarization training only. All basic ski skills are introduced, but further experience will be required for practical ski mobility.

(2) The Scout Skier Program is designed to instruct functional skiing to selected Marines from the training battalions aboard MWTC. These Marines are identified during the first phase of the training cycle that demonstrates an aptitude in skiing skills. Their missions may include route reconnaissance, trail breaking, flank security, mountain picketing, and being task organized as the maneuver element of an attack.

(3) The Military Ski Instructor Program is designed to instruct Marines on the functional and technical aspects of skiing in order to be ski mobile and teach basic skiing to Marines. The MSI is taught to plan and lead skiborne units on patrols and attacks. A Marine qualifies for this level by completing all instructional requirements for the Winter Mountain Leaders Course or Winter Instructor Qualification Course at MWTC.

(4) The Military Ski Examiner Program is designed to increase a Marine's level of technical knowledge, skiing skills and practical understanding of military skiing situations. They are also trained on grading ski tests. They will be tasked to train Marines to MSI level, administer the MLC/IQC (MSI) ski test, and assist in developing MSIs to the MSE level. An individual qualifies for this level by successfully completing WMLC/WIQC, attending all scheduled MSE preparation ski training, and by passing the four phases of the MSE Certification Examination: flat and uphill skills, downhill skills, teaching skills and a written test.

TRANSITION: Now that we've discussed the past and present Marine Corps ski training program, let's talk about the training and skills required to become a MSI skier.

2. (10 Min) **SKI TRAINING**. The ski training is geared for the first time skier and anyone in good physical condition should be able to meet the standards required by this course. Initially, ski training will be conducted without the additional burden of equipment. As a student, you must demonstrate that you can execute 100% of the MSI skills correctly and be able to instruct novice skiers on basic skills. Later in the course, you will be evaluated during a 5 – 10 kilometer movement over variable terrain carrying a combat load.

   a. Flat Ground Skiing Skills. The following skill must be passed in order to be certified as a Military Skier Instructor:

   (1) Basic Athlete Stance

   (2) Star Turn

   (3) Recovery from a Fall

   (4) Kick Turn

(5) Diagonal Stride

(6) Uphill Diagonal

(7) Double Poling

(8) Side Step

(9) Forward Side Step

(10) Uphill Traverse

(11) Herringbone

b. Downhill Skiing Skills. The following skill must be passed in order to be certified as a Military Skier Instructor:

(1) Downhill Run

(2) Downhill Traverse

(3) Step Turns

(4) Transition and Absorption

(5) Side Slip

(6) Kick Turn

(7) Gliding to Braking Wedge

(8) Wedge Turns

c. Teach Backs. The following skills must be passed in order to certified as a Military Skier Instructor:

(1) Class Management

(2) Site Selection

(3) Explanation and Technical Knowledge

(4) Demonstration

(5) Practical Application of Skills and Exercises

(6) Skier Diagnosis

(7) Progression Usage

TRANSITION: The Marine ski program is designed to make the Marine Corps self-sufficient in ski training and provide ski-mobile Marines to the FMF for employment on the winter battlefield.

PRACTICE (CONC)

a. Students will practice what was taught in upcoming field evolutions.

PROVIDE HELP (CONC)

a. The instructors will assist the students when necessary.

OPPORTUNITY FOR QUESTIONS (3 Min)

1. QUESTIONS FROM THE CLASS

2. QUESTIONS TO THE CLASS

    Q. What are the different ski programs offered to the Marine Corps?

    A. (1) Basic Military Skier Program

       (2) Scout Skier Program

       (3) Military Ski Instructor Program

       4. Military Ski Examiner Program

    Q. What skills are required to pass the Military Ski Program teach back?

       (1) Class Management

       (2) Site Selection

       (3) Explanation and Technical Knowledge

       (4) Demonstration

       (5) Practical Application of Skills and Exercises

       (6) Skier Diagnosis

       (7) Progression Usage

SUMMARY (2 Min)

   a. This period of instruction has been an introduction to the Marine Corps ski training program. It has covered the history and requirements for ski mobility, the different levels of ski expertise, and has outlined the training that you will receive here at MWTC.

   b. Those of you with IRF's please fill them out at this time and turn them in to the instructor. We will now take a short break.

UNITED STATES MARINE CORPS
Mountain Warfare Training Center
Bridgeport, California 93517-5001

WML
WMO
09/31/00

LESSON PLAN

SKIJORING

INTRODUCTION (5 Min)

1. **GAIN ATTENTION**. Skijoring is a method of moving skiborne troops efficiently with the use of over the snow capable vehicles. However, without proper training it can become tiring, slow, frustrating and even dangerous.

2. **PURPOSE**. The purpose of this period of instruction is to introduce the students to the over the snow movement technique known as skijoring, to include safety procedures, advantages and disadvantages, and performance. This lesson relates to tactical movement.

3. **INTRODUCE LEARNING OBJECTIVE**

    a. TERMINAL LEARNING OBJECTIVE. Given the necessary equipment and over a prescribed course, conduct skijoring operations, in accordance with the references.

    b. ENABLING LEARNING OBJECTIVES

    (1) Without the aid of references. list in writing two advantages of skijoring, in accordance with the references.

    (2) Without the aid of references, list in writing two disadvantages of skijoring, in accordance with the references.

    (3) Without the aid of references, list in writing four safety requirements for skijoring, in accordance with the references.

4. **METHOD / MEDIA**. The material in this lesson will be presented by lecture and demonstration. You will practice what you have learned during upcoming field training exercises. Those of you with IRF's please fill them out at the end of this period of instruction.

5. **EVALUATION**. You will be tested later in the course by a written and a performance evaluation.

TRANSITION: First, let's talk about the advantages of skijoring.

BODY (40 Min)

1. (2 Min) **ADVANTAGES**

    a. Less energy is expended by each skier if skijoring is done properly. It's similar to water skiing behind a boat. However, if you keep falling down and picking yourself up time after time, you'll be defeating the purpose.

    b. Movement of a skiborne unit can be expedited by skijoring. An over the snow vehicle is going to travel faster than a unit skiing. However, improperly trained skiborne troops can be the exception.

TRANSITION: Now let's talk about the disadvantages.

2. (3 Min) **DISADVANTAGES**

    a. Skijoring improperly can become tiring and slow. Thus, training is critical and can take a lot of time. Proper training is important not only for the skiers, but also for the drivers. Improperly trained drivers could cause serious accidents to skiers.

    b. Cold weather injuries can occur anytime in a cold weather environment. When Skijoring, the wind chill against a skier is going to be increased significantly. Ski troops must be familiar with proper clothing protection to prevent injury.

TRANSITION: Before we talk about the techniques of skijoring, we will discuss the safety requirements of skijoring.

3. (5 Min) **SAFETY REQUIREMENTS**. The following safety requirements must be adhered to:

    a. <u>Never</u> under any circumstances will a skier <u>tie into a skijoring rope</u>. This could cause the skier to be dragged over, around, or through obstacles and cause serious injuries.

    b. The <u>rate of speed</u> for a vehicle pulling experienced skiers <u>should not exceed</u> 25 mph. For inexperienced skiers, the rate of speed should not exceed 15 mph.

    c. Skiers should be <u>spaced at least 1/2 ski length apart</u> from the tail of the front skier to the tip of the following skier. This will give the skiers ample time to react to most situations, such as a fallen skier or rounding a corner.

d. A safety rider will always be employed. He should be experienced in skijoring and must have visual contact with skiers and communications with the driver. Anytime a group of Marines are skijoring, a safety rider is required and his tasks include:

   (1) Observation. Be situated where he can observe all skiers.

   (2) Communication. Be in communication with the driver at all times, whether by using a whistle, a cord, or another man.

   (3) Halt signal. Give the halt signal for all others to hear in case anyone falls, and most importantly, so the driver knows that he should halt.

   (4) Other signals. Have a signal that everyone knows for stopping, slowing, accelerating, etc.

TRANSITION: We now know the requirements to take, now we'll talk about preparing the vehicle for skijoring.

4. (5 Min) **PREPARING THE OVER THE SNOW VEHICLE**. The over the snow vehicles used at MWTC are the DMC 1450, LMC 1200, SUSV M-973 (BV-206), HUMMV equipped with MacTrack. The preparations we will be discussing will remain the same for these vehicles, the only exception is the load towing capacity which will be determined by the driver of the vehicle based upon the snow conditions, and also the position of the safety rider.

   a. Mirrors. Ensure that the mirrors for the driver are in correct alignment for best viewing to the rear of the vehicle.

   b. Back-up Horn. Ensure that the vehicle has a back-up horn and that it is functional.

   c. TC Hatch. The safety rider occupies the TC Hatch so he can observe the skiers.

   d. Towing Capacity. Due to the different towing capacities of the vehicles used here, the amount of skiers to be skijored will be determined by the driver. To attach the skijoring rope to any one of these vehicles remain the same as stated below:

      (1) Towing rope. Find the middle of the towing rope.

      (2) Clove hitch. Open the trailer hitch and place a clove hitch into it using the middle of the towing rope. A clove hitch should be the only knot utilized.

      (3) Trailer hitch. Tighten down the clove hitch and replace the top of the trailer hitch and ensure it is locked down to prevent the rope from slipping out.

      (4) Figure of eight loops. These should be tied on the end of each line of the towing rope for the last men.

TRANSITION: The vehicle is now prepared, next we'll get into the actual method to be utilized by the skiers.

5. (10 Min) **SKIJORING TECHNIQUES**

   a. <u>Skiers</u>. Each skier should line up on the outside of his prospected line. The first skiers should be at least 1 ski length behind the vehicle or at a vantage where the safety rider can observe him.

   b. <u>Proper Interval</u>. The proper interval of 1/2 ski length apart from each other should be utilized before hooking up.

   c. <u>To Hook Up</u>. Once the interval is appropriate and the skiers are ready to hook up, the forward skiers should hook up first, then the second set, etc., until the last skiers hook up.

   (1) <u>Ski poles</u>. To do this, place the ski pole handgrips on top of the tow rope. Next grasp the tow rope in front of the pole and wrap the rope around your ski poles and directly under the handgrips, forming a half hitch. Place the ski pole baskets behind your outside armpit to use as a rest.

   (2) <u>Grip the rope</u>. Do this with the inside hand.

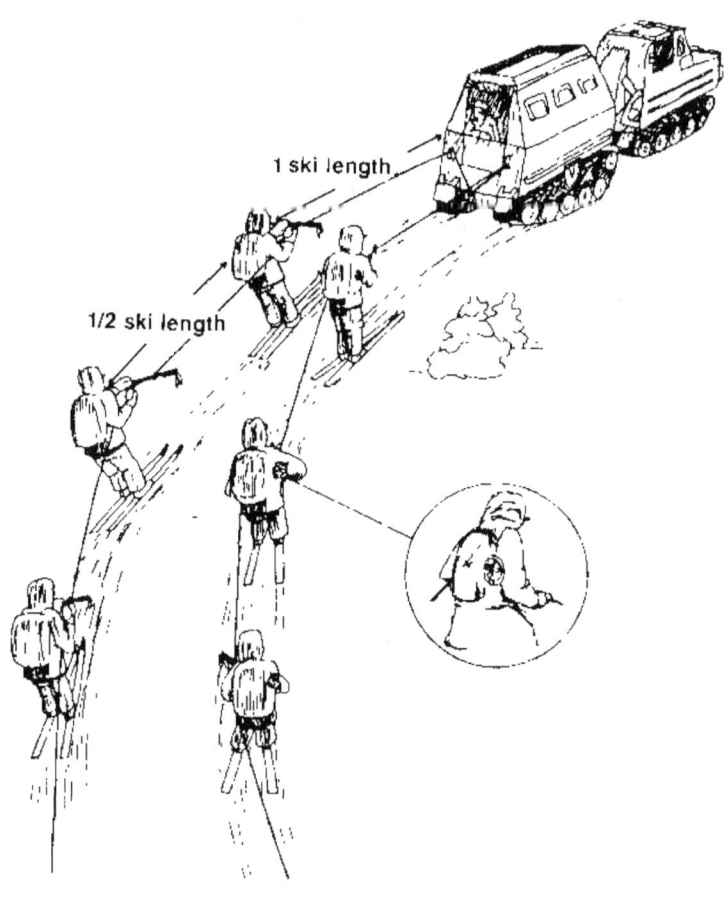

SKIJORING

15-4

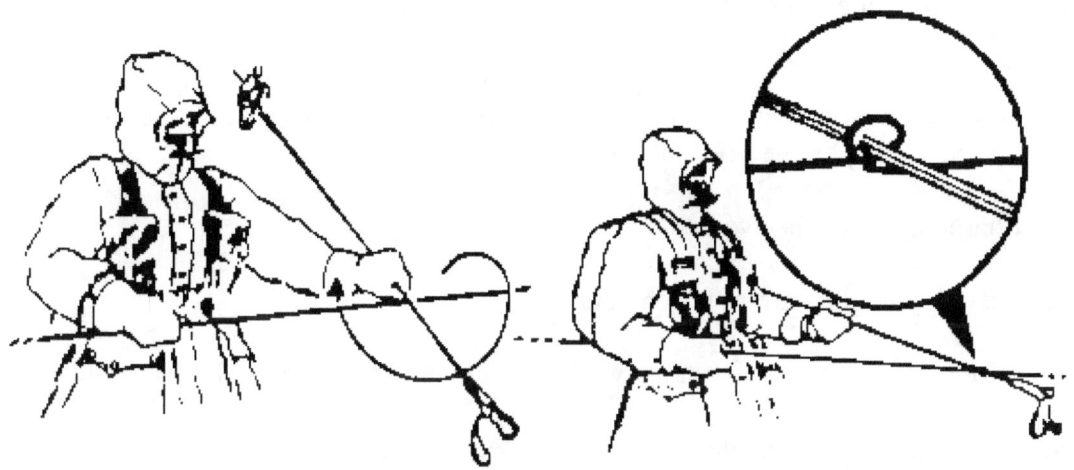

HOOKING UP TO SKIJORING LINE WITH SKI POLES

d. <u>Keep your skis parallel.</u> They should be about shoulder width apart, knees should be flexed in the basic ski stance.

e. <u>Last skiers.</u> The last skiers should ensure they keep the tension on the end of the rope by hanging on to the figure of eight loops. This will keep their ski poles in place.

f. If someone falls, everyone lets go of the rope and skis outboard.

g. Put experienced skiers in front and back of the rope.

h. Shuffle skis when starting to prevent skis from freezing to the ground.

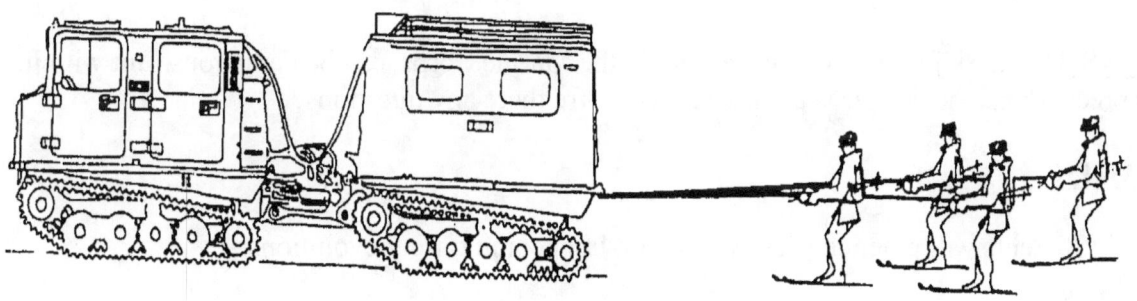

SKIJORING

<u>TRANSITION</u>: I now feel confident that everyone here has enough knowledge to successfully skijor, but let's cover a few special considerations.

6. (5 Min) **SPECIAL CONSIDERATIONS**

   a. Communication signals between safety rider and driver must be established.

15-5

b. The safety rider must be positioned to view all skijoring.

c. Start slowly with a consistent speed and stop gradually. Increase speed down gentle slopes and skiers will unhook before steep slopes. The decision to unhook for downhill grades will depend upon the skiers ability.

d. Sharp turns must be avoided.

e. Visibility might warrant fewer skiers.

f. Snow Conditions. In deeper snow, the vehicle might have to pull fewer skiers.

g. Skier's ability will naturally dictate the speed of the vehicle.

h. If sleds must be towed, it is best if they are towed separately for safety, and a safety rider will be utilized.

TOWING SLEDS

i. When negotiating turns or curves, ski to the outside of the turn, especially the rear skiers on the inside rope.

j. Ensure drivers are properly briefed.

TRANSITION: Skijoring, like any activity in the Marine Corps, can be dangerous. But with just a few precautions, the danger is greatly reduced. Are there any questions?

PRACTICE (CONC)

a. Students will practice what was taught in upcoming field evolutions.

PROVIDE HELP (CONC)

a. The instructors will assist the students when necessary.

OPPORTUNITY FOR QUESTIONS (3 Min)

1. QUESTIONS FROM THE CLASS

2. QUESTIONS TO THE CLASS

   Q. What are the advantages of skijoring?

   A. (1) Less energy is expended by the skier if skijoring is done properly.
      (2) Movement of skiborne units can be expedited by skijoring.

   Q. What are the disadvantages of skijoring?

   A. (1) Skijoring improperly can become tiring and slow.
      (2) Cold weather injuries can occur while skijoring.

   Q. What are the four safety requirements for skijoring?

   A. (1) The skier will NEVER tie into the skijoring rope.
      (2) The rate of speed for pulling skiers will not exceed 25 mph.
      (3) The skiers will be spaced at least 1/2 ski length apart.
      (4) A safety rider will ALWAYS be utilized.

SUMMARY (2 Min)

   a. We have discussed the advantages, disadvantages, methods, safety requirements, preparation of the vehicles and the special considerations of skijoring.

   b. Those of you with IRF's please fill them out at this time and turn them in to the instructor We will now take a short break.

UNITED STATES MARINE CORPS
Mountain Warfare Training Center
Bridgeport, California 93517-5001

                                                                                                     WML
                                                                                                     WMO
                                                                                                 09/18/02

LESSON PLAN

MARINE CORPS COLD WEATHER INFANTRY KIT

INTRODUCTION   (5 Min)

1. **GAIN ATTENTION**. Marines operating in a snow-covered environment may find themselves temporarily separated from their log train. If this is the case, then those Marines need a way to sustain themselves in harsh weather conditions and during combat. The Marine Corps has recently developed an answer to this problem. It is the Marine Corps Cold Weather Infantry Kit (MCCWIK).

2. **PURPOSE**. The purpose of this lesson is to introduce the student to the MCCWIK system. This lesson relates to the Winter Warfighting Load Requirements.

3. **INTRODUCE LEARNING OBJECTIVES**

    a. <u>TERMINAL LEARNING OBJECTIVE</u>. In a cold weather/mountainous environment, maintain a MCCWIK system, in accordance with the references.

    b. <u>ENABLING LEARNING OBJECTIVES.</u>

    (1) Given a MCCWIK, pack the sled for movement in accordance with the references.

    (2) Given a MCCWIK, maintain the MCCWIK in accordance with the references.

    (3) Given a MCCWIK in a cold weather mountainous environment, move with the sled while skiing or snow shoeing in accordance with the references.

    (4) Given an Extreme Cold Weather (ECW) tent, erect the ECW tent in accordance with the references.

    (5) Given an ECW tent, strike the ECW tent in accordance with the references.

(6) Given a squad stove, operate the squad stove in accordance with the references.

4. **METHOD / MEDIA**. The material in this lesson will be presented by lecture and demonstration. You will practice what you learn during upcoming field training exercises. Those of you who have IRF's please fill them out at the conclusion of this period of instruction

5. **EVALUATION**. You will be tested later in the course by performance evaluations.

TRANSITION: The first thing we need to discuss is the components of the Marine Corps Cold Weather Infantry Kit.

BODY (90 Min)

1. (5 Min) **COMPONENTS**. The Marine Corps Cold Weather Infantry Kit (MCCWIK) is a comprehensive collection of equipment that enables a four man fire team to operate in a cold weather environment. The following list are the components of the MCCWIK:

    a. (1) Fire Team Sled with Transport Bag.

    b. (1) Avalanche Probe with stuff sack.

    c. (2) Snow Shovels.

    d. (2) 1 Quart Thermos with spare stopper.

    e. (2) 33 oz. Fuel Bottles.

    f. (2) Funnels.

    g. (1) Cook Set.

    h. (2) Ski Wax Kits with stuff sacks.

    i. (4) Climbing Skins with stuff sacks.

    j. (1) Hatchet.

    k. (1) Whisk Broom.

    l. (1) Snow saw with sheath.

    m. (2) Squad Stoves.

    n. (1) Extreme Cold Weather Tent.

TRANSITION: Now that we talked about the components of the MCCWIK, let us now familiarize ourselves with the fire team sled.

2. (2 Min) **FIRE TEAM SLED**. The fire team sled is designed to hold all the group stores contained in the MCCWIK system. Each sled enables the equipment to be protected from the elements and offers a method to transport the equipment to different locations. Maneuvering the sled will be taught in the class TEAM SLED MOVEMENT.

TRANSITION: Now let us discuss the nomenclature of the fire team sled.

3. (5 Min) **NOMENCLATURE OF THE FIRE TEAM SLED**

   a. Hull. This is made of high-impact plastic with three runners used to keep the sled upright during movement. It has two clevises located in the front of the sled for attachment of the poles. The sled is also designed with a flanged top for securing the transport bag. There are three metal large D-rings located outside the hull. This is for assisting the movement of the sled by attaching traces. There are six metal small D-rings located on the sides of the hull for securing the transport bag to the hull.

   b. Transport Bag. This is made of coradora canvas. It has cinch cord on its rear corners and two plastic buckles on its front corners. The transport bag also has three adjustable compression straps with plastic buckles, two carrying handles, and a double zipper for top loading. This bag is where the group stores will be placed into for protection.

   c. Aluminum Pull Poles. There are a total of four aluminum poles. Two of these poles are "hooked" for the purpose of attachment to the clevise of the hull and have a small hole on the opposite end. The remaining two poles are have a snap button for attachment to the "hooked" poles and a small hole on the opposite end for assembly to the pull harness.

   d. Waist Harness. There are two different waist harnesses; the pull harness and the assist harness. The pull harness is equipped with a harness block with pin and lanyard with carabiner. This is the primary harness used to pull the sled. The assist harness is used when more than one Marine is needed to pull the sled. Both harnesses are constructed with nylon and have a plastic snap buckle.

TRANSITION: Now that we talked about the nomenclature of the fire team sled, let us now talk about its serviceability checks.

4. (5 Min) **SERVICEABILITY CHECKS OF THE FIRE TEAM SLED**. The following checks should be made before and during movement.

   a. Hull. The hull should be checked for any cracks or holes. Two clevises should be present and inspected for cracks or fracture lines. All metal D-rings should be present and not fractured.

b. <u>Transport Bag</u>. The transport bag should be free of rips or holes and the zipper functional. All three compression straps should be present and all plastic buckles inspected for cracks or breakage.

c. <u>Aluminum Poles</u>. The poles should be straight, not cracked or bent. All snap buttons should be functional.

d. <u>Waist Harnesses</u>. The harnesses should be inspected for possible rips or tears and the plastic buckles are not cracked or broken. The pull harness should be inspected to ensure that both harness blocks, lanyards, carabiners and pins are present and not cracked, ripped or fractured.

5. (5 Min) **ASSEMBLY OF THE FIRE TEAM SLED**

   a. <u>Transport Bag</u>

   (1) Place transport bag into the sled hull. Zipper and handles up, buckles toward bow of the sled hull.

   (2) Position cinch cord under flange located around top edge of the sled hull.

   (3) Stretch flap with female buckle end over front of sled.

   (4) Couple each set of buckles under their adjacent clevis.

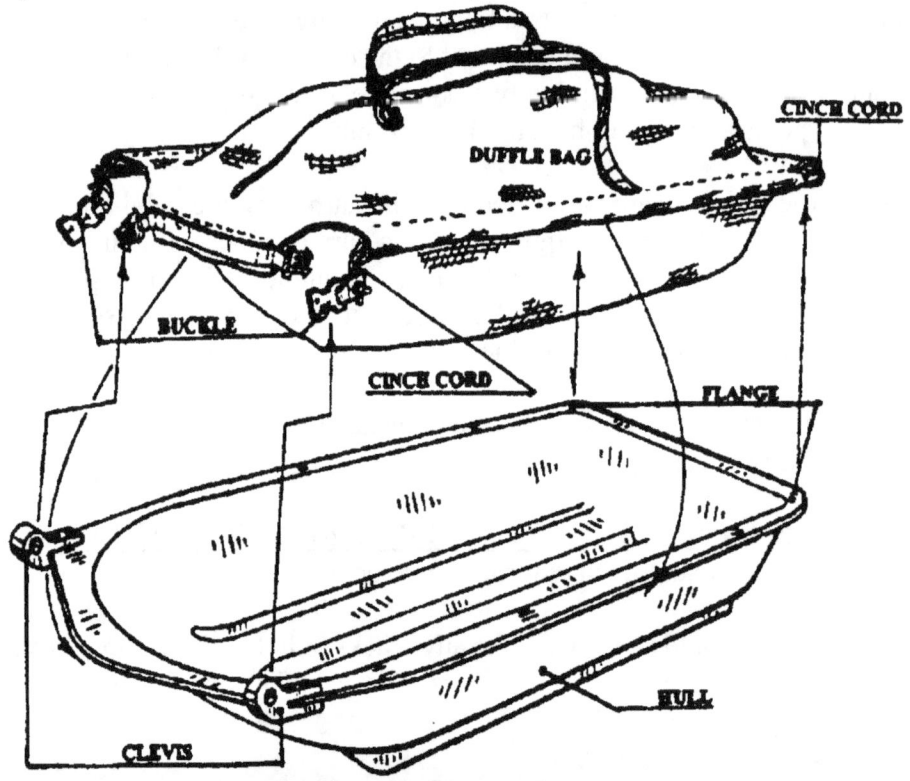

ASSEMBLING THE FIRE TEAM SLED

b. <u>Attaching Poles</u>

   (1) Place "hooked" end of pole through the hole on clevis while depressing snap button. Make sure bend is toward outside of the sled.

   (2) Assemble pole halves by depressing snap button into the larger sleeve aligning the snap button with the hole.

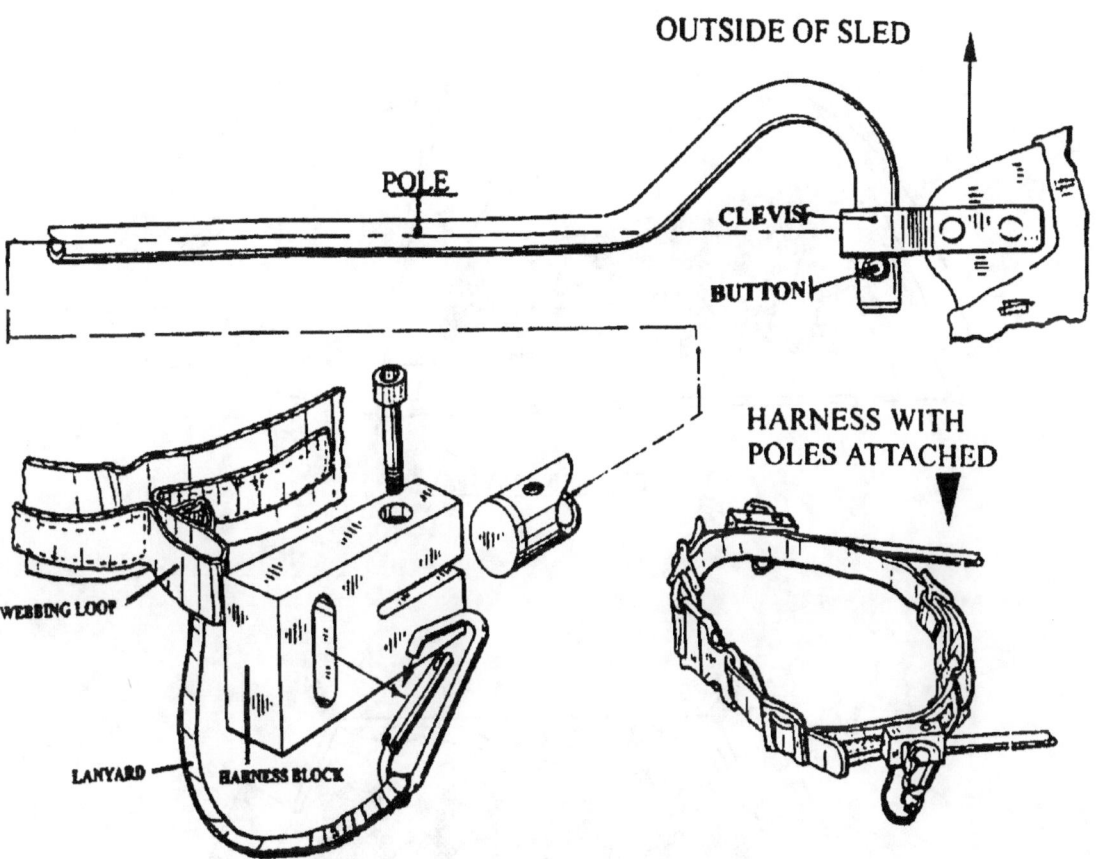

c. <u>Waist Harness</u>

   (1) The waist harnesses are adjustable in the front and rear to allow user to keep pull points directly at the sides of the hip.

   (2) Insert webbing loops through slot in harness block.

   (3) Place carabiner through the protruding web loop to the outside of the harness block.

   (4) Insert pole into the harness block and secure into place with pin.

<u>TRANSITION</u>: Next, we will discuss the extreme cold weather tent.

6. (5 Min) **CHARACTERISTICS AND NOMENCLATURE OF THE ECW TENT**

   a. <u>Characteristics</u>. The extreme cold weather tent was developed to replace the Norwegian tent sheets. It is lightweight and portable, weighing only 17 lbs. The tent is a self-standing dome-shaped, four-season design capable of holding four Marines within its 9852 square inches of floor space.

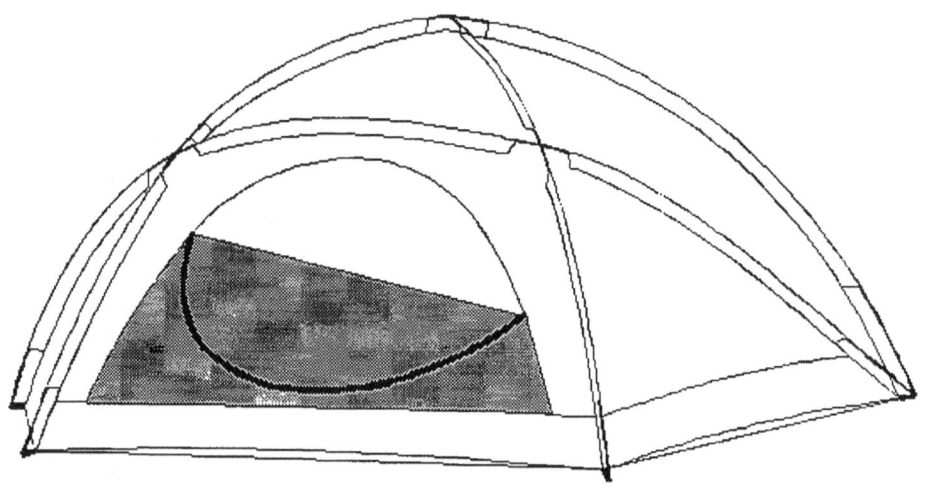

EXTERNAL VIEW

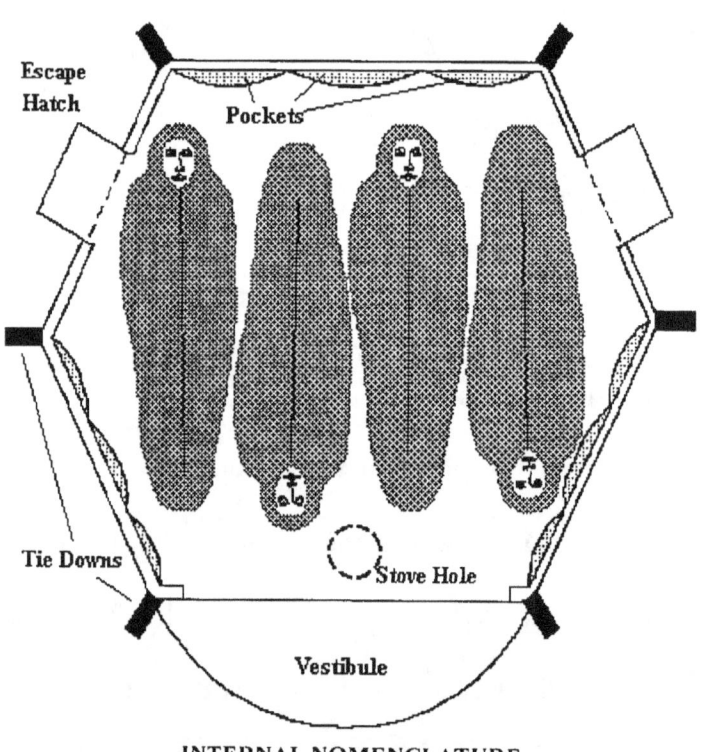

INTERNAL NOMENCLATURE

b. <u>Nomenclature</u>

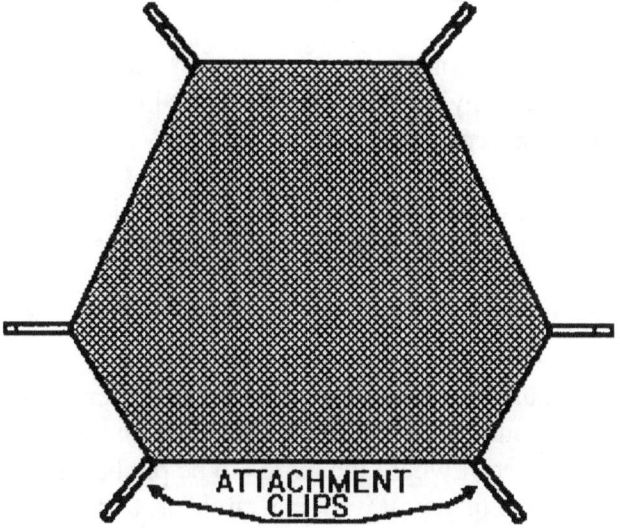

MESH DRYING RACK

(1) <u>Tent body</u>. Made of 3 ounce per yard urethane coated taffeta nylon. Inside the top of the tent is a mesh drying rack, and around the bottom are several mesh pockets for commonly used items. The entrance has a no-see-um mesh panel designed to keep bugs out. Later versions will have two openings spaced around the tent body that facilitate joining tents together for CP's or BAS's.

(2) <u>Flysheet</u>. The tent comes with two different flysheets; a woodland camo cover for forested areas and a white cover for snow covered terrain. These sheets are also made of nylon with a heavier urethane coating.

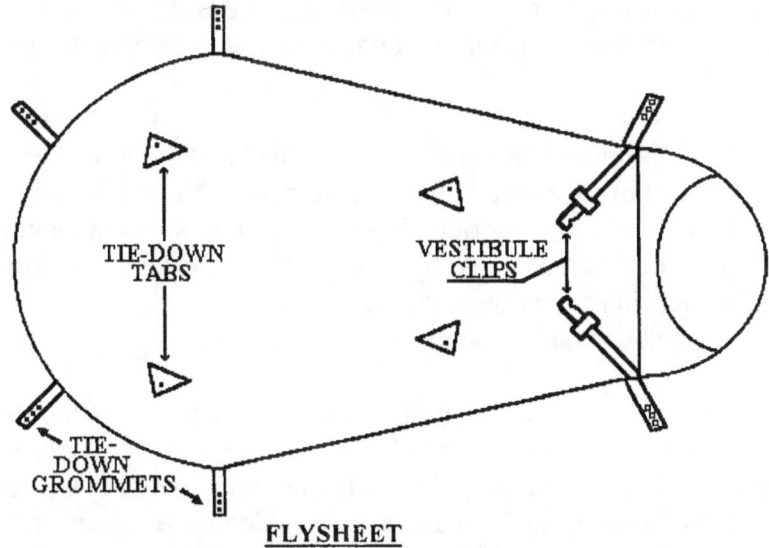

FLYSHEET

(3) <u>Poles</u>. The pole configuration used with this tent allows maximum use of floor space. The poles are comprised of nine sections of 7075 aluminum and are held together by shock cords which aid in connecting them when pitching the tent.

(4) <u>Accessory Kit</u>. Each tent comes with an accessory kit containing; 2 pole repair sleeves, 24 aluminum stakes, 12 nylon tie down cords, 12 line tighteners, a black foam spacer, and woodland colored repair tape 3" x 36".

<u>TRANSITION</u>: Now that we know the basic nomenclature and configuration of the extreme cold weather tent, let's take a look at how to pitch it.

7. (10 Min) **PITCHING THE TENT**

   a. <u>Clear an area</u>. Ensure that there is sufficient room for the tent (approximately 12 feet) by spreading it out on the ground and pulling the floor section tight. Another method is to have one man stand in the center of where the tent is to be pitched and hold a ski pole for a second Marine. The second man will walk in a circle around the first man using the ski pole to measure the pit radius, thus ensuring that the tent will have enough room. Once the circle is marked, should be dug down 4 to 6 feet. Depending on the tactical situation, if time is limited, you can pack down the snow to achieve some cover and concealment initially and then improve the position later by building up a surrounding snow wall.

   b. <u>Insert poles into sleeves</u>. Six of the poles go into the sleeves on the tent, with three being kept aside for the fly. The poles that form the triangle at the top of the tent should go in first, followed by the poles around the side. There are several grommets in each strap to adjust the tension of the tent. If the tent is too loose, snow and rain can accumulate.

NOTE: In pitching and striking, it is advisable to push rather than pull the poles so that the sections will remain engaged.)

   c. <u>Foam Spacer</u>. Attach spacer to snap located on the rear of the tent. The foam spacer is used to prevent the tent sheet from coming into contact with the tent, keeping water away from the tent.

   d. <u>Flysheet</u>. The fly is hooked onto the back of the tent and brought over the top, ensuring that the entrances on both are aligned. Insert the remaining poles into the sleeves and adjust for tension. Pull the front of the fly out away from the tent to attain maximum tension. Inside the fly are two straps that attach to the triangular buckles on each side of the entrance, use these to adjust tension and prevent the fly from blowing away. The fly is a very important part of the tent and tent performance is degraded without it.

   e. <u>Securing.</u> Use the tent stakes and guide lines provided to secure the tent. These tents, as with all tentage, are vulnerable to wind damage; therefore, it may be necessary to secure the corners prior to inserting the poles during pitching in high wind conditions. If you are pitching the tent in deep snow, it may be preferable to use "deadmen" to hold the tent down. Use all tie down points available depending on wind and tactical conditions. (Tying the tent down in ten different places may not be advisable if enemy attack is imminent.)

    f. Tactical considerations. The tent is designed with a light retention material, but it is not lightproof. Also, it may be possible to build a snow wall that not only shields light emissions, but also camouflages and protects the tent from snow as well. In deep snow it is best to dig down into the snow pack and keep a low silhouette. Camouflage with over-whites or netting.

TRANSITION: Now that the extreme cold weather tent is set up, let's discuss the procedures for striking and packing the ECW tent.

8. (3 Min) **STRIKING AND PACKING**. In order to strike the tent, perform the pitching instructions in reverse order. It is recommended that the tent fly be folded length ways into thirds. Before placing it in stuff sacks, roll tightly around the folded tent pole sections, squeezing trapped air out in the process.

TRANSITION: We are aware of the importance in maintaining our gear and equipment, so let's discuss how to maintain the ECW tent so that it can be used time and time again.

9. (5 Min) **MAINTENANCE**. The ECW tent requires very little maintenance.

    a. Cleaning. After each use, shake out loose debris. Sponge clean all dust and track marks.

        (1) If the fabric requires deeper cleaning, hand wash the tent in mild soap and warm water.

        (2) Air-dry the tent out of direct sunlight. Make sure that the fabric is completely dry.

        (3) Never store a tent damp, it will cause mildew and damage to the tent fabric.

    b. Tent Pole Care. From time to time, apply a thin layer of silicon lubricant to all parts of the poles. This is excellent protection against corrosion, prevents the poles from freezing together when they are very cold, and will make the joints work more smoothly in any weather.

    c. Seam Sealing. To ensure the waterproof of both the tent floor and flysheet, the seams must be thoroughly sealed.

    d. Zippers. Lubricate the zippers with a silicone spray to keep them running smoothly and to prevent freezing.

TRANSITION: As with all things, safety is very important, so let's discuss the safety considerations of the ECW tent.

10. (5 Min) **SAFETY CONSIDERATIONS**

a. Ventilation. With four people in the tent it can become a very foul atmosphere inside very quickly. To ventilate, open the door slightly, this also prevents condensation from forming inside the tent and humid air preventing the proper functioning of the drying rack.

b. Stoves. The place for cooking or melting snow for water is the vestibule. Great care should be taken when lighting a stove inside the tent because a flare up could be disastrous. If weather conditions prohibit cooking outside of the tent, then do so inside the vestibule. Cooking inside the tent itself can lead to fires or nasty spills on gear and other Marines resulting in burns. A small stove will heat the inside of the tent very quickly, but it will also consume all of the air in a sealed tent, resulting in asphyxiation. There have been numerous deaths caused by this. Also, you should never sleep in the tent while the stove is lit, this can have predictable results. If at all possible, do all cooking during the day as a light discipline technique.

TRANSITION: Now that we have discussed the ECW tent, let us talk about the team stove; the Coleman Peak 1 stove.

11. (5 Min) **NOMENCLATURE OF THE PEAK 1 STOVE**. The Coleman Peak 1 is a multi-fuel stove. It can burn white gas or kerosene. To be certain on how to operate this stove it is important to be able to identify the different parts.

   a. The following diagram corresponds to the numbered list:

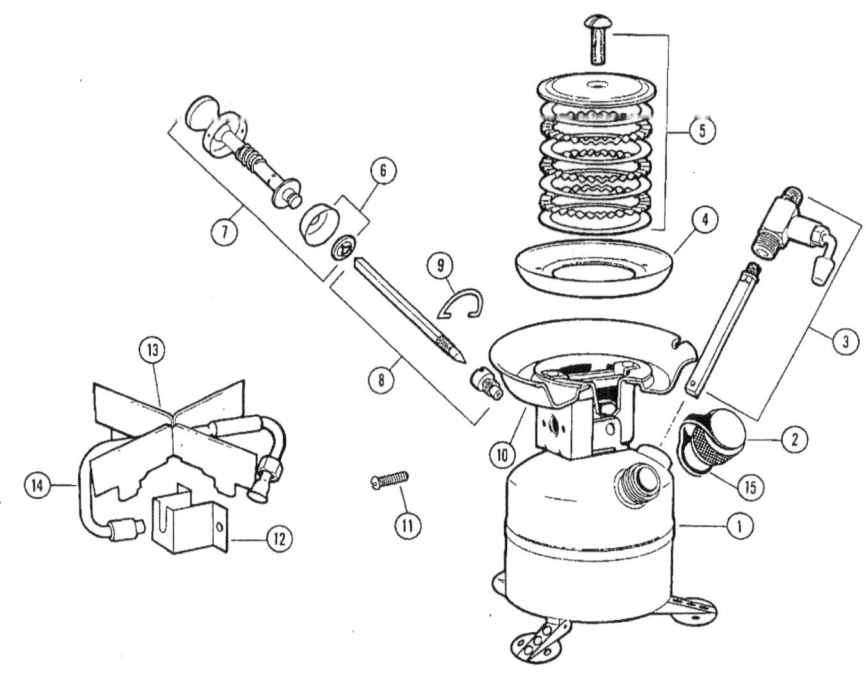

COLEMAN PEAK ONE

(1) Fount
(2) Filler Cap

(9) Clip for Pump Cap
(10) Burner Box Assembly

(3) Valve Assembly
(4) Burner Bowl
(5) Burner Ring Set
(6) Pump Cup
(7) Pump Plunger
(8) Air Stem and Check Valve

(11) Screw (Six)
(12) Generator Bracket
(13) Grate
(14) Generator Assembly
(15) Lanyard

TRANSITION: Let's now do a serviceability check for this stove.

12. (5 Min) **SERVICEABILITY CHECKS FOR THE STOVE.** In order for this stove to operate correctly, serviceability of the stove is needed.

    a. Burner Assembly. Make sure the screw is tightened down.

    b. Grate. Ensure that this is not loose, bent or damaged.

    c. Pump Cup. Make sure that this is not bent and seats well into the fuel tank. Lubrication may be necessary to provide efficient pressure.

    d. Pump Plunger. Check all the parts for cracks.

    e. Filler Cap. Ensure that the filler cap has a gasket and a tight fit to the fuel tank.

    f. Pump Cap Clip. Ensure that this is in place to hold the pump assembly stable while pressurizing the fuel tank.

    g. Generator Assembly. Check for kinks or fuel leaks.

    h. Valve Assembly. Ensure the threads are not damaged and that they fit properly into the fuel tank.

    i. Fount. Check for fuel leaks and that the proper, clean fuel is used in the fuel tank.

TRANSITION: Maintenance of this stove is required in order for it to operate proficiently.

13. (5 Min) **MAINTENANCE FOR THE STOVE**

    a. Parts. Ensure that all parts of the stove are tight. If this is not done periodically, you will have a stove falling apart on you and fail to operate properly.

    b. Carbon Buildup. By taking a small toothbrush and a small amount of fuel, removal of carbon from the following areas will be possible:

       (1) Burner head.

       (2) Outside area of the burner head.

(3) The entire stove itself due to spilt food.

    c. <u>Plastic or Rubber Parts</u>. If at all possible, keep the stove inside a pack or wrapped in clothing and out of the extreme cold until use to prevent possible cracking and damage.

<u>TRANSITION</u>: We are now ready to light the Peak 1 stove.

14. (5 Min) **<u>LIGHTING THE STOVE</u>**

    a. Ensure that the fuel tank has sufficient fuel in it. No more than 3/4 full to allow for pressurization.

    b. Place the stove on a level surface. <u>DO NOT TIP THE STOVE</u>.

    c. Be sure that the control knob is in the "off" position.

    d. Open the pump knob one turn counterclockwise.

    e. With the thumb over the hole in the pump knob, pump air into the fuel tank, <u>DO NOT OVER PRESSURIZE THE FUEL TANK</u>. If little or no resistance is felt, lubricate or replace the pump cup.

    f. Close pump knob firmly to the right.

    g. Hold a lit match to the burner bowl.

    h. Turn the black control knob counterclockwise to the "HI" position. If a yellow flame or liquid fuel appears in the burner, turn the control knob "OFF" and allow the flame to burn out excess fuel or allow it to evaporate before re-lighting.

    i. It may be necessary to re-pump the stove occasionally during use for full heat output.

    j. To regulate the heat, turn the control knob between "HI" and "LOW".

    k. To turn the stove off, turn the control knob fully clockwise to the "OFF" position and the flame will slowly extinguish itself.

<u>TRANSITION</u>: Now that we know about the team stove, let us discuss the rest of the MCCWIK components.

15. (5 Min) **<u>GROUP STORES</u>**. These are the remaining components of the MCCWIK system that we have mentioned earlier but did not discuss in full detail. These are stores that are required to operate for a prolonged period in a cold weather environment. Common sense should be used when deciding to survey worn out items. The fire team sled should not be used to haul individual gear, that gear is to be carried in the individual's pack. Additional items that may be placed in the sled are:

a. Extra fuel for the team stoves.

b. One case of extra MRE's / RCW's.

c. Candles or lantern.

d. Trash bags.

e. Crew Served Weapons.

f. Ammunition.

TRANSITION: Now let's discuss packing this gear in the sled.

16. (5 Min) **PACKING THE SLED**. There are no concrete rules as to how to pack the sled. However, there is one simple principle that should be adhered to: keep the center of gravity low, centered, and to the rear half of the sled. This principle will facilitate movement of the sled and help prevent it from tipping over or nose-diving into the snow when being pulled. In keeping with that principle, load the heaviest items first. In the standard group stores, the case of chow is the heaviest. It should be placed in the bottom, rear of the sled. Seldom used items (extra fuel, candles, and trash bags) should also be on the bottom, front half of the sled. The ECW tent should be in its stuff sack and placed inside the sled. All remaining items are then placed inside of the sled. The shovels and pioneer gear should be packed on top of all other components so that you can get to them easily during a movement or when first establishing a bivouac. Once all group stores are inside the sled, zip the canvas cover and tighten all compression straps to hold the gear securely in place.

TRANSITION: We have just covered the packing of the fire team sled. Now let's look at the Large Sled for casevac or crew-served weapons use.

17. (10 Min) **LARGE SLED FOR CASEVAC/CREW-SERVED WEAPONS**. The large sled is designed to haul a Marine on a back-board or stow any of the crew-served weapons with it's ammunition found in the infantry battalion. It is capable of being pulled by 1 Marine unloaded or 4 Marines loaded, carried by 6 Marines, helo/vertical lifted, and/or being towed by a snow mobile. It is being fielded; 2 per company, 1 per mortar (60 and 81mm), 1 per heavy machinegun, 1 per Javelin (optional depending on how many missiles are being carried by the team), and 1 per TOW (though manpacking a TOW is discouraged). It is similar in design to the Team Sled. We will now discuss the features of the Large Sled.

INSTRUCTOR NOTE: HOLD UP SLED AND SHOW ALL FEATURES UNIQUE TO LARGE SLED IN COMPARISON TO TEAM SLED.

PRACTICE (CONC)

a. Students will practice what was taught during upcoming field evolutions.

PROVIDE HELP (CONC)

    a. The instructors will assist the students when necessary.

OPPORTUNITY FOR QUESTIONS (3 Min)

1. QUESTIONS FROM THE CLASS

2. QUESTIONS TO THE CLASS

    Q. What are the four main parts to the fire team sled?

    A. (1) Hull
       (2) Transport Bag
       (3) Aluminum Pull Poles
       (4) Waist Harness

    Q. What principle should be adhered to when packing the fire team sled?

    A. Keep the center of gravity low, centered and to the rear half of the sled.

SUMMARY (2 Min)

    a. During this period of instruction we have discussed the Marine Corps Cold Weather Infantry Kit which includes the fire team sled, the extreme cold weather tent, the team stove and addition group stores.

    b. Those of you with IRF's please fill them out at this time. We will take a short break.

UNITED STATES MARINE CORPS
Mountain Warfare Training Center
Bridgeport, California 93517-5001

WML
WMO
08/16/01

LESSON PLAN

TEN-MAN ARCTIC TENT & STOVE

INTRODUCTION (5 Min)

1. **GAIN ATTENTION.** The 10-man arctic tent is the Holiday Inn of winter bivying. Great for peace-keeping ops, but not a tactical tent at 75 pounds.

2. **OVERVIEW.** The purpose of this period of instruction is to familiarize the student with the 10-man tent and Yukon stove, their nomenclature, set up, use and take down. This lesson relates to bivouac routine.

3. **INTRODUCE LEARNING OBJECTIVES**

    a. TERMINAL LEARNING OBJECTIVE In a cold weather/mountainous environment, bivouac tactically, in accordance with the references.

    b. ENABLING LEARNING OBJECTIVES

    (1) Given a 10 man tent in snow covered terrain, erect a 10 man tent in accordance with the reference.

    (2) Given a 10 man tent in snow covered terrain, strike a 10 man tent in accordance with the reference.

    (3) Given a Yukon stove, operate a Yukon stove in accordance with reference.

4. **METHOD/MEDIA** The material in this lesson will be presented by lecture and demonstration. You will practice what you have learned in an upcoming field training exercise. Those of you with IRFs, please fill them out at the end of the class.

5. **EVALUATION** You will be tested later by performance evaluation.

TRANSITION: Before we start erecting the tent, let's talk about nomenclature first.

BODY (50 Min)

1. (5 Min) **NOMENCLATURE AND SERVICEABILITY**

    a. Apex. This is the metal plate found in the middle of the tent.

    b. Stove Hole. This is the hole on the side of the tent with a rubber around it. The rubber should be serviceable and the covers for the hole should be rolled up and tied off when a stove is used.

    c. Front Door. The front door is the nearest the stove hole. All zippers should work and all buttons should be present.

    d. Back Door. The back door is opposite the front door and should always be secured. Ensure that the zippers work and that the buttons are present.

    e. Primary Lines. These are gold colored lines on the vertical seams of the tents. There should be one on each vertical seam.

    f. Intermediate Lines. These are the white lines on the horizontal seam between the primary lines. There should be a line between each of the primary lines.

    g. Secondary Lines. Secondary lines are white lines located above the primary lines on the vertical seams. There should be one secondary line above each primary line.

    h. Canvas. The canvas should be free of holes. Any holes should be patched up or sewn up.

    i. Tent Cover. This should also be free of holes and should have straps in order to secure the folded tent.

    j. Telescopic Pole. The tip of the telescopic pole must fit into the apex and be able to extend without collapsing when the weight of the tent is placed on it.

    k. Liner. The tent liner is white and should be free of holes. All holes must be sewn or patched. The liner doors must have zippers that work.

    TRANSITION: Now let's talk about a few considerations in selecting a suitable location for setting up this tent, a good bivouac site.

2. (2 Min) **SITE SELECTION**. There are a few basic criteria that should be considered in choosing a site in which to erect the ten-ma arctic tent. Forested areas in the mountains offer the best site. This area provides cover and containment for the Marines in the site. The trees also provide protection from the wind. Wooded areas also provide firewood and other materials for construction of defensive positions. If fires are built, the trees overhead will

help disperse the smoke. These are just a few principles in the selection of a site. For more guidance on site selection refer to WML.4.6 – Bivouac Routine.

TRANSITION: Let's now discuss the erection of the arctic ten-man tent.

3. (5 Min) **ERECTION**. In order to set up the ten-man tent, a level area must be created. If the snow is deep, it can be shoveled out or packed down. The diameter of the tent can be measured by using a ski pole held at arm's length. Two men will be needed: each man will hold one end of the ski pole. One man will position himself in the center of the prospected site for the tent, the other man will walk a complete circle, which as a result, will determine the diameter needed. The following steps are taken in order to erect the ten-man tent:

   a. The tent should be laid out in a <u>hexagonal shape</u>, with the front door at 11 or 1 o'clock from the wind, canvas side up.

   b. Primary Lines. Lay out the gold primary lines down the seams of the tent.

      (1) Stake Down. Take the tent pins and prepare to stake down these lines, or tie off to trees, or use deadmen to anchor. A dead man is made by using a log or object buried 2 or 3 feet in the snow, or by using a stick with a lead line stuck through a hole in ice if on a lake or river (the lead line is for cutting away).

   c. Telescopic Pole. One man with the telescopic pole extended, opens the door and enters the tent. He sets the pole in the ring at the apex of the tent and raises the pole to a vertical position. Use the tent cover as a base for the pole, so it does not sink in the snow. Do not twist the pole or it will collapse.

   d. Primary Lines. Once the tent is raised, the primary lines can be secured and tightened. Set in tent pins or dead men about 5 feet away from the tent on line with the seams. Secure opposite lines to ensure an even pull, in order to keep the pole straight (the pole man is still holding the pole, until all primary lines are secure).

   e. Stake down the corners and doors.

   f. Secondary lines. After the primary lines, the secondary lines can be staked out and secured. Do this about one foot farther out than the primary anchors.

   g. Inter mediate lines can now be staked down and tightened.

   h. Adjust all tent lines to ensure a tight and secure tent.

   i. Bottom Flap. Tuck the flap under and cover it with gear in wet-cold conditions. Fold flap out and cover with snow in dry-cold conditions.

   j. Several tents can be married together with the zippers to form CPs or BASs.

TRANSITION: Now that we have talked about erecting a tent, let's talk about striking now.

4. (5 Min) **STRIKING**. Many times it will be necessary to brush off any excess snow or frost from both the inside and the outside of the tent. Once this is done, the following steps are to be taken in order to strike the ten-man arctic tent:

   a. Remove the tent pins/stakes holding the tent lines down.

   b. Once the tent is laid out flat, close the door zippers and snaps, and tie up all lines securely.

   c. Now the tent is ready to be folded.

      (1) Lift the peak and one corner of the tent off the ground. Then bring the next corner.

      (2) Repeat placing and one corners together until the panels have been folded into an accordion pleat.

      (3) While bringing the corners together, make sure the inside liner is folded smoothly alongside the tent.

      (4) Lay the tent down, even out the folds, and place the tent lines into the tent.

      (5) Fold the top half down to the eave, then fold the bottom up and over the eave.

   d. Place folded tent in cover and strap-up tightly.

   e. Storage and Movement. Now the ten-man arctic tent is ready for storage or movement in the ahkio.

TRANSITION: Now that we have discussed the tent, let's talk about the Yukon stove.

5. (5 Min) **YUKON STOVE**. The Yukon stove is the primary heater of the 10 man tent. Individual squad stoves can be used, if there is no Yukon stove available. All Yukon stoves should be checked to ensure that they are SL-3 complete, serviceable and fitted properly.

   a. The following are the SL-3 parts:

      (1) M-1950 60,000 British Thermal Unit (BTU) stove body.

      (2) Five stove pipe sections.

      (3) Three guy lines.

      (4) Draft diverter.

(5) Burner plate and drip valve.

(6) Gas hose.

(7) Fuel can adapter with five-gallon gas can.

(8) Vent tube in two sections.

(9) Stove grate.

  b. Serviceability. Upon checking for all required parts, one must check its suitability for use.

    (1) Check that all five stovepipes fit together in sequence.

    (2) The #1 stovepipe must also fit snugly into the stove body.

    (3) The stove body itself has stove legs that swing down to make the stove level.

    (4) The draft diverter is also to be checked for a tight fit on the pipe, as well as the guy lines capable of being secured to it.

  c. Fitting.

    (1) The burner assembly should fit on top of the stove body with the retainer lugs holding it in place.

    (2) The gasoline hose should be checked for holes and the brass connections on both ends are tight and usable.

    (3) Check fuel can adapter and vent tube for a close fit.

TRANSITION: Now that we have all the parts and they are in working condition, let's now discuss the assembly of the Yukon stove.

6. (5 Min) **ASSEMBLY OF THE YUKON STOVE**. Once the Yukon stove has been checked for parts and condition, it can be prepared for operation. The steps in assembling it are as follows:

  a. #1 Stovepipe. Attach the #1 stovepipe to the stove body. This is done by twisting the stovepipe under the clasps surrounding the thimble hole.

  b. Stove Body. Swing out the stove legs from underneath the stove body.

    (1) The stove body should be level.

    (2) Once this is done, the stove can be put inside the ten-man tent.

c. Setting the Stove Inside

   (1) Position it so that the door and draft slide face out towards the tent door.

   (2) Install the #2 through #5 stovepipes together with the draft diverter and guy lines.

      (a) This section of stovepipe is best installed from outside the tent.

      (b) Attach by inserting the pipe through the stovepipe hole in the tent.

   (3) Secure the stovepipe by lashing the draft diverter guy lines to the ten-man tent. By doing this, both the stovepipe and tent will move together in strong winds, preventing a hot pipe from burning the tent.

d. Burner Assembly. It can be inserted once you have completed installing the pipe.

   (1) Insert the burner assembly into the burner hole on the top of the stove body.

   (2) Secure it by turning the retainers (spinnerets) until they catch under the burner hole.

   (3) Now the wire loop on the burner can be lowered onto the ends of the retainers (spinnerets).

   (4) The burner can now be rotated to ensure that the drip valve is offset to one side of the stove.

e. Fuel Can Adapter and Vent Tube. Assemble these two parts together as one part.

   (1) Insert it into the fuel can with the cam handle free.

   (2) Tighten down the adapter plug.

   (3) Once tight, press down the cam handle on the fuel can to seat the adapter and fuel can together.

f. Gas Hose. Attach the gas hose to the drip loop and the drip valve on the burner.

   (1) Make sure the hose does not come into contact with the stove.

   (2) To ensure this, run the gas hose over the drying line on the inside of the tent.

   (3) Check to ensure that the drip valve is closed.

g. Fuel Can. In order to ensure a smooth flow of fuel, the can must be elevated at least 3 feet higher than the stove.

(1) This can be done by lashing it to a tree or by making a tripod.

(2) Invert the fuel can outside into this position.

(3) Some fuel might leak out while this is being done, this is due to either the vent tube clearing out fuel, in which case it should stop within a few seconds, a bad seal/connection from the adapter to the fuel can or by a faulty or incorrectly inserted vent tube within the adapter.

(4) Wipe up the fuel immediately.

TRANSITION: Now the Yukon stove is assembled correctly, let's discuss the operation of the stove.

6. (5 Min) **OPERATION**

   a. The four steps taken to light the M-1950 Yukon stove are:

      (1) Open the stove door. Wipe up any excess fuel that has collected inside the stove, making sure the tent door is open and keeping your face away from the stove door.

      (2) Hold a lighted match under the edge of the burner plate.

      (3) Turn drip valve counterclockwise until 12-15 drops of fuel flow through the glass window.

      (4) Once lit, close the door of the stove and shut the draft slide.

   b. After the Stove is Lit

      (1) Adjust the drip valve to regulate fuel to heat the tent.

      (2) The gas should drop slowly through the glass window and not flow freely.

   c. To Turn the Stove Off. Turn the valve clockwise while watching the glass window until the fuel stops dripping.

   d. Gasoline and Diesel. It will burn 5 gallons of gasoline every 8-12 hours of operation.

   e. Wood. To burn wood, the burner plate cover must be closed and the grate inverted to allow space for a draft and ashes. The draft vent on the stove door should be open. When lighting the wood, EXTREME CAUTION should be used on how the wood is lighted. Try to avoid using any type of fuel, use a solid flammable material such as paper, dried vegetation, fabric, etc. Once the wood is lit, close the door and add wood accordingly for moderate temperature control within the ten-man tent. Periodically empty the ashes from the stove for efficient operating use from the stove.

TRANSITION: Now that the Yukon stove is working, let's talk about some safety precautions to ensure safe proper heating of the ten-man tent.

7. (5 Min) **SAFETY**. The following safety precautions should be observed:

    a. Never leave a stove running unattended.

    b. Turn the stove off if leaving, even for a short period of time.

    c. Make sure all fittings on the stove fit tightly.

    d. Keep the stove level at all times to spread an even flame on the burner plate. This can be accomplished by placing a piece of wood or an MRE sleeve under the stove legs and #1 stovepipe.

    e. Protect the gas hose from being pulled on or coming into contact with the stove.

    f. Do not use excessive force on the drip valve or the hose fittings. Too much pressure at these points will damage the threads and the stove will be useless.

    g. Check the rate of fuel flow at regular intervals. The drop in fuel will change the rate fuel in the burner plate burns.

    h. If the stove goes out accidentally, close the drip valve and wait for the stove to cool down before relighting.

    i. Store all fuel supplies outside the tent to prevent an explosion.

TRANSITION: Once the stove has been assembled and operated, on must be able to disassemble it.

8. (3 Min) **DISASSAEMBLY**. Once the stove has cooled, it can be dissembled. This is done in the following manner:

    a. Lower the gas can and right it.

    b. Disconnect the gas hose from the fuel can adapter and drip valve.

    c. Untie the guy lines from the draft diverter.

    d. Break down the stovepipe sections.

TRANSITION: Last of all, let's talk about the proper method of packing the Yukon stove.

9. (5 Min) **PACKING**. After the stove has been disassembled, it can be packed up and placed in the ahkio.

a. Method For Packing. Packing is basically the reverse of removing the parts which were inside the stove body before assembly.

   (1) Nest the sections of stovepipe inside the base section.

   (2) Place the draft diverter next to the stovepipes.

   (3) Place the pipes inside the stove body.

   (4) Place the burner assembly on top of the draft diverter.

   (5) Coil the gas hose and place the hose and fuel can adapter and vent tube next to the burner assembly.

   (6) Close the burner plate cover and the stove door and fold down the legs.

b. Storage or Shipment. The M-1950 Yukon stove is now ready for storage or movement.

TRANSITION: Proper care and maintenance of the stove is important, for your life may depend on it someday. Are there any questions?

PRACTICE (CONC)

   a. Students will practice what was taught in upcoming field evolutions.

PROVIDE HELP (CONC)

   a. The instructors will assist the students when necessary.

OPPORTUNITY FOR QUESTIONS (3 Min)

   1. **QUESTIONS FROM THE CLASS**

   2. **QUESTIONS TO THE CLASS**

      Q. What is the heat output of the Yukon stove?

      A. 60,000 BTU's

      Q. What are the draft diverter guy lines tied to?

      A. To the tent.

      Q. Where is the ideal area to erect the ten man tent?

      A. In a wooded area.

SUMMARY (2 Min)

a. This period of instruction we have discussed the 10 man tent and Yukon stove, how to set them up and take them down, nomenclature, and operation.

b. Those of you with IRF's, please fill them out and turn them in to the instructor at this time. We will now take a short break.

UNITED STATES MARINE CORPS
Mountain Warfare Training Center
Bridgeport, California 93517-5001

WML
WMO
08/16/01

LESSON PLAN

SPACE HEATER ARCTIC STOVE (SHA)

INTRODUCTION (5 Min)

1. **GAIN ATTENTION**. The Space Heater Arctic Stove is one of the stoves used by the Marine Corps to heat large areas such as the ten-man tent, GP tents, BAS's and so forth. By the end of this period of instruction you will have an understanding of the capabilities of this stove.

2. **OVERVIEW**. The purpose of this period of instruction is to familiarize the student with the Space Heater Arctic Stove, their nomenclature, set up, use and take down. This lesson relates to bivouac routine.

3. **INTRODUCE LEARNING OBJECTIVES**

    a. TERMINAL LEARNING OBJECTIVE In a cold weather/mountainous environment, bivouac tactically, in accordance with the references.

    b. ENABLING LEARNING OBJECTIVES

       (1) Given a Space Heater Arctic Stove, operate a SHA Stove in accordance with reference.

4. **METHOD/MEDIA** The material in this lesson will be presented by lecture and demonstration. You will practice what you have learned in an upcoming field training exercise. Those of you with IRFs, please fill them out at the end of the class.

5. **EVALUATION** You will be tested later by performance evaluation.

TRANSITION: Before we start erecting the tent, let's talk about nomenclature first.

BODY (30 min)

1. (5 Min) **Space Heater Arctic (SHA)**. The SHA stove is the primary heater of the 10-man tent. Individual squad stoves can be used, if there is no SHA stove available. All SHA stoves should be checked to ensure that they are SL-3 complete, serviceable and fitted properly.

    a. The following are the SL-3 parts:

        (1)   Heater body assembly.

        (2)   Six stove pipe sections.

        (3)   Stack cap assembly.

        (4)   Fuel supply hose.

        (5)   Fuel over flow hose.

        (6)   Stove grate.

        (7)   Gravity feed adapter

        (8)   Fuel can stand (optional).

        (9)   Thermoelectric fan (optional)

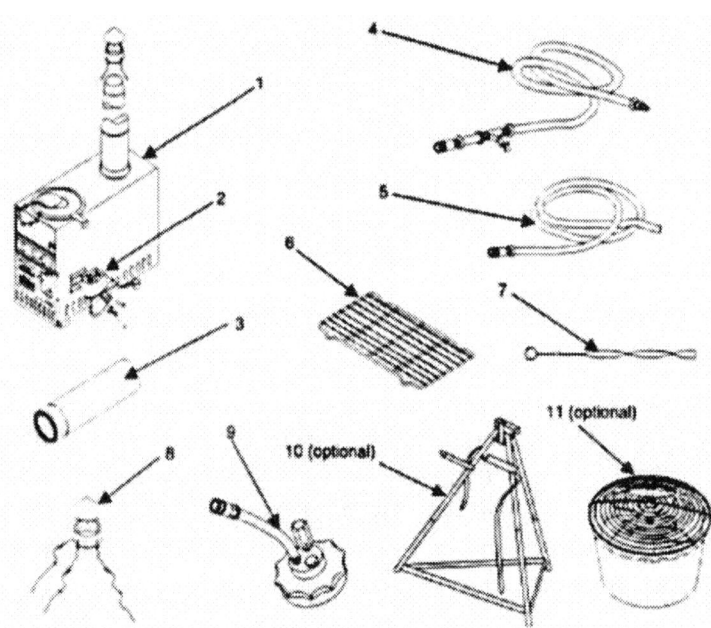

18 - 2

b. Serviceability. Upon checking for all required parts, you must check its suitability for use.

   (1) Check that all six stovepipes fit together in sequence.

   (2) The #1 stovepipe must also fit snugly into the stove body.

   (3) The stack cap assembly is also to be checked for a tight fit on the pipe

   (4) Check the guy lines to ensure proper connection to the stack cap assembly.

   (5) The fuel supply and fuel overflow hoses should be checked for holes and the connections on both ends are tight and usable.

   (6) Check fuel can adapter and vent tube for a close fit.

   (7) Check thermoelectric fan (if applicable) for free spinning blades by blowing into the bottom. Check it for dents, cracks, or holes.

   (8) Check remaining items for dents, cracks, holes, broken seams, and torn or ripped cordage or straps as appropriate.

TRANSITION: Now that we have all the parts and they are in working condition, let's now discuss the assembly of the SHA stove.

INSTRUCTOR NOTE: Have AI assemble the stove as you talk about the procedures.

2. (10 Min) **ASSEMBLY OF THE SHA STOVE**. Once the SHA stove has been checked for parts and condition, it can be prepared for operation. The steps in assembling it are as follows:

   a. #1 Stovepipe. Attach the #1 stovepipe to the stove body. This is done by placing the big opening into the Stack adaptor assembly on the Stove body.

   b. Stove Body. Set up the stove for the type of fuel you will burn.

      (1) Liquid: Place the burner door assembly (1) inside the doorframe (2). Located behind the front door (3). Ensure the burner grate is not installed (7). Close and latch (4) the front door (3)

      (2) Solid: Place the burner door assembly (1), smooth side down, over the burner assembly opening (5) until the rear edge (8) catches under the retaining clip (9). Ensure the retaining chain (6) is clear of the door. Install the solid fuel grate (7), legs down

(10), on the deck (11) of the upper heater area. Close the front door (3) and latch it (4).

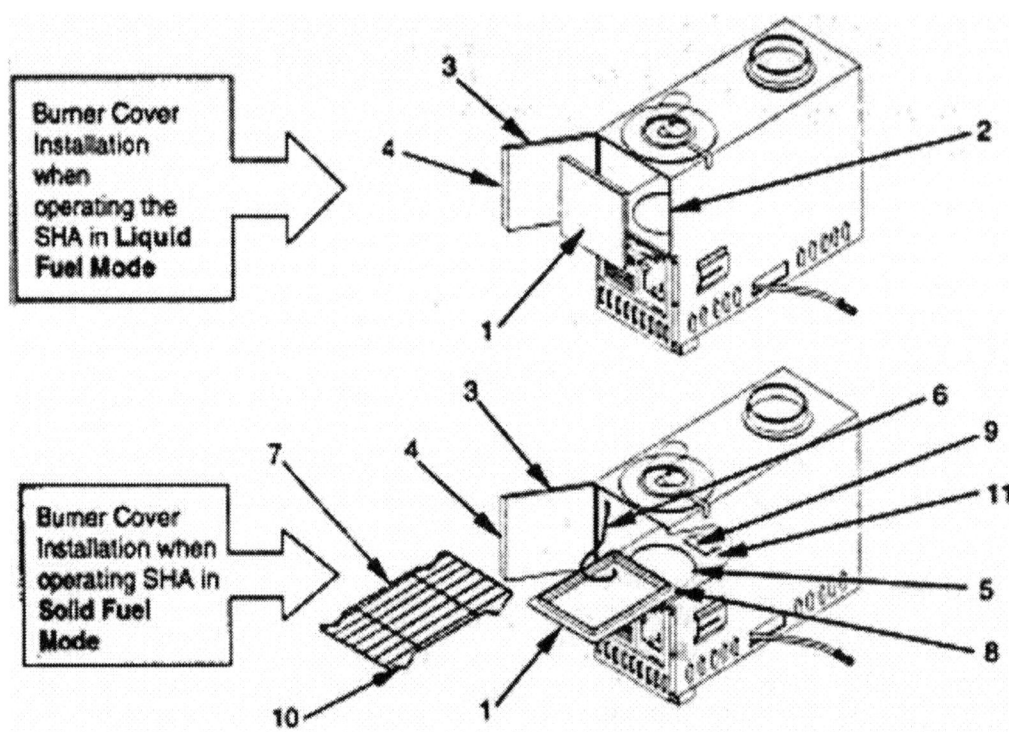

c. Setting the Stove Inside the tent.

(1) Position it so that the door and draft side face out towards the tent door and so that the stove body is level. A stove board may be used to help with floatation.

*NOTE: A stove board is a piece of wood approximately ½" x 1' x 2' that you must construct.

(2) Install the #2 through #5 stovepipes together with the draft diverter and guy lines.

(a) This section of stovepipe is best installed from outside the tent.

(b) Attach by inserting the pipe through the stovepipe hole in the tent, then down to the stove.

(3) Secure the stovepipe by lashing the draft diverter guy lines to the ten-man tent. By doing this, both the stovepipe and tent will

move together in strong winds, preventing a hot pipe from burning the tent.

***NOTE:** Further set up is not required to burn solid fuels.

    d. Fuel flow control valve. The fuel flow control valve can be inserted, for liquid fuel mode, once you have completed installing the pipe.

        (1) Insert the fuel control valve (16) into the slot (22) on the right side of the stove (as you look at the door). Slide the fuel control valve all the way back in the slot to prevent accidentally knocking it off.

        (2) Ensure that the fuel control valve is in the off position.

        (3) Attach the hose from the burner (31) that comes from the bottom of the stove to the bottom of the fuel control valve.

        (4) Attach the fuel overflow hose (30) to the right quick connect on the fuel control valve; and run it outside of the tent downhill to a safe runoff area.

        (5) Run the fuel line in from the outside of the tent were the fuel can is to be set up and connect it to the left quick connect of the fuel control valve. Ensure the T valve end is at the fuel control valve, not at the fuel can. Make sure the t valve is closed.

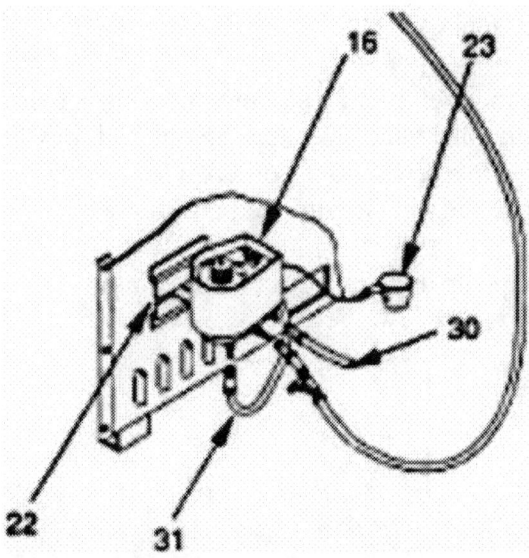

    e. Gravity feed adaptor. These parts come pre-assembled and should not require any assembly.

(1) Insert it into the fuel can by threading it on in-place of the fuel can lid. This will only work on a plastic fuel can.

(2) The fuel can should then be inverted and checked for leaks. If any are detected attempt to retighten or reseal the adaptor. If the leak is coming from the vent tube attempt to retighten. Do not over tighten the adaptor or vent tube or damage to the threads may occur.

f. Fuel Can (25). In order to ensure a smooth flow of fuel, the can must be elevated at least 3 feet higher than the stove.

(1) This can be done by lashing it to a tree or by making a tripod or using the optional tripod (24).

(2) Set the fuel can as far from the tent as the fuel hose will allow.

(3) (Optional) You may dig a small trench around the fuel can and stand to stop spilled fuel from running inside the tent.

(4) (Optional) Attach a piece of 550 cord or other string to the fuel supply hose at a point approx. 4 inches from the quick connect. This should be at a point still inside the trench. This will stop any fuel that leaks from running down the fuel line and running into the tent.

(5) Invert the fuel can outside into its position (i.e. on the tripod or lashed to a tree) and secure it. The optional tripod has two legs to set the can on. One goes through the handle and the other holds the can. Run the strap through the handle and secure it around the fuel can.

(6) Some fuel might leak out while this is being done. This is due to either the vent tube clearing out fuel, in which case it should stop within a few seconds or a bad seal. If there is a bad seal repeat step 2 of the Gravity feed adaptor section.

(7) Wipe up any spilled fuel immediately.

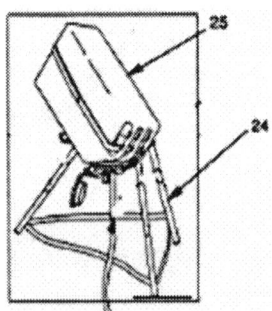

TRANSITION: Now the SHA stove is assembled correctly, let's discuss the operation of the stove.

3. (5 Min) **OPERATION**. To utilize the stove in solid fuel mode, burn your fuel as you would in any common wood stove. To burn wood or coal, draft vent on the stove door should be open. When lighting the wood, EXTREME CAUTION should be used on how the wood is lighted. Try to avoid using any type of liquid fuel. Use a solid flammable material such as paper, dried vegetation, fabric, etc. Once the wood is lit, close the door and add wood accordingly for moderate temperature control within the ten-man tent. Periodically empty the ashes from the stove for efficient operation.

   a. This section will deal with liquid fuel mode specifically. The stove should be set up for liquid fuel mode for these steps. The liquid fuels that this stove will burn are:
      1. JP-5
      2. JP-8,
      3. DIESEL (DF-A, DF-1, DF-2)
      4. Kerosene
      5. Jet A

   NOTE: Do not burn JP-4 or Regular Unleaded Gasoline.

   b. Rate of consumption is 6.08 fluid oz. per min. for DF-1 and DF-2 or 7.43 fluid oz. per min. for JP-5, JP-8 and DF-A. No rate is given for other fuels.

   c. Operating temperatures are from –60 degrees F to 50 degrees F.

   d. It operates at elevations up to 6,000 feet above sea level.

   e. Lighting the stove:

      1. Select the appropriate fuel.

      2. Lift the selector control knob (2) on the fuel control valve (1) and set the knob in the appropriate temperature rating according to the outside air temperature. There are two positions above and below -25°F. Pull the knob, rotate it to the desired setting and release it. Check it to make sure it is in the detent for the setting.

      3. Set the ON/OFF control (4) to the ON position.

      4. Set the fuel flow rate to high. Wait 2 to 5 minutes to allow the burner to fill with fuel.

      5. Shake or tap the fuel hoses to clear any air bubbles that might be trapped inside.

6. Priming the burner

   a. Open the heater lid (5)
   b. Hold the priming cup (attached to the fuel control valve) under the T fitting on the fuel supply hose. Open the T valve slowly and fill the priming cup with fuel and close the valve.
   c. Pour fuel into the bottom of the burner through the heater lid. If the temperature is below -25°F pour in an additional cup.

7. Take a piece of toilet tissue or paper and ball it up. Use the tissue to wipe up any remaining fuel from the priming cup. Do not discard the paper.

8. Light the paper on fire and drop it into the burner assembly. If required, use the cleaning tool to get the paper into the burner hole, were it can ignite the priming fuel. Make sure the burning paper remains in the burner hole and close the heater lid.

9. When the heater has warmed up sufficiently and begins to give off heat (approximately 5-10 minutes) gradually adjust the flow adjustment knob (3) to desired heat output.

10. To shut off the stove turn the knob on the fuel flow control valve to the off position. Allow the remaining fuel to burn off.

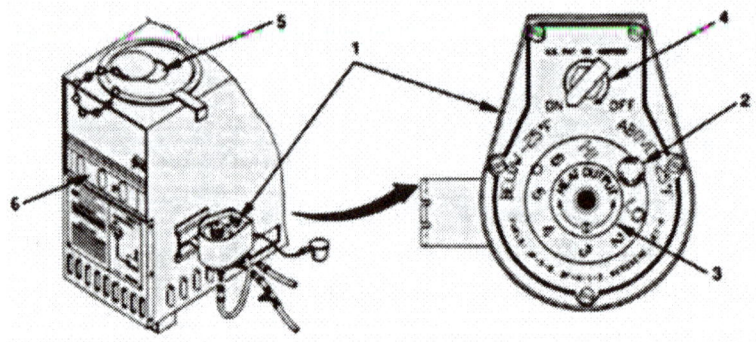

11. To refuel the stove:
    a. Shut the stove off and allow it to cool.
    b. Remove the fuel can from it's stand; and replace or refill.
    c. Replace the fuel can in the stand.
    d. Restart the stove.

12. (Optional) To operate the Thermal Electric Fan: remove the fan from the storage case. Take the site glass cover (not the lid assembly) and allow it to hang off the side of the stove. Place the TEF on the lid assembly. The heat will activate the fan.

TRANSITION: Now that the SHA stove is working, let's talk about some safety precautions to ensure safe proper heating of the ten-man tent.

4. (2 Min) **SAFETY**. The following safety precautions should be observed:

　　1. Never leave a stove running unattended.

　　2. Turn the stove off if leaving, even for a short period of time.

　　3. Make sure all fittings on the stove fit tightly.

　　4. Keep the stove level at all times

　　5. Protect the fuel hose from being pulled on or coming into contact with the stove.

　　6. Check the rate of fuel flow at regular intervals.

　　7. Store all fuel supplies outside the tent to prevent an explosion.

TRANSITION: Once the stove has been assembled and operated, on must be able to disassemble it.

5. (3 Min) **DISASSAEMBLY**. Once the stove has cooled, it can be dissembled. This is done in the following manner:

　　1. Lower the gas can and right it.

　　2. Disconnect the fuel hose from the fuel can adapter and stove.

　　3. Untie the guy lines from the stack cap assembly.

　　4. Break down the stovepipe sections.

　　5. Disassemble the fuel flow control valve

TRANSITION: Last of all, let's talk about the proper method of packing the SHA stove

6. (5 Min) **PACKING**. After the stove has been disassembled, it can be packed up and placed in the sled.

　　1. Method For Packing. Packing is basically the reverse of removing the parts, which were inside the stove body before assembly.

　　2. Nest the sections of stovepipe inside the base section.

　　3. Unlatch and open the front door; remove the burner cover assembly (1).

4. Install the solid fuel grate (2) with its feet up.

5. Stow the nested stack assembly (3) on the right side of the enclosure on top of the solid fuel grate.

6. Coil up the fuel overflow hose (4) and place inside the nested stack assembly.

7. Stow the lid assembly (5) against the left sidewall of the heater.

8. Coil up and store the fuel supply hose (6) between the lid assembly and the nested stack cap. Don't allow the connections or t-valve to come into contact with the site glass on the lid assembly or breakage could result.

9. Store the cleaning tool (7) on top of the solid fuel grate.

10. Replace burner cover assembly in the doorframe and shut the door.

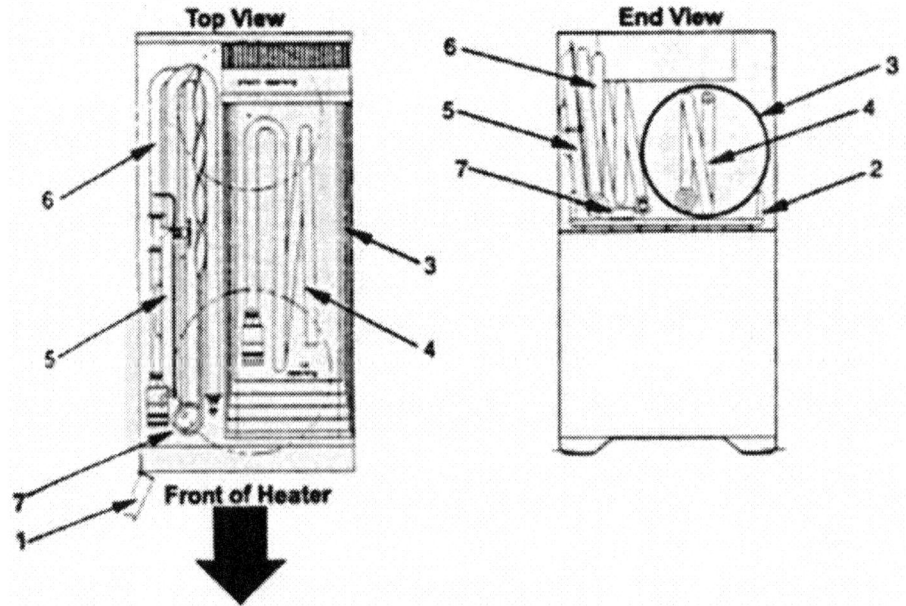

11. Press down on the top edge of the rear door, swing out and remove.

12. Wrap the tent lines around the stack cap (1) or stuff the lines inside the cap, and place on the upper right side of the rear storage enclosure (2).

13. Place the fuel flow control valve (3) in the lower left side of the rear storage enclosure.

14. Place the gravity feed adaptor (4) in front of the fuel flow control valve.

15. Install the rear door over the storage compartment by placing the bottom edge of the door in the slot along the lower edge of the rear storage compartment. While

applying downward pressure, press the rear door into the rear storage compartment doorframe until it clears the retaining pin. Release the rear door

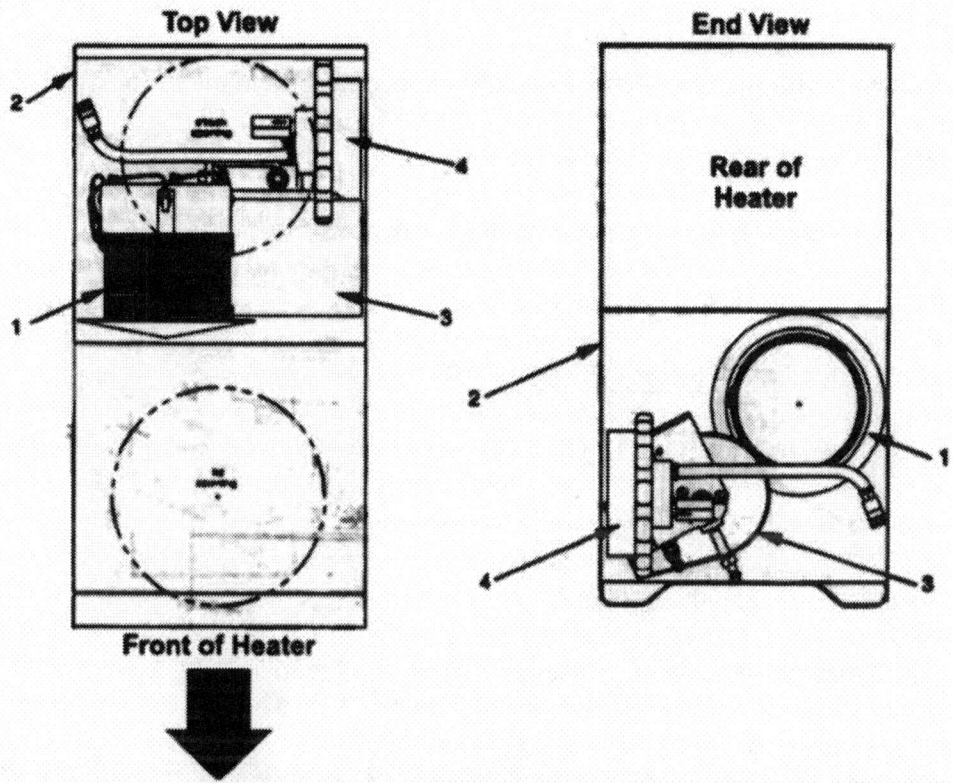

16. The SHA is now packed and ready for movement.

QUICK REFERENCE

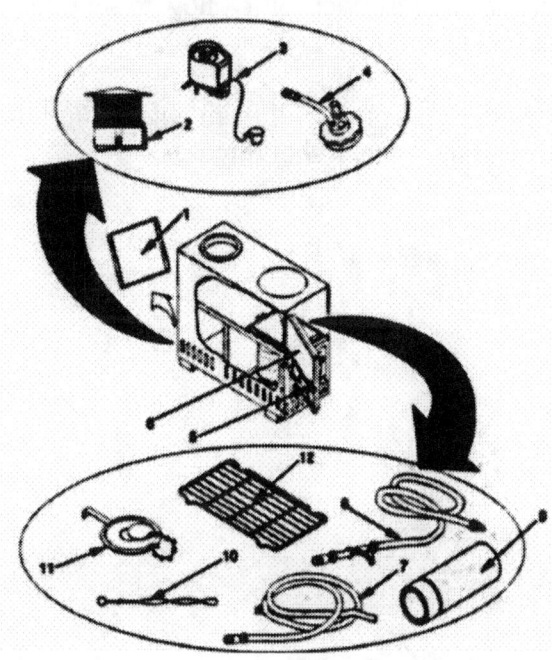

18 - 11

TRANSITION: Proper care and maintenance of the stove is important, for your life may depend on it someday. Are there any questions?

PRACTICE (CONC)

    a. Students will practice what was taught in upcoming field evolutions.

PROVIDE HELP (CONC)

    a. The instructors will assist the students when necessary.

OPPORTUNITY FOR QUESTIONS (3 Min)

    1.   QUESTIONS FROM THE CLASS

    2.   QUESTIONS TO THE CLASS

        Q. What is the heat output of the Yukon stove?

        A. 15,000 – 25,000 BTU/HR

        Q. What are the stack cap assembly guy lines tied to?

        A. To the tent.

SUMMARY (2 Min)

    a. During this period of instruction we have discussed the Space Heater Arctic Stove, its nomenclature, serviceability, how to set it up, operate it, take it down, and pack it.

    b. Those of you with IRF's please fill them out and turn them in to the instructor at this time. We will now take a short break.

UNITED STATES MARINE CORPS
Mountain Warfare Training Center
Bridgeport, California 93517-5001

WML
WMO
08/21/02

LESSON PLAN

BIVOUAC ROUTINE

INTRODUCTION (5 Min)

1. **GAIN ATTENTION**. Living in a winter environment is a far cry from a dream vacation, still, with a little work and common sense, miserable living conditions can easily be made bearable. Lessons learned from this class will take care of minor, petty chores, thus making living in this environment second nature and allowing you to focus your effort on defeating the enemy.

2. **PURPOSE**. The purpose of this period of instruction is to introduce the student to the basics of establishing a bivouac in a winter environment to include site selection considerations, steps in establishing a bivouac, and the specific and general use areas found in a bivouac. This lesson relates to the Marine Corps Cold Weather Infantry Kit, Defensive Positions and Field Fortifications, and Effects of Cold Weather on Infantry Weapons and Optics.

3. **INTRODUCE LEARNING OBJECTIVES**

   a. TERMINAL LEARNING OBJECTIVE. In a cold weather/mountainous environment, bivouac tactically, in accordance with the references.

   b. ENABLING LEARNING OBJECTIVES.

   (1) Without the aid of references, list in writing the eight bivouac site selection considerations, in accordance with the references.

   (2) Without the aid of references and in priority, list in writing the five steps of establishing a bivouac, in accordance with the references.

   (3) Without the aid of references, describe in writing the five specific use areas in a bivouac site, in accordance with the references.

4. **METHOD / MEDIA**. The material in this lesson will be presented by lecture and demonstration. You will practice what you have learned during upcoming field training exercises. Those of you with IRF's please fill them out at the conclusion of this period of instruction.

5. **EVALUATION**. You will be tested on this material later in the course by written and performance evaluations.

TRANSITION: Let's first look at where we will put our bivouac site.

BODY (60 Min)

1. (5 Min) **SITE SELECTION**. There are several factors to be considered in picking a good bivouac site. Higher Headquarters will assign you to a general sector, still leaving you with a lot of leeway in picking the specific site. Things to look for in that site, from most to least important are:

    a. <u>Not near or under a suspected avalanche site/runout zone</u>. It is quite conceivable that in such an area, an entire unit could be wiped out by one enemy soldier with an explosive charge.

    b. <u>A good defensive position</u>. In choosing between two good sites. pick the one that is more easily defended. Remember, *it is better to be miserably alive than comfortably dead.*

    c. Be sure that the area is <u>large enough to contain your whole unit</u>. In mountainous terrain, it will be difficult to find an area large enough to accommodate a whole company. Usually, platoons will have to set up near a centralized point to company integrity but still establish their own bivouacs with a 360 degree defense. As always, camouflage and concealment is essential; it will take only one misplaced tent to give away your position.

    d. <u>In a forested area</u>. There are several reasons for choosing a forested area:

       (1) Natural cover and concealment under the trees for tents, vehicles, and tracks.

       (2) Protection from the wind.

       (3) Readily available firewood and defensive position construction material if needed.

       (4) The trees can conceal and disperse smoke.

    e. If a forested area is not available, then choose nearby <u>depressions or knolls</u> and dig down.

    f. <u>Near an adequate water supply</u>. Snow can be melted, but this is a time and fuel consuming process and it may not be as clean as a running stream.

g. Generally, the best sites are on the <u>leeward sides of mountains</u>. Often just into or below the tree line offers an excellent site.

h. <u>Off the valley floor</u>. Cold air will settle during windless periods.

2. (15 Min) **ESTABLISHING THE BIVOUAC**. The order of establishing the bivouac is a very important consideration. If it is not properly adhered to, a unit will probably spend a lot of extra hours trying to establish it correctly. The following sequence should be second nature to a unit in a cold weather mountainous environment: security, track plan, defensive positions, living areas and specific use areas.

   a. <u>Security</u>. Security is always essential. After a march in a cold weather environment, the tendencies will be to set up tents, rest and eat with no consideration for security or concealment of the bivouac site. While a unit is waiting to enter the bivouac site, they should be staged in a good defendable position, which will offer protection from the elements if possible. SECURITY MUST ALWAYS BE STRESSED.

   b. <u>Track Plan</u>. While the unit leaders are doing their recon, they are going to be establishing a track plan. No unnecessary tracks are made. Good track discipline is going to be vital for the leaders so that the unit doesn't get confused and start making a major highway system around the bivouac site. A map study with a visual recon from afar will aid the leaders in knowing how they can set up their bivouac site to minimize confusion upon arrival.

      (1) <u>Jump-off Point</u>. This is where the trail to the bivouac site meets the approach trail. The jump-off point must not be detectable by the enemy. It should also be well concealed by such things as large trees, rivers, boulders or other natural obstacles that will make it hard to detect. Generally when making the jump-off point, leave the main trail at a right angle, or leave it heading back toward the original direction of march for best deception. It should be covered by fire from the defensive position.

      (2) <u>Dummy Track</u>. This is a deceptive trail that extends past the jump-off point on the main trail. Care should be used by the troops making this trail, so it fools the enemy into thinking that the unit has continued on the main trail. The main trail should extend well beyond the bivouac site to an area that can be used as a dummy bivouac site. When returning to the jump-off point, care should be exercised to ensure that your unit is not leaving marks to indicate that they were returning on this track. Some

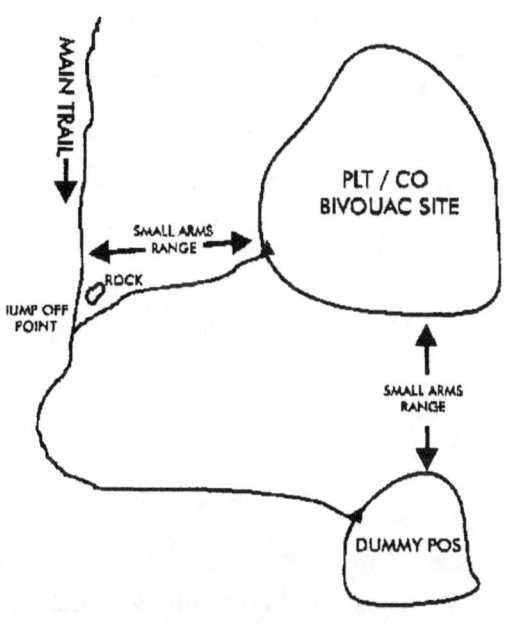

PLATOON / COMPANY TRACK PLAN

examples to aid in the return to the unit's position are:

(a) If on skis, return without using your ski poles.

(b) Reverse snowshoes and walk with them on backwards. A string may be attached to the tails to make walking easier.

(c) Looping the dummy track which will allow the men making the track to circle the track in front of the unit's actual defensive position, as well as provide the convenience to avoid retracing their tracks.

(3) <u>Dummy Position</u>. The end of the deceptive track can be made to look lived in. However, you don't want to alert the enemy to your presence in the area, so don't make it too obvious. If the enemy should find your trail, it is better that he should think you are in a dummy position, rather than being detected in the actual bivouac site.

(4) <u>Track plan in the bivouac site</u>. Every area in the bivouac site should be designated in the track plan.

(a) Central tracks should interlock everything so troops don't start making their own tracks.

(b) The defensive positions should be designated at this time.

(c) Before stamping out the tent sites, the leaders should have a good knowledge of the tent size. This will avoid having to enlarge the area by chopping trees, etc., or stamping an area for a tent that is to small and being tasked with recovering it.

(5) <u>Camouflage and concealment of the track behind you</u>. Camouflage is a continual process in the bivouac site. Here are some considerations for the track plan:

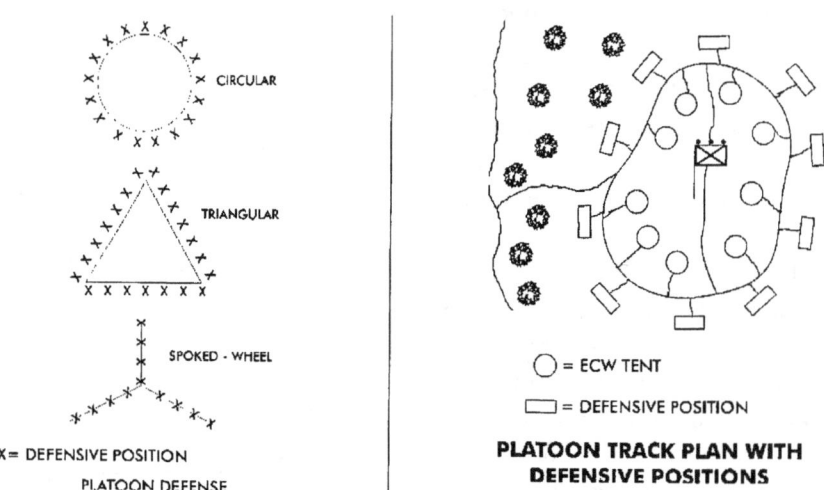

(a) Tracks should meander as much as possible avoiding straight lines. Straight lines are easier to detect because there are few things in nature that are perfectly straight.

(b) A track that is covered by trees will also be harder to detect from above. Use tree cover as much as possible.

(c) Re-emphasize track discipline for everyone in the unit. One or two stray tracks may give away your whole position.

c. <u>Defensive Positions</u>. The tactical situation will dictate the type of defense established. Some of the options are: circular perimeter, triangular perimeter, or a spoked-wheel perimeter. Each type of perimeter has its pros and cons. The unit commander must weigh his options and choose a type best suited for his needs. Individual positions should be constructed in accordance with DEFENSIVE POSITIONS AND FIELD FORTIFICATIONS. Here are some things that must be considered when establishing your defensive positions:

(1) Security must be maintained while constructing the positions.

(2) Positions are constructed so they can cover the whole perimeter.

(3) Automatic weapons are positioned to cover the jump-off point, dummy position, and likely avenues of approach.

(4) Your defensive positions must be on the outside of the perimeter just beyond the range of the noises generated from inside the perimeter, so that the noises do not hinder the sentry from listening to his front. Approximately 30 meters is a good rule.

(5) All positions, i.e. fighting holes, tents, heads, etc., will be connected by communications trenches.

(a) The walls of the trenches will be constructed at an angle and the edges will be rounded off so that they do not cast shadows.

(b) These trenches should be chest to shoulder deep to protect you from incoming fire.

(c) They will be constructed so that they will be afforded the best camouflage and concealment.

(d) The trenches should be made in a zigzag pattern to avoid receiving fire down the long axis.

(6) At a minimum, 2 men per squad should stand arctic sentry duty at a time on the squad's sector of responsibility.

(a) Arctic sentry duty consists of a double staggered watch to ensure that the sentries are fresh and alert at all times. For example: Two men are in fighting positions at 0200 hours; one of these men came on post at 0130 and the other at 0200. The first man will be relieved at 0230 and the second man will be relieved at 0300. This allows one of the sentries to be fresh at all times.

(b) In extreme cold temperatures, a fire watch may also be needed. His duties are:

1. Maintain communication with the sentries through wire or other means.

2. Prepare a hot wet for the sentries upon their return from post.

3. Alert the others in case of danger.

4. Be a snow-watch during storms to prevent the tents from collapsing.

d. <u>Living Areas</u>. Once defensive positions are identified and manned, the Marines may start the construction of their living areas. These living areas must be clearly marked during the leader's recon and are connected to all other positions by comm trenches. The actual construction and organization of these living areas will be discussed in depth later in this period of instruction.

e. <u>Specific Use Areas</u>. The final step in establishing the bivouac site is to designate and establish specific use areas.

(1) <u>Head area.</u> This will be centrally located but downwind of the living areas. It should not be so close to the living areas that people may get sick from it, nor so far away that people will not make the trek to use it in bad weather. If the tents are dispersed over a large area, more than one head area may be built. The head should be erected in a relatively sheltered area out of the wind, or erect some ponchos/tarps to protect the user from the wind. Heads can be constructed as follows:

(a) The sleeve of an MRE case, when lined with plastic bags, makes an acceptable toilet. However, in wet conditions, the cardboard may get soggy and collapse under the weight of a Marine.

(b) Plastic buckets or boxes, also lined with plastic bags, make an excellent alternative.

(c) If establishing a long-term bivouac, a toilet can be made by lashing a sturdy pole in between two trees at about knee level. This pole can be used to sit on while defecating. A second pole can be lashed to the back side of the trees to provide a back rest.

(d) Urinals may be placed outside the individual tents for easy access. Measures should be taken to avoid urinating anywhere besides the designated "piss tree".

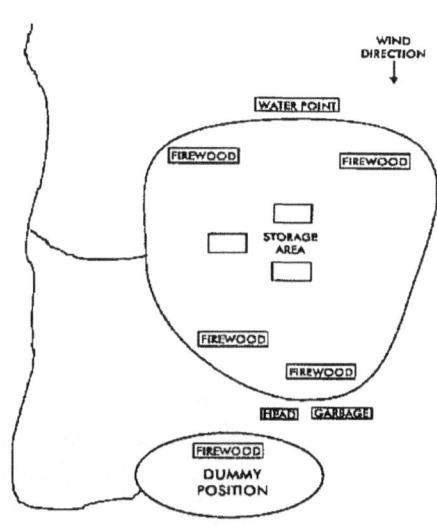

SPECIFIC USE AREAS

NOTE: Here at MWTC, all human feces will be double-bagged, clearly labeled as "human waste", and extracted down the mountain to the sanitation plant.

 (2) The <u>water point</u> is the next site designated. If it is a stream, it will be the furthest point upstream. If no stream is available, then a large, clean, sheltered snow bank must be identified and marked off. It should be located upwind and as far from the head area as possible. If chemical or biological agents have been used at anytime in the past, the whole snow bank must be tested for contamination before use.

 (3) The <u>garbage point</u> will be located next to your heads. All garbage will be retrograded without exception. In combat, the garbage could be disposed of in the dummy position.

 (4) <u>Storage points for the unit</u> are also designated. These areas are inside the perimeter and fall under the control of the company gunny / police sergeant. These areas are for the unit's excess gear and equipment such as; vehicles, rations, fuel, ammunition, communications equipment, extra skis, etc. Some special considerations are:

  (a) All gear will be protected from the elements to maintain serviceability.

  (b) Safety precautions must be taken. For example: fuel should be stored at least 25 meters from any flame or explosives/ammunition.

  (c) Points will be clearly marked to identify each point, and each man should know the location.

  (d) All gear/equipment should be dug down and properly camouflaged and concealed.

 (5) An area to gather <u>firewood and building materials</u> must be designated. Some key points to consider are:

  (a) Spread the cutting out so as not to give away your position by defoiling one part of the forest.

  (b) Do the entire cutting during daylight hours so that there are more natural noises to cover up your activities.

  (c) You may consider cutting and gathering your wood from the dummy position.

3. (10 Min) **ESTABLISHING THE LIVING AREAS.** To maintain proper bivouac discipline and unit efficiency in a cold weather environment, there are many things that need to occur when establishing your bivouac.
 a. <u>Tent Site</u>. To establish a tactical bivouac, the following steps should be taken:

(1) The tent site should, when possible, be located under overhanging tree limbs or near bushes. This provides anchor points for securing the tent, protection from the wind, and also helps conceal the tent from enemy observation.

(2) The tent site should be located at least 10 meters off of the main communication trench. This provides ample room for members of the tent to perform necessary functions (such as equipment maintenance) outside of their tent without blocking the access of the main trench line.

(3) When digging the tent down, ensure you pack down the snow to create a smooth, firm floor. This will help prevent restless nights.

(4) Ensure the pit you dig is large enough to allow you to walk around the exterior of the tent. This allows you to remove the snow that will build up on the roof of the tent during snow storms.

(5) Position the tent in the pit with the entrance at the downwind side. This will help reduce the wind blowing inside the tent every time the door is opened and closed.

(6) Build a snow wall around the perimeter of the tent with the snow removed from the pit. This snow wall will help protect the tent from the wind, conceal it from enemy observation, and limit the amount of light that escapes from the tent when entering/exiting at night.

(7) The tent trench should be dug with a sharp bend in it. This will prevent the possibility of the enemy shooting down the trench's axis and into the tent.

(8) Offset the tent's entrance from the comm trench. This will help enforce light discipline by reflecting the light from an opened tent door back into the tent vice down the length of the trench.

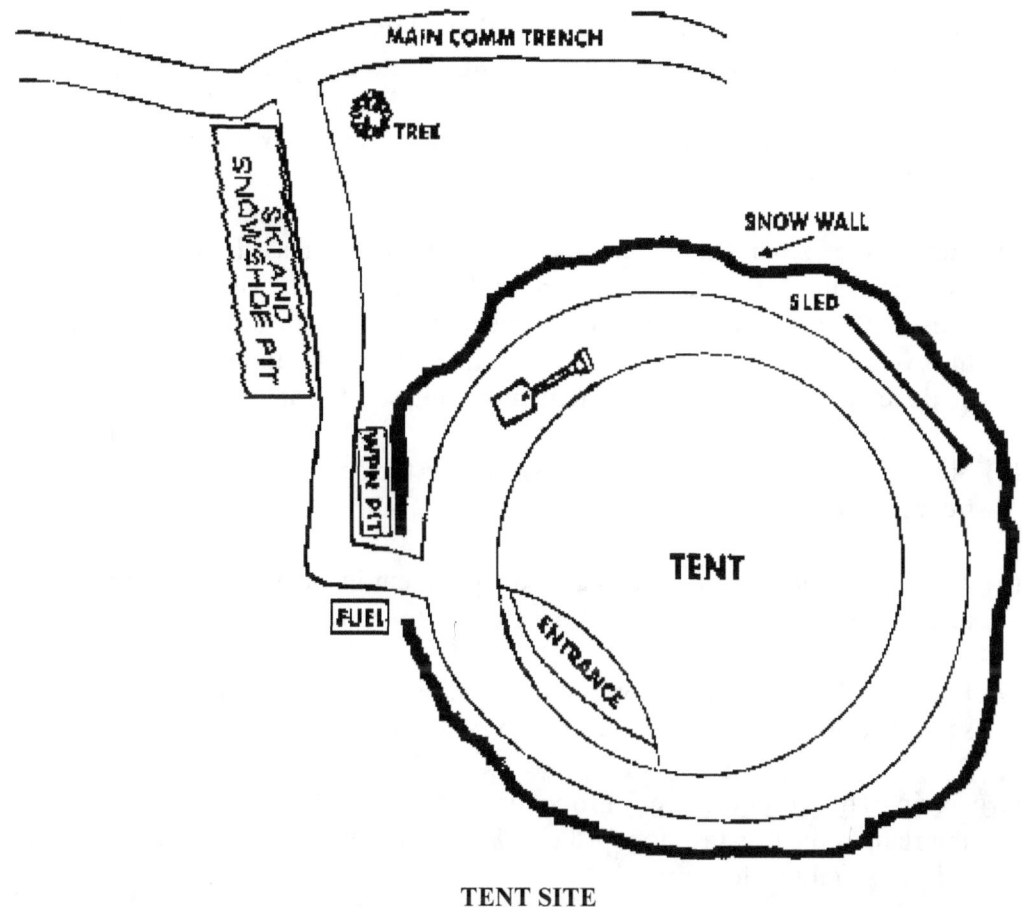

TENT SITE

b. <u>Organizing The Exterior Of The Tent</u>. A Marine unit hauls a lot of gear and equipment with them in a cold weather environment. Due to the size of the ECW tent, the majority of this gear must be left outside of the tent. To ensure no gear is lost or unaccounted for when displacing bivouac during periods of reduced visibility, a unit should have a set SOP as to where and how this excess gear is stored. The following is a recommendation as to how to store this gear.

(1) Ski Pit. Build a ski pit to the left of the tent entrance. The pit should be long enough to accommodate the skis while laying flat on the surface, and wide enough for all four pair of skis and poles. It only needs to be deep enough to allow the skis to be stored below the surface of the snow pack. Place pine boughs or branches on the floor of the pit to rest the skis on. Failure to place these branches will result in the skis actually freezing to the snow, forcing you to scrape the ice-buildup off before using the skis.

(2) Snowshoe Pit. Also on the left side of the entrance, build a pit for the team's snowshoes. This pit is constructed in the same manner as a ski pit.

(3) Weapons Pit. On the right side of the entrance, build a pit for the team's weapons (personal and crew served), and extra ammunitions. This pit is also constructed in the same manner as the ski pit. It is also necessary to cover these weapons and ammunitions with a poncho or some sort of tarp to protect them from the elements.

Some unit commanders may prescribe that all weapons are kept inside the vestibule of the tent instead of being left outside. If this is the case, strict attention must be paid to the effects of condensation as discussed in EFFECTS OF COLD WEATHER ON INFANTRY WEAPONS AND OPTICS. In extreme CW environments, the weapons will be stored outside regardless.

(4) Fuel Storage Area. On the left side of the tent, and at least 1 meter from the tent. Designate an area to store extra cans and bottles of fuel. It is suggested that these cans/bottles are dummy-corded to a tree branch or tent line to facilitate locating them in the event of heavy snowfall. To prevent a tent fire, all refueling of stoves/lanterns must be done outside away from the tent at the fuel storage area.

(5) Shovels. Ensure you leave your shovels near the door of the tent so you can keep your trenches free of new snowfall.

(6) Team Sled. Place all unused gear and equipment inside the sled, secure the cover, and place the sled on its side against the back of the tent.

(7) Piss Tree. Designate a tree or mound of snow as the tent team's piss tree. This tree should be about 5 meters or so from the entrance of the tent.

c. <u>Organizing The Vestibule Area.</u> Proper organization of the vestibule area of the tent is just as important as the organization of the exterior and interior of the tent. The following procedures should be followed:

(1) Cold Hole. Dig a rectangular shaped hole about 1 to 2 feet deep, in between the vestibule door and the tent door. This hole serves two purposes. First it provides a place to trap cold air and prevents it from drafting into your tent. Secondly, it provides a place to stand in when entering and exiting the tent.

(2) Packs. The team's packs should be left outside the main tent and stored in the vestibule. This creates more room inside the tent and prevents dragging excess snow inside the tent. All personal gear that is not being immediately used by the tent occupants should be stored inside their packs.

(3) Whisk Broom. A whisk broom should be inside the vestibule. This broom is used to brush off the snow from the occupants prior to their entry into the tent. This prevents the snow from melting inside the tent, creating pools of water that soak personnel and their gear.

(4) Trash Bag. Place a trash bag inside the vestibule. The tent occupants need to place their trash in this bag as soon as it is created.

(5) Stoves / Cooking. All cooking and melting snow for water should take place in the vestibule. This is to prevent the inadvertent spilling of a pot of water or chow on the inside of the tent. Great care should be taken when lighting the stove inside the

vestibule because a flare up could result in igniting the tent fly or tent body. The preferred method is to light the stove outside and then carry it into the vestibule. A squad stove will heat the vestibule and tent very quickly. However, it will also consume all the oxygen in a sealed tent, resulting in asphyxiation. To prevent this, leave the top of the door unzipped about 8 to 12 inches.

    d. <u>Organizing The Interior Of The Tent</u>. Four Marines in an ECW Tent can easily be compared to sardines in a can. In order to prevent temper flare-ups from the cramped living conditions, the following guidelines should be followed:

        (1) Sleeping Arrangements. The occupant's sleeping bags and mats should be laid out in accordance with the MARINE CORPS COLD WEATHER INFANTRY KIT.

        (2) Individual Gear. All individual gear that is not stored in the individual's pack should be kept neat, orderly, and staged in the individual's sleeping area. Do not allow your gear to invade into your fellow occupant's "space". When gear is not being used, it should be placed inside the individual's pack or WP bag to prevent it from getting wet.

        (3) Drying Wet Gear. In a cold weather environment, it is inevitable that Marines will get their gear wet. The ECW tent has a mesh drying rack in the roof of the tent. All wet gear should be placed on the rack to dry. However, depending on the amount of gear, the occupants may have to take turns drying their gear. Too much gear on the rack will hinder the flow of air, thus slowing down the drying process. Excessively wet gear should be wrung out outside to prevent excessive dripping on the occupants and their gear. A lit stove can enhance the drying of clothes immensely. Ensure that there is proper ventilation to prevent asphyxiation.

        (4) Lighting. To provide adequate lighting inside the ECW tent, the occupants can hang flashlights or chemlights from the drying rack buckles. Candles are not recommended for use inside the tent due to the fire hazard they create. Small lanterns work very well when hung from the drying rack. Lanterns, however, are a high fire hazard, and should be lit outside and brought into the tent once glowing.

        (5) Ventilation. The interior of the ECW tent quickly becomes rancid with the scent of unbathed Marines, dirty socks, wet gear, etc. To help alleviate this problem, open up all doors and escape hatches on a regular basis. To prevent asphyxiation, always leave the top of the door unzipped for 8 to 12 to facilitate the flow of fresh air.

        (6) Temperature. Temperatures inside the tent should be kept comfortably cool. This helps to conserve fuel and also helps maintain the acclimatization of personnel.

4. (10 Min) **<u>GENERAL TIPS FOR STAYING WARM</u>**. The following tips will help you stay warm.

    a. In general:
        a) Stay hydrated. Drink hot wets often.
        b) Take a vasodilator. (Garlic or Tobasco).

      c) Avoid vasoconstrictors. (Cigarettes or caffeine).
      d) Eat a lot of chow. Your body needs the fuel to stay warm.
      e) Wear a hat.
      f) Add an extra layer.

  b. Hands:
      a) Move your fingers to stimulate blood flow.
      b) Spin your hands violently in a wide circle in order to force the warm blood into the fingers.
      c) Pull your fingers out of the glove fingers into the palm. Make a loose fist so the fingers actually warm each other.
      d) Put on bigger/thicker gloves.
      e) Wear looser fitting gloves, tight gloves restrict blood flow.
      f) Change out of wet gloves as soon as possible.
      g) Never wear just the insulating inserts. Use the shells to keep them dry.
      h) If wearing gloves and cold, switch to mittens.
      i) Take off your gloves and place your hands in your armpits, or better yet in your buddy's armpits. Skin on skin contact is essential.
      j) Do not use vapor barriers on your hands. It is too easy to expose wet hands to cold temperatures and wind when the barriers are removed to gain dexterity.

  c. Feet:
      a) Move. Walk around.
      b) Change into dry socks.
      c) Put more insulation between you and the ground (stand on isopor mat, etc).
      d) Wear a vapor barrier. A trash bag works great.
      e) Put them on your buddy's stomach.

  d. Sleeping:
      a) Before going to sleep at night, eat a hot meal and drink a hot wet. Putting warm fluids/food into your body keeps you warm longer. Also, keep some snacks handy to eat during the night to maintain body heat. Remember, food is fuel.
      b) Place extra insulation between you and the ground. Extra mats, empty packs or pine boughs all add extra insulation.
      c) The more crowded a tent the warmer it is. If one man does not have a sleeping bag or has a lightweight sleeping bag, place him in the middle.
      d) Wear extra layers when sleeping. No wet clothing or Gore-Tex layers should be worn.
      e) Wear a balaclava or a hat and neck gaitor when you sleep.
      f) Canteens filled with warm water and placed in your sleeping bag make excellent hot water bottles that pre-heat your bag. Ensure the canteen caps are tightly secured or you may wake up with a wet sleeping bag. This also prevents your water from freezing over night.
      g) Fill your thermos with a hot wet every night before going to sleep. This will provide you with a hot wet to sip during the night and in the event you have to move in the middle of the night.

5. (10 Min) **GENERAL TIPS FOR LIVING IN THE COLD**. The following tips may help you adapt to living in the cold, not just survive it.

   a. Always keep your gear neat, orderly and packed in the event you have to displace suddenly.

   b. Take care of your personal hygiene needs (shaving, sponge bath, etc.) at night prior to going to sleep. This will reduce the removal of natural skin oils that help prevent CW injuries, i.e. windburn, sunburn, frostbite, etc.

   c. Keep all battery operated equipment (flashlights, hand-held radios, etc.) inside the sleeping bag with you if feasible. Leaving this gear laying on the tent floor over night will kill the batteries.

   d. The five gallon water jug should be stored upside down to prevent the freezing of the water at the pouring point. Insulation covers are available through the supply system.

   e. Handle all fuel outside the tent and always use contact gloves. Never handle fuel near a lit stove. In extreme cold temperatures, fuel spilled on unprotected skin will freeze the skin tissue almost immediately.

   f. Whenever there is a stove or flame source inside a tent, maintain a fire watch.

   g. Damp clothes (socks, glove liners, etc.) may be placed in the sleeping bag overnight to dry them. Do not put extremely wet clothes in the sleeping bags, as this will only make the bag wet. Wet socks and contact gloves should be wrung out and placed directly against your chest to dry.

   h. Headlamps are very useful in a tent.

   i. Have a sponge or rags handy to mop up spills or excess condensation. A whiskbroom is handy to knock excess snow from boots and gear.

   j. When melting snow, fill a bag with clean snow and bring it back to your tent.

6. (10 Min) **PULL POLE PROCEDURES.** When preparing to leave a bivouac site it is a tendency among Marines to begin tearing down too early, trying to ensure they are not the last teams. The goal of following these steps is to avoid standing around in the cold waiting for others to finish. Do not forget security when conducting pull pole.

   a. When planning a pull pole, give as much advance warning as possible so Marines can plan their schedules and ensure rest and readiness. Give a specific time, staging area, direction of movement, order of movement, camouflage pattern and over the snow mobility type.

   b. At least a half hour before the designated pull pole time the following should be accomplished. All personnel should have all gear packed up and ready to go. This includes all water containers topped off and skins already on the sled pullers' skis (if skiing). Any trenches dug to fighting positions should already be filled. Marines on security should be in buddy teams with combat loads and over the snow mobility with them. If it is extremely cold, these Marines will have to be rotated often. Marines still in the tents should be wearing what they will have on when they exit the tent save for a jacket. Marines should be sitting on sleeping mats with a stove burning waiting for the pull pole time.

 c. At 15 minutes prior to the pull pole time, the stove should be turned off, packs and sleds staged outside of the tent site. Sleeping mats should remain in the tent until the last moment in the event of postponement. All snow and ice should be removed from tent and tie down lines. Dig or cut deadmen out of the snow/ice.

 d. Five minutes to pull pole the isomats should be stowed. All equipment pits should be filled in and personnel should man their designated pull pole positions. These include the personnel that will strike and stow the tentage, stoves (if any) and who will fill in the tent site.

 e. On order, the tents will be dropped at the same time.

 f. When ready, stage at the designated spot in the order of movement. Bring in security and move.

TRANSITION: We've covered site selection, establishing the bivouac, establishing the living area, general tips for living in the cold, and pull pole procedures. Are there any questions?

PRACTICE (CONC)

 a. Students will practice what was taught in upcoming field evolutions.

PROVIDE HELP (CONC)

 a. The instructors will assist the students when necessary.

OPPORTUNITY FOR QUESTIONS (3 Min)

1. **QUESTIONS FROM THE CLASS**

2. **QUESTIONS TO THE CLASS**

 Q. What are the eight bivouac site considerations?

 A. (1) Not near or under a suspected avalanche site/run-out zone.
  (2) A good defendable position.
  (3) Insure that the area is large enough for the whole unit.
  (4) In a forested area.
  (5) Near an adequate water supply.
  (6) Off the valley floor.
  (7) On the leeward side of mountains, into or below the tree line.
  (8) Near depressions or knolls and dig down.

Q. What are the five steps of establishing a bivouac in order of priority?

A. (1) Security
   (2) Track plan
   (3) Defensive position
   (4) Living area
   (5) Specific use areas

Q. What are the five specific use areas in a bivouac site?

A. (1) Head
   (2) Water point
   (3) Garbage point
   (4) Storage points
   (5) Firewood/building materials

SUMMARY (2 Min)

a. During this class, we have discussed site selection and establishment of a bivouac to include specific use areas, the duties of the tent team, and shelter discipline techniques.

b. Those of you with IRF's please fill them out at this time and turn them into the instructor. We will now take a short break.

c. NOTE: A master Tent Team Leaders Checklist is attached as an appendix to this outline. Reproduce it/laminate it, as needed, for field use as pocket checklist.

TENT TEAM LEADER CHECKLISTS

SETTING IN CHECKLIST

OUTSIDE TENT
1. SECURITY POSTED, ROTATED
2. TRACK PLAN ESTABLISHED
3. DEFENSIVE POSITIONS COMPLETED
4. TENT SITE DUG OUT

5. TENT ERECTED CORRECTLY
6. TENT ANCHORED
7. ENTRANCE DOWNWIND
8. SLED STAGED, COVER CLOSED
9. SKI PIT BUILT/ARRANGED
10. PISS TREE DUG
11. TENT FREE OF SNOW/ICE

INSIDE TENT
1. CLOTHES BRUSH AT ENTRANCE
2. SLEEPING SPACE ALLOCATED
3. HOUSE DUTIES ASSIGNED:
    a. ROTATION OF STOVE MAN/PAD
    b. 2 MEN MOVING AT A TIME
    c. 15 MIN EARLY RISER
4. SENTRY ROSTER/ALERT STATE
5. COLD/COOKING HOLE DUG IN VESTIBULE
6. PACKS/WEAPONS ARRANGED IN VESTIBULE
7. ISOMATS LAID OUT

6. EXTREMITY CHECK, SELF/BUDDY
7. FIRE/SNOW WATCH W/ARCTIC SENTRY
8. WEAPONS CLEANED/LUBED
9. SKIS SCRAPED BEFORE PUT IN SKI PIT
10. SKI SKINS OFF/SLEEP ON TO REACTIVATE GLUE
11. NO GEAR ADRIFT (NOT MATS, STUFF DRYING)
12. SHAVE/HYGIENE BEFORE SLEEPING
13. LIGHT&NOISE DISCIPLINE
14. TRACK/CAMO IMPROVEMENT
15. ENSURE ALL EATING/DRINKING
16. PISSS TREE CHECK
17. ENSURE CLEANING COOKING UTENSILS
18. FIRE PRECAUTIONS
19. TENT FREE OF SNOW/ICE
20. SECURITY ROTATING
21. WORD PASSED

PULL POLE CHECKLIST

MINUS 15 MINUTES
1. STOVE TURNED OFF
2. PACKS PACKED/STAGED
3. SLEDS STAGED
4. ISOMATS LEFT OUT
5. SNOW/ICE REMOVED FROM TENT/LINES
6. DEADMEN DUG OUT/CUT AWAY

MINUS 5 MINUTES
1. ISOMATS STOWED
2. SLED/SKI/PISSTREE FILLED IN
3. TEAM MEMBERS OCCUPY PULL POLE POSITIONS

PULL POLE
1. ON ORDER COLLAPSE TENT
2. PACK TENT
3. ALL HOLES FILLED IN
4. WHEN READY, STAGE ON TRACK IN ORDER OF MOVEMENT

TENT ROUTINE CHECKLIST

1. STOVE ON, COOK ONLY OPERATES STOVE
2. MELT WATER, HOT WET FIRST
3. TOP OFF ALL WATER BOTTLES BEFORE SLEEPING
4. GEAR DRYING ROTATION
5. AIR OUT FEET, CHANGE SOX

UNITED STATES MARINE CORPS
Mountain Warfare Training Center
Bridgeport, California 93517-5001

WML
WMO
09/18/02

LESSON PLAN

SLED MOVEMENT

INTRODUCTION (2 Min)

1. **GAIN ATTENTION**. Moving in deep snow is never easy especially when pulling a sled. But it can be made easier and less strenuous if the sled is moved over a previously broken path. Therefore, the knowledge of trail breaking is essential.

2. **PURPOSE**. The purpose of this period of instruction is to introduce the student to the principles of trail breaking to include organization of the trail breaking party, equipment carried, and duties of the members. This lesson relates to the Marine Corps Cold Weather Infantry Kit (MCCWIK), all sleds, and Route Planning in Cold Weather Operations.

3. **INTRODUCE LEARNING OBJECTIVES**

   a. TERMINAL LEARNING OBJECTIVE. Given military ski or snowshoe equipment and sleds, move over snow covered terrain on skis or snow shoes while pulling a sled, in accordance with the references.

   b. ENABLING LEARNING OBJECTIVES.

   (1) In a cold weather/mountainous environment and given the necessary equipment and a designated distance, break trail, in accordance with the references.

   (2) In a cold weather/mountainous environment and given the necessary equipment execute the seven different formations in accordance with the references.

4. **METHOD / MEDIA**. The material in this lesson will be presented by lecture and demonstration. You will practice what you learn during the upcoming field training exercises. Those of you with IRF's please fill them out at the conclusion of this period of instruction.

5. **EVALUATION**. You will be tested later in the course by performance evaluation.

TRANSITION: Before a trail can be broken, a trail breaking party must establish the route and prepare the trail for the rest of the unit so that your movement is conducted smoothly.

BODY (35 Min)

1. (15 Min) **TRAIL BREAKING**. The purpose of trail breaking is to use a small body of troops to prepare a track or trail so that the main body can move as easily and as quickly as possible, arriving at their destination fresh and ready for follow-on missions. A trail breaking party is generally responsible for performing four tasks; these are reconnaissance and selection of the routed, navigation, preparation of the route, and acting as an advanced guard for the main body. Leaders should consider the following when deciding when or how to use a trail breaking party.

TRAILBREAKING

   a. Planning. Whenever a commander contemplates an over-snow movement that does not make use of existing tracks, he should automatically incorporate the use of a trail breaking party into the plan. Although the general route and the number of routes are determined by the commander overall in charge of the operation, the detailed selection will depend upon conditions underfoot, and the responsibility for this decision will rest with the leader of the trail breaking party.

   (1) The initial selection of a route is made based upon maps and aerial photographs, as well as any information that can be gathered from reconnaissance reports or local inhabitants. Factors to be considered are:

   (a) The tactical situation.

   (b) The main body's method of movement, and the equipment they will be required to carry.

   (c) The terrain.

   (d) Snow, weather, and light conditions.

   (2) The commander will also have to determine the size of the trail breaking party. This number could be as large as one third of the total force moving; a rule of thumb is that a

squad breaks for a platoon, a platoon breaks for a company, and so on. The final determination will depend upon a number of factors, to include:

    (a) The size of the main body, and the number of trails required to accommodate it.

    (b) The likelihood of enemy contact. A security force may need to accompany the trail breakers.

    (c) Anticipated difficulties in opening the route.

(3) The commander will have to determine how far in advance of the main body the trail breaking party should depart in time to reach the destination and to provide local security before the arrival of the main body. The trail breakers should not be allowed to get so far ahead of the main body that they move outside of the radius of available indirect supporting fires. They must also maintain contact with the main body so that the commander can be made aware of any changes to the route or tactical situation. When determining the trail breaking team's departure time, consider the following:

    (a) Type of terrain.

    (b) Weather, light, and snow conditions.

    (c) Number of trails to be broken.

    (d) Degree to which the trails need improvement.

    (e) Tactical situation.

b. <u>Organization</u>. Once a unit leader has been tasked to perform the trail breaking mission, he must organize the party and assign duties within the team. The leader should make every effort to preserve the tactical integrity of the element selected to break the trail. The order of march and organization of the trail breaking party is as follows:

(1) Breaker. This is the point position within the trail breaking party. He breaks the initial trail in the direction indicated by the section leader. He will attempt to travel the easiest route possible. In deep snow and steep terrain, this Marine will tire quickly so try to rotate him frequently.

(2) Straightener. Straightens curves and improves the direction of the trail.

(3) Party Leader. Selects the routes, navigates, and rotates tasks within the team. He will position himself to best control the team.

(4) Right Cutter. Clears obstruction on the right side of the trail. Might expand and level out the trail if necessary to clear the route for unit pulling sleds.

(5) Left Cutter. Clears obstruction on the left side of the trail. Might expand and level out the trail if necessary to clear the route for unit pulling sleds.

(6) Trail Packers. Remainder of the party constitutes the packing team. They improve the trail by filling in depressions, flatten the trail where it is uneven, and mark the route.

EXAMPLES OF TRAILBREAKING DUTIES

c. Equipment. Besides organizing the personnel within the trail breaking team, the party leader must ensure that the breakers are properly equipped. The following equipment should be carried:

(1) Party Leader. Compass, map, route card and/or overlay.

(2) Breakers/Straighteners. Hatchet or machete and wire cutters each.

(3) Cutters. Hatchet or machete each.

(4) Packers. Shovel each, as well as trail marking material.

d. Rotation of Duties. Some trail breaking tasks are harder than others. To avoid exhaustion and fatiguing of the trail breaking team, the party leader must rotate the duties regularly.

e. Marking the Trail. A trail will usually need to be marked, especially if it crosses existing trails or if it will be used during periods of limited visibility. Any of the following methods may be used, as long as the main body is aware of what to look for.

(1) Branches on trees and bushes broken in a predetermined manner.

(2) Flags, sticks, or guiding arrows placed in the snow.

(3) Markers made of rags or colored paper tied to trees.

(4) Cairns of snow or small piles of brush.

(5) Chem lights may be used, but should be shielded from enemy observation. Remember the chem lights, when cold, will illuminate weakly.

(6) Beverage base powder from MRE's can be utilized.

   f. <u>Multiple Trail Breaking</u>. Often the tactical situation will dictate that several trail breaking parties be employed so that other formations besides a column can be used. Each trail breaking party should conduct themselves tactically and follow the proper techniques and procedures for trail breaking.

<u>TRANSITION</u>: Now that we've covered trailbreaking, let's discuss sled pulling and movement.

2. (10 Min) **SLED PULLING AND SLED FORMATIONS**. At times, a commander may determine that the Marines must have their sleds immediately available regardless of their location. If the terrain or weather prevent transportation assets accompanying the Marines, they will have to pull their sleds along with them. Pulling sleds over snow-covered terrain is a time consuming and labor-intensive task. Because of this, commanders must allow for more time during the planning phase. Consideration should be given to the following:

   a. <u>Proper equipment.</u> Ensure that each sled team is equipped with the following items:

(1) Two waist harnesses.

(2) One 9' trace.

(3) 60' of cordage.

   b. <u>Movement techniques</u>. Select the movement technique most appropriate to the conditions and, most importantly, to the Marines level of training. Towing a sled while on skis demands a high level of proficiency. Snowshoes are usually the wiser choice insofar as movement techniques are considered. If skis are used, the personnel doing the actual towing should be equipped with climbing skins; wax will often not provide the necessary traction and will wear off quickly. If skins are not available, snowshoes are the logical choice.

SNOWSHOE MOVEMENT WITH A SLED

c. <u>Towing arrangements</u>. Utilize the correct towing arrangement for the terrain. Personnel may have to disconnect their trace from the sleds and reconnect it at a different attaching point as the terrain changes. There are different configurations for flat ground, ascending, descending, and moving side-slope.

d. <u>General tips</u>. The following apply regardless of movement technique:

(1) One team member can assist the sled-puller by pushing the sled from the rear with his poles.

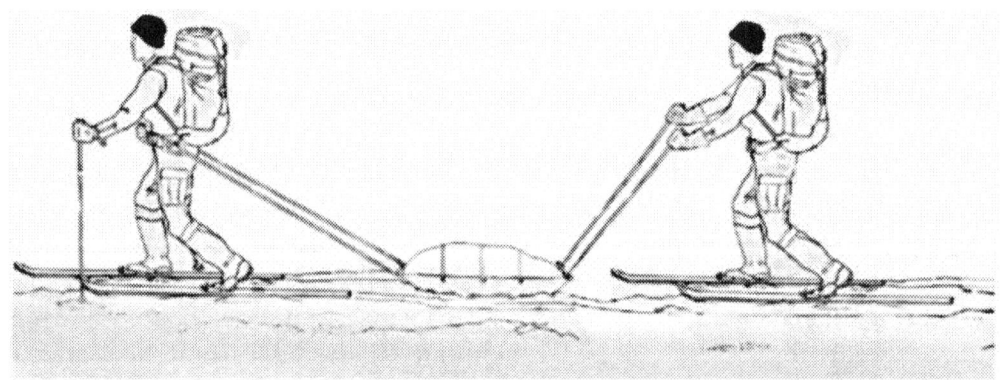

ASSISTING THE SLED PULLER

(2) On moderate slopes, have one Marine attach a trace from him to the sled and ski behind the sled while assisting the sled puller.

MOVING DOWN A MODERATE SLOPE

(3) Attach a trace to the sled when traversing a moderate slope. A Marine uphill from the sled attaches the trace from himself to the both sides of the sled. This will maintain balance to the sled and prevent it from sliding out.

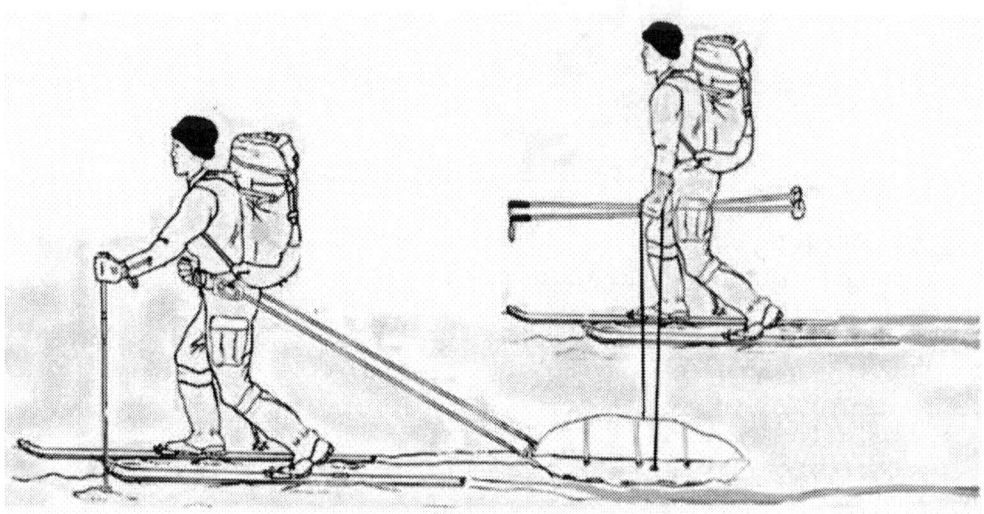

TRAVERSING A MODERATE SLOPE

(4) Sled-pulling teams must be rotated frequently to avoid exhaustion.

(5) A belay rope may be used to assist moving the sled on extreme slopes.

(6) Ascending steep slopes may require two sled pullers with assistance from a third using ski poles.

ASCENDING STEEP SLOPES

TRANSITION: Adequate trail breaking and sled movement is essential for the mobility of a unit if we are to accomplish our mission as Marines. Are there any questions?

PRACTICE (CONC)

a. The students will practice what was taught in upcoming field exercises.

PROVIDE HELP (CONC)

a. The instructors will assist the students when necessary.

OPPORTUNITY FOR QUESTIONS (3 Min)

1. **QUESTIONS FROM THE CLASS**

2. **QUESTIONS TO THE CLASS**

    Q. Who are the members of a trailbreaker party and their standard equipment?

    A. Breaker　　　　　　Wire Cutters
        Straightener　　　　N/A
        Section Leader　　　Map, Compass, Overlay or Route Card
        Right Cutter　　　　Hatchet or Machete
        Left Cutter　　　　 Hatchet or Machete
        #1 Packer　　　　　Shovel
        #2 Packer　　　　　Shovel
        #3 Packer　　　　　Shovel, Marking Material

    Q. What is the order of march for a trail breaking party?

    A. Breaker
        Straightener
        Section Leader
        Right Cutter
        Left Cutter
        #1 Packer
        #2 Packer
        #3 Packer

SUMMARY (2 Min)

   a. We have covered the organization of a trail breaking party, its equipment, order of movement and the duties of each man.

   b. Those of you with IRF's please fill them out at this time and turn them in to the instructor. We will now take a short break.

UNITED STATES MARINE CORPS
Mountain Warfare Training Center
Bridgeport, California 93517-5001

WML
WMO
08/15/01

LESSON PLAN

LIGHT AND NOISE DISCIPLINE IN A WINTER ENVIRONMENT

INTRODUCTION   (5 Min)

1. **GAIN ATTENTION**. Light and noise travels great distances in cold temperatures. If you are not aware of this, you may compromise your unit and its mission. Using the techniques discussed in this class should aid you in avoiding any type of compromises.

2. **PURPOSE**. The purpose of this period of instruction is to introduce the student to the factors that make noise and light discipline imperative in a cold weather environment. This lesson relates to establishing a bivouac.

3. **INTRODUCE LEARNING OBJECTIVES**

    a. TERMINAL LEARNING OBJECTIVE. In a cold weather mountainous environment and given cold weather clothing and equipment, execute light and noise discipline in accordance with the references.

4. **METHOD / MEDIA**. The material in this lesson will be presented by lecture and demonstration. You will practice what you have learned during upcoming field training exercises. Those of you with IRF's please fill them out at the conclusion of this period of instruction.

5. **EVALUATION**. You will be tested later in the course by performance evaluations.

TRANSITION: Let's look at some light problems which occur during various bivouac functions.

BODY (25 Min)

1. (5 Min) **LIGHT DISCIPLINE CONSIDERATIONS IN A BIVOUAC SITE**

    a. General.

    (1) Insulator. Snow acts as a natural insulator which bounces sound waves off it's surface.

    (2) Reflector. Since snow is actually crystallized water, it reflects light tremendously, especially at night.

    b. Light. Various tasks and duties carried out during bivouac routine can produce enormous amounts of light in the winter environment.

    (1) Lighting stoves creates a great light hazard.

        (a) Notice the flash of light as the stove was lit. If the stove must be lit after dark, care must be used to light the stove with the tent vestibule closed.

    (2) Care must be taken when entering or leaving a tent, because it can create a light hazard.

        (a) When entering or leaving an extreme cold weather tent, you should step between the tent and the vestibule and close the flap behind you before opening the next flap or place a poncho over the doorway as a second flap.

        (b) Pack snow around the base of the extreme cold weather tent, taking special care around the vestibule area, to keep light from stoves, flashlights, and lanterns from escaping.

        (c) To keep light discipline in a survival shelter you must double up a poncho to create a double flap.

    (3) Flashlights are also a light hazard. As you can see, flashlights, even with red lenses/blue lenses can be seen a great distance. Their use should be restricted to emergency use only. Chemlight sticks can be used in the track plan to reduce the possibility of compromising your position.

    (4) Cigarette smokers are also light hazards. Personnel around the tent, or on sentry routine should not have a lit cigarette.

TRANSITION: We've just seen light infractions created in a bivouac area. Now let's listen to noises created while in a bivouac area.

2. (5 Min) **NOISE DISCIPLINE CONSIDERATIONS IN A BIVOUAC SITE**

a. <u>Noise</u>. Various tasks and duties carried out during bivouac routines, also carry noise surprisingly far in a winter environment.

   (1) Care must be taken when entering or leaving the tent. The zipper should be opened or closed slowly to prevent amplifying the noise.

   (2) Another important function is the preparation of chow and melting snow/ice for water.

   (3) One of the greatest noise discipline infractions is talking. Voices must be kept low or not exist at all.

   (4) Loosen the cap on the one gallon fuel cans to help equalize the pressure.

   (5) The more organized your bivouac, the less noise you are likely to make.

TRANSITION: Let's listen now to some sounds made while on the move.

3. (5 Min) **NOISE CREATED BY TROOPS ON THE MOVE**. While all of it cannot be helped, some of it can be reduced.

   a. <u>Skis or Snowshoes</u>. As a man moves on skis or snowshoes, he creates noises from them clacking together, poles may strike the skis or snowshoes, and also, the crunching of snow from the skies or snowshoes can be created if the snow conditions are such. Care must also be taken by the individual to keep his skis or snowshoes apart, this is especially so when nearing possible enemy activity.

   b. <u>Falling.</u> Not only because the noise created by hitting the snow, but skis and gear banging into each other.

   c. <u>Radio Operators</u>. Care must be observed by radio operators that their handsets are turned down, or in whisper mode, so that only they can hear them.

TRANSITION: Noise on the move can be your greatest enemy in the hushed environment of winter in the mountains. Are there any questions?

PRACTICE (CONC)

   a. Students will practice what was taught during upcoming field evolutions.

PROVIDE HELP (CONC)

   a. The instructors will assist the students when necessary.

OPPORTUNITY FOR QUESTIONS (3 Min)

1. QUESTIONS FROM THE CLASS

2. QUESTIONS TO THE CLASS

   Q. What are three things we can do to avoid lighting up the bivouac area at night?

   A. (1) Have only one door of the extreme cold weather tent/vestibule open at a time.
      (2) Pack snow around the base of the tent.
      (3) Offset the door of the vestibule from the track plan access.

   Q. What are three things that we can do to prevent excess noise from compromising our position?

   A. (1) Use care and open tent door zippers slowly.
      (2) Voices must be kept low, or not at all.
      (3) Avoid falling.

SUMMARY (2 Min)

   a. This period of instruction has been on noise and light discipline in a cold weather environment, specifically lights and noises in a bivouac site, and noises made during movement.

   b. Those of you with IRF's please fill them out at this time and turn them in to the instructor. We will now take a short break.

UNITED STATES MARINE CORPS
Mountain Warfare Training Center
Bridgeport, California 93517-5001

WML
WMO
09/31/00

LESSON PLAN

CAMOUFLAGE, COVER, AND CONCEALMENT

INTRODUCTION (5 Min)

1. **GAIN ATTENTION**. When fighting over snow-covered terrain, a Marine's ability to accomplish his objective may be greatly influenced by his knowledge of camouflage, concealment, and deception. This class will not qualify you as an expert in this field, but it will give you a basic understanding of these three elements as they apply in cold weather and over snow-covered terrain.

2. **PURPOSE**. The purpose of this period of instruction is to introduce the student to the fundamentals of camouflage, concealment, and deception. This will include the different methods and materials used to camouflage both yourself and your position. This lesson relates to defensive positions and tactical bivouac.

3. **INTRODUCE LEARNING OBJECTIVES**

    a. TERMINAL LEARNING OBJECTIVE. In a cold weather mountainous environment, execute cold weather mountain operations, in accordance with the references.

    b. ENABLING LEARNING OBJECTIVES. In a cold weather mountainous environment, execute camouflage, cover, and concealment techniques, in accordance with the references.

4. **METHOD / MEDIA**. The material in this lesson will be presented by lecture. You will practice what you have learned during upcoming field exercises. Those of you with IRF's please fill them out at the conclusion of this period of instruction.

5. **EVALUATION**. You will be evaluated later in the course by performance examination.

TRANSITION: First, let's look at some general problems encountered in snow covered terrain.

BODY (50 Min)

1. (5 Min) **GENERAL**

    a. In snow-covered terrain the stark contrast between light and dark emphasizes any item which does not blend naturally with its surroundings. Furthermore, every movement by vehicles or dismounted troops leaves readily identifiable tracks in the snow which can provide detailed intelligence to an enemy. Before any movement, track planning must receive high priority. Nature may assist by covering tracks with newly fallen snow or by providing a storm with which movement can be concealed but these occurrences cannot be counted upon. Therefore, a careful plan balancing a minimal track pattern with a track deception plan must be instituted. In this regard, camouflage, concealment, and deception from air observation is a major concern.

    b. In snowy terrain, backgrounds are not necessarily all white. Rocks, bushes, trees, and shadows make sharp contrast with the snow. Snowy terrain in wooded regions, when viewed from the air, reveals a high proportion of dark areas which may impact on the overall camouflage plan.

    c. Firing of weapons, vehicle exhausts, stove vents, and breathing will, in extreme cold, cause local ice fog or vapor clouds which can be readily seen although the weapons, vehicle, tent, or man is well concealed. Smoke from fires may hang immediately above a position if there is no wind to disperse it. If green or wet wood is used for fires, its odor can also give important information to an enemy. It may be necessary to move weapons frequently, shut off vehicle engines or leave vehicles in rear area to reduce these telltale signs. On the other hand, deception or concealment may be enhanced by deliberately creating vapor clouds or smoke clouds. In this regard, heat sources or any type are of major note if the enemy has a thermal detection capability. Dummy positions and deception will be discussed later.

    d. When working in high Northern or Southern latitudes the (depending on hemisphere and season) amount of daylight can materially increase or decrease the time available for reconnaissance. Example: At 63°N latitude on December 21 there is 4 hours and 8 minutes of daylight and on June 21 there is 19 hours and 52 minutes of daylight with the rest of the day being dusk. Considering the above, light discipline takes on major importance as reflection of light off of snow can be seen at great distances.

        (1) Lights should be equipped with red or diffuse lenses.

        (2) Tents and shelters should be placed in defilade so that light escaping from doors and vents has less effect.

        (3) Snow walls must be built with an offset entry.

(4) Tents and shelters should be sealed. Stove holes, vents, rips, etc., should be draped with cloth, tape, or branches. Tent sheet doors covered with either extra tent sheets or ponchos.

TRANSITION: Now, let us discuss how an individual Marine would properly conceal himself from possible detection by the enemy.

2. (10 Min) **INDIVIDUAL CONCEALMENT**. Individual concealment requires the individual to be familiar with the various types of terrain over which he will pass and to know how to adjust his clothing to adapt to his surroundings and how to move through an area in the least obvious manner. The individual must also utilize individual camouflage paint and expedient measures as appropriate.

CAMOUFLAGED MARINE

   a. Clothing Combinations. By using a combination of these items, four different color combinations can be attained as listed below.

      (1) All green.

      (2) All white.

      (3) White over green.

      (4) Green over white.

b. Terrain/Clothing Alternatives. The following examples of different types of terrain and the best suited clothing combinations are considered the basics upon which the individual must build using his own initiative and imaginations.

   (1) Thickly wooded areas. These areas consist mainly of secondary growth coniferous or deciduous trees with thick underbrush. An all green clothing combination is normally best these conditions.

   (2) Low brush or light scrub areas. These areas are often found at and above the tree line or in hilly areas with poor soil. In most cases an open snow background predominates and a combination of white over green is usually found to be suitable.

   (3) Forested areas. These areas are covered with primary growth, coniferous and deciduous, of varying density, with little underbrush. The normal clothing combination here is green over white.

   (4) Above tree line. Except in very mountainous regions where rock faces and large areas of talus may interrupt the universal whiteness, those areas above the tree line form a vast unremitting expanse of whiteness. The only relief to this whiteness is offered by different textures in the snow where it has been compressed by the wind or smoothed by the sun and, of greater tactical importance, by shadow. An all-white clothing combination is the most suitable and care must be taken to ensure that weapons and other equipment items are similarly camouflaged. Even the straps of an ALICE pack, if not camouflaged, will stand out clearly at great distances. Only with experience will the individual become fully proficient in the art of camouflage above the tree line, but the following points are worth remembering. On bright days, a man may be more difficult to spot when he has the sky as a background, because the sky, in these conditions, will usually be darker than the terrain. In the same vein, backgrounds of unbroken snow should be avoided as snow-covered ground reflects many times more light than bare ground and any deficiencies in camouflage will be exaggerated proportionally.

   (5) Mixed surroundings. In mixed surroundings frequent changes of camouflage may become necessary. However, to avoid unnecessary changes, individuals should be trained to avoid areas of local growth and dark outlines. In addition, camouflage changes must be unit-wide to aid in friend or foe identification. Over-whites must be quickly available to change as the terrain changes.

TRANSITION: Now that we've covered how an individual can conceal himself from the enemy, how about his gear?

3. (10 Min) **WEAPONS AND EQUIPMENT CONCEALMENT**. Small items of equipment are relatively easy to camouflage with good results being achieved by the use of matte white point or white tape. White tape is particularly useful for camouflaging webbing equipment, although some types of tape have a tendency to crack and peel at low temperatures. Care must be taken when camouflaging weapons, whether with paint, tape, or garnish, to ensure that the material used does not interfere with the working parts or the cooling system. This camouflaging can also provide insulation from the cold for the individual user. The methods of

concealing larger items, tents and vehicles, as stated below, apply in principle to other large pieces of equipment.

a. <u>Weapons</u>. Camouflage weapons using the same principles as in other environment. The color scheme is unique to a cold weather environment.

   (1) T/E infantry weapons must be camouflaged to match the predominate terrain and vegetation of your operation area.

   (2) Previously mentioned materials can be used on a variety of weapons.

   (3) Ice fog must considered for all long range weapons. Secondary and even tertiary positions may be used to "shoot and move."

b. <u>Tents and Vehicles Below the Treeline</u>. The specific methods used depend largely on the intensity of the tree cover, the materials available, and whether white or dark colors predominate in the area. If the region is thickly wooded the tent (or vehicles) can be camouflaged by thickening the area around it with branches after it has been sited. Care must be taken not to disturb the snow cover or the trees being thinned. In other instances, where white predominates, the main object is to break up the shape of the tent and the easily recognized shape of a vehicle. This can be accomplished first by digging the tent or vehicle into the snow, providing both cover and concealment, the by draping white camouflage netting, clothing, or other materials over the tent.

c. <u>Tents and Vehicles Above the Tree Line</u>. In instances when units are required to operate above the tree line, their tents and vehicles will be visible for great distances owing to their marked contrast to the uniformly white background. Many camouflaging options are available to include but not be limited to the following:

   (1) Dig tents into snow drifts. Smooth the excavated snow as possible to erase the change in texture and make it conform to the surrounding drift.

   (2) White camouflage netting and garnish should be available for use.

   (3) In some cases, white parachutes are also useful as tent and vehicle camouflage sheets.

   (4) As a last resort, the liner from a ten-man tent could be removed to use as camouflage over the outer. However, the texture of the liner is different from the snow and the liner must be cut to allow proper draping over guy lines.

d. <u>Vehicles</u>. In addition to the above principles, the following items are significant when considering the camouflage of vehicles.

   (1) Vehicles should be prepared for operations in this environment by pointing in accordance with the current editions of Marine Corps Order P4750.3 with matte paint in irregular patterns. It should be realized that as terrain changes, so must the

camouflage, so the most universal paint scheme should be used with expedients used in specifically needed instances.

    (2) In addition to painting, each vehicle should be equipped with an all-season camouflage net to be used as needed.

    (3) Whenever possible, a vehicle should be parked so that its shadow falls on a bush, interrupting the straight lines of its own shadow.

    (4) In wooded areas, lean-to's or snow shelters can be built to provide cover and concealment for vehicles.

    (5) In cold conditions, consideration must be given to the exhaust from vehicles since it will form ice fog or vapor clouds that are easily detected.

  e. <u>Camouflage Materials</u>. Since white is the predominate color in winter, snow becomes a most important camouflage material. By intelligent use of camouflage clothing and equipment together with what nature makes available, effective individual and group camouflage can be achieved.

    (1) Improvised camouflage clothes can made from sheeting, tape, whitewash sacking or painted canvas. White paper, when wet, can be applied and allowed to freeze on all kinds of surfaces. Snow thrown over an object helps to increase the camouflage effect.

    (2) White paint has many uses in winter camouflage. Weapons, vehicles, skis and sleds can be effectively painted with a white non-glossy type paint.

    (3) In some cases white smoke may be used to help the camouflage plan. As in any use of smoke, wind conditions must be carefully considered.

    (4) Camouflage face paint, white and loam color combination, may be applied to exposed areas of the face and hands to blend effectively with the snow cover.

WARNING: Frostbite signs may be covered by camouflage paint.

    (5) White zinc oxide ointment may be used in place of cammie sticks for several reasons. It has a sun block factor, is easier to apply and remove for changing conditions, and it is easier to perform buddy checks for frostbite without the danger of erasing part of a frostbitten face.

    (6) The use of camouflage nets and parachutes is effective in concealing tents and vehicles. Careful use of garnish is especially important in this environment.

    (7) The use of natural vegetation is highly recommended, but as in any environment, the age of the cut vegetation and collection procedures are important. Vegetation must be changed before it begins to wilt and the individual must be very careful to selectively

gather vegetation so as to not make his positions obvious because of stripped areas or indiscriminate tracks.

TRANSITION: We have now learned how to conceal ourselves and equipment, next let us discuss how to conceal our positions.

4. (10 Min) **CONCEALMENT OF POSITIONS**. All defensive positions must be camouflaged, whether they are dug into the ground, into the snow or are built above the snow level. They should be connected with well camouflaged communication trenches and should be sited under brush or tree line when possible. The following points should be considered when building and camouflaging defensive positions in the snow.

   a. The sides and ends of all trenches should have pronounced slopes and rounded edges to reduce shadows.

   b. When vegetation is gathered, it is important to selectively gather small amounts from varied areas so as to not leave obvious stripped areas.

   c. The clearing of fields of fire must be done carefully due to the obvious tracks left. Again, areas should not be stripped of vegetation.

CAMOUFLAGED POSITION

   d. If there is sufficient snow cover, trenches should not be dug to ground level as grass, leaves and dirt will be mixed with the excavated snow which will make these areas very obvious, especially from air observation.

   e. The constant use of trails and trenches will glaze and dirty these avenues. New snow should be added frequently to improve the overall camouflage plan.

   f. The snow excavated from positions should be smoothed out and not left in humps and uneven piles which cast obvious shadows.

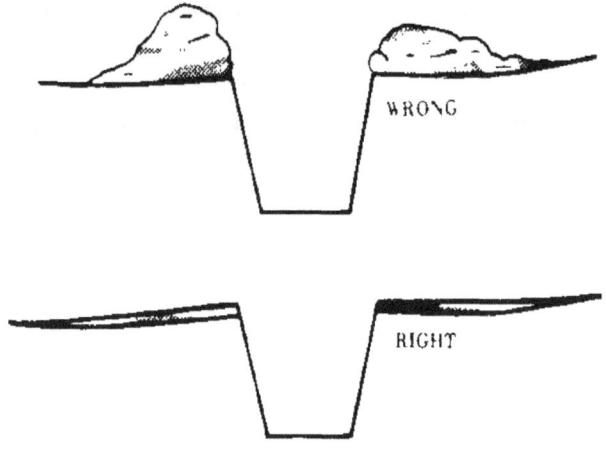

SNOW REMOVED FROM A TRENCH

g. Positions should be selected such that the least amount of modification to the natural surroundings is required.

h. Overhead cover and concealment should be considered in all cases.

CAMOUFLAGED SHELTER AND OP

i. Approaches to positions should be made under the concealment offered by trees or bushes behind snow drifts or slopes and in shaded areas. They should also lead through the position to one or more dummy positions. Trails should never end at a camouflaged position.

j. In selecting positions it is important to consider the background against which the position will be constructed. Appropriate inspection and camouflaging is necessary. The individual must consider the attacker's view of the positions.

k. Positions should be camouflaged as much as possible before they are occupied.

TRANSITION: Now that we've learned how to conceal our positions, let's discuss infrared deception.

5. (5 Min) **INFRARED DECEPTION**

   a. Environmental Effects. The winter environment has a significant effect on the, thermal infrared signatures of targets, enhancing them unless weather conditions are degraded (such as during a snow storm). High thermal contrasts result when low background temperatures become coupled with high temperatures produced by body heat, engines, mechanical friction, and the human need to heat the immediate surroundings for comfort and survival. These contrasts are further enhanced when background temperatures rapidly change because of radiational cooling. Poor thermal contrasts exist when snow is falling.

   b. Thermal Backgrounds

      (1) Terrain. A deep, undisturbed snow cover will present a uniform and clutter free background to a thermal infrared sensor, provided the snow is deep enough to completely cover a large area. When we now put rocks, trees, and bushes into the picture, we get a contrast due to the absorbing effects these objects have to thermal heat throughout the day. As the day cools down into the evening, the snow pack temperature is less than that of the exposed objects. This causes the contrast in the thermal infrared sensors. Such a contrast in the background creates problems for an infrared munition because it generates numerous potential false targets.

      (2) Personnel. The individual soldier is particularly vulnerable to infrared surveillance in winter because of the large difference between body temperature and a cold background. Materials with low thermal emittance characteristics may eventually be used in the manufacture of clothing to mitigate the problem, but for now, a Marine must rely on the creative use of existing equipment and natural materials to hide from surveillance. The picture below illustrates the use of materials gathered from nature and in every Marines pack to conceal a fighting position or bivouac from infrared surveillance. It blends with the background both visually and thermally.

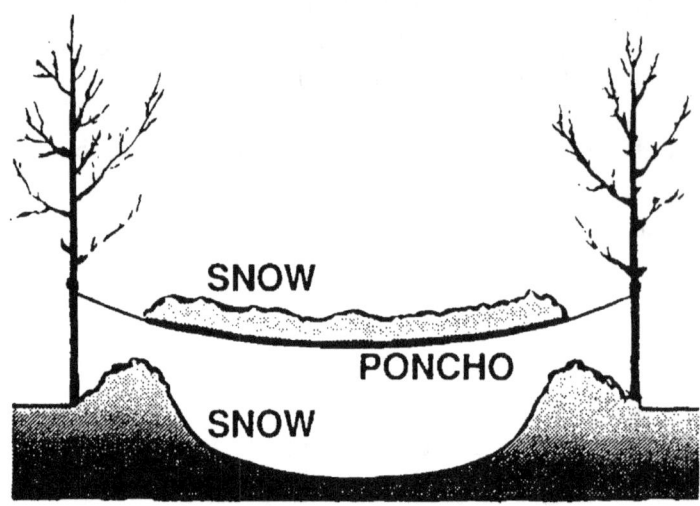

OVERHEAD CONCEALMENT

(3) Internal Combustion Engines. The same problem exists for internal combustion devices that are used by Marines in the field, namely that thermal contrasts are greater in winter than at any other time of the year because of lower background temperatures. Nets can effectively visually camouflage field generators and such, but they do not effectively hide or disrupt their signature. There are some means of deception being developed, but as of now, we have none.

(4) Tents. Tents are highly vulnerable in winter to infrared targeting and surveillance owing to the operation of personnel heaters.

(5) Personnel Heaters. Personnel heaters also increase the overall thermal signature of armored vehicles in winter, making them more vulnerable to infrared targeting.

(6) Thermal Inversions. Thermal inversions are common in winter on clear, calm evenings, especially when a snow cover is present. They can cause rapid radiational cooling of the snow surface and large thermal gradients in the air near the surface. Armored vehicles contrast sharply with a natural background during an inversion owing to their capacity. Contrasts are greatest for vehicles left exposed in low lying areas.

(7) Obscurants. The natural obscurant most frequently associated with winter is falling snow. Infrared radiation will not penetrate falling snow any better than visible radiation. Even in light to moderate amounts, falling snow significantly reduces the target detection ranges of thermal infrared systems. Ice fog is a winter obscurant that significantly reduces visibility. It is difficult to see through with the naked eye. Thermal imagers could be used to provide effective security against intruders in ice fog.

TRANSITION: We must be particularly aggressive in our protective measures against thermal imagery devices in a winter environment, especially when the possibility exists that the enemy possesses the means to use these types of detection devices. Are there any questions?

PRACTICE (CONC)

    a. Students will practice what was taught in upcoming field evolutions.

PROVIDE HELP (CONC)

    a. The instructors will assist the students when necessary.

OPPORTUNITY FOR QUESTIONS (3 Min)

1. QUESTIONS FROM THE CLASS

2. QUESTIONS TO THE CLASS

    Q. What are three causes of localized ice fog or vapor clouds?

    A. (1) Weapons.
       (2) Vehicle exhaust.
       (3) Stove vents.

    Q. What is the camouflage used when in a very thickly wooded area?

    A. An all green clothing combination.

    Q. What are the two important points when camouflaging a defensive position in the snow?

    A. (1) The sides and ends of all trenches should have pronounce slopes and rounded edges to reduce shadows.
       (2) The snow excavated from positions should be smoothed out and not left in humps or uneven piles which cause obvious shadows.

SUMMARY (2 Min)

    a. This period of instruction has been on camouflage, cover, and concealment and has dealt with personal, unit, and equipment camouflage, deception, types of camouflage clothing and various forms that it is worn in.

    b. Those of you with IRF's please fill them out at this time and turn them in to the instructor. We will now take a short break.

UNITED STATES MARINE CORPS
Mountain Warfare Training Center
Bridgeport, California 93517-5001

WML
WMO
08/15/01

LESSON PLAN

DEFENSIVE POSITIONS AND FIELD FORTIFICATIONS

INTRODUCTION (5 Min)

1. **GAIN ATTENTION**. When we're cold and tired after a long ski march to our newly-assigned location, the last thing that we want to do is begin construction of a defensive position. However, it may be that position that we worked on until near exhaustion that is the determining factor in repelling the enemy come sunrise. It may be this class that enables you to build that very position.

2. **PURPOSE**. The purpose of this period of instruction is to introduce the student to the basics of defensive positions in a cold weather environment. This includes site selection, characteristics of defensive positions, types of positions, building materials, and placement of obstacles. This lesson relates to bivouac routine.

3. **INTRODUCE LEARNING OBJECTIVES**

    a. <u>TERMINAL LEARNING OBJECTIVE</u>. In a cold weather mountainous environment, properly utilize defensive positions and field fortifications, in accordance with the references.

    b. <u>ENABLING LEARNING OBJECTIVE</u>.

    (1) In a cold weather/mountainous environment, construct defensive positions, in accordance with the references.

    (2) Without the aid of reference and given a list of building materials, state in writing the minimum thickness needed for each building material for a defensive position for protection from small arms fire in accordance with the references.

4. **METHOD / MEDIA.** The material in this lesson will be presented by lecture and demonstration. You will practice what you have learned during upcoming field training exercises. Those of you with IRF's please fill them out at the conclusion of this period of instruction.

5. **EVALUATION**. You will be tested later in the course by performance evaluation.

TRANSITION: Now that we know our purpose behind this class. let's discuss the site that we should work on.

BODY (50 Min)

1. (5 Min) **SITE SELECTION**. The first thought to building your fighting hole is, "Where should it go?" This is most likely going to be decided by your superiors to cover likely avenues of approach, but this still leaves the individual Marine a lot of leeway to work with. Here are some basic considerations to take into account:

    a. Where you can best cover your assigned sector with good fields of fire.

    b. On the high ground, it is nearly impossible to conduct an attack over icy slopes, or wading through knee to waist high snow uphill.

    c. Covering likely avenues of approach, i.e., frozen streams, lakes, ravines, tree lines, etc.

    d. Covering natural obstacles, i.e., avalanches, windfall, etc.

    e. In a location where you can incorporate natural strengthening features such as rocks, trees, fallen timber, small knolls, or depressions. Not only will this mean less work, and offer natural cover and concealment, but these positions will be immensely stronger then just snow fortifications.

    f. In shaded areas, not only to give you concealment, but also to prevent the sun from melting away your protection.

    g. With you back to the prevailing winds, not only will this keep you more comfortable and more alert, but also gives you the advantage of seeing better and further; whereas the enemy has to look into the wind. When this is not always possible, bend your positions to provide some slight relief, so that you and your buddy can look half right into the wind; or half left, so as not to take on the full force of the wind.

TRANSITION: Let's now look at other considerations.

2. (5 Min) **OTHER CONSIDERATIONS**. These terrain features will probably be utilized by a platoon or company-sized unit.

a. <u>Snow-covered and Icy Slopes</u>. Slopes are normal obstacles that troops and vehicles encounter in a mountainous environment. But, when slopes are snow/ice-covered, the loss of sufficient traction can occur and movement will become more difficult.

b. <u>Windfalls</u>. Strong winds or supporting arms may knock down many trees in a wooded area. These trees are known as windfalls. They are very effective obstacles when covered with snow, especially to personnel wearing snowshoes or skis.

c. <u>Lakes and Streams</u>. Not all natural obstacles are equally effective in the winter as in the summer. Normally, bodies of water are considered natural obstacles but, under winter conditions, the ice which forms may turn these former obstacles into excellent avenues of approach. This illustrates an important reason for re-evaluating defensive positions before winter arrives.

d. <u>Avalanches</u>. An avalanche makes an excellent obstacle for blocking roads and passes. Since it occurs in a cold weather/mountainous environment where there are few natural avenues of approach, avalanches can have a far reaching influence in combat with the correct timing and location. If it is possible to predict in advance where an avalanche could occur, then we can use this against the enemy.

<u>TRANSITION</u>: Now let's take a look at some characteristics.

3. (5 Min) **CHARACTERISTICS**. As with any fighting position, you want to build down to offer a low profile. If the snow is shallow, dig a regular fighting hole into the ground. This can be difficult, but very strong. Care must be taken not to scatter the dirt across the snow, as this may give away your position.

   a. <u>Camouflage</u>. The best way to camouflage your fighting hole is to build it similar to a nearby snow bank. If the surrounding area has powdered snow, so should yours. Ice, branches, or nothing at all. Nothing should give away your position as being different from the surrounding area.

   b. Your position should not contain straight, sharp lines, as nothing in nature is ruler straight. So round off your edges and weave your trails.

INDIVIDUAL FIRING POSITION

c. The bottom of your fighting position should be kept camouflaged. Do not dig completely down to the dirt and leave it exposed, but fill in the bottom periodically with fresh snow. Dirt will show very obviously in aerial photographs, and even normal foot traffic in snow will stain the trench bottoms, unless they are renewed with fresh snow. MRE boxes can be used to insulate your position for camouflage, and keep your feet warmer.

d. The best time to build your fighting hole is just before any snowfall. This will help cover your major construction, and you will only have to redo minor work.

e. Never walk directly in front of your position. If you walk in front, you have just given your location away.

TWO MAN DEFENSIVE POSITION

    f. Overhead cover can be applied by building a log roof, and burying it with dirt and snow. However, make sure your firing port is not too large because large a snow bank with a hole in the middle could be a "dead" give away.

DEFENSIVE POSITION WITH GOOD OVERHEAD COVER

TRANSITION: Let's look at the three general categories of fighting positions and a few cold weather considerations.

4. (5 Min) **TYPES OF POSITIONS**. There are three general categories of fighting positions.

    a. <u>Primary Position</u>. This position accomplishes the mission assigned to you by covering your assigned sector of responsibility. Obviously, this position will receive your attention first, and you should continue to improve on it for as long as it is occupied. It should demonstrate your knowledge of KOCOA and be situated on:

        (1) Dominating terrain concealed from enemy observation and direct fire.

(2) Have excellent observation and fields of fire.

(3) Cover likely avenues of approach and obstacles.

PRIMARY FIRING POSITION FOR A NON-MOBILE TOW SYSTEM

b. <u>Alternate Position</u>. Even the best primary position must sometimes be made of snow. You may find your protection degraded by the impact. You may also find your vision impaired and position disclosed by flash, firing smoke, carbon residue, or even ice fog. In either case, you will need one or several nearby alternate positions where you can move to carry out your mission. This should be the second position constructed.

c. <u>Supplementary Positions</u>. These positions are prepared to accomplish missions other than your primary mission. These positions may be constructed to cover an open flank, protect a boundary, provide reinforcing fires to another sector, or deny enemy penetration, but may have a higher priority than alternate positions. Their priority of construction will be designated by the unit commander.

TRANSITION: Next we'll take a look at the building materials used in a cold weather environment.

5. (5 Min) **BUILDING MATERIALS**. There are several different materials that can be used to build your fighting position with varying degrees of effectiveness in stopping the enemy's fire. When properly constructed, even positions made only of snow can protect you. The following data is subjective with respect to conditions.

a. <u>Newly Fallen Snow</u>

   (1) Minimum thickness: 13 feet

   (2) Construction: Cut into the back of a large snow bank or piled up snow.

b. <u>Packed Snow</u>

   (1) Minimum thickness: 7 feet

   (2) Construction: Stomp down snow in MRE boxes to form blocks.

c. <u>Frozen Snow/Water - Snowcrete</u>

   (1) Minimum thickness: 4 feet.

   (2) Construction: Pour water onto packed snow and allow it to freeze.

d. <u>Ice</u>

   (1) Minimum thickness: 3.5 feet.

   (2) Construction:

      (a) Make ice blocks in empty ammo cans or line MRE boxes with trash bags and fill with water.

      (b) Cut ice blocks out of lakes and streams. This not only builds your position but denies an avenue of approach.

e. <u>Ice-Crete</u>

   (1) Minimum thickness: 1 foot.

   (2) Construction:

      (a) Mix soil and water in empty ammo cans or lined MRE boxes and allow it to freeze.

      (b) Build log forms and fill with dirt. Pour the water into the forms and allow it to freeze.

<u>TRANSITION</u>: Now we should discuss a few of the notable features of field fortifications and the emplacement of obstacles around them.

6. (10 Min) **FIELD FORTIFICATIONS**. Although defensive positions made out of snow and other mixtures, i.e., ice, ice-crete etc., have suitable dampening capabilities when used properly, you may want to enhance those capabilities with other natural materials.

   a. <u>Rocks, Logs, and Dirt</u>. Just because you are operating over snow-covered terrain is no reason to neglect the normal defensive materials. If the situation permits, positions dug directly into the ground with overhead cover will provide the best protection. Remember, 8 inches of logs/dirt is required to stop mortar and artillery fragments.

   b. <u>Sandbags</u>. Filled with snow, dirt, or rocks and frozen make it harder to penetrate and is an expedient way to make packed snow.

FIGHTING POSITION USING SANDBAGS

   c. <u>Simple Snow Fortifications</u>. These can stop concentrated small arms fire for a period of time and even heavy machine gun fire temporarily.

   d. <u>Wooden Logs</u>. It is possible to build your position out of wooden logs. This is usually done in conjunction with your regular fighting position. It requires a lot of work but is very effective. There are three basic types of log walls that we will discuss:

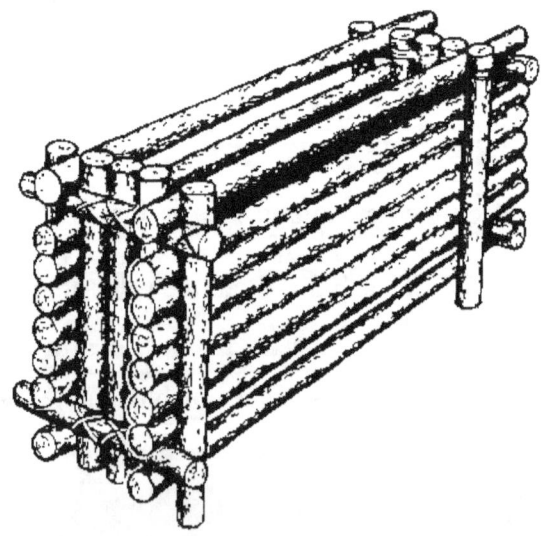

EXAMPLE LOG FORM

(1) <u>Natural tree wall</u>. Find any two sturdy trees approximately eight feet apart. Stack logs on the enemy side until they stand approximately chest high. Pile snow, rocks, and branches against these logs to hold them against the trees.

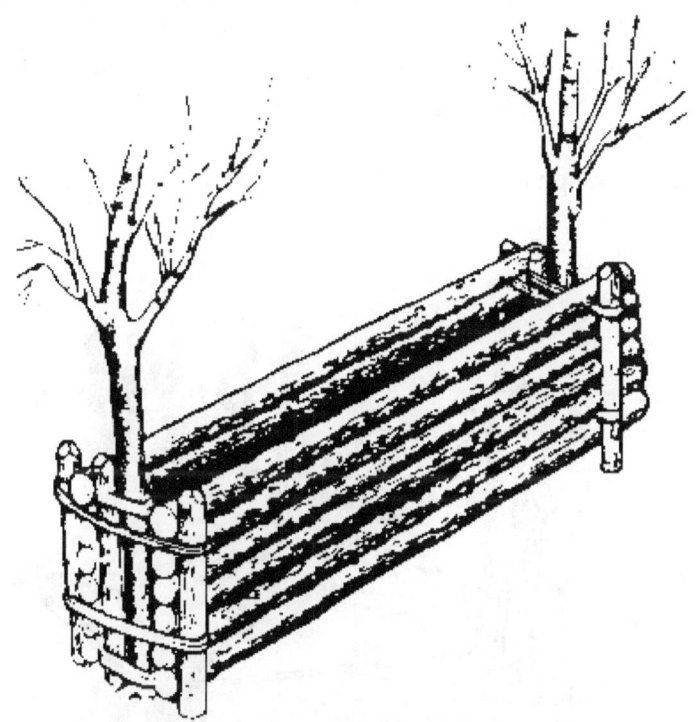

TREE WALL ON BOTH SIDES OF TREES

TREE SUPPORTED WALL

(2) <u>Artificial anchor wall</u>. It may be necessary to construct a wall when there are no suitable anchors to hold the wall up. This method is very effective but care must be taken for site selection and camouflage. To construct:

   (a) Drive two logs into the ground approximately eight feet apart and three feet deep.

   (b) Dig a trench on the enemy side of the two logs about one foot deep and eight feet long. Lay a log on this trench. This will act as a "dead man".

   (c) Connect cordage from the ends of the "dead-man" to the top and bottom of the two original logs.

   (d) Stack logs up on enemy side of the two original logs.

   (e) Pile snow, rocks, and branches up on the enemy side of the stacked logs to hold these logs in place and to make the position look like a snow bank.

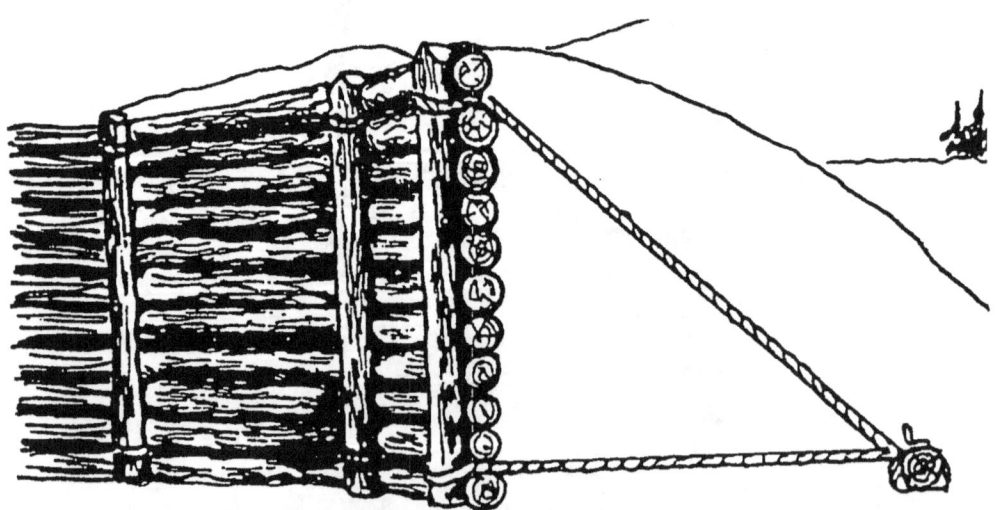

ANCHOR SUPPORTED WALL

(3) <u>Tripod wall</u>. Another method of building a fighting position that will provide some overhead cover is a tripod wall. This wall again is constructed when there are no suitable anchors to hold the wall up. To construct:

(a) Lash three logs together with cordage to form a tripod.

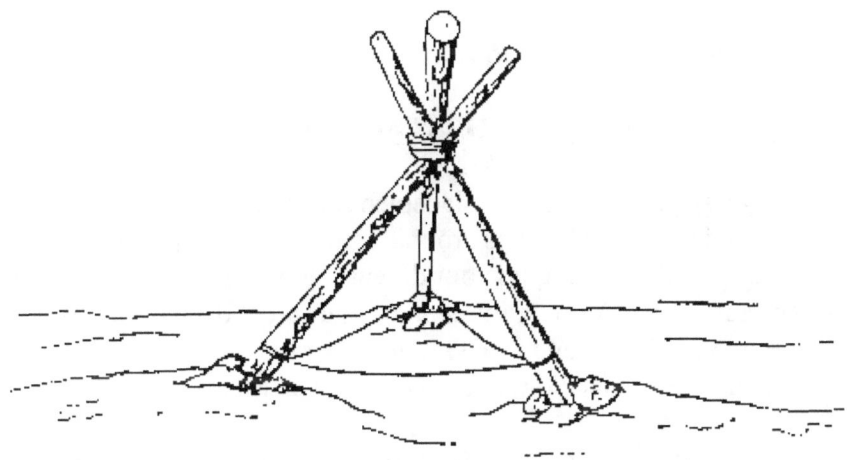

(b) Position the tripod legs so that two of the legs are facing the enemy.

(c) Stack logs up on the enemy side of the tripod legs.

TRIPOD SUPPORTED WALL

(d) Pile snow, rocks, and branches up against the stacked legs to hold them in place and to make the position look like a snow bank. Care must be taken to camouflage the "apex" of the tripod.

NOTE: These positions are designed for two men. Each man fires from the outside edges of the position. An important factor in defensive positions is "continuing actions", always improving the positions.

TRANSITION: Last of all we will look at emplacement of obstacles.

7. (5 Min) **EMPLACEMENT OF OBSTACLES**

   a. Natural Obstacles. The actual placement of natural obstacles is not feasible. You cannot move windfalls or avalanches around your defense to provide obstacles. Therefore, it is important, prior to setting up your defense, to recognize the natural obstacles and use them to your advantage. An overall recon of your area for natural obstacles will dictate where you set up your PDF's and FPF's. Remember that natural obstacles should channelize troop movement and enhance the effectiveness of artificial obstacles.

   b. Artificial Obstacles. There are numerous types and they are limited only to your imagination. Here are some examples of good obstacles:

   (1) Barbed-wire. There are many types of artificial obstacles used under summer conditions which are also suitable for winter use. Barbed-wire makes an effective obstacle in soft, shallow snow. Triple concertina is especially effective since it is easy to install, in addition to it being difficult to cross. However, as the snow gets deeper and more compact, a point is reached where it is possible to cross the barbed-wire on top of the snow.

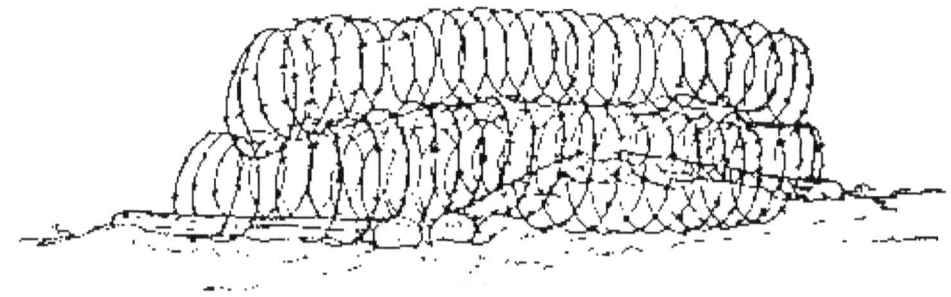

BARBED WIRE

   (2) Tangle foot. This is one of the most common summer type of obstacle. It is constructed by driving poles or logs into the snow, leaving approximately one foot showing above the snow, the "zigzag" barbed-wire form pole to pole. The disadvantage to this is that, if a large snowfall covers it up, it becomes useless.

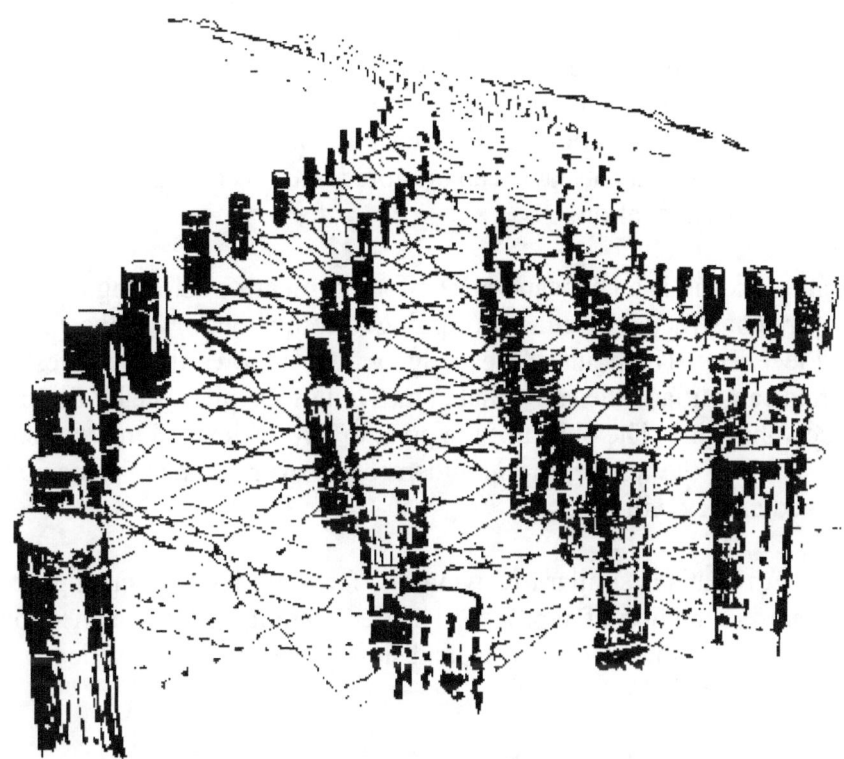

TANGLE FOOT

(3) <u>Lapland fence</u>. The lapland fence uses a floating type of anchor point or one which is not sunk into the ground. Poles are used to form a tripod. The tripod is mounted on a triangular base of wood. Six strands of barbed-wire are strung on the enemy side of the fence, four strands along the friendly side and four strand along the bottom. As the snow becomes deeper, the tripods are raised out of the snow to rest the obstacle on top of the newly fallen snow. The bottom of the tripod and the base wires give enough floatation to prevent the fence form sinking into the snow.

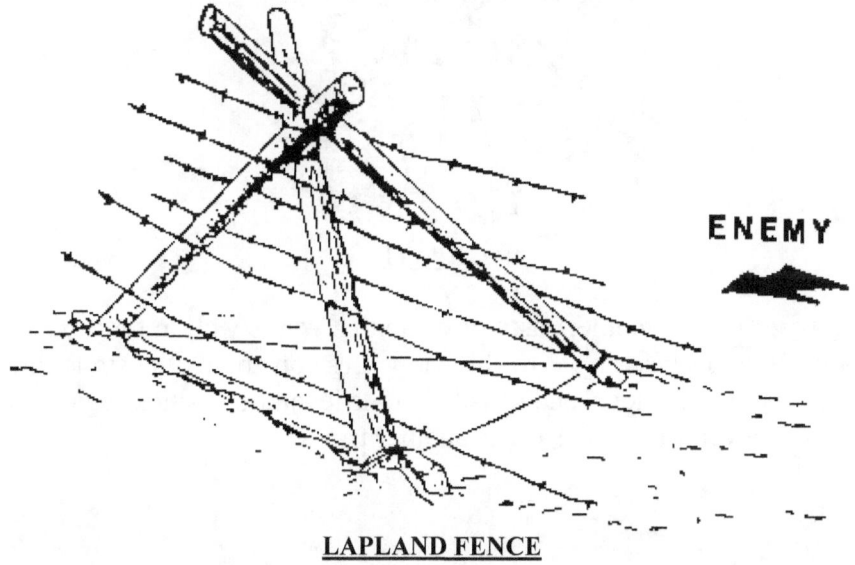

LAPLAND FENCE

(4) <u>Knife rests</u>. Knife rests are portable, barbed-wire fences which can supplement obstacles partially covered by snowfall. They are constructed by tying two wood poles at their center and should be approximately four feet long. The "X's" are then lashed to a ten feet to twelve feet pole. This forms a framework to which barbed-wire is fastened on all four sides. This obstacle can be stored until needed and then easily transported to the next desired location.

(5) <u>Ski pit</u>. A ski pit is a hasty and very effective means of slowing down skiborne troops. They are constructed by cutting (digging) a wedge out of a slope approximately 2 to 2-1/2 feet deep. The point of the wedge will point toward "friendlies". Care must be taken to camouflage the snow that comes out of the pit. Usually the snow is thrown on the downhill side. The object of the pit is to have skiers ski into the pit, catch their tips in the point of the wedge, flipping themselves over. You can place punji-sticks where you predict they will fall.

(6) <u>Abatis</u>. An abatis is similar to a windfall. Trees are felled at an angle of 45 degrees to the enemy's direction of approach. The trees should be left attached to their stumps to retard removal along trails, roads, and slopes. An abatis can inflict casualties to skiers and damage vehicles.

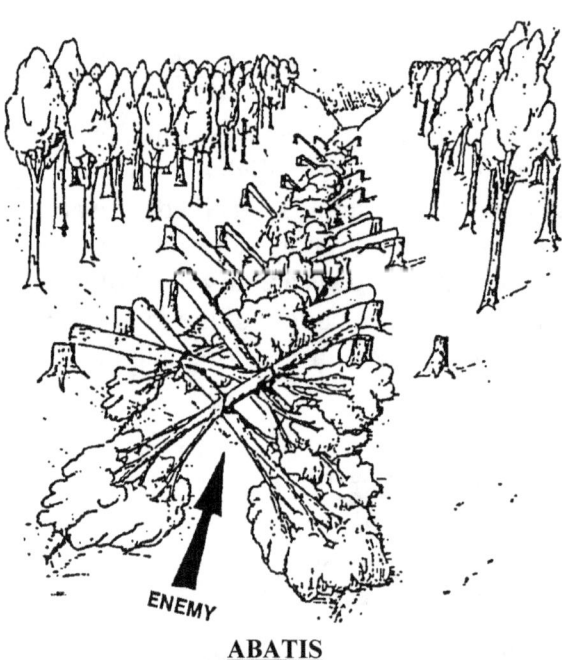

ABATIS

(7) <u>Trip wire</u>. This is another hasty obstacle that works well in forested areas on ski-borne troops. It works best on a downhill slope. It's constructed by stretching wire from tree to tree about throat level and possibly ankle level. Punji-sticks again will help reinforce the ankle level trip wire. Here is another trip wire:

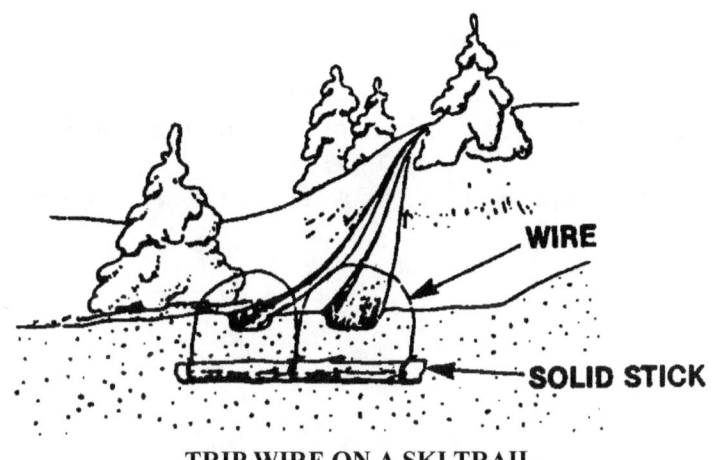

TRIP WIRE ON A SKI TRAIL

<u>TRANSITION</u>: Other devices can be constructed, their vileness only limited by your imagination. Are there any questions?

PRACTICE (CONC)

    a. Students will practice what was taught in upcoming field evolutions.

PROVIDE HELP (CONC)

    a. The instructors will assist the students when necessary.

OPPORTUNITY FOR QUESTIONS (3 Min)

1. **QUESTIONS FROM THE CLASS**

2. **QUESTIONS TO THE CLASS**

    Q. What is the minimum thickness required to stop small arms fire in newly-fallen snow?

    A. 13 feet

    Q. What are some examples of artificial obstacles?

    A. (1) Barbed-wire
        (2) Tangle foot
        (3) Lapland fence
        (4) Knife rests
        (5) Abatis
        (6) Ski pit
        (7) Trip wire

SUMMARY (2 Min)

    a. During this period of instruction we have discussed cold weather defensive positions including, building materials, site selection, and natural and artificial obstacles.

    b. Those of you with IRF's please fill them out at this time and turn them in to the instructor We will now take a short break.

UNITED STATES MARINE CORPS
Mountain Warfare Training Center
Bridgeport, California 93517-5001

                                                                                            WML
                                                                                            WMO
                                                                                           09/06/02

LESSON PLAN

ROUTE PLANNING IN COLD WEATHER OPERATIONS

INTRODUCTION   (3 Min)

1. **GAIN ATTENTION**. Movement in a cold weather environment is in itself difficult; however, when combining the inexperience of Marines on skis or snowshoes, changes in snow condition, tactical situation and mountainous terrain, movement becomes a monumental task. To move a large unit across varied terrain requires more than just words of encouragement or demands by higher authority. It requires positive leadership on the small unit level and a keen sense of technical skill in route selection.

2. **PURPOSE**. The purpose of this period of instruction is to familiarize the student with the requisite knowledge to plan and execute proper route selection. Including planning for additional time required for movement due to additional equipment and terrain considerations.

3. **INTRODUCE LEARNING OBJECTIVES**

    a. TERMINAL LEARNING OBJECTIVE. In a cold weather mountainous environment, plan a route, in accordance with the references.

    b. ENABLING LEARNING OBJECTIVES.

    (1) In a cold weather/mountainous environment, using the route planning considerations, plan a route, in accordance with the references.

    (2) In a cold weather/mountainous environment, execute tactical movements, in accordance with the references.

(3) Given a blank route card, map and a scenario, prepare a route card in accordance with the references.

(4) Given overlay paper, map and a scenario, prepare a route overlay in accordance with the references.

4. **METHOD / MEDIA**. The material in this lesson will be presented by lecture and demonstration. You will practice what you have learned during upcoming field training exercises. Those of you with IRF's please fill them out at the conclusion of this period of instruction.

5. **EVALUATION**. You will be tested later in the course by performance evaluation.

TRANSITION: To be able to plan a route, we must first understand exactly what a route is. Let's also look at the definition of selection.

BODY (30 Min)

1. (2 Min) **DEFINITION**. By Webster's Dictionary, a route is "a course, way, or road for travel or shipping." Similarly, Webster has defined selection as "the act or an instance of choosing in preference to another of others." As you can see then, route selection is nothing more than the logical or systematic approach of determining one course over another. As a small unit leader or planner, we take on an awesome responsibility to ensure that the route we choose is the best. There are unfortunately an infinite number of routes across one piece of terrain. However, there is only one logical or "common sense" route, if all the necessary precautions and considerations are taken.

TRANSITION: Let's look at some route considerations that must be seriously analyzed during the planning of a route.

2. (8 Min) **ROUTE CONSIDERATIONS**. In the summer there are very few variables in route selection. Perhaps we can get by with considering only the terrain and tactical situation. However, to do so in the winter will surely cause disastrous results. There are several factors which must not be neglected!

   a. The Eight Route Considerations are:

   (1) Terrain

   (2) Weather

   (3) Avalanches

   (4) Snow conditions

   (5) Group's ability

(6) Tactical situation/mission

(7) Equipment

(8) Time

b. Each Consideration Contains Its Own Respective Meaning

(1) Terrain

(a) Contour. The lay of the land in the mountains will have a great deal to do with the way in which we negotiate that piece of real estate. To move directly up or down a slope without regard to gradient could be disastrous. Unnecessary cold/heat casualties could result, not to mention the fact that your Marines will not be effective to fight once on or at the march objective. Routes curve, following a contour line, rather than going straight on an azimuth. Contouring also reduces slope angle of the actual movement.

(b) Natural lines of drift. As their name indicates, these are terrain features that tend to draw a unit into them, due to the ease of movement they provide. All things, water, animals, and even humans, tend to take the path of least resistance whenever given the opportunity. But, to allow yourself to be drawn into this situation, could be leading your unit straight into an enemy ambush. A good leader in this environment will set up an ambush near one of these natural lines of drift, because he knows how easily a less experienced leader can allow the terrain to dictate his route.

(c) Altitude. What is the high and low altitude mark for the route? Will this be a time factor or clothing factor due to the lapse rate?

(d) Tree line. Above or below the tree line during the route? Deciduous or evergreen trees?

(e) Vegetation/rock type. Predominant micro-terrain type(s).

(2) Weather. The weather is perhaps the greatest hazard in a mountainous environment for the untrained and unprepared unit. Weather factors are temperature, visibility, precipitation and the wind velocity and direction; all of which will hamper your movement. Bad weather causes even the simplest tasks to become a burden, particularly security awareness. The long nights and short days in extreme northern or southern latitudes will account for lack of visibility.

(3) Avalanches. The avalanche hazard in the area of operations must not be taken lightly. Moving across such a critical slope without proper prior reconnaissance could be adverse to the attainment of a unit's goal. Because avalanche initiation is a feasible means by which our adversary can destroy our forces, movement on or near critical slopes must be carefully planned. Advance initiation by supporting arms is an option.

Engineer assets ill need to be forward to clear the MSR. In the IPB process, an avalanche overlay will be needed for the MCOO in determining go/no go mobility corridors. Collect essential elements of intel in order to make a hazard analysis, like; local avalanche maps, local knowledge, current winter weather patterns, last 2 weeks weather, historical trends and forecast.

(4) <u>Snow conditions</u>. Changing snow conditions can greatly enhance or deteriorate a skier's ability to negotiate a particular route or segment of that route. The constant and sometimes unpredictable changes in the snow conditions make it very difficult for combat laden Marines to utilize a particular route; especially in areas which are windswept or crusted because of temperature changes. The anticipation of merging to and from varying snow conditions may cause Marines to become rigid, thereby busting their ass constantly. What is the depth? Is it dry, wet, crusted or frozen? Is it early, mid, or late season conditions? What is the slope aspect and angle? These answers will determine your over-the-snow mobility selection when balanced with the group's ability.

(5) <u>Group's ability</u>. As we all remember, being aware of the group's ability plays a major factor in cold weather and mountainous operations. In ski or snowshoe training, Marines must be continually challenged with falls and recovery emphasis. However, to move a unit across diversified terrain without regard to degree of slope, snow conditions, temperature, etc. (all of which influence ability) could be detrimental to both morale and mission accomplishment. If the unit has not yet even mastered the wedge turn, then multiple traverses up or down a slope will be necessary using kick turns. Remember as a leader to not base the difficulty of the route on your ability, but to your unit's ability. What is the unit's morale? Are they acclimatized? What is the physical conditioning of the unit? What is they're training status; cold weather experience, navigation, night ops, etc.?

(6) <u>Tactical situation/mission</u>. Although independent of each other, the tactical situation and/or the mission effect route selection in a similar manner. Under normal training circumstances, unit leaders and commanders tend to neglect the tactical scenario while overburdened with the difficulties of moving the unit from one march objective to another. Route selection is very difficult in these circumstances, since moving on the valley floors and out of the cover and concealment of trees, is easier and faster for inexperienced troops in a snow-covered mountainous terrain. As leaders, we must ensure that Marines are not silhouetted on crests and are not caught out in the open valley floors. Our particular concern then becomes route selection for multiple columns, mountain pickets, and cross-compartment movement. Speed vs. security will be balanced with the threat condition in route selection.

USING TERRAIN AND SURROUNDINGS TO CAMOUFLAGE YOUR TRAIL

(7) <u>Equipment</u>. The equipment and the degree of training which Marines have received in the use of that equipment, will directly influence route selection. What equipment is available?…skis, snowshoes, sleds, ropes for skijoring, BV 206s, pack animals, snowmobiles, etc. Is the unit trained on any of it? As a general rule, the more equipment that you use, the easier the route must be. When Marines are pulling sleds, the slope gradient cannot be much more than 30 degrees if traversing that slope. To do so would only allow the sleds to roll over, or slide sideways. To negotiate excessively steep terrain, which should have been avoided, may require belaying sleds to the bottom or by extensively traversing with ropes (it can become impossible). Choosing sleds or no sleds and the over all weight of equipment will determine how aggressive a route one can choose.

(8) <u>Time</u>. When route selecting, we tend to look at a piece of terrain in relation to the amount of time which will be required to negotiate it. Considering all the factors mentioned thus far, picking a route which may appear to temporarily take us away from our march objective and ultimately converge with our original axis of advance, may in the long run gain time. The tendency however, is to sacrifice proper route selection to gain time, when in actuality time is lost due to subsequent problems perhaps manifested as a casevac. Appreciate time requirements. Tempo of ops is relative to the

enemy, not the clock. Time precedence missions are usually unrealistic in the winter. They can quickly lead to a disaster as a unit loses cohesiveness as the straggler column grows.

TRANSITION: We've looked at route considerations, now let's discuss the actual route planning phase.

3. (7 Min) **ROUTE PLANNING TOOLS**. All the considerations already discussed must be considered for proper planning of any size unit. Failure to consider the above could mean disastrous results in unit movements. When planning for unit movement in a temperate zone, axis of advances, boundaries, and other control measures are routinely established without regards for weather conditions, and certainly not track discipline. In snow-covered mountainous terrain, staff planners must consider any and all factors which influence mobility. These considerations must then be analyzed through map/photo studies and ultimately depicted by an overlay or other description. Let us look at the varied planning assets and documents, which are influenced by the route selection considerations.

   a. The four route planning tools are:

   (1) Map/photo reconnaissance

   (2) Orders/Overlays

   (3) Time-Distance Formula/Commander's Log

   (4) Route Cards

   b. Each route planning consideration has its own meaning:

   (1) Map/Photo Reconnaissance. Before any route description or overlays can be established, a thorough map/photo study must be conducted. Realizing that this type of study is normally done during temperate climatic operations, the need to only discuss those items peculiar to a cold weather/mountainous environment exists. The prevailing weather patterns that influence the snow pack can be correlated with terrain aspects in a map study. For instance, knowing that our weather generally comes from the W/SW, we can conclude that most north and east facing slopes will have the greatest deposition of snow, and subsequent formations of cornices. For planning purposes, we must be careful when establishing routes which cross sharp ridgelines oriented with northeasterly facing slopes. Convex and concave slopes can be determined. Another common area of neglect through improper map reconnaissance is cross-compartment movement. Realizing that the tactical situation and the mission at hand may dictate establishing cross-compartment movement to contact, attention should be directed toward skirting avalanche prone slopes. From standard 1:50,000 or 1:25,000 maps, critical slopes can easily be determined by looking for a few key indicators such as type of slope (either convex or concave), gradient of the slope, and vegetation on that slope, i.e., obvious cuts in the tree line. Finally, when establishing routes utilizing a map or

photo, wider than normal boundaries will have to be established, or a greater degree of flexibility, allowing the unit leader more freedom to select his route. The large contour interval in mountains can hide many obstacles. Aerial photos show actual vegetation type and coverage, as well as the hidden obstacles in between contour lines. However, snow can obscure linear terrain features. An actual flyover will give you a good feel for the difficulty of the terrain.

(2) <u>Orders/Overlays</u>. When writing the operational order, particular attention should be directed towards allowing the commander of the subordinate units plenty of room for freedom of action; not only in the freedom of maneuver, but also for route selection. On the overlay, critical slopes and avalanche paths need to be identified, since they may be part of the adversary's barrier plan as well as a natural hazard. Include the TDF (time-distance formula) in the margin of the overlay. There are numerous cold weather specific considerations to be plugged into an op order, this will be covered in Cold Weather Patrolling.

(3) <u>Time-Distance Formula (TDF)</u>. This formula is designed to be a guideline and should not be considered as the exact amount of time required for your movement. The TDF is made for acclimated troops on foot in the summertime and/or on skis with skins or snowshoes in the wintertime. The TDF will vary based on unit size, physical conditioning, experience, load carried, angle of slope, snow conditions, surface conditions, etc. A set of TDFs are used for planning as follows:

(a) Mountain Leaders/Patrols with combat load and high level of mountain/winter experience - 3kph + 1 hour for every 300 meters ascent + 1 hour for every 800 meters descent.

(b) Assault Climbers/Scout Skiers/Patrols with combat load – 2kph + 1 hour for every 300 meters ascent + 1 hour for every 800 meters descent.

(c) Company/Bn movements with combat load – 1kph + 1 hour for every 300 meters ascent + 1 hour for every 800 meters descent.

Each route should show a completed TDF, however, commanders must be patient since the actual execution across that route may include hidden difficulties. This is especially true of routes across streams, roads, or across compartments. Logging and comparing estimated and actual times will enhance your ability to estimate the time it takes your unit to move in mountains and/or snow. The TDF totals are marked on the overlay, as well as the route card.

Commander's Log. The Commander's Log is the most accurate method for estimating time for a unit's movement. However, it requires experience in the operational area with current personnel (personnel turbulence, cohesion fills negate old information base). A Commander's Log is made via patrol/route debriefs and transferring route card information in order to see at a glance which units are moving at what rates in a particular type of terrain, weather, and load conditions.

INSTRUCTOR NOTE: A blank copy and a completed example copy of Commander's Log are attached at the back of the outline. The blank is to be used as a master copy for reproduction back at their units.

    (4) <u>Route Cards</u>. A route card is a written description of the route. The description needs to be good enough to enable you to follow the route without a map. Filling out a route card forces a Marine to do a detailed map study, which will aid his situational awareness throughout the route. Two copies are made. One for higher, which can be an aid in rescue if one becomes lost. The other should be carried separately from the man with the map.

ROUTE CARD (front)

UNIT I.D.	UNIT COMMANDER	NUMBER OF PERSONNEL	DATE AND TIME	MAP REFERENCE	
LEG	MAG AZ	DIST Meters	GRID 6 dig	EST TIME	ACT TIME	ELEVATION GAIN/LOSS	DESCRIPTION OF GROUND

TOTAL DISTANCE - _____ = _____ TIME
TOTAL ELEVATION GAIN - _____ = _____ TIME
TOTAL ELEVATION LOSS - _____ = _____ TIME
    = TOTAL ESTIMATED TIME - _____

LOG ACTUAL ROUTE TIME TOTAL - _____

NOTE: As many lines as needed for route legs are used, with totals at the bottom.

ROUTE CARD (back)

ESCAPE ROUTE(S): _____

WEATHER FORECAST:
    ALTITUDE, HI PT _____ LO PT _____

TEMPERATURE, HI _____ LO _____

WIND, DIRECTION _____ SPEED _____

FREEZING LEVEL _____

SNOW LEVEL _____

MOON PHASE _____

CLOUD COVER _____

REMARKS:

INSTRUCTOR NOTE: A blank copy and a completed example copy of the Route Card are attached at the back of the outline. The blank is to be used as a master copy for reproduction back at their units.

TRANSITION: After the macro-terrain route selection is planned, the micro-terrain route selection must be considered.

4. (5 Min) **GENERAL CONSIDERATIONS for MICROTERRAIN SELECTION**. All of the planning in the world cannot make up for poor performance. It is the actual Marine on the ground, who selects the route to either success or annihilation via micro-terrain choices. Here are some general considerations when selecting micro-terrain:

   a. Terrain Negotiation. Once we have determined where we are going, the physical route must be selected by the lead element.

      (1) By traversing slopes early on, we eliminate the necessity need for sudden gains or losses in elevation, which will ultimately add several minutes or hours to a previously short move.

      (2) It may be more efficient to contour around an object i.e., slope, draw or finger, than to travel in a straight line. This will prevent unnecessary loss or gain in elevation.

      (3) When descending steep slopes, utilize the "plunge-step" by removing skis or snowshoes. If pulling a sled, place the sled in front of you while descending.

   b. Kick Turns vs. Wide Turns. When descending steep terrain, the tendency is to traverse, then kick turn, then traverse, and kick turn through the completion of the slope. This technique is fine, provided we are not pulling sleds. Since performing kick turns while pulling sleds is very difficult, we should try to make wide, sweeping turns on benches, or uphill from large trees or pockets of vegetation. This prevents gaining too much speed, heading uncontrollably downhill. Finally, when conducting long traverses, try to remain just inside the tree line whenever possible. By doing so, you remain in generally softer snow and will not have the problems of negotiating crusted, windswept snow.

    c. <u>Narrow Depressions</u>. These include such things as creek beds, fallen logs, or similar obstacles. The preferable technique is to cross the obstacle in a snow bridge. If a snow bridge is not available, a few options exist. One option is to side step into the depression, and then side step out. Another is to build a snow bridge by shoveling snow into the appropriate areas. This is recommended if a snow bridge does not exist, and a large force must cross the obstacle.

    d. <u>Frozen Waterways/Lakes</u>. Any frozen body of water makes an ideal avenue of approach, as well as a large obstacle. Another concern is the fact that the tracks are very difficult to conceal unless there are several overhanging trees, or the areas crossed are shaded the majority of the time. Finally, the banks around these areas are frequently very steep and may be difficult to ascend.

<u>TRANSITION</u>: Remember, a good plan brilliantly executed is better than an excellent plan poorly executed. Are there any questions?

PRACTICE (CONC)

    a. Students will practice what was taught in upcoming field evolutions.

PROVIDE HELP (CONC)

    a. The instructors will assist the students when necessary.

OPPORTUNITY FOR QUESTIONS (3 Min)

1. <u>QUESTIONS FROM THE CLASS</u>

2. <u>QUESTIONS TO THE CLASS</u>

    Q. What are the eight route selection considerations?

    A. (1) Terrain
       (2) Weather
       (3) Avalanches
       (4) Snow conditions
       (5) Group's ability
       (6) Tactical situation/mission
       (7) Equipment
       (8) Time

    Q. What are the route planning tools?

A. (1) Map/photo reconnaissance
   (2) Overlays/orders
   (3) Time-distance formula
   (4) Route Cards

SUMMARY (2 Min)

a. This period of instruction has discussed the selection and planning of routes in a cold weather environment.

b. Those of you with IRF's please fill them out at this time and turn them in to the instructor We will now take a short break.

COMMANDER'S LOG

FOR TIME-DISTANCE ESTIMATION IN SNOW-COVERED TERRAIN
(attach route overlay and/or route card, if possible)

Fill out a copy for every route done by the whole unit or any subordinate unit. The more sheets accumulated, the more accurate the time-distance estimation will become.

UNIT:

DISTANCE (map):

ELEVATION GAIN, TOTAL:

ELEVATION LOSS, TOTAL:

WEATHER CONDITIONS:
(winds, precipitation, humidity, day/night)

TEMPERATURE (high and low):

ALTITUDE (high and low):

SNOW CONDITIONS:
(depth, hardness/flotation, dry/wet, etc)

OVER-THE-SNOW MOBILITY EQUIPMENT:
(skis, snowshoes, sleds, combat boots, VB boots, ski/march boots, skins, wax, etc)

REMARKS:

ROUTE CARD

UNIT ID		UNIT CMDR		# OF PERS		DTG		MAP REF
CP/ LEG	MAG AZIMUTH	DIST. meters	GRID 6 digit	ELEV GAIN	ELEV LOSS	EST TIME	ACT TIME	DESCRIPTION

TOTALS: ELEV.   ELEV.   EST.   ACT
DISTANCE_____GAIN_____LOSS_____TIME_____TIME_____

UNITED STATES MARINE CORPS
Mountain Warfare Training Center
Bridgeport, California 93517-5001

WML
WMO
09/17/02

LESSON PLAN

COLD WEATHER PATROLLING

INTRODUCTION (5 Min)

1. **GAIN ATTENTION**. Due to the nature of the terrain in a snow covered mountainous environment, and the increased need for accurate and current intelligence, patrolling has an increased importance in cold weather operations.

2. **PURPOSE**. The purpose of this period of instruction is to familiarize the student with those aspects of patrolling peculiar to a cold weather mountainous environment, including possible missions, composition, planning considerations, preparation, and patrol bases. This lesson relates to offensive operations, defensive operations, and route planning in cold weather operations.

3. **INTRODUCE LEARNING OBJECTIVES**

    a. TERMINAL LEARNING OBJECTIVE. In a cold weather/mountainous environment, conduct a patrol, in accordance with the references.

    b. ENABLING LEARNING OBJECTIVES.

    (1) In a cold weather mountainous environment, use cold weather patrolling techniques, in accordance with the references.

    (2) Without the aid of references, conduct patrolling operations in a cold weather mountainous environment, in accordance with the references.

    (3) Without the aid of references, operate from a patrol base, in accordance with the references.

    (4) Without the aid of references, and given a rifle, skis/snowshoes in a snow covered terrain, demonstrate the firing positions, in accordance with the references.

(5) Given a winter warning order matrix blank and a scenario, write a warning order in accordance with the references.

(6) Given a scenario, write a patrol order in accordance with the references.

(7) Without the aid of references, list in writing the things a patrol leader can do to prolong the time a man can stay warm and alert in an ambush position, in accordance with the references.

4. **METHOD / MEDIA**. The material in this lesson will be presented by lecture and demonstration. You will practice what you have learned in upcoming field training exercises. Those of you with IRF's please fill them out at the conclusion of this period of instruction.

5. **EVALUATION**. You will be evaluated later in the course by performance exercise.

TRANSITION: Let's first start by discussing where you begin with a patrol, the estimate of the situation.

BODY (70 Min)

1. (10 Min) **THE ESTIMATE OF THE SITUATION**. As in a temperate climate, a patrol leader must make an estimate of the situation. Although the basics are similar, some additional considerations must be addressed in a cold weather mountainous environment.

    a. Mission. Along with recon and combat patrols, there are various additional applications for patrolling that are peculiar to a cold weather environment. Some possible examples are:

        (1) Recon of the battle area and particular targets, while establishing a presence forward of the FEBA/FLOT.

        (2) Harassing the enemy lines, depriving them of shelter and rest.

        (3) Deep penetration by ski and/or helicopter to destroy logistic supply lines.

        (4) Installation of observation posts and radio relay/retrans sites.

        (5) Picketing of high ground on the unit's flanks during the advance.

        (6) TRAP mission; recovery of downed aircraft/troops and equipment due to the extreme weather in the mountains.

    b. Enemy. A normal SALUTE report is still the standard format used with some additional considerations peculiar to this environment. A good method of determining these is to evaluate your own unit. The human nature aspect, is more likely than not, going to influence how the enemy is dealing with the elements. This plays a crucial part in predicting how the enemy will select a route, when you are aware of how natural lines of drift tend to influence a unit.

   (1) The type of mobility the enemy is utilizing and their ability level.

   (2) The unit's discipline.

   (3) The unit's morale.

   (4) The enemy's last known action?

   (5) The units re-supply capabilities.

c. <u>Terrain and Weather</u>. It is very important for the leader to as well as gather as much information as possible about incoming weather patterns, do a thorough study of the terrain he will be moving into.

   (1) Terrain. At high elevations, cold weather and even snow can be encountered during all times of the year. Mobility will become extremely difficult due to these factors, and very taxing to the endurance of the unit. The wide variety of terrain encountered will affect speed of movement, concealment, and security.

      (a) Above the tree line, the exposed terrain makes movement and security more difficult. Use micro terrain and shadows as much as possible to make observation from the enemy more difficult. Movement during the hours of darkness should be done whenever possible.

      (b) Below the tree line, movement and bivouac areas can be concealed by vegetation.

      (c) Danger areas tend to be larger in the mountains, and due to the decreased mobility of a unit, must be given special consideration.

   (2) Weather. The uncertainty of weather in the mountains requires leaders to be flexible in their planning. Severe weather can move in suddenly and last for several days. Always plan for the worst case scenario.

      (a) Cold temperatures and high winds. These affect your Marines, their weapons, and equipment. Individual response time is slowed, and your time schedule must reflect this.

      (b) Snow cover. Affects the rate and mode of movement. Terrain can be affected if avalanche conditions exist. Snow depth and consistency may change considerably during the patrol and may greatly affect movement.

      (c) Visibility will become reduced during storms.

(d) In the high latitudes, you will experience about four hours of daylight from November to February. Mid-December has only a few hours of twilight per day.

d. <u>Troops and Fire Support</u>. A cold weather mountainous environment makes using some troop and fire support assets very limited under these conditions. The leader must take into account how much support to expect from available assets.

   (1) Troops. The size of the patrol will depend on the mission, and members should be employed in arctic buddy teams.

   (a) Cold weather patrols are generally larger due to the gear requirements, i.e. clothing, casevac gear, and substantial firepower assets must be spread loaded amongst the members of the patrol. Keep the gear list limited to only which is essential to the mission. Anything in addition to this will hamper your mobility.

   (b) Personnel must possess the physical capabilities required to accomplish the mission under extremely arduous conditions. If in doubt of an individual, leave him in the patrol base.

   (2) Fire Support. These assets are going to be limited by the same conditions that your patrol is experiencing. Their reaction time may be slower, due to the cold and may have difficulties operating their equipment. Ammunition may be less effective against certain targets.

   (a) Air support may be grounded due to weather.

   (b) Artillery may not be able to displace and may have a slower response time. Also terrain may mask portions of your area of operations.

   (c) Mortars, although an effective weapon in this terrain, are limited by their firepower and range.

   (d) Rockets. Method of target engagement may be limited in extreme cold due to ice fog.

   (e) Small arms may be limited in extreme cold because of reduced rate and range.

   (f) Grenades/Smoke must be pre-rigged with a floating device prior to leaving the patrol base.

e. <u>Time Space and Logistics</u>. This is where a leader will notice that this type of environment can hamper his planning ability the most.

   (1) Time will be affected more than anything else.

   (a) March rate will be affected by terrain, weather, visibility, type of mobility, danger areas, and avalanche conditions.

- (b) Any simple task should be expected to take longer due to slower response times and difficulties handling gear in the cold.

- (c) Security is a consideration that will be affected to a large degree if your movement requires you to establish mountain pickets, or if you have to use overwatch techniques. Speed, as a form of security, can be effective in some situations.

(2) Space. In the mountains, distances on the map seem far more obtainable than what they actually are due to the elevation lost and gained. A useful tool is the time/distance formula taught in MOUNTAIN SAFETY. Careful, detailed planning must be done using natural contours along a given route to avoid falling victim to moving along an azimuth.

(3) Logistics. Unusual weather and terrain conditions make problems of supply, casualty evacuations, transportation, and services more difficult and more time-consuming. More time must be allowed for moving supplies and troops because of the environment. The capacity of any support element to provide adequate logistic support may be the determining factor when evaluating the feasibility of an operation. Leaders must always be prepared to alter the plan.

TRANSITION: Now that we've covered the estimate of the situation, let's discuss how to organize the patrol.

2. (2 Min) **PATROL COMPOSITION**. The exact size and composition of the patrol will depend on the mission; however, the "arctic buddy system" should be used. The buddy system should be used for security, chow, bivouac routine, etc. Generally, arctic patrols are larger to carry more gear, provide additional firepower, and assist in trail breaking and casevacs. Using sleds on patrols should be avoided if possible.

TRANSITION: Now that we are tasked organized, let's talk about special planning for the cold weather.

3. (5 Min) **PATROL PLANNING**

   a. Patrol leaders considerations. Planning must be precise, extremely detailed and continually improved upon. In a cold weather mountainous environment, the planning and preparations for a patrol are similar to those in temperate climates with some additional considerations. A Winter Warning Order matrix blank and a Patrol Order bullet format sheet with winter considerations are included at the back of this outline.

   (1) Wax designation and skins, if skis are used.

   (2) Shelter and equipment to be carried (must be dependent on mission of patrol).

   (3) Trail breaking party designated and briefed on operation, if used.

(4) Immediate action drill considerations for the type of mobility.

(5) Establishment of ORP's and patrol bases.

(6) Skijoring operations, if available, are used.

(7) Communication considerations.

(8) Heliborne considerations, i.e. load considerations, ahkio huddles, if available are used.

(9) Route considerations. A poorly selected route will jeopardize the mission of the patrol. Therefore, it is important that route selection be a priority in the planning of the patrol. Refer to Route Planning Considerations in Cold Weather.

(10) Accountability of personnel. Due to skill level of a unit operating on skis/snowshoes, the possibility of personnel being separated could occur. It is essential then to increase head counts, and always employ the buddy system, selecting personnel of the same skill level if possible.

TRANSITION: After our planning, we must prepare for the patrol.

4. (5 Min) **PATROL PREPARATION**. Preparation of equipment and personnel must be thorough. The hostile environment places more dependence on everyone's equipment and fellow Marines. Orders should be complete and concise using aerial photos, terrain model, and route cards. Every phase of the patrol must be rehearsed.

   a. Camouflage

      (1) Unit designation prescribed for use with over-whites.

      (2) Ensure they are clean.

      (3) White tape should be firmly applied to all 782 gear and weapons.

   b. Immediate Action Drills. It is not the purpose of this outline to designate specific SOP's for IA drills. However, it is essential to rehearse immediate action drills for each mode of mobility until the reactions become second nature.

NOTE: What makes the difference in this environment is conducting drills on skis/snowshoes on varying terrain with appropriate firing positions and crawls until the unit leader gets the results he desires.

      (1) Break Contact. The conduct of the drill remains the same, however, the firing positions must be taken into consideration. The initial reaction of the personnel will be to drop down and return fire in order to suppress engaging unit. But, due to snow consistency,

Marines may drop into the prone position and be submerged into the snow, unable to return fire. Also getting back up out of the prone position can be tiring and slow causing the marines to be in the kill zone longer. Marines should lean or kneel behind cover, and could use their rucksacks as a firing platform if out in the open. Depending on how weapon systems are being carried, reaction time will become a critical factor to the success of the break in contact.

    (a) Once the Marines skis/snowshoes are in the opposite direction of the enemy in the withdrawal, all firing will be done with the skis/snowshoes in that direction and adopting a suitable firing position from this stance. Such as stemming one ski out to the side and firing to the rear. This will allow effective speed or movement and fire at the same time.

(2) <u>Hasty Ambushes</u>. The ultimate goal of the drill is the same as in a temperate climate. However, much more precaution must be taken in the emplacement and positioning of the Marines into the ambush site due to the mobility being used when they apply to the situation of the ambush. Always be aware of your own tracks, these may just tip off the enemy of your intention.

    (a) If the ambush situation is that, an enemy force is tracking your patrol during movement, in order to establish an effective ambush it may be necessary to "button hook" into the ambush site. This will set up an effective means of deceiving a tracking enemy into the kill zone and at the same time, maintain the units track discipline. The button hook should encompass terrain that favors the ambushing unit in security and observation of the oncoming enemy.

    (b) If the ambush situation is that an enemy is to the flanks or front, it may be advisable to go foot borne into the ambush site to ensure noise discipline and avoidance of obstacles on skis/snowshoes. If the goal of the ambush is harassment, this would probably not be a good course of action. Rapid withdrawal from the ambush site would be difficult.

    (c) The ambush site should facilitate an efficient avenue of withdrawal should a superior enemy force be encountered. This route should allow for rapid downhill movement in order to put as much distance between the ambush force and the enemy.

(3) <u>Counter-Ambush</u>. The drills for near and far ambush remain the same. The following considerations should be noted:

    (a) Firing positions, along with placing weapon systems into action, should be rehearsed prior to patrolling operations.

    (b) Also, it is important to note the dampening effect of snow on explosives and smoke devices, so proper preparation of this ordinance prior to the conduct of the patrol should take place.

(4) <u>Encountering an Enemy Force while Skijoring</u>.

    (a) All drivers should be aware of possible ambush sites, and accelerate through these sites. If fire is received, skijoring Marines will present as low as a silhouette as possible.

    (b) If the vehicles are attacked and rendered inoperable, then apply counter-ambush drills to clear the enemy force.

    (c) Crew-served weapons can be mounted in the VC hatch for fire support.

TRANSITION: Let's discuss how we will move along that route tactically.

5. (5 Min) **PATROL BASES**. Deceptive actions are mandatory in a snow-covered environment. Make use of jump-off points, deceptive tracks, dummy positions and hasty ambushes.

    a. <u>Jump-off point</u>. Is made on slopes or in dense woods where it is possible to hide the real track and where enemy pursuit will have such a high speed of travel that it is difficult for him to discover where the real track is, as well as any booby traps. The deceptive track should be made as far away as time permits.

    b. <u>Button Hook</u>. Is the preferred method of setting in a track plan to a patrol base. This technique allows for the track to be observed and covered by fire during the occupation of the position. (See illustration)

    c. <u>Deceptive tracks</u>. Used to mislead the enemy from your jump-off position. They should be located in an area where you can observe them and cover them with small arms fire from the patrol base.

    d. <u>Dummy position.</u> Make a position located at the end of the deceptive track as a decoy, to draw the enemy to this position, realism should be emphasized as much as possible. Erect tentage, antennas, etc. It should be continuously observed and located within small arms fire from your position.

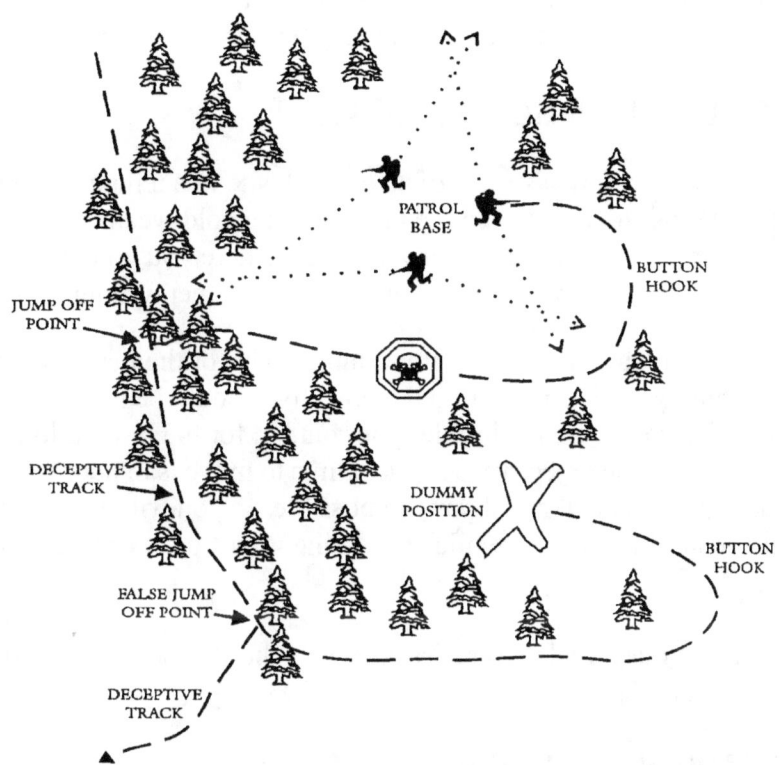

PATROL BASE PLAN

e. During occupation of the patrol base minimal activity should occur. Therefore, it may not be advisable to dig comm trenches and defensive positions. All food preparation, water procurement, weapon maintenance, etc. should occur in a covered area.

f. Due to the possibility of patrol base becoming detected and engaged by a superior enemy force, a contingency plan for escape must be planned, practiced, and implemented. Impeccable bivouac routine and security are the best means of preventing this possibility from becoming a reality. Here are some of the guidelines to follow:

(1) All equipment when not in use will be in the Marines pack, i.e., sleeping bag, stove, fuel, extra clothing, etc.

(2) Due to the importance of equipment to Marines in a cold weather environment, patrols operating out of a patrol base should take precautions prior to leaving, in case the patrol base becomes compromised and the patrol has to be self sufficient. Considerations such as:

   (a) Increasing the number of combat loads being carried by the patrol. A guideline should be for each buddy team to have one combat load.

   (b) All gear and equipment not going on the patrol should be packed prior to going out on patrol and cached outside of the patrol base.

TRANSITION: Due to the harsh environment that cold weather may present, it is important to discuss some considerations to have when conducting ambush operations.

6. (3 Min) **AMBUSH SITE RECONNAISSANCE**

    a. When conducting a reconnaissance of an ambush site in a temperate environment it is possible to get close to, or even on the position. In a cold weather environment this is almost impossible to do. As you are aware, everywhere you go in this environment, your tracks follow. There are obvious consequences of the enemy finding these tracks.

    b. The route to the ambush site is very important, again for the main reason of the tracks you will leave behind you. The route into the objective should be such that the chance of the tracks being discovered are negligible. The route in for the last 1 kilometer or so will be at a 90 degree angle to the objective. The ideal time to move is in poor weather during a heavy snowfall at night. This will not always be possible, so you will have to find terrain that will allow you to enter the position while concealing your tracks from all but a chance encounter.

TRANSITION: During your initial reconnaissance, you should choose a tentative position for the ORP, avenues, E&E routes, etc.

7. (5 Min) **THE AMBUSH POSITION**

    a. The site for the ORP needs to be well chosen; it needs to be able to afford concealment from enemy observation and cover from fire. It should also be reasonably close to the Ambush Line (supposing the ambush was on a linear feature). In the ORP, a warming tent may need to be set up. The ideal shelter is the 4 man tent vestibule. In an area ambush, there may be more than one ORP, so more than one warming tent will be required.

    b. A decision must be made on whether the ambush party should take ski's into the ambush position or leave them at the ORP. The decision making will involve answering some of the following questions:

    (1) Is it a near of far ambush?

    (2) Does the ambush party have to move through heavy vegetation to get to the position?

    (3) Is the route to the objective downhill or uphill?

    (4) Is the route over easy terrain which affords good concealment into the position?

NOTE: The above questions should already have been answered by your reconnaissance element.

    c. If you do not use skis, how do you get to the position?

    (1) Posthole with or without carrying skis.

(a) If you do take skis to the ambush site, they may have to be bundled to prevent any excess noise being made, or they may even be towed to the position.

(2) Use snowshoes with or without poles.

(a) If you do not carry ski's and use snowshoes, ski's may have to be left at a location easily identifiable, so that the returning ambushers can find their own ski's quickly.

TRANSITION: Once we are in the ambush position, we must consider our actions.

8. (5 Min) **ACTIONS IN THE AMBUSH POSITION**. The method of occupation into the ambush position doesn't change for the most part. When moving into a position in snow covered terrain, particular care must be taken to avoid tracking up the area. Choose an approach which provides as much concealment as possible from enemy observation. The firing positions inside the ambush site should be out of sight from the enemy, appearing natural and undisturbed. Additionally, they must provide adequate cover from enemy fire.

   a. The success of the ambush forces concealment is governed by good use of the terrain; however, varying types of snow conditions can hamper or help your concealment. The following are factors which should be considered:

   (1) Powder snow will allow you to easily create a fighting trench position in about 10-15 minutes, which when smoothed out will give you concealment from enemy observation and some protection against small arms fire.

   (2) Hard compact snow will make it difficult to dig a fighting position and will take considerably longer, 15-30 minutes. Hard compact snow provides excellent protection against small arms fire.

   (3) Hard frozen ice will be almost impossible to make a fighting position out of.

   b. In most ambushes, the attacker will load the firepower in his favor by laying claymores, and booby traps near the boundaries of the kill zone. The employment of such devices is for the most part impossible to do in an ambush site in a snow-covered environment, because of the tracks that are left by the installers. If using, check all explosives inside a warming tent or in the patrol base, if possible. Waterproof any electric or time fuse connections with plastic bags and duct tape. Place dry rags inside the plastic to absorb moisture. Waterproof fuse igniters.

   c. There are a number of things a patrol leader can do to prolong the time a man can stay warm and alert in an ambush position. These are:

   (1) On reaching ORP, all men put on all their warming layers.

   (2) Select an ambush site out of the wind. Situation permitting, you may dig into fighting positions, adding a one man sleeping bench along the back.

(3) Ensure all men have a dry isopor (insulating) mat to lie on. The thicker the insulation pad(s), the better.

(4) Ensure each man has a full, hot thermos. Don't use heat sources in the ambush site to make hot-wets. Buddy teams will only use 1 thermos at a time to maximize the heat maintained in the drink.

(5) Always employ Marines in buddy teams on 50% alert.

(6) Toes and hands need to be kept moving at all times to maintain circulation, because the extremities will be the first part of the body to get cold.

(7) Rotate the Marine back to the warming tent, if he is getting near the stage where he is spending more time trying to stay warm than alert (tactical situation permitting). He may be experiencing the effects of hypothermia/frostbite, unfortunately, if one Marine is in that state, others will soon follow.

(8) One sleeping bag per buddy team can be used for hot-bagging to warm-up for the man *not* on security.

(9) Set a no-show time for extract. A Marine can only lie so long on an isopor mat in the cold before he starts to become hypothermic.

TRANSITION: Another consideration for our ambush is the choice of weapons.

9. (5 Min) **WEAPONS**. When the signal to fire is given, everyone's weapon should operate at that precise moment in time. If it does not, the weight of fire may not be enough to destroy the enemy.

   a. A company in Norway was tasked with an ambush 6 clicks from their position. They moved out at last light and got into position at about 2330. On the way there, they had to move through a valley in which the temperature was much colder than the area they had left. About three hours into the ambush, the enemy was sighted, the signal to fire (a MG next to the company commander commencing fire) didn't happen. The weapon would not fire, the second MG also malfunctioned so the company commander attempted to give the signal with his personal weapon, which also failed to operate. The company commander then shouted FIRE. His company, who saw the enemy and heard the order, tried to engage, but only 4 of the 120 weapons functioned. On the subsequent investigation, it was found that the weather conditions had caused the problems. On the ski march to the ambush site it had been snowing lightly, the temperature just below freezing. The temperature in the valley that they had moved through was around -12°F. It was surmised that the snow on the weapons melted slightly and got into the working parts, the drop into the valley caused the melted snow to freeze, jamming the weapons.

   (1) With the new and improved lubrication that is available, the above problem may be a thing of the past. But keep it in mind just the same.

b. It is worth remembering that the propellant used in all weapon systems is affected by the cold, and the burn time of that propellant is noticeably slower.

c. The use of light to illuminate the enemy works very well in snow covered terrain as the light is reflected off of the snow. However, care must be taken where the light is placed in the sky, as the additional magnification afforded by the snow can highlight your position just as easily.

TRANSITION: As soon as the assault is completed and the enemy is either dead or dying, we should move onto reorganization before the enemy can send more victims to your meat grinder.

10. (2 Min) **SEARCH TEAMS**. Once the enemy has been taken on and killed, there are two courses of action:

   a. The first is to bug out and leave the area.

   b. The second is to search the enemy and pick-up any information that could be useful.

TRANSITION: We must keep in mind that the theory behind warfighting is not the seizure of hills or towns. It is the destruction of the enemy.

11. (3 Min) **WITHDRAWAL**. If the decision is to stay and search, consider the following:

   a. Once the search teams are back, the main body can begin to move back to the ORP.

   b. If their skis and packs are there, then they would put them on. If they carried their skis to the ambush position then they have various choices of where they are put on.

   c. The ORP party should take down the warming shelter, and have sleds ready to retrieve any wounded personnel.

TRANSITION: Ambushes can be the most cost effective or costly means of engaging the enemy, it depends on how well you plan and execute it. Are there any questions on what we've just covered?

12. (5 Min) **FIRING POSITIONS**. The three firing positions used in snow-covered terrain are: Standing, Kneeling, and Prone.

   a. Standing Firing Position

      (1) Direction of fire. Point either the left or right ski/snowshoe in the direction you want to fire. Right hand shooters will have left ski/snowshoe towards the direction of enemy; left handed shooters will have right ski/snowshoe towards the direction of the enemy. Plant the opposite ski/snowshoe outward, edging the inside of the ski/snowshoe to form a half herringbone.

(2) Ski poles. If using ski poles, remove the straps from wrist. Next, plant one ski pole on either side of the ski/snowshoe, pointing toward the target approximately 12 inches forward of the ski/snowshoe bindings. Cross the top of the ski poles forming an X. The straps of the ski poles may be interlocked to provide a more stable platform.

(3) Weapon. Unsling your weapon and assume a firing position, bending the rear knee forward towards the ski poles. Place the weapon in the cradle of the X formed by the ski poles. Place the non-firing hand on the X of the ski poles, crouch forward, aim and fire.

STANDING POSITION

b. Kneeling Firing Position

(1) Direction of Fire. Point either right or left ski/snowshoe in the direction you want to fire. Plant the opposite ski/snowshoe outward, edging the inside part forming a half-herringbone. Bend the rear forward and rotate the ankle until the knees touches the snow.

(2) Ski poles. If using ski poles, remove the straps from your wrist, and plant your ski poles as for the standing position. Form an X with the ski poles approximately eye level with your non-firing hand.

(3) Weapon. Unsling your weapon, and assume the firing position placing the weapon in the cradle of the X formed by the ski poles. Grasp under the weapon with the non-firing hand, just behind the X formed by the ski poles, lean forward, aim and fire.

KNEELING POSITION

c. <u>Prone Firing Position</u>

   (1) <u>Ski poles used as elbow rest</u>

   (a) <u>With ski/snowshoes on assume a prone position</u>. Do this by forming a herringbone, and then bending both knees forward until they touch the snow. Once the knees make contact with snow, grasp both poles in the center with non-firing hand, fall forward, and break the fall with the ski poles.

   (b) Place the ski poles perpendicular to the direction of fire approximately 1-1/2 feet to your front.

   (c) Bring your weapon to the prone firing position with your elbow resting on the ski poles.

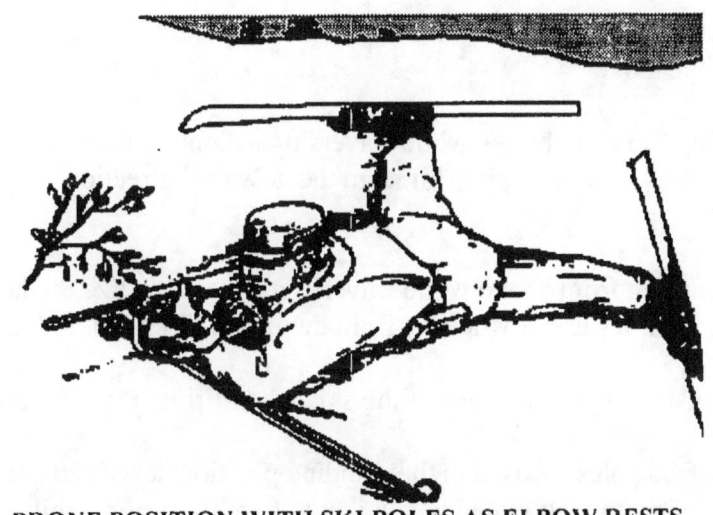

PRONE POSITION WITH SKI POLES AS ELBOW RESTS

(2) <u>Skis/snowshoes used as weapon supports</u>

　　(a) Assume a prone firing position with the skis/snowshoes off.

　　(b) Place the skis/snowshoes perpendicular to the direction of fire approximately 3 feet to your front. Place your ski poles at your side.

　　(c) Bring your weapon to the prone firing position with the forearm resting on the skis as in any prone supported firing position.

PRONE POSITION WITH SNOWSHOE SUPPORT

<u>TRANSITION</u>: Now that we have learned the different types of positions, let's discuss the carriage of the weapon while skiborne.

13. (5 Min) **CONSIDERATIONS FOR FIRING**

a. When moving downhill it's easier to turn into your strong side and return fire, vice trying to shoot on the move.

b. If receiving fire from below while traversing a slope, utilize the sitting position when returning fire. It's easier when hit from the downhill direction to sit uphill and face and fire downhill.

c. If receiving fire from above while traversing a slope, utilize the kneeling position when returning fire. It's easier when hit from the uphill direction, to kneel uphill and fire.

d. Always ensure to set the edges of the skis when firing; this will prevent slippage.

e. The use of ski poles, as used in the standing position described earlier, is not practical for a quick-fire technique. It would be quicker to come to a stop, while simultaneously bring the rifle to bear, squatting down using the thighs for support. This method is much quicker and provides a much smaller target for the enemy.

PRACTICE (CONC)

a. Students will practice what was taught in upcoming field evolutions.

PROVIDE HELP (CONC)

a. The instructors will assist the students when necessary.

OPPORTUNITY FOR QUESTIONS (3 Min)

1. QUESTIONS FROM THE CLASS

2. QUESTIONS TO THE CLASS

    Q. What are the four additional measures taken when establishing a patrol base?

    A. (1) Jump-off point
       (2) Button hook
       (3) Deceptive tracks
       (4) Dummy position

    Q. Where should a jump-off point be made?

    A. On slopes or in dense woods where it is possible to hide the real track and where enemy pursuit will have a high speed of travel that it is difficult for him to discover where the real track is.

SUMMARY (2 Min)

a. During this period of instruction we have covered cold weather patrolling, its missions, planning, organization, how to conduct ambushes and occupy patrol bases, and firing positions in the snow.

b. Those of you with IRF's please fill them out at this time and turn them in to the instructor. We will now take a short break.

PATROL ORDER
WINTER CONSIDERATIONS

ORIENTATION:
    Terrain model should include avalanche prone slopes.
    Avalanche conditions (aspect, angle, elevation, location, severity, etc).
    Full weather report (including; wind direction and speed, temperatures, precipitation).
    Snow conditions for travel (depth, wet, dry, crust, or frozen, over-the-snow mobility).

SITUATION:
    Enemy – include both foot and vehicle over-the-snow mobility assets/capabilities.
    Friendly –
    Attachments –

MISSION:

EXECUTION:
    Intent –
    Concept of Ops –
        Scheme of Maneuver – Route overlays should have avalanche prone slopes marked and the time distance formula totals annotated.
        Fire Support Plan –

    Tasks – There are many winter specific collateral duties.
        Avalanche search teams; probe line, marker, shovelers, transceiver team, hasty team, etc.
        Trail breaking team and rotation.
        Sled team rotation, if sleds used.
        Litter teams, augmentation litter teams.
        Snow pit analysis/rutschblock team.
        Ice recon team.

    Coordinating Instructions –
        Initial camouflage pattern for movement.
        Over-the-snow mobility selection including; sleds or no sleds, skins or wax type, snowshoes or skis below or out of snow line.
        Transceiver check, when and where.
        Priorities of work in a winter bivouac.
        Track discipline for travel and bivouac.
        When and where weapons will be cleaned/bores punched.
        Pull pole time.
        Shelter selection (tent, fly only, bivy bag only).
        Final inspection includes ski/snowshoe binding adjustment, thermos/canteens topped off, camy pattern check, pack and sled check

ADMIN AND LOGS:
    Type of ration resupply, water/no water resupply, fuel resupply.
    How protect POWs from elements.
    How conduct resupply from MSR to position, if applicable.

COMMAND AND SIGNALS:
    Hand and arm signal modification for cold weather clothing, if needed.

COMBAT ORDERS

Student Handout

1. **Fundamentals**

It is not only the commander's will which is decisive in war, but his manner of expressing that will. Commanders use orders to express their will and translate their decisions into actions. MCRP 5-2-A (Operational Terms and Graphics), defines an order as "a communication, written, oral, or by signal which conveys instructions from a superior to a subordinate. In a broad sense, the term order and command are synonymous. However, an order implies discretion as to the details of execution whereas a command does not." Combat orders are the second step in the three steps of action.

      a. Decision
      b. Communicate
      c. Execute

An experienced combat leader phrased it this way:

> ***The essential thing is action. Action has three stages: the decision born of thought, the order or preparation for execution, and the execution itself. All three stages are governed by the will.***
>   -General Hans Von Seekt

In essence, combat orders express the will or intent of the commander. They must be brief, clear, and definite. A decision, however promising, will probably fail if the commander cannot express it in an order.

The stages of action defined above are similar to the famous Boyd cycle--observe, orient, decide, act (OODA) loop. The key point being that the orders process is continuous. It begins when we receive or decide a mission and ends when the mission is complete. **(Never mistake the orders process as merely the development of the order itself.)**

A commander's ability to deliver orders corresponds directly to his tactical skill. If the commander makes an accurate estimate of the situation and arrives at a definite tactical decision, then he typically issues an effective order. Conversely, a commander who cannot make a decision will not produce an effective order.

Delivering combat orders, like tactical decision making, is an art. To be effective, commanders must frequently practice making decisions and articulating orders. Since orders express the commander's will, they should reflect the personality of the commander. A competent commander avoids highly formalized formats or lengthy order procedures. They limit his flexibility. Often, he must individualize orders to best match the abilities of those who receive it. (For more information on this aspect of combat orders see Von Schell's Battle Leadership, pages 11-12.) A commander will provide a more detailed order to inexperienced or unfamiliar subordinates than he would to those with whom he knows and trusts.

Mission tactics are crucial to the art of combat orders. The author of **Battle Leadership** describes them this way:

> *....we use what we term "mission tactics;" orders are not written out in the minute detail, a mission is merely given to the commander. How it shall be carried out is his problem. This is done because the commander on the ground is the only one who can correctly judge existing conditions and take proper action if a chance occurs in the situation. There is also a strong psychological reason for these "mission tactics." The commander who can make his own decisions within the limits of his mission, feels responsible for what he does. Consequently, he will accomplish more because he will act in accordance with his own psychological individuality. Give the same independence to your platoon and squad leaders.*

Of course, there will be situations where more detailed control is necessary and mission orders may not be practical. This is especially true for smaller, less experienced units, or units that have not had time to become cohesive. More detailed control is often applied during peacetime live fire exercises when precision is more important than flexibility. Use mission orders whenever the situation allows. Never assume, however, that they are a license to avoid careful thought or relax discipline.

Sun Tzu says that speed is the essence of war. Orders must also be timely. General Patton's observation "that a good plan violently executed now is better than a perfect plan executed next week" bears repeating. In describing some lessons learned by the U. S. Army in the First World War, General George C. Marshall wrote:

> *In studying the examples of the orders issued to our troops in France several important points deserve consideration in determining the relative excellence of the orders issued. It is frequently the case that what appears to have been a model order was actually the reverse, and a poorly and apparently hastily prepared order will often be erroneously condemned. Many orders, models in their form, failed to reach the troops in time to affect their actions, and many apparently crude and fragmentary instructions did reach front-line commanders in time to enable the purpose of*

higher command to be carried out on the battlefield. It is apparent that unless an order is issued in time for its instructions to percolate down throughout the organization sufficiently in advance of an engagement to enable each commander to arrange his unit accordingly, that order is a failure, however perfect it may appear on paper. Our troops suffered much from the delays involved in preparing long and complicated orders due to the <u>failure of the staff concerned to recognize that speed was more important than technique</u>.

According to MCDP 5 ***Planning***, "the more urgent the situation, the greater need for brevity and simplicity." Remember that an effective combat order is much more than merely passing information. You must convey your will. **A good order is as much inspiration as information**. The confidence and enthusiasm in which you deliver your order is as important as the order itself.

2. **Types of orders.** Combat orders are distinguished from administrative orders by their purpose, and tactical action. There are several types of combat orders. The most common are the warning order, operation order, patrol order, and fragmentary order. All definitions are from MCRP 5-2-A, Operational Terms and Graphics.

   a. A **warning order** is "a preliminary notice of an order or action which is to follow". Its primary purpose is to allow subordinate units to prepare while commanders continue planning or decide on a course of action. It is really nothing more than a heads ups. We teach that a warning order must contain four essential elements: the situation, mission, general instructions and special instructions. Don't delay issuing a warning order because you don't have all the information you would like to have.

   b. An **operations order** is "a directive issued by a commander to subordinate commanders for the purpose of effecting the coordinated execution of an operation." They are used by commanders at all echelons. They are issued orally or in writing. When written they can be only a page or two long or as thick as a phone book. When dealing with small units (squad, platoon, company) these orders are properly referred to by the action they intend. Some examples are attack, defense, withdrawal, or movement orders. You may hear them referred to as five-paragraph orders. Although such a reference is made in at least one doctrinal publication, you will want to be sure to distinguish the action intended by the order from the format used to organize the order. For example, say "attack" and "defense" order instead of five-paragraph order. An overlay order consists of operational graphics outlined on acetate. A matrix order is a method of communicating a written order in concise form and is generally used at company and higher levels in the operating forces. Operations orders often have several **annexes** that contain specific details on fire support, communication, heliborne movement, and so forth. Annexes are common in battalion and higher orders, but uncommon in typical platoon attack orders.

   c. A **patrol order** is an order given to a patrol.

   d. A **Fragmentary** or **"frag" order** is "an abbreviated form of an operation order, usually issued on a day-to-day basis, that eliminates the need for restating information contained in a basic operations order. It may be issued in sections." Frag orders are often necessary due to enemy countermoves. Expect frag orders in most operations. Remember "no plan survives contact with the enemy." Frag orders typically contain the mission (paragraph II) and execution (paragraph III). They also contain any other parts of the order that have changed since you issued the original order. Frag orders are an important technique to keep orders short. In general, Frag Orders contain changes to the previously given orders and consequently the information communicated does not repeat information that remains the same.

3. **The Spectrum of Combat Orders**

Conceptually, just as there is a spectrum of conflict, there is also a spectrum of combat orders. Combat orders come in a variety of forms. They range from a few hasty instructions shouted by a squad leader in the heat of battle to a phone book sized written operations order for a Joint Task Force signed by the Joint Force Commander and every type of combat order in between.

4. **Formats**

Several nations have agreed to use the format contained in standard agreement (STANAG 2014). Most Marines will recognize it as SMEAC, the five-paragraph operations order format. It is used throughout the U. S. Armed Forces, NATO, and elsewhere overseas. Nearly all types of combat orders are based on all or part of the five-paragraph format. Memorize this format. But, never forget that it is the action that follows your order that counts, not the format itself. A short, simple order that conveys your will is superior to a lengthy, complicated order. Do not allow your decision to become lost in a series of paragraphs, subparagraphs, alpha numerics, and acronyms. **Content is always more important than format.**

> *Standard order formats expedite understanding, prevent omissions, and facilitate ready reference. <u>However, content, clarity, and conciseness are more important than format.</u> Slavishly following a prescribed format can result in rigid form and unimaginative content not consistent with the unique requirements of each situation (MCDP 5, p90).*

5. **Some techniques**

   a. **Whenever possible give your orders orally, in person**. Much of the communication is nonverbal. You can better communicate to your subordinate leaders in person when you look them in the eye. Written orders are best reserved for larger units whose size preclude oral orders, or when recording operations for historical reasons.

b.   **Keep orders short and to the point.** Combat is extremely stressful. Your Marines will probably be exhausted, scared, and either too hot or too cold. Their attention span will be short. They will not listen to orders that are too long or complicated.

c.   **In rare occasions, issue your order to all hands.** At the company level and below you may have the opportunity to issue orders to the entire unit at once. This is a good method of communicating your will to all hands. Often the tactical situation will not permit the use of this technique.

d.   **Use active voice, direct language that conveys confidence.** Avoid vague terms, qualifiers, or gratuitous phrases. Terms such as "conduct a rehearsal if you can manage it", "attack vigorously", "radioman maintain radio communication" only serve to dilute the clarity and energy of your order. Strive to avoid them.

e.   **Give subordinate leaders enough time to prepare and issue their own orders.** Use the half rule or one-third rule. For the half rule, divide in half whatever time you have to prepare and issue your order. Use the first half yourself and give the second half to your subordinate leaders. Later in your service, as you assume commands of larger organizations, you may use the two-thirds rule. Divide your available time in thirds. Use one-third yourself. Give the remaining two-thirds of the time to your subordinates to prepare. Orders groups work well in many situations.

f.   **Designate an orders group.** Your orders group should contain all your subordinate unit leaders, key billet holders, and the leaders of all attached units. For example, a rifle platoon orders group would contain the platoon sergeant, platoon guide, radioman, squad leaders, and leaders of all attached units. Make sure all members of your orders group know they are in your orders group.

g.   **Don't always expect your subordinate leaders to come to you.** Another method of issuing orders is to visit each of your subordinate leaders personally and deliver your order individually. This may be the only method possible if you are under fire.

h.   **Use visual aids.** If possible, issue your order on the terrain where you will act. If that is not possible, issue your order over a model, dry erase board, or butcher block paper. If you're in the field you can build a terrain model. You can use wire, string, yarn, colored chalk, laminated index cards and even small plastic toys as means of graphically communicating your order.

i.   **Give a short orientation before you begin your order.** Your orientation should cover key terrain, tactical control measures, current location, expected enemy positions, direction of north, direction of attack, and other pertinent information. You can use 3x5 cards with grid coordinates of assembly areas, objectives, targets, and other tactical control measures on them. If you give your subordinate leaders a chance to copy the grids during the orientation, you can shorten your order by not having to dictate the grids. Offer your orders group the opportunity to ask questions about the orientation before you begin your actual order.

j.   **Tell your orders group to hold their questions until you have completed your order.** By holding all questions until the end, you eliminate questions that interrupt the flow of your order or may be answered later in the order.

k.   **Avoid excessive formality or informality when issuing orders.** Use a natural conversational tone and tempo per the attached examples. There is no need to announce each paragraph and subparagraph when giving your order. You can stay with the format without resorting to this time consuming, excessively formal habit. Similarly, you should never allow your orders to degenerate into a casual dialogue. Your order should convey confidence and authority. Those who hear your order should know that you are giving an order.

l.   **Take charge.** Never permit eating, sleeping, talking, or any sort of distraction during your order. Conduct a roll before you begin your orientation to ensure all key personnel are present. Tell subordinate leaders where you want them positioned to receive the order. Insist that they take notes.

m.   **Use backbriefs.** Backbriefs are an efficient method of verifying that your order is understood. A backbrief is when you ask questions about your order to those who have just received your order. In Michael Shaara's *Killer Angels* (page 93), Union General Reynold employed the briefback technique effectively on day one at Gettysburg when he gave an order to a captain.

n.   **Prioritize your order development.** In combat, time is crucial. You will probably not have time to give as thorough an order as you would like. **As you estimate your situation (METT-T), issue a warning order to allow your unit maximum prep time.** Then work the enemy situation, mission, and execution. Leave less critical portions of your order for last.

o.   **Don't read your order.** Naturally, you may refer to notes when delivering your oral order, but you must not read it. Focus on the eyes of the Marines you are about to order into harms way, not your notes. It is difficult to inspire confidence in your decision if you are reading it. If your understanding of the enemy situation is so weak or your scheme of maneuver so involved that you can't brief them without reading, then you probably need to rethink your decision. It is too complex. If possible, issue detailed information like grid coordinates, checkpoints, target numbers, frequencies, and call signs in writing before issuing your oral order. This is not a matter of turning an oral order into a written one, just a technique to keep the order itself focused and brief. Dictate data only as a last resort. Never dictate your enemy situation, mission, intent, scheme of maneuver, fire support plan, or tasks. Look at your Marines when you give these critical portions of your order.

        p.   **Tell your Marines why**. Of the who, what, where, when, and why of your mission and task statements, the why is most important. If Marines know why they are acting, they are better equipped to respond to unanticipated situations and fleeting opportunities.

6.   **The attached readings**

    Appendix A is the order format we will use for Company level operations and below

    Appendix B is the Sample Platoon Commander's Order.

    Appendix C is the Sample Squad Leader's order (Derived from the Platoon Commander's Order in Appendix B).

    Appendix D provides additional guidance for using the five paragraph order format. Remember, not every order will have every item of information mentioned in their appendix.

    Appendix E is a schematic that demonstrates how information from highers orders is reformatted in our order. It can be useful when you receive a timely and complete order. It is not much use in a dynamic tactical situation or when the order you receive is obsolete or incomplete.

APPENDIX A

FIVE-PARAGRAPH ORDER FORMAT
FOR COMPANY-LEVEL OPERATIONS

A map and/or terrain model orientation is normally given prior to issuing the order.

1. **Situation**
    a. <u>Enemy forces</u>
        (1) SALUTE (Composition, disposition, and strength), (size, activity, location, unit, time, and equipment).
        (2) DRAW-D (Capabilities and limitations to defend, reinforce, attack, withdraw, or delay).
        (3) EMPCOA (Enemy most probable course of action which of the capabilities above (DRAW-D) is most likely and/or dangerous).
    b. <u>Friendly forces</u>
        (1) Higher's mission and intent.
        (2) Adjacent unit missions (task and purpose).
            (a) Left
            (b) Front
            (c) Right
            (d) Rear
        (3) Supporting unit's (type of support GS, GS, or ATTACH, location, POF).
    c. Attachments and Detachments (date and time effective).

2. **Mission**
Task to be accomplished and purpose (who, what, when, where, and why).

3. **Execution**
    a. Commander's intent (relative to the friendly/enemy/terrain).
    b. <u>Concept of operations</u>
        (1) Scheme of maneuver. (Concise, plan to accomplish mission in general terms)
        (2) Fire support plan. (Purpose and how it supports SOM, direct and indirect)
    c. Tasks (subordinate element missions: main effort, supporting efforts, and reserve). (Includes units attached to you.)
    d. Coordinating instructions (identify and discuss instructions that are common to two or more elements).

4. **Administration and Logistics**
    a. Administration. (Bad guys, bandages)
    b. Logistics. (Beans, bullets, and batteries).

5. **Command and Signal**
    a. Signal.
    b. Command.
        (1) Location of key leaders.
        (2) Chain of command (command succession).

Any questions?   *The time is now.*

APPENDIX B

SAMPLE PLATOON COMMANDER'S ORDER

GENERAL SITUATION: You are the 1st Plat Cmdr, Company C, 1st Bn, 6th Marines. For the past three days, the enemy has been conducting a series of ambushes along the Bn's Main Supply Route (MSR), Rte 610, in an effort to disrupt our supplies and LOC

ORIENTATION. We are in the AA, here. This intermittent stream is our LD. Our Aslt Pos is here on the N side of this draw. This is CO OBJ A and BN OBJ 1. On call targets are here. Our sector is characterized by cross compartmentral terrain with a mix of deciduous and evergreen vegetation throughout. Visibility varies with up to several hundred meters on top of the fingers and down the long axis of major draws. This north-south finger dominates the center of our sector. Route 610 is to our south. Route 644 is our platoon's eastern boundary and this north south stream is to the west. Cannon Creek runs east-west through our sector just south of our tentative Aslt Pos. Any questions on orientation? OK, hold the rest of your questions until the end of the order.

1. SITUATION
   A. ENEMY
      (1) SALUTE. Platoon (-), hasty-D, vic GC 823671 (Bn Obj 1) --- Also a sqd (+) digging in on the high ground vic GC 816659 (Co Obj A). Elements of the CRF, spotted by Bn scout/sniper team at 0600 this morning. AK-47s, RPK MG's, Light mortars?? Bn S-2 suspects the squad (+) is a forward outpost for the Plat(-).

      (2) DRAWD. Defend initially. May reinforce Co Obj A (footmobile) with up to a squad, but would likely take an hour. Positions at Co Obj A and Bn Obj 1 are not currently mutually supporting.

      (3) EMPCOA. Squad(+) on Co Obj A will defend but if their pos becomes untenable they'll w-draw to the N and attempt to link up with the Plat(-) on Bn Obj 1. The EN on Co Obj A may have an SP/LP his front, but I believe his focus remains narrow to the S. CRF will target own pos and likely CFF to cover his w-drawl. If we're disorganized in our console, he may use mortars & MG's to support a X-atk..

   B. FRIENDLY
      (1) Higher
         (a) Mission. Our Co is Bn ME. At 1500, Co C will ATD EN vic Co Obj A & Bn Obj 1, IOT deny EN the ability to interfere with our MSR.
         (b) Intent. Maintain a secure route for our resupply convoys to travel on the MSR.
      (2) Adjacent
         (a) There is no one to our front. The scout/sniper team withdrew at 0800 this morning.
         (b) 2d plat is currently to our rear and is the Co ME. At 1530, 2d plat will ATD EN vic of Bn Obj 1, IOT deny EN the ability to interfere with our MSR.
         (c) 3d plat remains in the Co AA and is the Co Reserve. They are prepared to assume the mission of 2d plat.

      (3) Supporting. 81's in GS of Bn loc (805637). POF to Co C, B, then A. 60's in GS of Co loc (815646). POF to 2d plat, 1st, then 3d.

   C. ATTACHMENTS/DETACHMENTS - none

2. MISSION. Our plt is a SE. At 1500, 1st plat will ATD EN vic Co Obj A IOT prevent the EN from influencing the ME's atk on Bn Obj 1.

3. EXECUTION

   A. COMMANDER'S INTENT. I see the EN CV as his narrow frontal focus. We'll exploit this weakness by effective SptArms & numerically superior force rapidly closing on the EN flank. WE MUST PREVENT THE ENEMY VICINITY COMPANY OBJECTIVE A FROM INTERFERING WITH THE COMP ME ATK ON BN OBJ 1.

   B. CONCEPT OF THE OPERATION
      (1) Scheme of Maneuver. Flanking attack, 3 squads abreast, 1 as the ME and 2 as SEs, DOA is N. Move from AA to AP...at 1500, we'll LD and travel N to the AsltPos. Vic of AsltPos I'll call for 1 min of supp fire on the obj. As fires lift, we'll assault from E to W through the obj and consol oriented NW. BPT assume mission of Co ME.

      (2) Fire Support Plan. Purpose - suppress EN on Co Obj A IOT allow our plat to close with and destroy him. In the vic of the AsltPos, I'll call for 1 min of supp fire on the Obj, (AF2400 at GC 816659). CO has an on-call tgt on Bn Obj 1, (AF2401 at GC 823671).

   C. TASKS
      (1) 1st SQD. ME. At 1500, ATD the EN center 1/3 Co Obj A IOT prev EN vic Comp Obj A from influencing CO ME atk on Bn Obj 1.
      (2) 2d SQD. At 1500, ATD the EN left 1/3 Co Obj A IOT prev EN vic Comp Obj A from influencing ME atk on Bn Obj 1.
      (3) 3d SQD. At 1500, ATD the EN right 1/3 Co Obj A IOT prev EN vic Comp Obj A from influencing ME atk on Bn Obj 1.

D. COORDINATING INSTRUCTIONS
    (1) DOA - NORTH
    (2) TOA- 1500
    (3) Base Unit- 1st Sqd
    (4) Formations for movement
        AA to ATK Pos- Plat Column; order of mvt: 1st, 2d, 3d
        ATK Pos to LD to Aslt Pos- Plat Wedge; order of mvt: 1st-center, 2d-left, 3d-right
        Aslt Pos to Obj- On Line; order of mvt: 1st-center, 2d-left, 3d-right
    (5) Consolidation. Form a plat 180 oriented to the NE. 1st Sqd - center 1/3, 2nd Sqd - left 1/3, 3d Sqd - right 1/3. Adjacent Sqds tie in physically. Flank Sqds refuse your flanks.
    (6) Tactical Control Measures
        AA- GC 814646
        Atk Pos- GC 814648
        LD- intermittent stream running generally NW/SE through GC 814649
        Aslt Pos- GC 814656
        Co OBJ A- GC 816659
        Bn OBJ 1- GC 823671
    (7) Time Line
        Brief Backs from Sqd Ldrs-   0930
        Sqd Ldr's Time(orders/rehearsals/inspections)-   0945
        Platoon inspection-   1300
        Platoon rehearsals-   1330
        Move to Atk Pos-   1430
    (8) MOPP Level- 0

4. ADMINISTRATION & LOGISTICS
    A. ADMINISTRATION
        (1) WIA/KIA: Self aid, buddy aid, Corpsman Aid; urgent or priority cas will evac to the PltSgt who will evac to Co GySgt at the Co AA. Walking wounded during the assault proceed up to Comp Obj A.
        (2) EPWs: 5 S & T; evac through the Plat Guide who will evac to Co XO at the Co AA.
    B. LOGISTICS
        (1) AMMO: 180 rds per M-16. 400 rds per M-249. 8 HE grenades per M-203.
        (2) Chow/H2O resup after ME's atk complete. Every Marine LD with basic load.

5. COMMAND & SIGNAL
    A. SIGNAL

Call Signs and Freqs:	Freq	Call Sign
Co Tac-	37.95	A1T
1st plat-		L1T
2d plat-		B2A
3d plat-		C6E
Wpns plat-		Y7H
Bn Tac-	42.25	R5B
Inf Mort Net-	46.50	V3C

The CEOI will be carried in the radio operator's left breast pocket.
    Consolidate: (Primary) - Green Star Cluster   (Alternate) - whistle and voice
    Cease Fire: Primary- White Star Parachute   (Alternate) - messenger and voice
    Challenge & Password: SPIT/ TREE   Running: Taco Bell   Alt: 7

    B. COMMAND
    Co Cmdr, 1st Sgt will be with 2d plat; Co XO, GySgt with 3d plat
    Plat Cmdr will be with 1st sqd; Plat Sgt will be with 2d sqd; Plat Guide will be with 3d sqd
    Succession of Command: 1st sqd ldr, Plat Sgt, Plat Guide, 2d sqd ldr, 3d sqd ldr.
    Time is now 0900. Are there any questions?

APPENDIX C

SAMPLE SQUAD LEADER'S ORDER

(DERIVED FROM THE PLT CMDR'S ORDER ISSUED IN APPENDIX B)

<u>GENERAL SITUATION</u>: You are the 1st SqdLdr, 1st Plt, C 1/6. For the past three days, the EN has been conducting a series of ambushes along the Bn's Main Supply Route(MSR), Rte 610, in an effort to disrupt our supplies and lines of communication.

ORIENTATION. 1st fire team sits here, second team on the left, third on the right. Break out your note taking gear, orient your maps and listen up. We are in the AA, here. This intermittent stream is the LD. Our Aslt Pos is here on the N side of this draw. This is Co OBJ A and Bn OBJ 1. On call tgts are here as well. Our sector is characterized by cross compartmented terrain with a mix of hardwood trees and thick undergrowth. Visibility varies with up to several hundred meters on top of the fingers and down the long axis of major draws. This N-S finger dominates the center of our sector. Route 610 is to our S. Route 644 is our platoon's eastern boundary and this N-S stream is to the west. Cannon Creek runs E-W through our sector just S of our tentative Aslt Pos. Any questions on orientation? OK, hold the rest of your questions until the end of the order.

1.   SITUATION
        A.   ENEMY
               (1)   <u>SALUTE</u>. Plat(-), hasty D, vic GC 823671 (Bn Obj 1) --- Also a sqd(+) digging in on the high ground vic GC 816659 (Co Obj A). Elements of the CRF, spotted by Bn scout/sniper team at 0600 this morning. AK-47s, RPK MGs, light mortars? Bn S-2 suspect the sqd(+) is a forward outpost for the Plat(-).

                (2)   <u>DRAWD</u>. Defend initially. May reinforce Co Obj A (footmobile) with up to a sqd, but would likely take an hour. Positions at Co Obj A and Bn Obj 1 are not currently mutually supporting.

                (3)   <u>EMPCOA</u>. Squad(+) on Co Obj A will defend but if their pos becomes untenable they'll w-draw to the N and attempt to link up with the Plat(-) on Bn Obj 1. The EN on Co Obj A may have an SP/LP to his front, and may be narrowly focused to his S. CRF will tgt his own pos & likely CFF to cover his w-drawl. I believe his MGs are oriented down the draws to his immediate front. If we are disorganized in our consol, he may use mortars and MGs to spt a X-atk.

        B.   <u>FRIENDLY</u>
               (1)   <u>Higher</u>
                      (a)   <u>Mission</u>. Our plt is a SE. At 1500, 1st plat ATD EN vic Co Obj A IOT prevent EN from influencing ME's atk on Bn Obj 1.
                      (b)   <u>Intent</u>. CV is his narrow frontal focus. He'll become disorganized if confronted by sptg arms & a numerically superior force rapidly closing on his flank. PREVENT EN VIC COMP OBJ A FROM INTERFERING WITH THE COMP ME ATK ON BN OBJ 1.
               (2)   <u>Adjacent</u>
                      (a)   There is noone to our front. The scout/sniper team withdrew at 0800 this morning.
                      (b)   <u>Left</u>. (2nd Squad) At 1500, ATD EN left 1/3 Co Obj A IOT prevent EN from influencing ME's atk on Bn Obj 1.
                      (c)   <u>Right</u>. (3rd Squad) At 1500, ATD EN right 1/3 Co Obj A IOT prevent EN from influencing ME's atk on Bn Obj 1.
                      (d)   <u>Rear</u>. (2d Plt) Co ME. At 1530, ATD EN vic Bn Obj IOT deny EN ability to interfere with our MSR.
                      (e)   <u>Rear</u>. 3d plat remains in the Co AA and is the Co Reserve. BPT assume mission of 2d plat.

               (3)   <u>Supporting</u>. 81s in GS of Bn loc (805637). POF to Co C, B, A. Comp 60s in GS of Co loc (815646). POF to 2d plat, 1st, then 3d.

        C.   ATTACHMENTS/DETACHMENTS - none

2.   MISSION. Plt ME. At 1500, 1ST Sqd ATD EN center 1/3 Co Obj A IOT prevent EN from influencing the ME's atk on Bn Obj 1.

3.   EXECUTION

        A.   <u>COMMANDER'S INTENT</u>. I see the EN CV as his narrow frontal focus .We'll exploit this weakness by use of effective SptArms & rapidly closing on his flank. WE MUST PREVENT THE EN VIC CO OBJ A FROM INTERFERING WITH THE COMP ME ATK ON BN OBJ 1.

        B.   <u>CONCEPT OF THE OPERATION</u>
               (1)   <u>Scheme of Maneuver</u>. Our plt conducts a flanking atk...we are the center squad. We'll atk w/ 3 teams abreast, 1 as the ME and 2 as SEs. DOA is N. Move from AA to AP 1st in Plt column. LD at 1500 as center of plt wedge & atk N to AsltPos. Vic of AsltPos, we'll come on line while plt cmdr calls for 1 min of supp fire on the Obj. As the fires lift, we'll aslt from E to W through the Obj and consol oriented NW. BPT assume mission of Co ME.

               (2)   <u>Fire Support Plan</u>. Purpose - suppress EN on Co Obj A IOT allow our sqd to close with & destroy him. Vic AsltPos, the plt cmdr will call for 1 min of supp fire on the Obj, (AF2400 at GC 816659). Comp also has an on-call tgt on Bn Obj 1 (AF2401 at GC 823671).

        C.   <u>TASKS</u>
               (1)   <u>1st FT</u>. ME. At 1500, ATD EN center 1/3 of 1st Sqd's sector of Co Obj A IOT prevent EN from influencing ME's atk on Bn Obj 1.

               (2)   <u>2d FT</u>. At 1500, ATD EN left 1/3 of 1st Sqd's sector of Co Obj A IOT prevent EN from influencing ME's atk on Bn

Obj 1.

(3) <u>3d FT</u>. At 1500, ATD EN right 1/3 of 1st Sqd's sector of Co Obj A IOT prevent EN from influencing ME's atk on Bn Obj 1.

    D. <u>COORDINATING INSTRUCTIONS</u>
        (1) DOA - NORTH
        (2) TOA- 1500
        (3) Base Unit- 1st FT
        (4) <u>Formations for movement</u>.
AA to ATK Pos- Plat Column (order of mvt: 1, 2, 3) Squad column (order of mvnt: 1, 2, 3)
ATK Pos to LD to Aslt Pos- Plat Wedge (order of mvt: 1-center, 2-left, 3-right) Squad column (1, 2, 3)
Aslt Pos to Obj- Plat On Line (order of mvt: 1-center, 2-left, 3-right) Squad on Line (1-center, 2-left, 3-right)
        (5) <u>Consolidation</u>. On consolidation, we will form a plat 180 oriented to the NE. 1st Sqd will have the center 1/3 **[with 2nd FT on the left, 1st FT in the center and 3rd FT on the right]**, 2nd Sqd the left 1/3 and 3d Sqd the right 1/3. Adjacent Sqds tie in physically. Flank Sqds refuse your flanks.
        (6) <u>Tactical Control Measures</u>
AA- GC 814646
Atk Pos- GC 814648
LD- intermittent stream running generally NW/SE through GC 814649
Aslt Pos- GC 814656
Co OBJ A- GC 816659
Bn OBJ 1- GC 823671
        (7) <u>Time Line</u>
            Brief Platoon Commander-   0930
            FT Ldr prep Time
            My order 0945
            Rehearse
                Actions in Obj Area   1030
                IA Drills   1100
                Consolidaton/BPT's   1130
                Actions in Obj Area   1200
            Chow/FT Ldr Prep   1230
            Squad Gear & Knowledge Inspection   1245
            Platoon inspection   1300
            Platoon rehearsals   1330
            Move to Atk Pos   1430
        (8) MOPP Level- 0

4.   <u>ADMINISTRATION & LOGISTICS</u>
    A. <u>ADMINISTRATION</u>
        (1) <u>WIA/KIA</u>: Self aid, buddy aid, Corpsman Aid; urgent or priority casualties routed through me and then evac to the PltSgt who will evac to Co GySgt at the Co AA. Walking wounded during the aslt proceed up to Comp Obj A and link up with us after the aslt.
        (2) <u>EPWs</u>: 5 S & T; process through me - I will send to Plat Guide who will evac to Co XO at the Co AA.
    B. <u>LOGISTICS</u>
        (1) AMMO: 180 rds per M-16   400 rds per M-249.   8 HE grenades per M-203
        (2) Chow/H2) resupplied after ME's atk is completed. Every Marine cross LD with basic load.

5.   <u>COMMAND & SIGNAL</u>
    A. <u>SIGNAL</u>

Call Signs and Freqs:	Freq		Call Sign
Co Tac-	37.95		A1T
1st plat-			L1T
2d plat-			B2A
3d plat-			C6E
Wpns plat-		Y7H	
Bn Tac-	42.25	R5B	
Inf Mort Net-	46.50	V3C	

The CEOI will be carried in the Plt RTO's left breast pocket.
        Consolidate: (Primary) - Green Star Cluster   (Alternate) - whistle and voice
        Cease Fire: Primary- White Star Parachute   (Alternate) - messenger and voice
        Challenge & Password: SPIT/ TREE   Running: Taco Bell Alt: 7

    B. <u>COMMAND</u>
Co Cmdr, 1st Sgt will be with 2d plat; Co XO, GySgt with 3d plat
Plat Cmdr will be with us... Plat Sgt will be with 2d sqd.... Plat Guide will be with 3d sqd
Succession of Command: 1st FTL, 2nd FTL 3rd FTL

                Time is now 1025. Are there any questions?

APPENDIX D

DEVELOPING THE ORDER

1. The six troop-leading steps (BAMCIS) are the sequence by which a leader receives, plans, and executes his mission. Combat orders are developed along with the troop-leading steps. The development of the combat order begins at the receipt of the mission and does not end with combat, but continues throughout and after the fight in anticipation of the next mission. It includes the techniques by which orders and instructions are organized, sequenced, and transmitted from leaders to subordinates. The combat order is a continuing process with accomplishment of the mission as its main goal. Among combat orders, there are: the **Warning Order, the Five-Paragraph Order,** the **Operation Order,** and the **Fragmentary Order**. Orders generally adhere to the five-paragraph (SMEAC) format though each will differ due to time available and information available or required.

2. Once the leader completes the tactical plan, he issues his order to his subordinates. A five-paragraph order gives subordinates the essential information needed to carry out the operation. It sets forth the **S**ituation, the **M**ission, the plan and method of **E**xecution, **A**dministration and logistics, and **C**ommand and signal information. This format is commonly referred to and remembered by the acronym SMEAC. The order converts the leader's plan into action, gives direction to the efforts of his unit, and provides specific instructions to subordinate elements.

3. Outlined below is the format used for combat orders. At the rifle company level and below, orders are most commonly issued orally with the aid of a **terrain model**.

ORIENTATION. Prior to issuing an order, the unit leader orients his subordinate leaders to the planned area of operation using a terrain model, map, or when possible, the area of operation. The purpose of the orientation is to simply orient subordinates prior to the issuing of the order. Keep the orientation simple and brief. Orientations typically include:

- **Direction of north**
- **Present location (grid)**
- **Unit objectives (grid, terrain feature and designator)**
- **Key features and their potential effects on your mission to include:**
  - **Land forms (hill, valley, finger, draw, depressions, etc.)**
  - **Streams, rivers and lakes (names and general direction of flow)**
  - **Roads (names and general direction)**
  - **Firebreaks, trails and power lines**
  - **LZs, and beaches (grid and designator)**
  - **Vegetation and its potential effect on the mission (forest, jungle, desert, etc.)**
- **Boundaries outlining your planned area of operation**
- **Weather forecast and its potential effect on terrain, personnel, equipment, and mission**
- **Astronomical data that is applicable to the mission (BMCT, sunrise, sunset, EECT, lunar and tidal data)**
- **Time zone (when applicable)**

1. SITUATION. The situation paragraph contains information on the overall status and disposition of both friendly and enemy forces. The information provided is that deemed essential to the subordinate leader's understanding of the current situation. The situation paragraph contains three subparagraphs: Enemy Forces, Friendly Forces, Attachments and Detachments.

   a. **Enemy Forces**. Information about the enemy contained in this subparagraph should be the culmination of intelligence provided by higher headquarters and information gathered (facts and assumptions) which pertain to the accomplishment of the mission. Analysis of the enemy is conducted during your estimate of the situation (METT-T). The Enemy Forces subparagraph has three subparagraphs within it.

   (1) **SALUTE**. This information is usually obtained directly from your higher commander's order and should be tailored to your subordinates so that it is meaningful and relevant. This subparagraph provides information on such things as known and suspected enemy locations, current/recent activities, what type of unit the friendly force is facing, (i.e., light infantry, mechanized, armor, T-55 & BTR-60 equipped), the strength estimate with respect to equipment, personnel and support capabilities (mortar, artillery, air, NBC, recon, patrols, etc.). A helpful acronym to remember when developing this paragraph is SALUTE. In relation to enemy forces, this acronym stands for: **S**ize of the enemy force, their **A**ctivity, last known **L**ocation, **U**nit type/designation, **T**ime the enemy was last

observed, and <u>E</u>quipment they possess.

          (2)   <u>DRAW-D</u>. This subparagraph should highlight what courses of action the enemy is capable of executing and how much time it would take to execute a given course of action. Also discussed are possible enemy weaknesses and vulnerabilities. Particular attention should be given to the enemy's capability to project combat power with respect to time, space, and resources (e.g., the enemy is capable of a foot mobile counterattack on Platoon Objective ALPHA from the east with a squad size element within 30 to 45 minutes.) An acronym to assist you in determining the enemy's capabilities and limitations is DRAW-D, which stands for: <u>D</u>efend, <u>R</u>einforce, <u>A</u>ttack, <u>W</u>ithdraw, and <u>D</u>elay. There is no requirement to mention every action the enemy might possibly take, only those that are likely.

          (3)   <u>Enemy most probable course of action</u>. (EMPCOA) A concise statement of the enemy's most probable action <u>within your assigned zone, sector or objective</u>. What are the enemy's objectives and how the enemy will likely fight the battle and react to friendly actions? How are enemy forces deployed? Determine this by analyzing the possible courses of action open to the enemy, and inform our subordinate unit leaders of what you expect the enemy to do during your mission execution. In determining the enemy most probable course of action you consider the elements of DRAW-D again, though you must go one step further. Anticipating the enemy most probable course of action includes combining pertinent intelligence provided by higher combined with other facts and assumptions about the enemy. Considering the enemy most probable course of action is <u>an essential element</u> in the development of your scheme of maneuver.

    b.   <u>Friendly Forces</u>. Information contained in this subparagraph is obtained directly from your higher commander's order. It contains the missions and locations of higher, adjacent, and supporting units, and the next higher commander's intent for the operation. Information should be limited to that which subordinate leaders need to know to accomplish their assigned mission. It can be remembered by the acronym HAS and includes in order:

        (1)   <u>Higher</u>. The location, mission and <u>intent</u> of the next higher unit (for a squad leader's order, the platoon's mission and the platoon commander's intent).

        (2)   <u>Adjacent</u>. The mission and location of units to your left, right, front and rear having effect on your mission, as well as units tasked with a reserve mission. *These units often provide security for your own.* Listed is the unit providing security, their mission and general location.

        (3)   <u>Supporting</u>. *Nonorganic* units providing fire support or combat service support are addressed here. Listed are the units providing support, the location of the supporting unit, the command relationship (DS, GS, etc.), priority of the support and the unit being supported, if known.

    c.   <u>Attachments and Detachments</u>. *Nonorganic* units attached, and/or organic units detached from the issuing unit by higher headquarters are addressed here. The unit and effective time of attachment/detachment are given. If there are no attachments or detachments state "none."

2.   <u>MISSION</u>. The mission statement is a clear and concise statement of what the unit is to accomplish. The mission statement is derived from the leader's mission analysis. It expresses the unit's primary <u>task</u> and <u>purpose</u> by addressing the "five Ws" -- Who (unit), What (task), When (time), Where (grid), and Why (purpose ...in order to...) for the mission assigned. The mission statement should also include the type of operation (attack or defend) and the control measures that will be used (such as "objective" and "battle position"). <u>The mission statement is the heart of the order</u>, and as such is always stated in full and must stand alone without references to any other document except a map. The task describes the action to be taken while the purpose describes the desired result of the action. Of the two, <u>the purpose is predominant</u>. While the situation may change, making the task obsolete, the purpose (short term intent) is more permanent and continues to guide our actions.

3.   <u>EXECUTION</u>. The execution paragraph contains the "how to" information needed to conduct the operation. This paragraph consists of Commander's Intent, Concept of the Operation, Tasks to subordinate unit leaders, and Coordinating Instructions.

    a.   <u>Commander's Intent</u>. The commander's intent is a vision provided to subordinates which enables them to act in a changing environment in the absence of additional orders. It describes the commander's longterm purpose of the operation with respect to the relationship among friendly forces, the enemy and terrain. <u>At the tactical level, intent is conveyed throughout the order because the commander personally drafts and delivers the order; he can therefore emphasize key points that he believes are vital to the success of the mission.</u>

    b.   <u>Concept of the Operation</u>. The concept of operation includes the scheme of maneuver, fire support plan, and when applicable, the employment plan of other combat multipliers such as obstacles used in the defense.

(1) <u>Scheme of Maneuver</u>. Using a graphic, sketch, or terrain model, the leader explains his plan to accomplish the unit's assigned mission. It should be described in general terms without identifying specific units. Brief the scheme of maneuver in logical sequence; begin at your current location and brief your unit's actions through completion of your mission. For an offensive operation the scheme of maneuver includes: <u>form of maneuver, planned distribution of forces (including main effort and supporting effort), direction of movement, tactical control measures, and consolidation</u>. For a defensive operation the scheme of maneuver includes: defensive technique, planned distribution of forces, general direction of fires/location of planned engagement areas and security plan. When applicable, reserve forces are also briefed.

(2) <u>Fire Support Plan</u>. Describes how fire support will be used to <u>complement the scheme of maneuver</u>. The fire support plan ties in directly with the scheme of maneuver. Organic, attached and supporting indirect fires may be included. In some instances, this paragraph can also be used to describe how direct fires will be used to support the scheme of maneuver. Included in the fire support plan may be:

- the <u>purpose/concept</u> of fire support and how it will integrate with and support the scheme of maneuver.
- the <u>priority</u> of fires and when priority shifts within the unit
- the <u>location</u>, description and target designation of preplanned targets that support the scheme of maneuver.
- the location of firing units (if not already covered in friendly forces subparagraph)
- <u>permissive</u> and <u>restrictive</u> control measures on the use of fires
- <u>allocation</u> of targets (i.e., in the offense - priority targets; in the defense - final protective fires)

c. <u>Tasks</u>. The specific missions to be accomplished by each subordinate element of the unit will be listed in a separate numbered subparagraph, including reserves. <u>Task statements are your subordinate unit's mission statements, and as such, should be written in the same manner as any mission statement.</u> Just as your mission statement from higher, your subordinate task statements should answer the "5 Ws," Who, What (task), When, Where, and Why (purpose/in order to ...) for the missions you assign. When tasks are multiple, they are itemized with subparagraphs. If there is a priority or sequence of accomplishment, it is stated. When a subordinate unit is designated the main effort, state it in the tasking statement. Anticipated (...be prepared to...) missions should be included (i.e., pursuit, defense). Subordinate unit tasks should be listed in a logical sequence (i.e., from start to finish; or most important to the least important missions, followed by anticipated missions).

d. <u>Coordinating Instructions</u>. Coordinating instructions are those specific instructions and tasks that tie the plan together. Included are details of coordination and control applicable to two or more units in the command. Items commonly addressed in coordinating instructions include:

1. Order of movement and planned formations during movement
2. Consolidation, reorganization, counterattack plan
3. Movement into the defense (used for deliberate occupation of the defense only).
4. Location/grids for tactical and fire control measures--these control measures should also be depicted on your terrain model/operational graphic. Examples include: check points, phase lines, release points, battle positions, SP/LP, TRPs, etc. When briefing tactical control measures, point out the location on your terrain model and then give grid coordinates.
5. Target precedence--assigned to specific units/weapon systems to provide guidance on what targets to engage when multiple targets are presented.
6. Security plan--the plan to provide early warning/protection to the unit to prevent surprise upon enemy contact.
7. Engagement and disengagement criteria and instructions
8. Priority of work (used for defensive operations)
9. Reporting requirements
10. Rules of engagement (if applicable)
11. MOPP level
12. Planning and execution time lines

4. <u>ADMINISTRATION AND LOGISTICS</u>. This paragraph contains all the information

necessary for subordinate units to coordinate their resupply, recovery of equipment, and evacuation of wounded and prisoners. This paragraph addresses the "*FIVE* Bs" -- BEANS (chow), BULLETS (ammunition), BATTERIES (COMM/NVG), BAND-AIDS (MEDEVAC) & BAD GUYS (EPWs) and is divided into two subparagraphs.

    a.   <u>Administration</u>

        (1)   Medical evacuation plan for wounded
        (2)   Enemy prisoners of war (EPW) handling procedures and evacuation plan

* Admin subparagraphs should outline POCs at your level and at least one level up. Specific instructions such as when to evacuate casualties and location of collection point are also included.

    b.   <u>Logistics</u>

        (1)   Initial issue and resupply plan (ammo, chow, water, batteries)
        (2)   Any other logistical concerns to include transportation, etc.

* Also included in logistic subparagraphs should be who is responsible for drawing, who gets special gear and any POCs necessary for coordination.

5.   <u>COMMAND AND SIGNAL</u>. This paragraph contains instructions and information relating to command and communications (control) functions. It contains two subparagraphs--signal and command.

    a.   <u>Signal</u>. Specifies the signal instructions for the operation. Include both the <u>primary</u> and <u>alternate</u> signal plans as well as methods of communication in priority (example: primary means of communication is land line, alternate is radio, then messenger). Also included are the times when the signal plan changes.

        (1)   Communication plan to include primary and alternate call signs/frequencies (CEOI index number if applicable) should be specified as well as time of change.
        (2)   Visual signals required to coordinate the concept of operations (examples include: signals to commence, shift, and cease the support by fire; signal for displacement of the support by fire force; signals to commence, and cease the FPF; signal to break contact).
        (3) Challenge/Password (primary & alternate)/running password and time of change.
        (4) Brevity codes and code words

  b. <u>Command</u>. Identifies your location and the location of other leaders as required.

  (1) Location of the higher commander. (key leaders)
  (2) Your location before, during and after the battle.
  (3) Succession of command (i.e., sqd leader, 1st fire team leader, etc.)

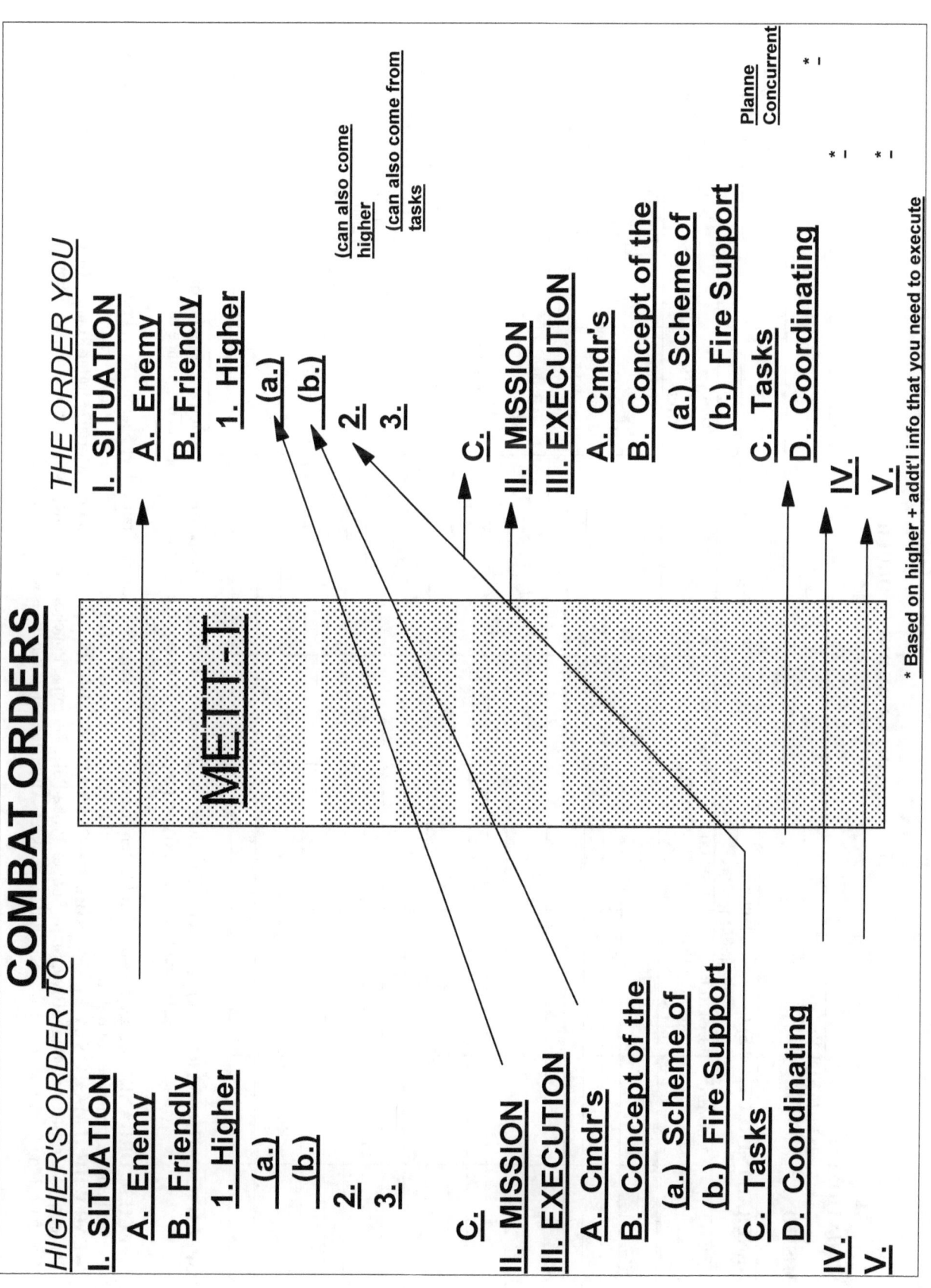

WINTER WARNING ORDER

A. SITUATION: 1) ENEMY: _____
                2) FRIENDLY: _____

B. MISSION: _____

C. GENERAL INTRUCTIONS: Overwhite Pattern: _____ Shelter Type: _____ Chows/man: _____
                Over snow Mobility: _____

1. NAME	2. CHAIN OF CMND	3. GENERAL ORGINIZATION (ELEMNTS)	4. SPECIFIC ORGANIZATION (TEAMS)	5. DUTIES	6. ARMS, AMMO AND EQUIPMENT	7. GEAR COMMON TO ALL	TIME SCHEDULE			
							When	What	Where	Who
						☐ Pocket Items		Draw Ration		
						☐ Assault Load		Draw Fuel		
						☐ Combat Load		Draw Wpns		
						☐ Transceiver		Draw Comm		
						☐ Over Whites		Test Comm		
						☐ Helmet		Draw Ammo		
						☐ VB Boots		Draw Ord		
						☐ Ski March		FSupt. Cord		
						☐ Gaiters		For.Unit Cord		
						☐ LBE w/ H2O		Chow		
						☐ First Aid Kit		Patrol Order		
						☐ Daypack		Init Insp		
						☐ Large Pack		Rehearsal		
						☐ Skis w/skins		Final Insp		
						☐ Ski Poles		Pull Pole		
						☐ Snow Shoes		Fill in old pos.		
						☐ Goggles		Test Fire		
						☐ Headlamp		Transceiver Ck		
						☐ Overboots		T.O.D.		
						☐ Shovel, pers		T.O.R.		
						☐ Thermos		Debrief		
						☐ Sling Rope & Carabiners				

D. SPECIFIC INSTRUCTIONS: a) APL: is in charge when I am gone. You will supervise patrol preparation and drawing of equipment. Adhere to schedule. b) Element Leaders supervise preparation of respective elements and report compliance to APL.
    c) _____ will construct terrain model.

E. SPECIFIC ORGANIZATION TEAMS: Aid/Litter Team(s), Hasty Avalanche Search, Transceiver Search Team, Probe Team, Shovel Team, Marker, Ice Recon Team, Rutchsblock/Snow Analysis Team, Trail Breaking, Navigation Team,

F. GEAR WHICH MAY ONLY BE CARRIED BY CERTAIN MEMBERS: MG, Bulk Ammo, Mortar, Mortar Round, Rope, Chemlites, Radio, extra Battery, NVGs, Sleeping Bagw/Bivy, Snow Shovel, Probe Pole, Thermos, Poncho, Fuel Bottle, Pyro, grenades, Binos, Team or Large Sled, Ski Litter, Stove, Compass, Map/Protractor/Pens, WaxKit, Litter, Cookset, Altimeter, Snow Analysis Kit, Ice Auger, Snow Saw, Map, Route Card, Rope, Hatchet, Wire Cutter,

UNITED STATES MARINE CORPS
Mountain Warfare Training Center
Bridgeport, California 93517-5001

WML
WMO
09/05/02

LESSON PLAN

ICE RECONNAISSANCE AND BREACHING

INTRODUCTION (5 Min)

1. **GAIN ATTENTION**. In the combat arms MOS, a Marine may be required to cross a frozen body of water. Accurate information must be determined to cross safely.

2. **PURPOSE**. The purpose of this period of instruction is to introduce the student to ice reconnaissance and breaching to include route selection for crossing, how to lay demolition charges for breaching, and how to establish ambushes for the enemy. This lesson relates to Tactical Movement.

3. **INTRODUCE LEARNING OBJECTIVES**

    a. TERMINAL LEARNING OBJECTIVE. In a cold weather mountainous environment, conduct ice reconnaissance and breaching operations, in accordance with the references.

    b. ENABLING LEARNING OBJECTIVE

    (1) In a cold weather/mountainous environment, prepare tools and special equipment required for ice reconnaissance, in accordance with the references.

    (2) In a cold weather/mountainous environment, select a route to cross ice, in accordance with the references.

    (3) Given the necessary equipment and a frozen water obstacle, determine the thickness of the ice in accordance with the references.

    (4) Given a frozen water obstacle, cross the ice in accordance with the references.

    (5) In a cold weather/mountainous environment, prepare demolition charges for breaching ice, in accordance with the references.

(6) Without the aid of reference, state in writing the minimum thickness of ice needed to support a man on foot in accordance with the reference.

(7) Without the aid of reference, state in writing the minimum thickness needed to support a man on skis/snowshoes in accordance with the reference.

4. **METHOD / MEDIA**. The material in this lesson will be presented by lecture and demonstration. You will practice what you learn during upcoming field exercises. Those of you with IRF's please fill them out at the conclusion of this period of instruction.

5. **EVALUATION**. You will be tested later in the course by performance evaluation.

TRANSITION: In winter, swamps, rivers, and lakes become an asset to movement, in summer a liability. Detailed engineer reconnaissance is necessary to determine the ice thickness before use for any vehicle movement or as airstrips. Frozen waterway, (lakes, streams, bays), only require snow removal and possible strengthening of the ice in places and are affected by slopes only at the entrance and exits of the waterway. Roads and airstrips over lakes, streams, and bays are constructed only after intensive, detailed reconnaissance of ice conditions. Thickness is only one of many factors affecting the ability of ice to support loads, the entire route must be checked by personnel qualified to interpret ice characteristics (i.e. Reconnaissance, Combat Engineers, Winter Mountain Leaders).

BODY (70 Min)

1. (5 Min) **RECONNAISSANCE**. Pre-check points, for consideration to check for ice routes are:

    a. Check for

        (1) Mines.

        (2) Ice obstacles.

        (3) Tank traps on ice.

        (4) Demolitions under ice.

        (5) NBC evidence.

    b. Actual Reconnaissance Preparation

        (1) Measure the thickness of:

            (a) Ice and its formation.

            (b) Snow on the ice.

(c) Snow on the banks.

(2) Determine

(a) Ice on banks. How ice is attached to the banks? Does it hang over the water?

(b) Slope of banks.

(c) Width and depth of water.

(d) Current speed and slowest current area.

(e) Weight capacity of ice.

(f) Theoretical growth of ice.

(g) Best routes for main and secondary crossing.

(3) Forthcoming work. Will it be a large one?

(4) Natural material. Is it available for reinforcement?

c. Tools/Special Equipment Required

(1) Ice-measuring rod.

(2) Ice auger/axe-bar.

(3) Chisel/spud.

(4) Ice saw.

(5) Weighted depth cord.

(6) NBC detectors.

(7) Probes.

(8) Mine detectors.

(9) Belay rope.

(10) Thermite grenades.

(11) Axe.

(12) Ski poles or staff.

(13) Demolitions.

NOTE: Gear carried depends on availability and need.

TRANSITION: Using this information, we can now select a proper crossing site, and routes.

2. (5 Min) **THE CROSSING ROUTE**. The best condition for ice crossings exists where the stream is broad and straight, with banks that are relatively low or have a gentle slope and are easy to prepare. It should also have a fairly even bottom, well-defined channels and a water flow that is reasonably deep with a low, uniform velocity. These conditions favor the formation of strong thick ice suitable for ice bridges. Of particular interest is the effect of a small run in stream on the outside curve of the main stream. Observation have shown that where a main current curves at near a right angle, and a secondary current enters perpendicular to or into the main current, there is a slowing of the entire movement of that particular area freezing from bank to bank, capable of supporting troops on snowshoes or skis.

   a. The Route

      (1) Designate the proposed route, approaches, exits, and sounding holes with bags, ice blocks, snow piles, stakes, tape, etc., depending on tactical consideration.

      (2) Determine the width of the water obstacle.

      (3) Determine if any dams are upstream.

      (4) Determine tidal rise and drop for bays and length of high tide.

      (5) Along the axis, cut and drill holes through the ice every 3-5 meters by the banks, and every 10 meters in the channel.

      (6) Sketch rivers profile on special engineer paper.

      (7) Determine alternate routes.

   b. Special Considerations

      (1) Immediately adjacent to the shore, the ice formation is thin and weak and more likely to develop cracks than ice in the center of a frozen stream. Depending upon the gradient of the river bed and the thickness of the ice near the shore, it is generally safer to maintain a route near the shore if the ice rests upon the river bottom.

      (2) Where an under-ice current of water flows under a large ice area, the ice in contact with the current is subject to a greater variation in temperature in a given time, and therefore thicker than the ice in adjacent areas.

      (3) Shallow water ice is usually thinner than deep water ice.

(4) Good quality ice is clear and free from bubbles and cracks. In a body of water containing clear and cloudy ice, the clear ice will frequently be thinner then the cloudy ice.

(5) Muskeg lakes contain a great deal of vegetation whose decomposition retards freezing and results in weak ice.

(6) Flooded snow when frozen, produces "slush ice" which is white and may contain air bubbles. Slush ice has a load carrying capacity approximately 1/4 less than that of prime natural ice.

(7) During freezing weather the thickness of ice is increased by removing its snow cover.

(8) Ice that remains unsupported after a drop in the water beneath it, has little strength. Reservoirs and lakes with runoffs are examples of this.

(9) During extremely cold weather, cracks caused by contraction of the ice may be enlarged by heavy traffic. In Norway, some rivers and lakes are controlled by flood bates and will be opened by home guard units upon attack.

(10) In spring the main body of ice can be traveled over, if water is on the ice surface for a limited time only. Pot holes demand extra caution.

(11) A reinforced ice crossing close to a summer crossing site that uses floating equipment, should be located downstream of the summer crossing site to minimize the danger of damage to the bridging equipment during the thaw.

TRANSITION: What about fords in winter operations?

3. (5 Min) **FORDS**. Broad flood-plains with sandbars and shifting water produce weak or unsupported ice with open water areas and difficult working conditions. Such locations will probably require a combination of an ice crossing, conventional floating bridges/winter ford. The fording of streams on foot is advised in summer months, or during other seasons to times when weather and water conditions are favorable.

   a. Summer. The variation in stream velocity and depth are important considerations, particularly in streams and rivers from glacial or ice-cap areas. Fording of such streams may be feasible only during certain hours of the day, usually early morning, when volume and velocity of the water are at a minimum. Precautions are necessary to prevent the use of the ford at unfavorable times and to provide assistance when difficulties arise during actual crossing of equipment or personnel.

   b. Winter. The fording of streams in the winter should be avoided because of the difficulties encountered in the crossing, and the effects of water on personnel and equipment when the ambient temperatures are very low. Some streams, particularly those flowing in broad flood-plains in which valley icing occurs, have open channels that continually shift about the valley during early winter. Valley ice is treacherous because a shifting stream leaves sheets of ice unsupported. Equipment breaks through such ice and is difficult to recover. The sides of open water channels are frequently steep and water is generally deep. Passage, even by heavy tractor equipment, is hazardous. If such fording operations become necessary, the route should be marked with care, all unsupported ice should be removed, and special precautions should be taken to provide safe passage through active stream channels. Continual reconnaissance should be maintained upstream from the ford to determine probable shifts in the water channel. The position of the active stream channel can sometimes be controlled by upstream damming and diversion. The whole water channel should be kept clear to a point below the ford.

TRANSITION: Now, let's be more specific. What kind of ice is safe to cross?

4. (10 Min) **ICE**. Ice is classified in three general types: Salt water ice, fresh water ice, and land ice. These types have widely different characteristics and occur in different areas, and present different problems. All may be used for construction.

   a. Salt Water Ice. Salt water ice is more resilient than fresh water ice and will bend before it will break. Salt water ice first forms in crystals within the layer of salt water affected by convection. These crystals give salt water oily or opaque appearance. Salt water ice is classified by its concentration as: Pack ice, drift ice, fast ice, and ice foot. Sea ice older than one year is much stronger than young sea ice. Young sea ice must be at least 1.66 times the thickness of old sea ice to carry the same load.

      (1) Pack and drift ice. Pack ice is formed on the open sea, remains packed together, and is influenced by currents and winds. When pieces of ice break off the pack they are called drift ice. Pack ice seldom grows thicker the 5-7 feet in the first year, but has attained thickness' of 12 feet in the polar basin where pack ice is perennial. Telescoping and piling up of ice flows can create ice thickness' of up to 125 feet around the edges of the pack. Although the interior of the pack is solid and relatively smooth, it is usually not smooth enough for airstrips without some grading. Currents and winds constantly fracture pack ice leaving lanes or leads of open water. When the leads freeze, they provide sites of smooth ice which make good airstrip locations. The length of such an airstrip, must be doubled the length of the normal maximum takeoff for run for the fully loaded aircraft.

      (2) Fast ice and ice foot. Fast ice is either attached to the shore (land-fast) or otherwise confined so that it does not drift. An example of fast ice, is ice in a bay or lagoon. The portion of fast ice that is attached to the shore is called the ice foot. Fast ice may be attached to the ice foot or be separated fro the ice foot by a crack. Fast ice rises and falls with the tide but the ice foot remains fast to the shore. The more irregular the shoreline and the greater the number of islands in the area, the wider the fast ice will be. Fast ice provides better movement routes and emergency landing fields than does pack ice.

   b. Fresh Water Ice. Fresh water ice begins to form on lakes and rivers under normal conditions, from 3-5 weeks after the daily temperature drops be 32°F. Rates of formation and types of ice formed vary tremendously. Fresh water ice is generally more uniform and stronger than sea ice.

(1) Lake ice. Lakes generally freeze with a smooth surface, and as the ice thickness increases no crystalline structure shows, and the surface retains it's smooth, dry polished appearance. Lake ice is generally weak in the areas of streams, inlets, springs, or outlets. Decaying vegetation on the bottom of a lake may give off air bubbles, which slow ice formation and create weak ice.

(2) River ice. Ice that forms on wide, slow-moving rivers frequently has the same smooth appearance as lake ice. Warm weather and wind may; however, create a rough surface which will remain rough throughout the winter. This ice is filled with air bubbles. In super cooled water (below 32°F), ice will form around solid particles in fine, spicular, sharp pointed crystals in loose spongy masses called frazil ice or slush ice. Frazil ice floats upwards and can accumulate to great thickness under ice sheets and become an integral part of the river ice.

(3) There are four types of ice found on inland rivers and lakes. They are as follows:

   (a) Blue ice is by far the best type. Normally the color is light blue or green in shallow areas and black over deep water. In some cases, where the water depth is less than 3 feet the ice will be clear and the bottom visible. A few cracks may be visible, but are not a sign of weakness if they run in the same direction as the current. These cracks are caused by ice contraction in extreme cold; no bubbles should be present.

   (b) Chandelier ice is normally encountered in the spring. Water covers the surface of the ice during formation due to surface melting or an upstream breakup, which floods the surface. Candle ice is formed when this water percolates/melts through the remaining ice to reach the water below. The ice then appears as a series of icicles or "candles". Because the horizontal strength of the ice has been weakened there is no cohesion for strength. This ice is dangerous to cross even though it may be 5-6 inches thick.

   (c) Rotten ice can be encountered at any time. It can be caused by a thaw or by incomplete freezing. In wintertime it can be caused by bogs, rotting vegetation and sewers; indicates the presence of contamination of some type. Generally it is dull and chalky in color and is very brittle. Rotten ice has no strength and should not be used.

   (d) Unsupported ice occurs when there is a space between the ice and water. It is normally found in areas where the water table has fallen due to tidal action, etc. Unsupported ice can be detected by cutting a hole in the ice. If the water rises less 3/4 of the way up the side or does not rise at all, then the ice is unsupported and should be avoided when possible.

NOTE: The strength of the ice depends upon the following factors: ice structure, purity of water, freezing process, cycles of freezing and thawing, crystal orientation, temperature, ice thickness, snow cover, water current, underside support, and age.

    (4) Normally, fresh water does not freeze to a thickness greater than 8 feet in a single season. In lakes, the normal ice depth by late March are from 3 1/2 feet to 6 feet, depending on winter temperatures.

       (a) The following will speed up freezing:

          1. Low stable temperatures.

          2. High wind-chill factor.

          3. No snow cover.

          4. No current.

       (b) The following will retard freezing:

          1. Fluctuating temperature.

          2. Fast current.

          3. Snow cover.

          4. Salt water/other impurities.

   c. The amount of ice required to support men and vehicles with the proper distance apart:

ICE SAFETY TABLE

ITEM LOADED	WEIGHT IN TONS	ICE THICKNESS NEEDED cm / in	DISTANCE APART NEEDED Meters / Feet
Man on skis/snowshoes	0.1	3 / 1.2	5 / 15
Man on foot	0.1	5 / 2	5 / 15
Horse	0.2	10 / 4	5 / 15
Horse drawn sled/cart	1	15 / 6	15 / 50
BV 206	2	15 / 6	15 / 50
APC	3	20 / 8	15 / 50
5 ton truck & load	7	30 / 12	20 / 70
Bulldozer	5	25 / 10	15 / 50
Tank	50	90 / 36	25 / 80

TRANSITION: Now that we know what type of ice is safe to cross, let's discuss ice crossing.

5. (5 Min) **ICE CROSSING**

    a. <u>Safety Precautions</u>. There are six safety precautions to take prior to crossing an ice-covered body of water.

        (1) Loosen bindings on skis and snowshoes, if so equipped.

        (2) Remove wrist loops of ski poles.

        (3) Sling pack and weapons onto one shoulder.

        (4) Only expose one to the danger at a time or until weight factor is determined.

        (5) First groups of individuals should be belayed by ropes. Mandatory belay if over rapid flowing water.

        (6) Clothing should be worn snugly. All ties tied securely - wrist straps, waist straps, collars, trouser cuffs, etc. This gives buoyancy if breakthrough occurs, and reduces cold shock.

    b. <u>Self-Rescue Techniques</u>. If an individual or group breaks through the ice on an ice-covered body of water, carry out the following self-rescue techniques:

        (1) Remove the unnecessary gear such as packs, weapon, and skis. Throw onto the ice.

        (2) Retain your ski poles and use these as daggers to drag/push yourself across the ice.

        (3) Use your ski poles as described above to extract yourself from the water. Remain flat, do not stand up near hole.

        (4) <u>Carry out re-warming process immediately!!</u> The usual result of sudden immersion is severe hypothermia and shock. This is the most important step in the self-rescue technique.

TRANSITION: Now that we have discussed ice reconnaissance, let's move on to talk about ice breaching.

6. (2 Min) **TACTICAL USES**

    a. Deny the enemy use of a frozen river or lake as an avenue of approach or MSR.

    b. Protect defensive position on a lake or river-line.

    c. Ambush troops and vehicles, or helicopters using a frozen body of water.

TRANSITION: Now, let's discuss the organization of the ice breaching team.

7. (3 Min) **TEAM ORGANIZATION**. The ice breaching party is organized into three teams:

    a. Reconnaissance Team.

    b. Demolition Team.

    c. Initiator.

TRANSITION: Now, let's talk about the duties of each team.

8. (10 Min) **TEAM DUTIES**

    a. Reconnaissance Team. Conducts an ice reconnaissance to select the appropriate site and determine the demolition requirements and reports this information to the demolition team. Furthermore, the reconnaissance team goes ahead of the demolition team to carry out the following tasks:

    (1) Prepares the holes in the ice by chopping, drilling, or blasting.

       (a) Chopping can be done with a piton hammer, axe-bar, etc.

       (b) Drilling is done with small, tamped charges placed on the surface of the ice in a partially excavated hole.

    (2) The holes should larger than required to pass the main charges through. When chopping through the ice, the darker water will be seen before you break through.

    (3) At this point, clear out the hole to its desired diameter, then attempt to quickly clear the remaining ice, before the hole fills up with water.

    (4) The reconnaissance team then moves off the ice.

    b. Demolition Team. Having prepared the charges in a concealed position on a single detonating cord chain, or each charge individually prepared for electric detonation, the demolition team brings them onto the ice by sled, and places one charge by each hole.. One team member remains with each charge and suspends the charge through the hole using a stake to bridge the hole, and hold the charge in place. Once this is accomplished, the demolition team follows the reconnaissance team off the ice. In cold weather, the following points should be noted in connection with demolitions.

    (1) At sub-zero temperatures C-4 can become brittle. The preparation of charges should be done in warm surroundings if the explosive has to be molded. TNT demolition blocks are a suitable alternative to C-4 due to TNT being water resistant and freeze proof. TNT, however, cannot be molded.

(2) Detonating cord becomes very brittle in the cold and may not explode its whole length unless all kinks or bends are removed very carefully. There should be no sharp bends or knots.

(3) Electric detonation by means of a series circuit is an effective alternative to using detonating cord in extreme cold weather. This type of detonating process should be used on small gaps/widths.

(4) When working with explosives and accessories a complete sense of touch is required. Due to the cold, the preparation of explosive charges will take twice as long as normal in the cold. No man should work for more than 30 minutes before going into a warm tent to warm his hands. Contact gloves can increase the 30 minute period.

(5) The effect of a charge in snow is considerably lessened. However, when used in a pattern charge on ice, explosives are very effective. If temperatures remain between 14°F and -4°F, a blown hole in ice will be passable for light vehicles in 14 days and heavier tracked vehicles in 6 weeks.

(6) Improvising in a combat situation may be needed to accomplish a given mission. If C-4/TNT are not present, The use of other explosives such as bangalore torpedoes, M-15 anti-tank mines, and cratering charges can be utilized effectively.

c. Initiator follows the remainder of the group, checks the charges and initiates the charges if non-electric blasting is being used. The time fuse should be sufficiently long to ensure that the initiator clears the ice before detonation. If electric blasting caps are used, the initiator completes the firing circuit. This method is used when detonation is time-sensitive, or is to be on command at a future time, but if it is used, the charges will have to be reexamined periodically.

TRANSITION: Let's now go over the Ice-Breaching Formula.

9. (5 Min) **ICE-BREACHING FORMULA**

   a. Weight of Charge: $W = 1.4 \times T^3$ power.

   b. Depth of Charge: $D = .6 \times T$

   c. Radius of Crater: $R = 6.56 \times W^{1/3}$ power.

   W - pounds TNT, T - ice thickness, D – depth, R - radius

   d. If the water is over 8 feet deep, charges should be suspended 4 feet under the ice. If the water is less than 8 feet deep, charges are suspended halfway between the ice and the bottom. The charges may have to be weighted in river currents to keep them stationary at the proper depth. The charges should be set up on a continuous detonation cord trunk line with each charge being wrapped with det cord and taped. This ensures positive detonation, and quick assembly as each charge is measured off for proper separation. The charges are suspended from the det cord by wrapping it around the stake bridging each ice hole. For this reason, the proper length between charges should be the separation distance, plus twice the suspended distance, plus 2 feet for wrapping the charge and the stake.

TRANSITION: Now let's discuss water obstacles and ice ambushes.

10. (5 Min) **WATER OBSTACLE MAINTENANCE**. In cold weather, the water obstacle will begin to refreeze soon after breaching. This can be delayed by removing the shattered blocks of ice from the water or pushing them under the downstream ice in a river. The open water can then be covered with a tarpaulin, or plastic to insulate it from the colder air. This will inhibit ice formation. When examining the obstacle after breaching, be careful to avoid large sympathetic cracks which may have opened nearby.

TRANSITION: Next, we'll discuss ice ambushes.

11. (5 Min) **ICE AMBUSHES**. There is sometimes a need to blow ice without marking the snow on top of it. Ambush sites are typical of this requirement. This may be done by employing the following technique:

    a. Near the bank, remove a section of ice 1.5 x 1 meters in size.

    b. Cut a number of saplings up to 10 meters high and trim the branches.

    c. Attach explosives to these saplings and connect the detonator cord.

    d. Slide saplings through the hole in the ice with their explosives attached. Lash the end of one sapling to front of the next until the length of ice which is to be blown is covered. The charges will be kept afloat and flush against the ice through the buoyancy of the saplings.

    e. Blow the charge from the bank as required.

TRANSITION: Obviously, if this is to be an ambush, it will not be blown until the enemy is on the ice or lured to a position that puts them at a disadvantage.

PRACTICE (CONC)

   a. The students will practice what was covered in field training exercises.

PROVIDE HELP (CONC)

   a. The instructors will provide help as required.

OPPORTUNITY FOR QUESTIONS (3 Min)

1. QUESTIONS FROM THE CLASS

2. QUESTIONS TO THE CLASS

    Q. What is the minimum thickness of ice needed for a man on foot?

    A. 2 inches

    Q. How do you cross an ice-covered surface?

    A. (1) Loosen bindings
       (2) Remove ski pole loops
       (3) Sling pack and rifle over one shoulder
       (4) Wear clothing snugly
       (5) Deep well dispersed
       (6) Belay individuals as necessary

    Q. Why do we use ice breaching?

    A. (1) Deny the enemy the use of an ice MSR.
       (2) Protect a river-line defense.
       (3) Ambush enemy forces crossing frozen bodies of water.

SUMMARY (2 Min)

   a. What we have just discussed is ice breaching and reconnaissance to include types of ice, crossing techniques, how to establish demolitions for destroying ice avenues of approach and ice ambushes.

   b. Those of you with IRF's please fill them out at this time and turn them in to the instructor We will now take a short break.

UNITED STATES MARINE CORPS
Mountain Warfare Training Center
Bridgeport, California 93517-5001

WML
WMO
09/05/02

LESSON PLAN

COLD WEATHER NAVIGATION

INTRODUCTION (5 Min)

1. **GAIN ATTENTION**. Navigation in an cold weather environment poses it's own set of problems. Heavy snowfall will cause features to blend with each other, compass reaction may be slower, and handling of maps/air photographs will be awkward and difficult.

2. **PURPOSE**. The purpose of this period of instruction is to familiarize the student with those aspects of navigation that are peculiar to a cold weather environment. Including problems, methods, measuring distance and direction. Also covered will be some hints for navigators to make a difficult task a little less so. This lesson relates to Cold Weather Patrolling.

3. **INTRODUCE LEARNING OBJECTIVES**

    a. TERMINAL LEARNING OBJECTIVE. In a cold weather mountainous environment, execute cold weather mountain operations, in accordance with the references.

    b. ENABLING LEARNING OBJECTIVES. In a cold weather/mountainous environment, execute cold weather navigation techniques, in accordance with the references.

4. **METHOD / MEDIA**. The material in this lesson will be presented by lecture. You will practice what you have learned in upcoming field training exercises. Those of you IRF's please fill them out at the conclusion of this period of instruction.

5. **EVALUATION**. You will be tested later in the course by performance evaluation.

TRANSITION: Navigation can be difficult or simple. Being aware of problems that may be encountered in a cold weather/mountainous environment will aid the navigator in movement.

BODY (50 Min)

1. (5 Min) **NAVIGATION PROBLEMS**. For Marines operating on a cold, snow-covered environment, almost every task they must perform becomes more difficult. Land navigation is no exception to this rule. Long nights, fog, snowfall, blizzards, and drifting snow all drastically limit visibility. An overcast sky and snow-covered ground will create a condition of reduced visibility, which make it difficult to recognize ground features. Prior to beginning a movement, consider the following factors that may affect your ability to find the way to the objective.

    a. In cold weather regions long periods of darkness, snowfall, fog, and wind-driven snow may drastically reduce visibility. At times, overcast skies above open, snow-covered ground may produce a condition known as "whiteout", where the surrounding terrain appears to blend into the sky; terrain features disappear, and depth perception becomes impossible.

    b. Deep snow may completely cover tracks, trails, streams, and improved roads, making them indistinguishable from one another or completely concealing their presence from the observer.

    c. Lakes and ponds, when snow-covered, may be confused with area of open ground. If a body of water appears to have an easily recognizable and distinct shape on your map, this shape may not be so easily recognized when you look at the same body when it is blanketed with snow.

    d. Drifting snow may hide small depressions which appear on your map, or it may change the appearance of small hills by collecting on the leeward side, making the hill appear to be larger or differently shaped than the map indicates.

    e. Aerial photographs taken during the winter may be difficult to read because of their lack of detail and the absence of discernible relief or contrasts.

    f. Except where they cover population centers, maps of cold regions are notoriously inaccurate and outdated. In general, the cold regions of the world are sparsely populated, and are characterized by very limited road networks and a lack of man-made structures. This has meant limited demand for detailed surveys, and makes surveying on the ground difficult, at best. Only recently, with the development of aerial and especially satellite surveying techniques, has this situation begun to improve.

    g. Handling maps, compasses, and other navigational instruments in extreme cold temperatures is difficult with the bulky hand-wear necessary to protect against cold weather injuries. Battery-operated global positioning systems (GPS) may become unreliable or inoperable in the cold weather. The "plugger" GPS will not operate in temperatures below –4°F unless it is kept warm. On an extremely cold day, it may become too cold to operate quickly enough to cannot obtain an accurate position, even if it is carried beneath ones insulating layer of clothing, and only exposed to the cold when needed.

h. In polar regions, there are often wide, open coastal plains or the dense Boreal forests. The lack of terrain features, or total concealment by dense forest, may make terrain association difficult or impossible.

i. As you approach the earth's magnetic poles, compass deflection (expressed as the grid-magnetic angle) increases. The further north or south you move, the more your ability to accurately determine direction will be affected, to the point that when you reach either of the magnetic poles, the needle of a compass will continually rotate about its axis.

j. Cold regions are characterized by many magnetic ore deposits that, if encountered, can cause large deviations from the grid-magnetic angle listed on maps for that area. Prior to beginning a movement consider the following factors, which will affect your ability to find your way to your objective.

TRANSITION: Now since we have been made aware of problems, it is time to look at some navigational techniques..

2. (10 Min) **NAVIGATIONAL TECHNIQUES**. The same methods used to navigate in other parts of the world are used in cold regions; there is no special technique that is used only in the snow or cold. However, certain techniques work better than others in the conditions of both terrain and weather that are characteristic of cold regions.

   a. Direction Method. This is the normal method of navigating using a compass and map. The following are factor to keep in mind when employing this method:

   (1) Compasses.

       (a) Standard liquid filled compasses may become sluggish. Dry filled compasses are more useful in the arctic.

       (b) Metal objects, such as ski poles or weapons will affect the direction the compass needle indicates.

   (2) Steering marks.

       (a) These are well defined points used in maintaining direction when traveling. It is easier to take a bearing on an aiming mark with a compass and then march to it, than it is to continually refer to the compass.

       (b) By day, lone trees, mountains or personnel may be used as steering aids.

       (c) By night, often the only steering marks are stars. A star near the horizon, with a bearing within 2 degrees of your compass course, may be used. It will be good for up to 30 minutes when you are heading north, but only for 15 minutes south. After these intervals choose another star. In latitudes under 70 degrees, when traveling north the Polaris star makes a good aiming point. Since it's bearing is usually only about 1 degree from true north, and is no more than 2-1/2 degrees away from it. In higher latitudes the North Star is too high in the sky to indicate good direction. A guide may be needed at points where men could take a wrong turn.

b. <u>Dead Reckoning</u>. Dead reckoning is a navigational technique that you can use to determine your location by accurately and continuously plotting where you have been. Specifically, dead reckoning consists of plotting and recording a series of courses, each measuring distance and direction from a known point. Dead reckoning is the best technique to use in areas where terrain association is difficult, either due to a lack of terrain features or to your inability to see or identify the existing features.

(1) The following equipment is needed to accomplish this technique:

(a) Appropriate map(s) or aerial photograph(s) of a known scale.

(b) Compass.

(c) Protractor.

(d) Route Card.

ROUTE CARD

UNIT I.D.	UNIT COMMANDER	NUMBER OF PERSONNEL	DATE AND TIME	MAP REFERENCE

LEG	MAG AZM	DIST	GRID	EST TIME	ACT TIME	ELEVATION GAIN/LOSS	DESCRIPTION OF GROUND

(e) Pace counter.

(f) Known distance measuring cord.

(2) Dead reckoning consists of:

(a) Selecting the route to your objective.

(b) Plotting it on a map or an aerial photograph.

(c) Filling out a route card.

(d) Determine an accurate pace count for the method of movement you will employ, i.e. foot, snowshoe, or ski.

(3) While navigating by dead reckoning, you must do the following:

(a) Trust the compass.

(b) Maintain an accurate pace count.

(c) Adjust the plotted route as required to negotiate obstacles.

(d) Record azimuths, distances, times, adjustments to the route, and any pertinent notes in the log. The log should be constantly be updated during movement.

(e) Whenever possible, verify your location by terrain association or resection. Remember, errors are cumulative!

c. Altimeter. The compass merely points the direction to magnetic north. The altimeter merely gives the elevation. But by monitoring the elevation and checking it against a topographic map, Marines can keep track of their progress, pinpointing their location, and find the critical junction in their route. Keep in mind the following when utilizing an altimeter:

(1) The altimeter should be calibrated at a known position before and during the march. This will keep your altimeter current and accurate.

(2) Barometric air pressures will affect the accuracy of the altimeter's reading. As air pressure rises, the altimeter reading will be lower and vice-versa. This does cause havoc when navigating, but it can assist you in predicting future weather changes.

(3) Try to keep the altimeter at a relatively constant temperature by carrying it next to your body. This will help prevent any possible false readings due to temperature exposure.

TRANSITION: One important aspect to remember while applying cold weather navigation, is any one of the three techniques in distance measuring.

3. (5 Min) **DISTANCE MEASURING**. Distance can be measured by pacing or by using a measuring cord.

a. <u>Pacing</u>. When the pacing method is used, the pace must be checked against a measured distance over terrain similar to in which will be encountered during movement. Also keep in mind any aids to the movement that might be employed, i.e. snowshoes or skis. The pace count will be affected by the following:

   (1) Slopes.

   (2) Surface composition/snow density.

   (3) Head/tail winds.

   (4) Weight of clothing and equipment.

   (5) Stamina.

   (6) Level of proficiency (usually only applies if wearing skis).

NOTE: Since it is virtually impossible to maintain a normal pace while breaking trail, pacemen should travel with the trail element of a unit. This will allow them to move on a well-broken trail where they can make an accurate count.

b. <u>Cord Method</u>. Although skis are the fastest, most energy-efficient, and least tiring method of dismounted movement (for Marines with a high level of proficiency) in snow-covered terrain, it is very difficult to maintain an accurate pace count. The following is a technique for determining distance using the cord method:

   (1) The lead man marches off in the desired direction trailing the cord behind him while carrying markers.

   (2) The rear man jerks on the cord when the lead man reaches the end, signaling the leader to place the first marker. The lead man then puts a marker down and starts off again. The rear man follows insuring to keep the cord taut.

   (3) When the rear man comes to the marker, he stops, jerks the cord, and picks up the marker. The lead man stops, checks for snags, and then drops the second marker.

   (4) This procedure is repeated until the required distance has been measured.

   (5) When the rear man has all of the markers, the lead man must wait for him to catch up and either retrieve all of the markers or they can switch roles.

c. <u>Tick off features</u>. Tick off features are features that are used to "tick off' on you route card as you pass by. Keep in mind that poor visibility may hamper this method.

<u>TRANSITION</u>: Now let us discuss some top tips for the navigator.

4. (5 Min) **HINTS FOR THE NAVIGATOR**. Remember the object of navigation is not only to get from point A to point B, but to get there by the best possible route, in keeping with the tactical situation, and with the minimum amount of delay and fatigue.

   a. Keeping the compass warm will speed up the taking of bearings.

   b. When no aiming marks exist to your front, march on a back azimuth. Your aiming mark may be a natural feature to your rear, or an artificial aiming mark left behind by you.

   c. When visibility is poor only close in aiming marks will be seen. Under these conditions the navigator should try to pick up further aiming marks along the correct bearing as he approaches each one. This can only be done accurately when the route to each mark follows a straight line. Frequent compass checks should be made to ensure that the correct bearing is being maintained. The compass should also be set for night marching.

   d. Distance Measurement

      (1) Pace counting is useful only if you know the length of your pace.

      (2) Remember to convert paces to meters.

      (3) Measuring distance by, a known distance length of cord, is more accurate than pacing.

TRANSITION: Let's see what's needed before the march.

5. (5 Min) **BEFORE THE MARCH**

   a. Planning

      (1) Before the march begins, the route should be planned in detail from route reconnaissance, maps, and air photographs. The easiest route, consistent with tactical demands, should be chosen.

      (2) Route cards should be completed before the march begins. The leader of the trail breaking party must be thoroughly briefed.

      (3) Not only leaders and navigators, but all Marines should know the route. During the march, the men should be kept informed of their progress and at each halt should be told the distance yet to be marched and how long it should take to get there, plus any changes to the route that are made.

      (4) Camouflage and security must be upheld at all times.

TRANSITION: Are there any questions?

PRACTICE (CONC)

a. Students will practice what was taught in upcoming field evolutions.

PROVIDE HELP (CONC)

a. The instructors will assist the students when necessary.

OPPORTUNITY FOR QUESTIONS (3 Min)

1. QUESTIONS FROM THE CLASS

2. QUESTIONS TO THE CLASS

   Q. What are the three methods of measuring distance using cold weather navigation methods?

   A. (1) Pacing
      (2) Cord method
      (3) Tick off features

   Q. What are four areas where cold weather navigation problems exist?

   A. (1) Deep snow
      (2) Magnetic disturbances
      (3) Air photos
      (4) Handling maps, compasses, and other navigation tools with bare hands

   Q. What are the methods assisting you in cold weather navigation?

   A. (1) Direction Method
      (2) Dead Reckoning
      (3) Altimeter

SUMMARY (2 Min)

a. This period of instruction has discussed those aspects of navigation that are affected by a cold weather environment such as map reading, finding location, pacing and determining distance, and route planning.

b. Those of you with IRF's please fill them out at this time and turn them in to the instructor We will now take a short break.

UNITED STATES MARINE CORPS
Mountain Warfare Training Center
Bridgeport, California 93517-5001

WML
WMO
08/15/01

LESSON PLAN

EFFECTS OF COLD WEATHER ON INFANTRY WEAPONS AND OPTICS

INTRODUCTION (4 Min)

1. **GAIN ATTENTION**. Every type of environment has a definite effect on the operation of your weapon, dust in the desert, or mud/rain in the jungle, etc. This is particularly true in a cold weather environment. Since effective operation of your weapon is absolutely essential to both the success of the mission, and to personal survival, an understanding of how the cold will effect your weapon, and the techniques used to counteract these effects, will be of importance.

2. **PURPOSE**. The purpose of this period of instruction is to familiarize the student with the effects that cold weather will have on weapons systems. This includes; individual and crew-served weapons, mechanical and electrical systems, and procedures to alleviate these problems. This lesson relates to cold weather tactical operations.

3. **INTRODUCE LEARNING OBJECTIVES**

    a. TERMINAL LEARNING OBJECTIVE. In a cold weather mountainous environment, operate and maintain your assigned weapon system efficiently, in accordance with the references.

    b. ENABLING LEARNING OBJECTIVES

    (1) Without the aid of references, list in writing the six problems with weapons in general encountered in a cold weather environment, in accordance with the references.

    (2) Without the aid of references, state in writing the temperature ranges of CLP, LAW, and LSAT in accordance with the reference.

(3) Without the aid of references, select from a list, the ways to compensate for ice fog in accordance with the reference.

(4) Without the aid of references, state in writing how to compensate for slope angle when judging target range sight data in accordance with the references.

(5) Without the aid of references, state in writing how backblast is effected by cold in accordance with the reference.

(6) In a winter environment and given a weapon, maintain the weapon in accordance with the references.

(7) In snow-covered terrain, given a weapon, skis, ski poles, engage targets while skiborne in accordance with the references.

4. **METHOD / MEDIA**. The material in this lesson will be presented by lecture and demonstration. You will practice what you have learned during upcoming field training exercises. Those of you with IRF's please fill them out at the conclusion of this period of instruction.

5. **EVALUATION**. You will be tested by written and performance evaluation.

TRANSITION: Before we can fully understand the effects that cold weather will have on weapons, we must know how the cold will effect the individual materials from which weapon and their accessories are made. So let's first take a look at the effects of cold on materials.

BODY (55 Min)

1. (5 Min) **EFFECTS OF COLD ON MATERIALS**. In general, most materials used in the construction of weapons will became more brittle and less able to handle stress or impact without breakage. Some materials will be severely effected such as metals, plastics and rubbers; while others such as glass (fiberglass) and ceramics (ceramic fiber, like kevlar), will be virtually uneffected.

    a. Metal, Plastic and Rubber. All will become brittle and more susceptible to damage as temperature drops, thus necessitating careful handling.

    (1) Steel. Impact strength of steel will drop slowly until it reaches a critical temperature, when there will be a very drastic drop. This critical temperature can vary widely, depending on the quality, and heat treatment of the steel.

    (2) Volume or dimensional changes can occur. In cold weather, metal components of weapons will contract. This can be particularly important in close tolerance parts, such as bearings. During firing, these components will heat up and expand rapidly and possibly unevenly. This will increase the chances of breakage or malfunction.

b. Lubricants and Other Liquids as in leveling bubbles may become thick, gummy, or frozen.

c. Propellants will burn slower.

d. Electrical Power. Batteries of all types deliver less power at low temperatures. Dry batteries should be stored above 10°F and should be gently warmed before use. Lithium batteries perform best and last longer when kept cool, rather than frozen. Optimum storage temperature is 35°F or slightly cooler.

TRANSITION: Let's now look at what kind of problems will be encountered by weapons in general.

2. (5 Min) **PROBLEMS AND SOLUTIONS PERTAINING TO WEAPONS IN GENERAL**. Several problems will be common to most weapons available to the infantryman in the field. They are:

   a. Sluggishness. This is normally caused by improper lubrication.

      (1) For all weapons, lubrication Cleaner, Lubricant, and Preservative (CLP) will be used. CLP is good in cold climates from -35°F and above. After cleaning a weapon with CLP, apply a light coat to provide extra lubrication and corrosion protection.

      (2) Temperatures. 0 to -65°F, Lubricant Arctic Weather (LAW) should be used as CLP will freeze at -35°F. If no LAW is available, the weapon should be fired dry. Lubricant Sub-Artic Temperature (LSAT) will freeze at 0°F.

      (3) To prepare a weapon for cold weather operation, the weapon should be stripped of all lubricants and rust preventatives. Then a light coat of CLP should be applied.

      (4) Sluggishness may also be caused by ice or snow in the firing mechanism. If this is the case, the weapon must be cleaned, dried, and lubricated with a light coat of CLP.

   b. Breakage and Malfunction. The higher rate of breakage and malfunction in cold weather, is normally caused because of the factors we discussed earlier; however, malfunctions can be caused by snow and ice.

      (1) Breakage. This normally occurs during the first few rounds from the uneven expansion of metal parts due to the rapid temperature rise. Firing slowly or in short bursts, at the sustained rate initially, will give the weapon time to warm up and alleviate the problem. When automatic weapons must be fired at the rapid rate (such as in an ambush) plan for a minimum of 2 automatic weapons and one to break upon engagement.

      (2) Malfunctions. Malfunctions, due to ice and snow, can be reduced by careful handling and use of muzzle covers. Be particularly cautious when moving through forested or brushy areas to avoid getting ice or snow in moving mechanisms.

c. <u>Condensation</u>. Condensation forms on weapons when they are brought from cold environment into a heated shelter. The moisture in the air condenses on a surface, leaving a film of moisture. This film of moisture can freeze when taken outside again, causing malfunctions.

   (1) This condensation or sweating will continue for approximately 1 hour in a heated shelter. After an hour, the weapon should be cleaned and lubricated.

   (2) To prevent or reduce this effect, weapons should be stored outside of the tent in the rifle pit. If the tactical situation doesn't permit the weapons to be stored outside, weapons can be stored in the vestibule of the ECW tent. Ammunition (mags, LBV, belts, etc.) should be stored outside or in vestibule, too.

d. <u>Visibility</u>. In a cold weather environment, falling or blowing snow, whiteouts, and greyouts, may severely hamper visibility, and the ability to deliver accurate fire. Nights are longer in winter and the higher in latitude, the longer the night.

   (1) Ice fog. In extreme cold, -35°F and below, and where conditions of little or no wind exist, (less than 3 mph), ice fog may form upon firing of the weapon. Ice fog forms when moisture in the air is crystallized when a projectile leaves the muzzle, and travels down range. Ice fog can hover over the weapon and along the path of the projectile for several minutes, obscuring vision and pinpointing the weapon's position. Weapons must then be displaced to an alternate position.

e. <u>Emplacement</u>. Crew-served weapons requiring some type of base or platform for firing will need special consideration. Emplacement of a weapon on snow, ice, or frozen ground, may result in breakage, or inaccuracy because of sinking, or the inability to absorb shock. Emplacements relating to particular weapons will be discussed in the section pertaining to that weapon.

f. <u>Reduced Velocity and Range of Projectiles</u>. As temperature drops, so does the muzzle velocity, and thus the range of projectiles. This is because of a change in both internal and external ballistics.

   (1) Internal ballistics. This occurs inside the weapon; burning rate of propellant decreases, thus decreasing the rate of gas expansion, and in turn decreasing speed of projectile down the bore. Closed breech weapons gases are contained within a specific volume.

   (2) External ballistics. This occurs after the projectile leaves muzzle. Decreased muzzle velocity reduces stability of the projectile once it leaves the muzzle, possibly severe enough to cause projectiles to tumble. At longer ranges, this further reduces velocity, range, and accuracy. Colder air is denser than warmer air which increases drag on the projectile, further reducing range.

   (3) Weapons should be test fired to establish BZO and new range cards.

(4) Slope angle considerations. Gravity effects the trajectory of a round only in the horizontal distance traveled. This is true when firing either uphill or downhill. This causes overestimation of range (line of sight distance to target), the shooter hits high whether firing uphill or downhill. To compensate for the effect for slope angle; use the map distance for sight settings, estimate range in the horizontal plane if a map is unavailable, aim at 6 o'clock or less and adjust from impact. Laser range finders give line of sight range, so the sight data will always be less. The steeper the slope angle, the greater the sight data estimation problem.

SLOPE ANGLE, EXAMPLE:

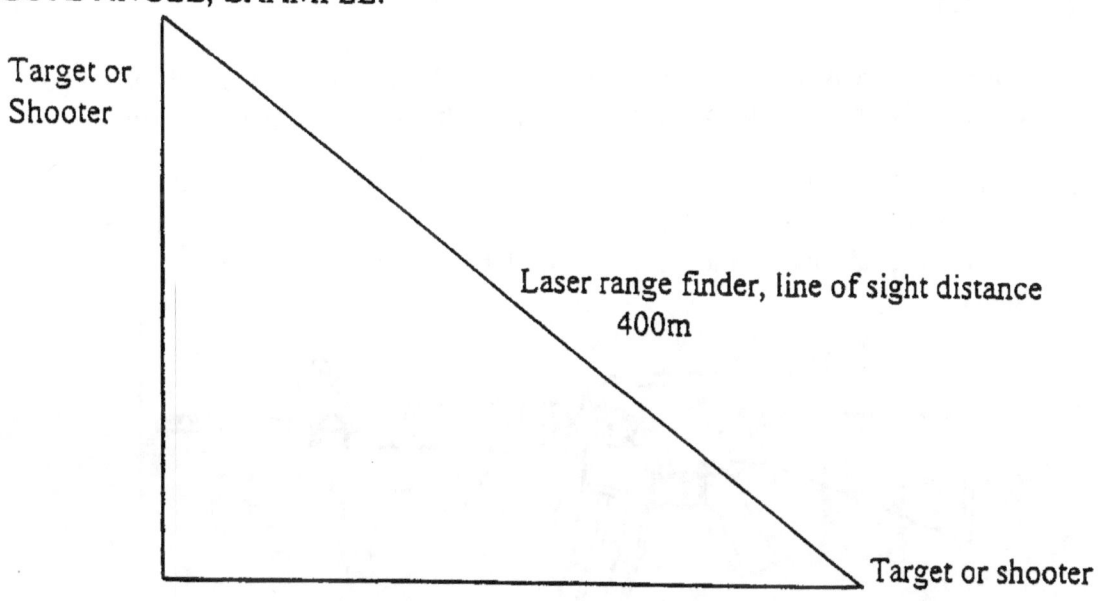

TRANSITION: Let's now take a look at some specific problems with the following weapons.

3. (15 Min) **PROBLEMS AND SOLUTIONS PECULIAR TO INFANTRY WEAPONS**. In addition to the problems encountered with weapons in general, each particular weapon will have its own peculiar problems. The following problems are by weapons organic to an infantry battalion, and require proper techniques to combat these problems:

   a. <u>M9 9mm Pistol</u>

      (1) Damage or breakage to moving parts due to rapid warming or ice and snow in the weapon (firing pin, extractor/ejector).
      (2) Difficulty operating and firing weapon while wearing mittens or gloves.

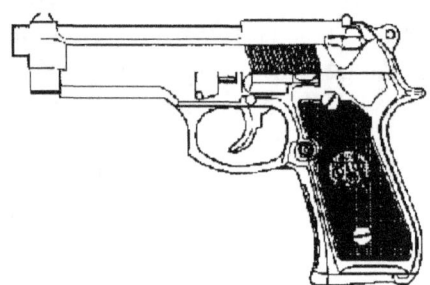

   (3) Magazine freezing in magazine well.

b. <u>M16A2 Rifle</u>

   (1) Damage or breakage to moving parts due to rapid warming or ice and snow in the weapon (firing pin, extractor/ejector). Fire at the slow rate to allow warm up.

   (2) Condensation in the buffer tube will hamper the shock absorbing ability, which may result in breakage and retarded recoil, which can result in the omission of the cocking step in the cycle of operation. Seal the outlet hole with wax or gum and wipe out buffer tube regularly.

   (3) Ice fog. Although rifles will create ice fog, an individual can adjust his position easily and as frequently as necessary. Plan and prepare extra firing positions.

   (4) Rifle preparation; use white cloth tape for camouflage and to reinforce handguards and butt stock, select muzzle cover (must be able to shoot through it for immediate action) and carry spare (tape also works well), select magazine well cover (a mag works easiest), ensure ejection port is closed at all times, open trigger guard for use with mittens/thick gloves (can leave closed but unlatched for safety).

c. <u>M249 Squad Automatic Weapon</u>

   (1) Damage or breakage to moving parts due to rapid warming or ice and snow in the weapon.

   (2) Increased breakage, malfunction and ice fog compared to the M-16A2 in part due to increased rate of fire.

   (3) Attach ski pole baskets or assault snowshoe, etc. to bipod legs for flotation in snow.

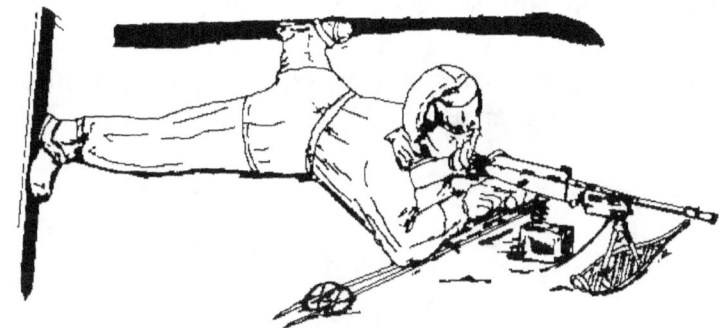

(4) Protect the ammo belt! Do not move with a belt in the feed cover, it is too easy to get a jam from snow/ice. Protect the feed cover and use a muzzle cover.

d. <u>M203 Grenade Launcher</u>

(1) Breakage. The M-203 is not very susceptible to breakage except the handguards in extreme temperatures. However, the rule of careful handling should still be observed, particularly in relation to sights.

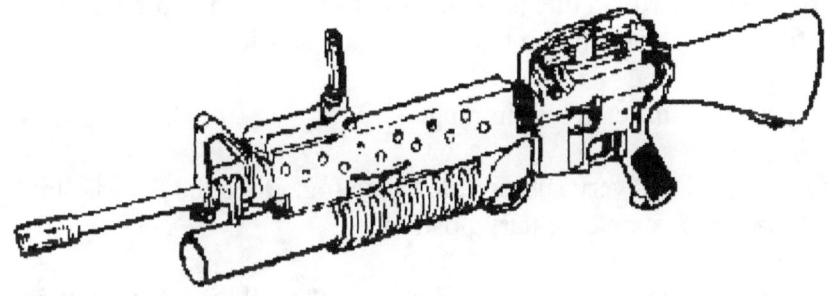

(2) Freezing. May freeze around slide.

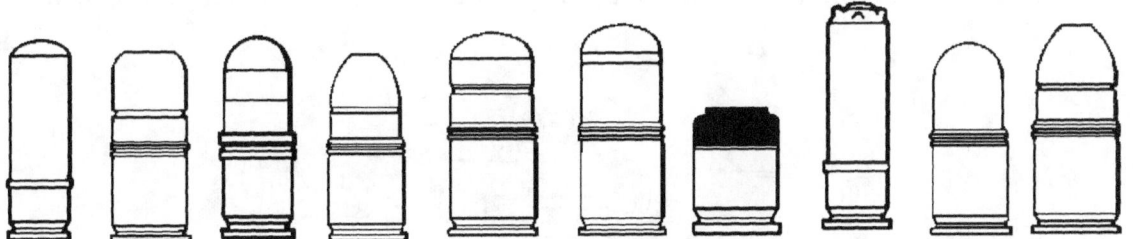

(3) Effectiveness of round. Will be reduced because of the dampening effect of snow, and an increased number of duds due to the impact fuse not detonating in soft snow. Grenadiers should fire into trees to obtain airbursts, if possible.

e. <u>AT4</u>. Components that are effected in cold weather are:

(1) Sights. Plastic sights become brittle and break easily.

(2) Rubber end caps. Become brittle and damage easily.

(3) Propellant. Propellants burn slower and less efficiently in the cold, as a result of this missile velocity is slightly decreased.

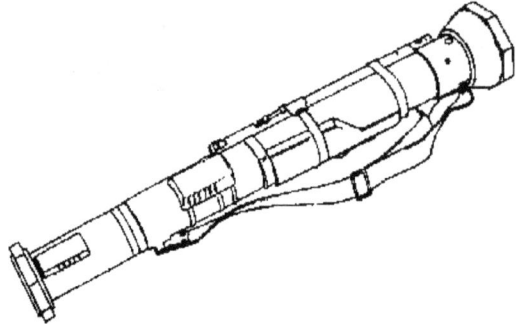

(4) Hangfires are not noticeably increased in this environment.

(5) Cocking handle. Snow and ice build up on cocking mechanism can make the AT4 difficult to cock.

(6) Backblast area is tripled due to several factors (unburned propellants, snow/ice projectiles, lack of vegetation to absorb it, air density, etc.).

(7) There is not a significant loss in range.

(8) Ice Fog. The AT4 is very susceptible to ice fog. Once the AT4 is fired, the individual should move to a supplementary position.

f. **M153 SMAW**: Components that are effected by the cold weather include:

(1) Propellant. Propellants burn slower and less efficiently in the cold, as a result of this missile velocity is slightly decreased. When firing at temperatures below freezing, set the range drum selector switch to the blue mark to compensate for range drop in the cold.

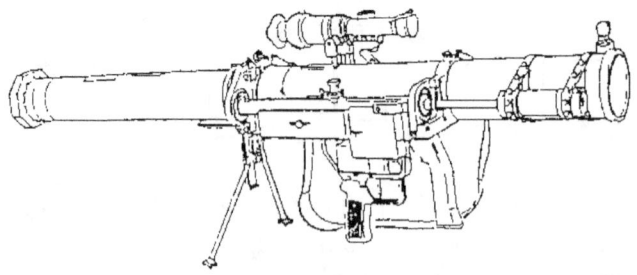

(2) Firing Mechanism. Ice and snow tend to collect in the firing mechanism and can be difficult to break out.

g. <u>Grenades</u>. Grenades are not susceptible to breakage, but do have certain problems peculiar to them.

   (1) Casualty radius. Reduced effective casualty radius because of the dampening effect of snow.

   (2) Platform. Smoke or gas grenades must be tied to a platform to prevent sinking into the snow.

   (3) Hand gear must be dry when handling grenades to prevent skin/glove from freezing to them.

   (4) Procedures for throwing must be modified when using hand gear in the following manner:

      (a) Hold near neck of fuse to insure a positive grip.

      (b) Spoon should rest between first and second knuckles of thumb to insure sensitive feeling.

h. <u>Mines</u>. Mine warfare, in a cold weather environment, calls for special precautions to be successful.

   (1) As little as six inches of snow can virtually neutralize the effects of antipersonnel and anti-armor mines.

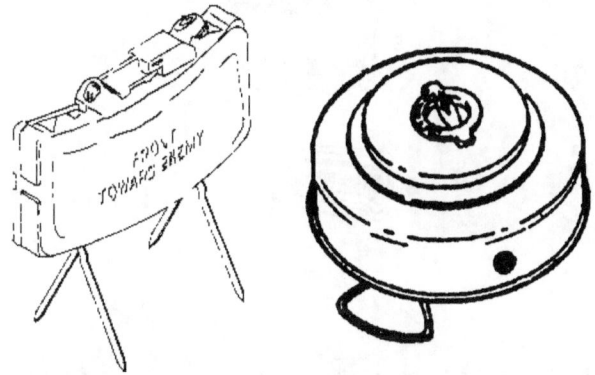

   (2) Digging in mines becomes nearly impossible on frozen surfaces.

(3) To prevent mines from sinking into deep snow, crossed branches or sticks can provide a useful platform. Also mines may be stacked to account for some sinking.

(4) A large amount of snowfall in a single night can make mines placed prior to that ineffective.

i. <u>Demolitions</u>. The cold effects the ability to employ demolitions effectively. Handling becomes a team effort, due to the requirement for wearing gloves.

   (1) Plastic explosives (C-4) becomes very hard, making insertion of a blasting cap difficult unless done indoors prior to use.

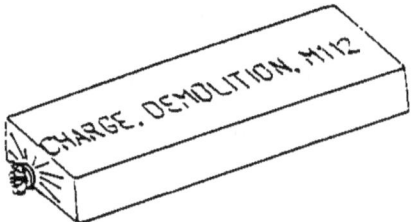

   (2) In extreme cold, C-4 has been known to simply shatter from the explosion of the blasting cap rather than to detonate.

   (3) TNT blocks are excellent for use in cold weather for projects like hasty fighting or firing positions due to the water resistance.

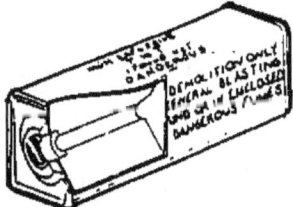 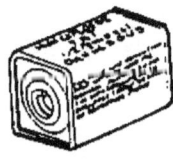 

   (4) Detonation cord becomes stiff and is difficult to tie and will break easily in extreme cold weather.

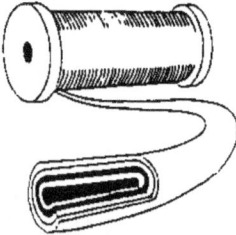

   (5) Time fuze tends to maintain its curl when extremely cold. Unless done inside a warm shelter, uncurling usually results in breaking the fuze.

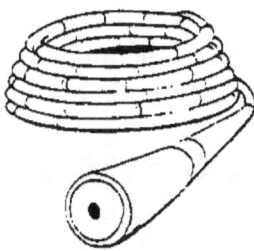

(6) Condensation also contributes to the increased chance of misfires. Misfire and hangfire waiting times should be doubled in a cold environment.

j. <u>Lasers</u>. Laser sighting equipment is effected by visibility problems, such as ice fog and blowing snow, common in a cold weather climate. Lasing a snow or ice covered target may result in refraction of the laser up to several miles off target. There is a documented case of a missile tracking a refracted signal in Norway which landed off-base, not just off-target.

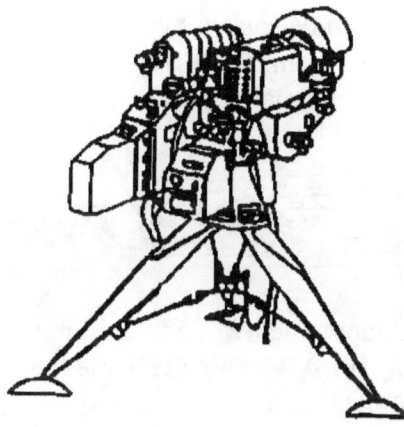

(1) The HHLRF can only view for about 300 meters in heavy ice fog or snow due to light refraction. The human eye can see approximately 200 meters further in the same conditions.

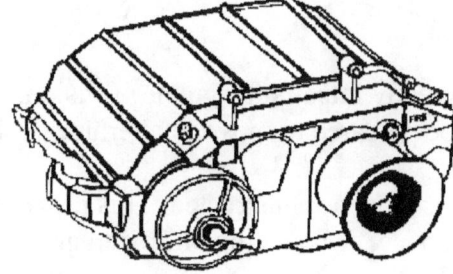

(2) When the weather is clear, cold, and dry, the HHLRF will be accurate to within one meter at a ten thousand meter range.

TRANSITION: Individual weapons will not be the only weapons a Marine will have to contend with. Crew served weapons will be extremely important to any mission, and therefore, the infantryman should know the problems and limitations of these weapons.

4. (10 Min) **PROBLEMS AND SOLUTIONS PECULIAR TO CREW SERVED WEAPONS**. Because of the inherent complexity of crew served weapons, i.e., mortars, heavy machine guns, anti-armor weapons, etc., a greater number of problems will arise in cold weather operations. The complexity of operations, and subsequent problems encountered, are far too difficult to cover in any great detail in one class. However, the major difficulties, and solutions, will be discussed in this section. For detailed operating instructions in a cold weather environment, you should refer to the particular TM pertaining to each individual weapon.

    a. <u>Machine Guns M60, M240G, M2, MK19</u>

        (1) All machine guns have a high rate of breakage due, in part, to the cold and their high rate of fire. Gun crews should carry extra gear and bolt components.

            (a) A short, violent recoil can be caused by slower burning propellants or by the freezing of the buffer mechanism. This shortened recoil results in the increased probability of breakage and malfunction.

            (b) If temperatures reach down below -35°F, all internal components should be lubricated with LAW. If LAW is not available the weapon should be fired dry. For the MK-19, LSAT freezes at 0°F, so CLP or LAW is then used.

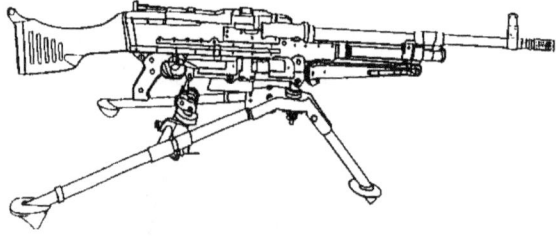

            (c) Since ice fog greatly impairs the gunner's vision, and identifies the gun's position, crews should be prepared to move to alternate positions.

            (d) Firing platforms must be constructed. This prevents the gun from sinking into snow, sliding on ice, or bouncing on frozen ground.

                1. Different materials may be used. MRE boxes, sandbags, sleds (buffed and anchored), or evergreen boughs may be used to prevent sinking in snow or to cushion ice, and frozen ground.

2. Ski pole baskets may be wired to the bipod legs of medium machine guns for flotation.

3. A sled may be used to provide a stable, yet mobile platform for a tripod mounted gun.

(e) Never place hot barrels directly on snow or ice.

1. Warping or cracking will result due to the sudden temperature changes.

2. Barrel may also sink out of sight, necessitating a delay in changing barrels.

b. <u>Ammunition</u>

(1) Ammunition must be kept free of all snow and ice, and should be checked frequently.

(2) Ammunition should be stored in its original container, raised off of the ground 4-6" and covered with ponchos, tents or any other material that will protect it from the snow.

c. <u>Mortars 60mm, 81mm</u>

(1) Seating the gun. Yokes and baseplates have been known to crack at temperatures below -25°F and on uneven rocky or frozen ground.

(a) When possible select a position that has vegetation on the ground or at the minimum under the baseplate.

(b) Use two men on the bipod when seating on ice or at high angle elevation.

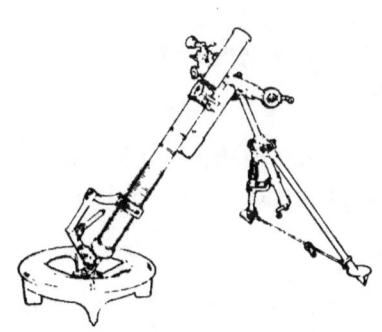 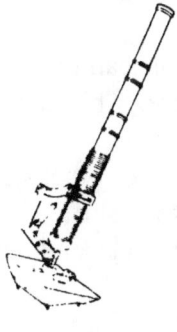

(c) Dig to the ground, if possible. If the ground is not frozen, seat normally. If it is frozen, break through the frozen layer with picks, shaped-charges, demolitions, burning the surface to thaw it, etc. If you cannot break through, fill sand bags with rocks, snow, ice etc. to buff the baseplate, or use any field expedient material. It should be thick enough to provide shock absorption between the baseplate and the frozen ground but not so much as to cause bouncing. If the snow is too deep to dig down, a flotation base must be built on the snow for the mortar. Pack the snow and place logs, or any suitable material, firmly on the snow. Then place snow-filled sand bags on top of the logs. This is the least preferred method. Digging to the ground and breaking it up is the most preferred method. Mortars can be set in on ice (frozen lakes, rivers) very easily, just by using a pick to chip the ice. Leave the chipped ice in the hole for the baseplate, it fills in around the fins providing excellent support.

BRUSH UNDER THE BASEPLATE

(2) Sighting and fire control leveling vials will take longer to adjust due to the liquid being thicker when cold.

(3) Care must be taken to avoid breathing on the sights, as this will cause them to fog up.

(4) Aiming stakes should not be pounded into frozen ground, or pried out of it.

    (a) A hole can be dug, stake emplaced, then water poured around base and allowed to freeze. This is the least preferred method

    (b) Rocks or blocks of ice/snow may be used, but melting of ice or snow may cause stake to shift. This is the chock method, the most preferred. An ammo can filled with rocks and dirt also works well.

NOTE: The length of time in a position will dictate the method used to secure the aiming stakes.

(5) Ammunition will be greatly effected.

    (a) VT fuzes are preferred in snow due to the severe dampening effect of snow. Malfunctions will occur in proportion with the severity of the weather. Airburst is preferred.

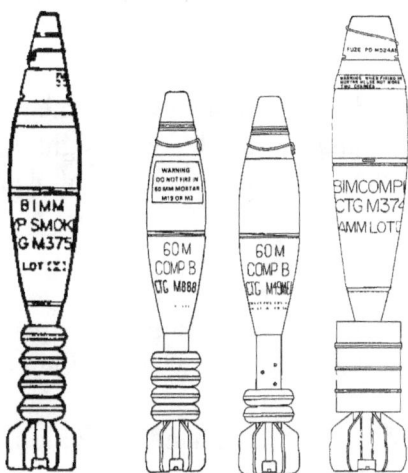

  (b) Tests conducted at -25°F experienced higher fuze failure at charge 0, but reliability increased when fired at a higher charge zone (1, 2, 3, or 4).

 (6) Contact gloves must be worn to prevent skin from freezing to metal.

  (a) Handwheels should be taped to provide better grip with gloved hands.

d. <u>Anti-tank Guided Missiles - TOW and Dragon</u>. All systems will operate effectively in cold weather down to –25°F which is the milspec criteria. Some problems which can be expected to be encountered are:

(1) During cold weather firing (10°F and below) there is a possibility of reduced reliability with the TOW missile due to blowout of the missile capstan block which connects the launcher to the missile guidance wire.

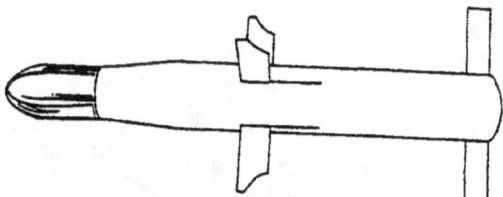

(2) When firing in cold weather the maximum effective range of the TOW missile is reduced to 3400 meters vice 3750 meters.

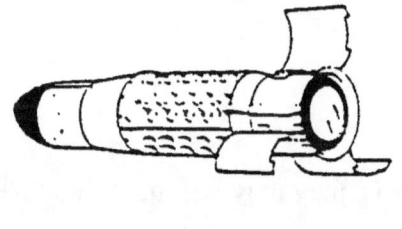
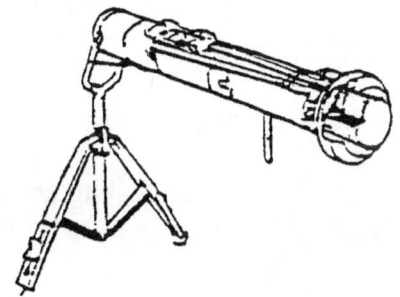

(3) The TOW, Dragon and Javelin can be operated at temperatures down to -25°F and stored at temperatures down to -65°F.

(4) Before firing the TOW, ensure that the night sight is turned on and focused as the missile guidance system (MGS) may select the night sight due to obscuration from ice fog or falling/blowing snow. (The gunner has no control over which sight is used).

(5) Keep in mind that the backblast area is tripled.

(6) Snow glare may create difficulties in the tracking of the missile to its target for the TOW and Dragon. Also wear eye protection when looking through the sight.

(7) Cover all exposed skin, including face, eye, hand protection, in case of still burning launch propellants as the missile exits the tube.

(8) The Javelin has a slight drop when fired in the cold, so do not fire from partial defilade or reverse slope positions.

(9) The Javelin and Dragon are easily man-portable and easy to emplace in snow. The TOW will be road bound with the HMMWV.

(10) Marines may be needed to augment sled teams when moving bulk missile reloads. However movement will still be slow and difficult.

5. (5 Min) **OPTICS**. Due to the delicate nature of optics, there are some special considerations to keep in mind when dealing with optics in a cold weather environment.

   a. <u>Night Vision Goggles</u>. When operating NVGs ensure:

   (1) Use the cold weather battery adapter.

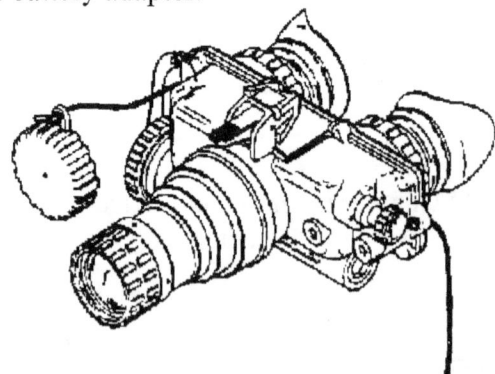

   (2) Use the de-misting shield.

   (3) Use extra care when dealing with moving parts, they may become brittle in extreme cold.

b. Thermal sights

   (1) A cold weather environment has a significant effect on the thermal infrared signatures of targets, enhancing them unless weather conditions are degrading (snowstorms etc.)

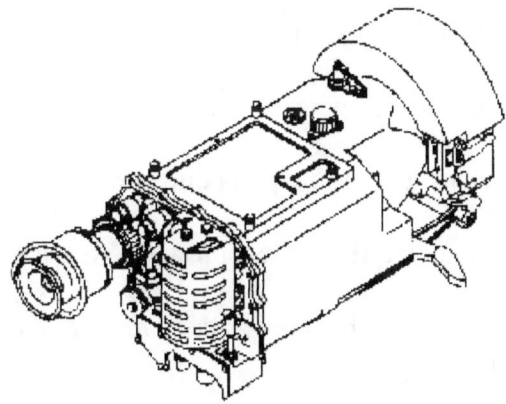

   (2) An undisturbed snow cover will present a relatively uniform and clutter free background to a thermal infrared sensor, provided that the snow is deep enough to completely cover a large area.

   (3) Thermal camouflage can be used to defeat these systems, such as reflective insulation (space blankets), one foot of snow on a poncho, etc. Note that movement (ski, foot, etc.) through snow will leave a thermal trail in the snow for a short time.

c. AN/PVS-4

   (1) Ensure that you use the daylight cover while on the move to protect from snow and ice from getting packed into the scope.

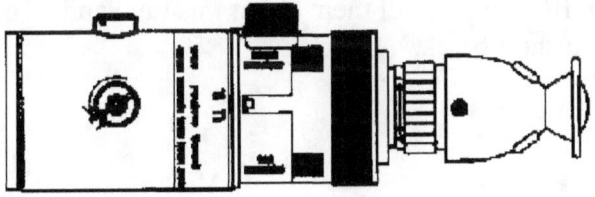

TRANSITION: In order to accomplish the mission, you must always keep these things in mind. Are there any question?

PRACTICE (CONC)

   a. Students will practice what was taught in upcoming field evolutions.

PROVIDE HELP (CONC)

   a. Instructors will assist the students when necessary.

OPPORTUNITY FOR QUESTIONS (3 Min)

1. **QUESTIONS FROM THE CLASS**

2. **QUESTIONS TO THE CLASS**

    Q. What are two main problems encountered with the M9 9mm pistol?

    A. (1) Damage or breakage to moving parts due to rapid warming or ice and snow in the weapon.
    (2) Difficulty operating and firing weapon while wearing mittens or gloves.

    Q. What are the six problems with weapons encountered in a cold weather environment?

    A. (1) Sluggishness.
    (2) Breakage and malfunction.
    (3) Condensation.
    (4) Visibility.
    (5) Emplacement.
    (6) Reduced velocity and range of projectiles.

SUMMARY (2 Min)

   a. During this period of instruction we have discussed several topics; the effects of cold on materials, general problems encountered by most weapons in the cold, problems that can occur with individual and crew served weapons, mines, demolition's, laser equipment, and optical devices. We also discussed some techniques used to counteract these problems.

   b. Those of you with IRF's please fill them out at this time and turn them into the instructor. We will now take a short break.

UNITED STATES MARINE CORPS
Mountain Warfare Training Center
Bridgeport, California 93517-5001

WML
WMO
09/04/02

LESSON PLAN

COMMUNICATIONS CONSIDERATIONS IN A COLD WEATHER ENVIRONMENT

INTRODUCTION (5 Min)

1. **GAIN ATTENTION**. Communicating in a cold weather, mountainous environment presents unique challenges that must be identified and overcome. Weather, geography, and altitude are all factors that can effect equipment, personnel, and communications organization. The key to success in combat is reliable, secure, rapid, and flexible communications. Commanders must understand cold's effects on their communications systems and their personnel, and know the procedures to counteract these effects.

2. **PURPOSE**. The purpose of this period of instruction is to familiarize communicators with the problem they will encounter in a cold weather environment, techniques that can use to combat these problems and recommendations for command group communication systems for foot mobile and mechanized winter mountain operations. This class relates to Wave Propagation, Antenna Theory, and Field Expedient Antennas.

3. **INTRODUCE LEARNING OBJECTIVES**

   a. TERMINAL LEARNING OBJECTIVE. In a cold weather mountainous environment, plan cold weather mountain operations, in accordance with the references.

   b. ENABLING LEARNING OBJECTIVE. In a cold weather/mountainous environment, demonstrate preventive measures for operating communication equipment in a cold weather environment, in accordance with the references.

4. **METHOD / MEDIA**. The material in this lesson will be presented by lecture and demonstration. You will practice what was learned during upcoming field training exercises. Those of you with IRF's please fill them out at the conclusion of this period of instruction.

5. **EVALUATION**. You will be evaluated by performance evaluation during the upcoming field exercises.

TRANSITION: Let's first take a look at some considerations for planning.

BODY (75 Min)

1. (5 Min) **PLANNING**. The primary communication planning factors that must be considered by battalion and company staff personnel during cold weather communication planning are:

    a. Communication equipment.

    b. Communication maintenance and supplies.

    c. Safety.

    d. The equipment load communications personnel must carry in addition to their required personal equipment.

    e. Communication plans.

    f. Additional personnel and equipment needed to man retransmission sites and to conduct mountain-picketing operations.

    g. Command group communication system configurations.

TRANSITION: The first consideration in using communications we will discuss, is the radio.

2. (10 Min) **RADIO**. Radios are the most common means of communicating. They are subject to many problems in the cold. The two major problems are reduced battery power and increased equipment failure.

    a. <u>Battery power</u>

    (1) Dry cell batteries lose efficiency and produce less power if not protected from cold weather.

    (a) Spare batteries. Store spare batteries inside heated shelters. They should be stored at a temperature above 10°F and gently warmed before use. Carry one spare set of batteries in a parka or trouser pocket between your body and your outside layer of protective clothing.

    (b) Snow. Never place the battery in the snow or unprotected against the shell of a sled. If snow covers the pressure release cover on the radio battery box, ice may form over it which may restrict it from air exchange.

    (c) Rotate batteries every 4 hours with a spare and label them with the amount of time used. Keep logs entries when batteries are changed.

(d) Always use cold weather batteries.

(2) <u>Lithium batteries</u>. Lithium batteries are superior to magnesium batteries in the cold. They are lighter, last longer, and perform best when kept cool (but not cold or freezing temperatures). These are hazardous material because of an explosive chemical reaction that may occur when not ventilated or when they are in contact with water or fire.

(a) Optimum storage temperatures are 35°F or slightly colder.

(b) These batteries should be serialized and will be accounted for during and after each exercise.

(c) Radio operators will keep the plastic bag that lithium batteries are issued in and repack the battery for turn-in when battery is depleted. This protects the battery from moisture.

(3) <u>Cold weather batteries</u>

Dry Cell	**Expected Life**	**Transmit/Receive Life in Hours**	**C/W Battery**
BA 4386	36-48 hrs	1/9	BA 5598
BA 30	24-48 hrs		BA 3030
BA 1588	18-36 hrs	1/9	BA 5588
No corresponding dry cell battery	72-96 hrs	1/9	BA 5590

b. <u>Material Failures</u>. When temperatures are below 10°F, radio equipment becomes brittle and is very susceptible to breakage.

(1) AT-271A (10 foot whip)

(a) Difficulty during movement in thick vegetation.

(b) Antenna may break if radio operator falls.

(c) Company/platoon radio operators should carry a spare AT-271A.

(2) RC-292. Coaxial cable, connectors, and antenna elements must have a thin coat of silicone lubricant spread on components.

(3) AN/GRA-39. The radio remote cables, and connectors must have a thin coat of silicone lubricant.

(4) H-250 or H-189. The handset cable, and connector, must have a thin coat of silicone lubricant.

(a) Press-to-talk button is subject to sticking from freezing. Company radio operators should carry a spare handset.

(b) Microphone. Moisture cover for handsets should be used to prevent moisture from freezing in microphone, or a plastic bag can be wrapped around the handset.

(5) Ice and snow. Keep radios, remote sets, telephones and cryptographic equipment off of the ice and snow.

(a) Radio remote antenna stations should use an ECW tent at the antenna station to keep the radio equipment warm.

(b) Fabricate insulated cold weather bags for radios, and equipment, if tent cannot be used.

(6) Condensation. Do not bring radio equipment from 0°F into warm tents above temperatures of 40°F, because the equipment will sweat, causing moisture to short the radio circuitry.

(a) Remove frost from the equipment, before bringing it into the tent.

(b) Operators must gradually warm equipment and batteries.

(7) Nighttime. Do not turn radio equipment off at night, if on line, and if needed for operation in the morning, unless equipment is in temperatures above 10 to 20°F.

(8) Operator maintenance. Because the polar regions are subject to disturbances, which affect radio reception, it is important to get the very best performance from radio sets. Operators must be intimately familiar with their equipment and should keep their radio equipment clean, dry, and where possible, warm.

(a) Always keep plugs and jacks clean.

(b) Antenna connections must be tight.

(c) Keep insulators dry and clean.

(d) Always remove snow and ice.

(e) Power connections must be tight.

(f) Motors and fans should turn freely.

(g) Knobs and control should operate easily.

(h) Keep batteries fresh, warm, and spares on hand.

(i) Install breath shields on all handsets.

(j) Coat cables and wires with silicone.

TRANSITION: The next topic we will discuss, will be wire communications. Wire is one of the most preferable types of communication in the cold, because there are less problems with actual communication; although, installing and maintaining the network can create several problems.

3. (5 Min) **WIRE (TELEPHONE)**

   a. Battery Power

   (1) BA 30 dry cell battery. It is used in field telephones, and is not a reliable battery in temperatures below 30°F.

   (2) BA 3030. This is the cold weather battery replacement.

   b. Material Failures. Like radio equipment when temperatures are below 10°F, telephone equipment materials become brittle, and are very susceptible to breakage.

   (1) Field telephone handset cables. Cables must have a thin coat of silicone lubricant.

   (2) TA 312 field phone. This provides the best wire communications; however, a microphone moisture cover must be installed in the telephone.

   (3) TA 1 sound powered telephones have a carbon element microphone which freezes, and needs to be kept warm and dry to operate.

   c. Planning Considerations

   (1) Wire line route maps should be drawn by the wire chief at the battalion headquarters alpha command, so wiremen have a recommended wire laying path to each company, and attachment.

   (2) Vehicles. Laying wire may take time unless a over-the-snow capable vehicle is available.

   (3) Helicopter and a wire dispenser case CY 1064A. Which holds five rolls of 1/2 mile communication wire. Allowing for a slack factor, the helicopter can lay about 1 1/2 miles of wire.

   (4) Crossing roads. Either bury the wire 6-12 inches down, or overhead 18 feet above the road.

   (5) Wire may be laid by over-the-snow mobile troops on snow shoes or skis using the DR-8: 15 lbs., MX306A: 25 lbs., or RL-159 and RL-27B (idiot stick): 70 lbs.

(6) FM 9-3. Gives guidance on laying wire for infantry units in relation to the time a unit will be stationary, and the distance between units.

(7) Plans. There are several plans that an infantry battalion can use t lay, and retrieve wire. One plan is as follows: The infantry companies move forward into position, when a company takes a position, the alpha command secures a position, and dispatches an over the snow vehicle to the companies position. During this time, the bravo command runs wire to the alpha command and other attachments. When the alpha command prepares the companies to move forward, the bravo command sends vehicles, and wiremen to receive the wire line route map, so wire retrieving can be started by the bravo command.

(8) Unit retrieving wire. The unit retrieving the wire must carry empty RL-159 and DR-5 reels to retrieve the wire for reuse.

TRANSITION: The most secure method of communication will be the messenger. I the cold, the messenger will have several problems to contend with, so before it is decided to employ a messenger, these things must be considered.

4. (5 Min) **MESSENGER**. This is the most secure means; however, it is limited by the terrain, weather, etc. Basically every man is a messenger, especially commanders that attend regular meetings. The use of messengers should be preplanned and they must NEVER travel alone.

   a. Problems Encountered. The following considerations in using messengers in a cold weather environment.

      (1) The enemy.

      (2) Personal survival. Messengers must have the proper equipment. An estimated time of arrival and return must be set and contingency plans made.

      (3) Transportation over the snow. Use an over-the-snow capable vehicle, if possible. Ensure the messenger is properly trained in the use of skis/snowshoes, familiar with the terrain and arctic navigation.

      (4) Wild animals.

TRANSITION: The most basic type of field communications is visual communications. We will begin by discussing the problems inherent to this type of communication in a cold weather environment.

5. (5 Min) **VISUAL COMMUNICATION**. Visual communications are an accepted method in most situations, but in cold weather environment, it can be rendered ineffective by blowing snow, such as whiteouts. Visual signals should be prearranged, and in your operations order.

a. <u>Air Panel Markers</u>. Air panel markers contain one set of white and black markers, with 13 markers per set. Each marker is 2 feet wide, and 3 feet long. The black marker will show up well against a snow background.

b. <u>Fluorescent Panel Markers</u>. One set contains 60 red and yellow, 18 inches wide, by 26 inches long markers. These markers are excellent for ground to air signals, by tactical air control party teams.

c. <u>Semaphore Flags</u>. Red and white are used on land, and red and yellow are used on water.

d. <u>Pyrotechnics</u>. Red and green colors can be most easily seen against a snow-covered background. A red signal is the International Signal for Distress or emergency.

TRANSITION: Let's now discuss audio communications, and considerations for employment in a cold weather environment.

6. (5 Min) **AUDIO**. Sound signals are satisfactory only for short distances, and their effectiveness is greatly reduced by battle noise. There are three types that are used in a cold weather environment. Other sound signals can be devised through the use of your imagination. They must be kept brief and simple to prevent misunderstanding.

   a. <u>Whistles</u>

   b. <u>Sirens</u>

   c. <u>UIQ 10 Loudspeaker</u>. Can project sound over a large area. A deception plan using tape recordings of mechanized vehicles, can be projected by a two-man team in front of the enemy location.

TRANSITION: To assure our equipment carries out its purpose, we must ensure proper maintenance techniques are performed.

7. (5 Min) **MAINTENANCE**. Performance of preventive maintenance is essential to ensure the proper operation of communication equipment.

   a. <u>Limited Technical Inspection (LTI)</u>. All communications equipment to be used must have a second echelon LTI performed on it before going to the field.

   b. <u>Daily Preventive Maintenance (PM)</u>. By operator. Personnel operating communication equipment in the field must perform daily PM.

   c. <u>Communication Contact Team</u>. Should attach to the log train to provide maintenance support for the infantry companies and attachments.

   d. <u>Maintenance Personnel</u>. The communication contact team, and headquarters groups must have maintenance personnel attached, and sufficient amount of pre-expended bin items and supplies, such as handsets, coaxial cable, connectors, whip and base antennas, silicone lubricant, plastic bags, duct tape, dry cloth, erasers, pencils, special cold weather electrical tape, and batteries.

TRANSITION: Safety while installing and operating communications equipment is most important. If safety is not practiced, we create the possibility of injury to ourselves and others.

8. (5 Min) **SAFETY**. The four safety precautions in operating communication equipment in cold weather are:

   a. Do not touch metal parts on communications equipment with bare hands if temperatures are below 0°F to 10°F.

   b. Construct antennas to be windproof.

   c. Ensure that the HF equipment is grounded properly.

   d. If the MRC vehicle is operating constantly, check the exhaust system to ensure proper ventilation.

TRANSITION: Great equipment is no good unless it can be transported to where it is needed.

9. (5 Min) **EQUIPMENT LOAD**

   a. <u>Foot Mobility</u>. During foot mobile operations, the battalion/company communication personnel may carry the following communication equipment in addition to the required cold weather pack:

QUANTITY	**EQUIPMENT**	**WEIGHT**
One	PRC 77 with spare battery	23 lbs.
One	KY 57 with spare battery	6 lbs.
One	KY 38 with spare battery	27 lbs.
*One	RC 292 antenna	45 lbs.
One	DR 8 with handle 1/4 mile reel of wire	18 lbs.
One	MX 306 1/2 mile wire	25 lbs.
One	RL 159 with handle 1 mile reel of wire	73 lbs.
One	TA 312 field telephone	10 lbs.
One	TA1 field telephone	3 lbs.
One	PRC 104 with spare batteries	25 lbs.
One	PRC 75 with spare battery	10 lbs.
One	PRC 68 with spare battery	4 lbs.
One	PRC 18 with spare battery	27 lbs.

* The weight of the RC 292 antenna can be reduces to 20 pounds by not carrying the 12 metal mast sections. Parachute cord 70 feet in length must be carried so that the RC 292 can be tree-topped.

b. Methods of reducing the communication equipment load on the individual Marine are:

   (1) Log trains. Utilize the log trains to provide re-supply of batteries, wire, preventive maintenance material, and maintenance support for the interchanging of equipment that is inoperable.

   (2) Spread load communication equipment and cold weather equipment between the Marines of each command group tent team.

c. <u>MRC 138 vehicles</u> must carry their pioneer gear: a pick, sledge hammer, two grounding stakes (4-6 feet in length) and salt.

<u>TRANSITION</u>: Let us now talk about the actual battalion communication plan.

10. (5 Min) **CONFIGURATIONS**. Staff personnel must know the radio circuits that are used for each operation. Depending on the task organization, a regiment or battalion operation at MCMWTC may use the following radio circuits:

RLT RADIO CIRCUITS/BATTALION RADIO CIRCUITS
RLT Link A/TAC VHF Battalion TAC 1 VHF
RLT Link B/Command HF Battalion TAC 2 VHF
RLT Link D/Intel VHF 81 mm mortar VHF
RLT FSC VHF Artillery COF VHF
Artillery Command/Fire Direction VHF TACP Local VHF
TAR/HR 1 I-IF Company TACs VHF
TAD/HD 1 UHF Platoon TACs VHF
LZ Control VHF LZ Control VHF
CSSD Request VHF/HF
MCMWTC SAR/MEDEVAC VHF

NOTE: The Battalion Tactical 2 radio net is by doctrine a VHF circuit. In a mountainous environment, battalions should establish the Battalion Tactical 2 net as an HF circuit. If a high frequency Tactical 2 net is used, additional PRC 104 radios for foot mobile communications must be requested from higher headquarters. The battalion communication platoon table of equipment is authorized only five PRC 104 radios and three MRC 138 radio vehicles which is not sufficient to provide additional high frequency communications for the Battalion Tactical 2 net.

<u>TRANSITION</u>: Let's now talk about a mobile communications system, concentrating specifically on that vehicle which we will most likely use in combat.

11. (10 Min) **MECHANIZED COMMUNICATIONS SYSTEMS**

a. Mechanized vehicles may provide additional equipment not identified in the infantry regiment T/E. All vehicles must be winterized. They also provide heated areas and as additional mobility capability.

b. <u>MRC vehicles</u>. When MRC vehicles remain off in temperatures below 0°F for over 4-5 hours, operators must allow 10-15 minutes at a constant idle for the vehicle to warm. Once the MRC vehicle is operating, allow 5-10 minutes for the High Frequency (HF), Very High Frequency (VHF), or Ultra High Frequency (UHF) mobile radios to warm up. If your radio does not key out, the problem may be that the radio set is not warm enough. You should start the MRC vehicles every 1-2 hours for 5-10 minutes to prevent freezing of vehicle, and radio components.

c. <u>AAV C-7</u>. The C-7 will provide a capability to move in the unfrozen waters of fjords and streams. If snow is considerable, the AAV will be roadbound. However, these vehicles are relatively mobile and displace rapidly.

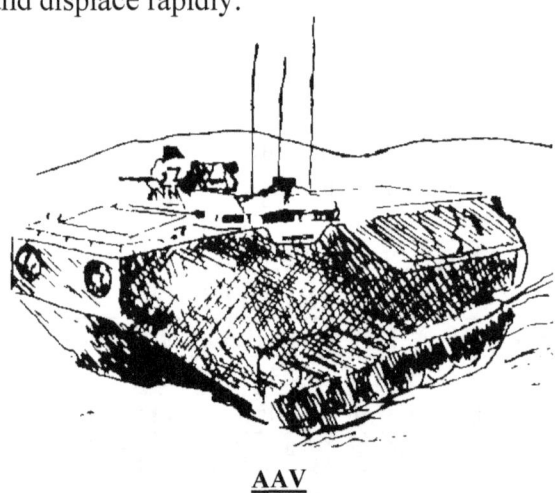

AAV

d. <u>LAV</u>. The LAV, equipped with chains, is fast and mobile both on and off the roads in moderate snow depths. Self-recovery capabilities of LAVs make them ideal for quick displacement. In deep snow and on ice roads, they will be roadbound.

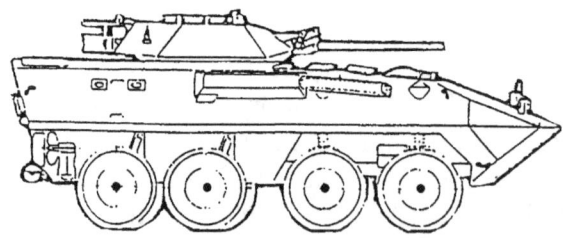

LAV

e. <u>BV-202 / BV-206 C2 Variant</u>. These vehicles provide exceptional COC/CP displacement capabilities and their off-road capability is unsurmassed. The BV-202s, which are dedicated to the USMC, are pre-positioned in Norway. They come with drivers/signalmen who can be used to augment USMC needs, especially in maintaining communications with allied units. Radios used in the BV-206 C2 variant will need to be provided from the unit's T/E. Mechanized infantry operations in the mountainous terrain of the MCMWTC, similar to NATO operations in Norway, should be supported by BV-206 over-the-snow vehicles. Using the BV-206 vehicle, Combat Operations Center (COC) for the forward, alpha, bravo, and log trains can be configured to best meet the characteristics of the BV-206 as shown.

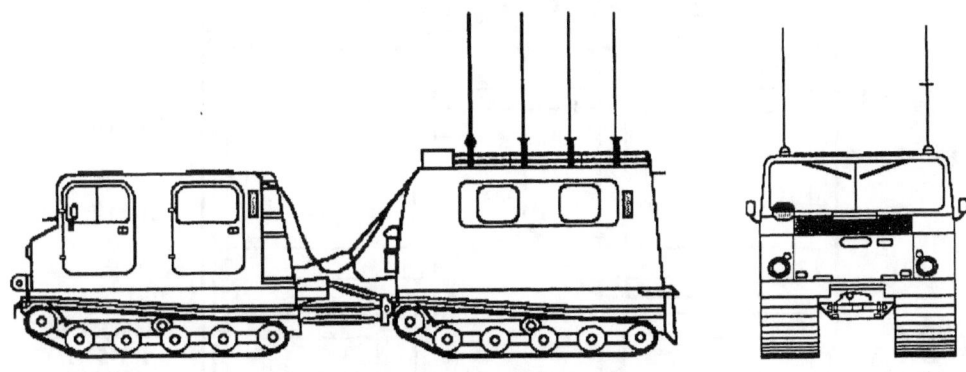

BV-206 SIDE AND FRONT VIEWS

(1) The BV-206 troop carrier vehicle is not configured with communications equipment. The antennas constructed on the rear cab of the BV-206 roof are recommended for the battalion alpha and bravo command group communication system configurations.

(2) <u>Front Cab Radio Configuration</u>

    (a) The two sets behind the commander and driver were removed and two RT-254 radio sets with KY-57 cryptographic mounts were installed. Two sections of 2x4's were bolted to the seat frame to provide a base for the mounts. The RT-524 mounts were wired to the BV-206 vehicle battery to power the radios.

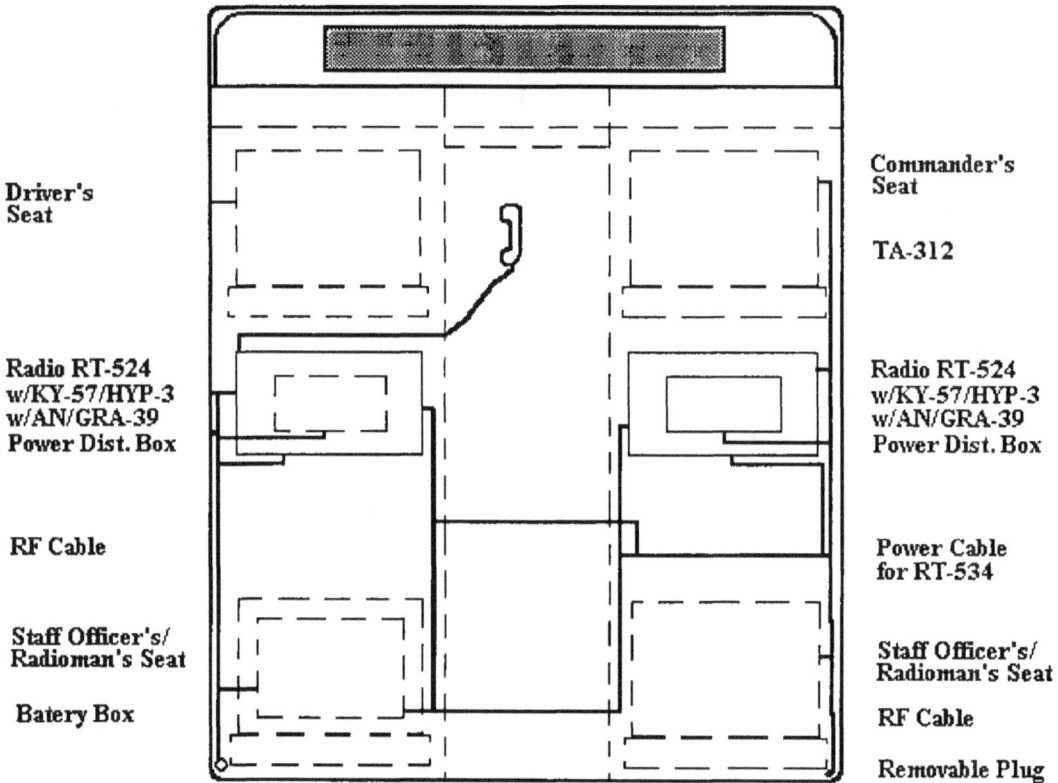

(b) The antenna system for the two radio sets consists of two MX-6707 matching units mounted on the rear cab roof. 15 feet of RF cable was used to connect each radio set to the MX-6707. The RF cable was run through the two removable plugs in the rear corners of the front cab t the MX-6707 matching units.

(c) One TA-312 telephone line was installed from the front cab to the rear cab serving as the intercom system.

(3) Rear Cab Radio Configuration

(a) The wooden seat on the side opposite the door entrance was removed and a 2x4 inch, ten (10) foot wooden mount for eight (8) radio/remote sets was constructed and bolted into the rear vehicle frame. A 2x4 inch, ten (10) foot wooden seat was constructed and bolted into the vehicle frame above the wooden mounts for the eight (8) radio/remote sets. The radio and cryptographic sets were secured in the wooden mounts by bungie cords. This permitted easy access to set frequencies, change batteries and troubleshoot the equipment.

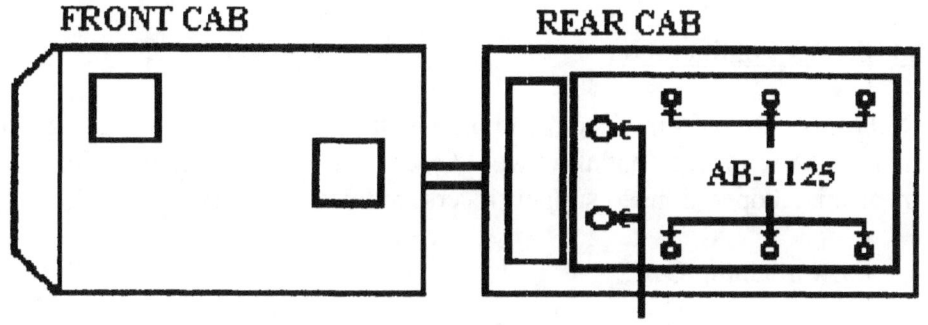

MX-6707
REAR CAB ANTENNA CONFIGURATION

(b) The antenna system for the six (6) radio sets was constructed using six (6) AB-1125 antenna stakes mounted on the rear cab roof. Fifteen (15) foot sections of RF cable were used to connect each radio to the AB-1125 stakes. The RF cable was run under the wooden mounts for the radio sets through the two (2) removable plugs in the corners of the rear cab. The cables should be marked on both ends to facilitate troubleshooting. The six (6) AB-1125 (TRC-166 base antennas) should be mounted to the rack on the rear cab within metal U-bolts. Additionally, six (6) AT-271 ten (10) foot whip antennas are needed to attach to the AB-1125 base antennas to provide radio communications.

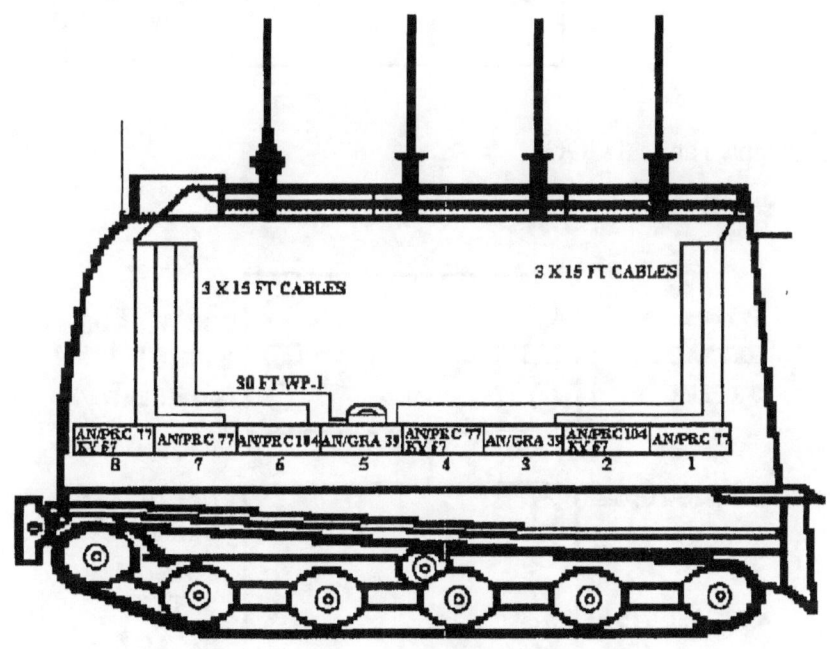

RADIO INSTALLATION IN REAR CAB

(c) One TA-312 telephone line was installed from the front cab to the rear cab which served as the intercom system.

(4) <u>Radio Circuits Configuration</u>

   (a) Radio circuit requirements depend on a battalion or regiment's organization and mission. A reinforced battalion would ideally use four BV-206 vehicles during mechanized operations to support its communications plan.

   1. Forward/Jump COC.

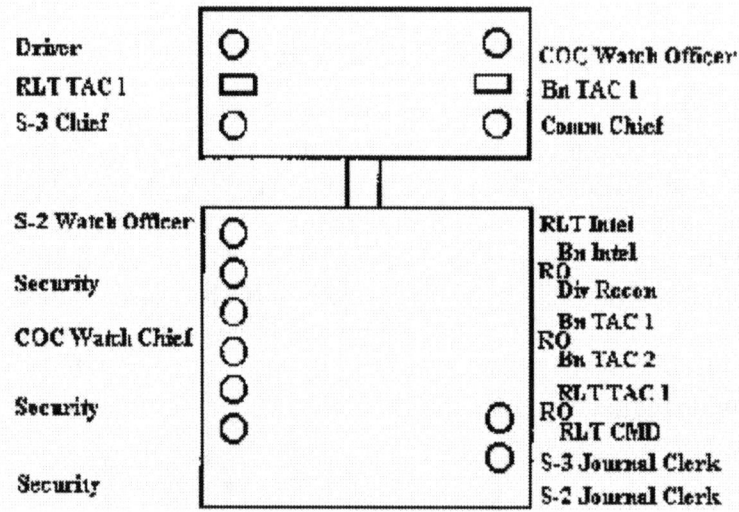

   2. Alpha/Tactical COC.

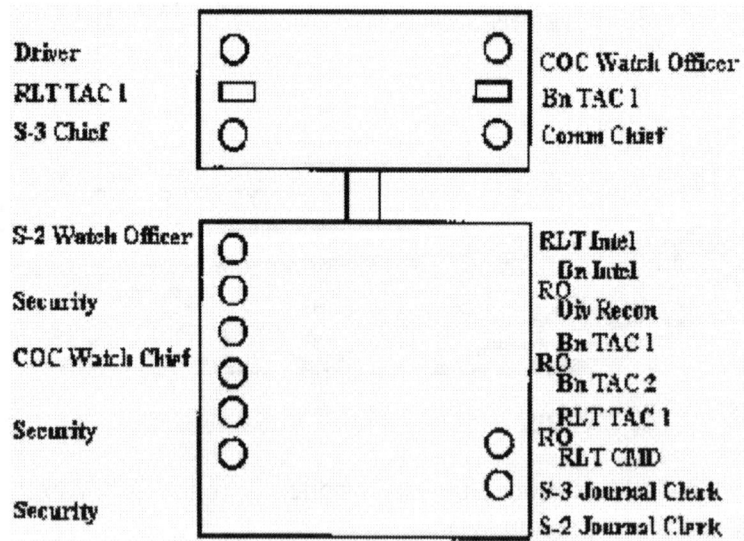

29-14

3. Bravo/Main COC.

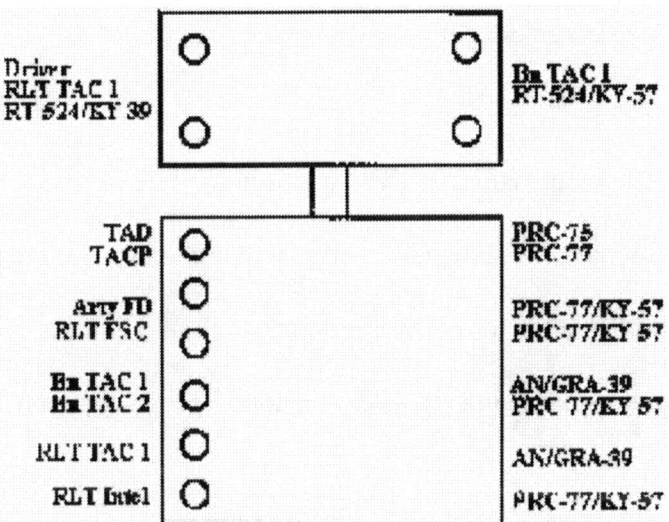

4. Log Trains.

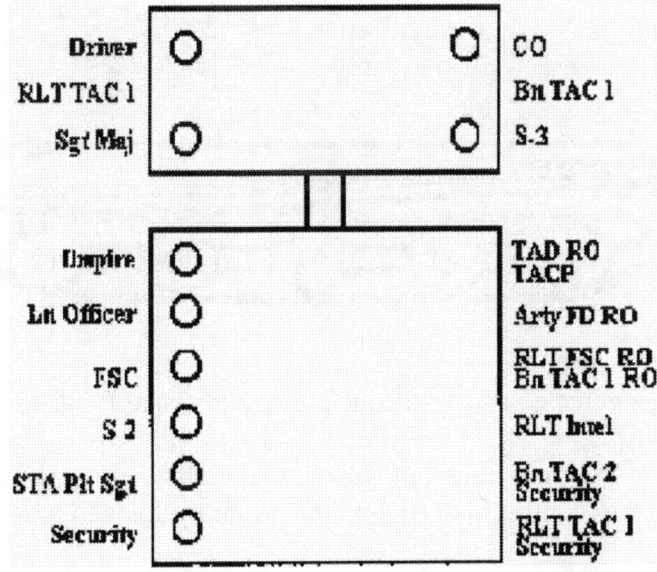

(b) If four BV-206 vehicles are not available, the battalion modifies its communications plan. Depending on the task organization of the battalion and the number of BV-206 vehicles available, the battalion radio circuits can be configured to best meet their operational doctrine.

(5) Chase BV-206

    (a) The chase BV-206 vehicle will carry additional communication technicians, radio operators, wiremen, and security personnel. The chase vehicle will support the COC BV-206 vehicle with:

        1. Tentage.

        2. Provide security for the COC configuration.

        3. Additional communication equipment and maintenance items.

        4. Radio operators to relieve the actuals on radio circuits.

        5. Wiremen to run and retrieve telephone lines to company positions.

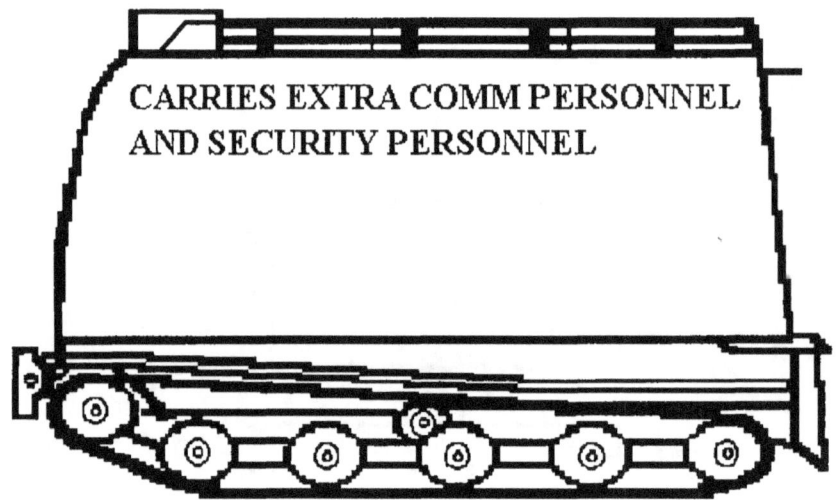

    (b) The chase vehicle is recommended if an additional BV-206 is available for use.

(6) Following are diagrams of recommended battalion COC using a ten man tent. Again, it is emphasized to configure the communications systems to best meet the battalion's requirements.

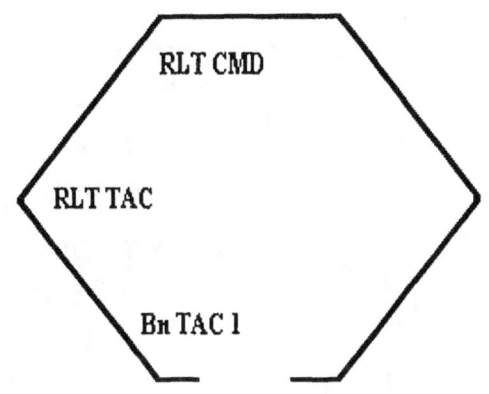

TENT CONFIGURATION FOR FORWARD/JUMP COC

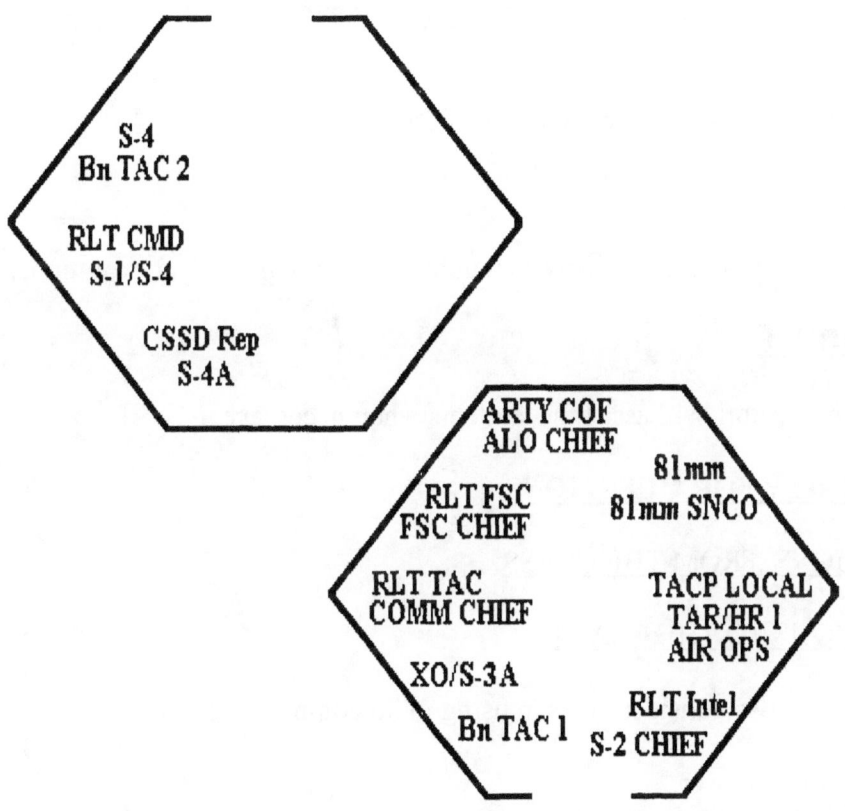

TENT CONFIGURATION FOR MAIN/BRAVO COMMAND

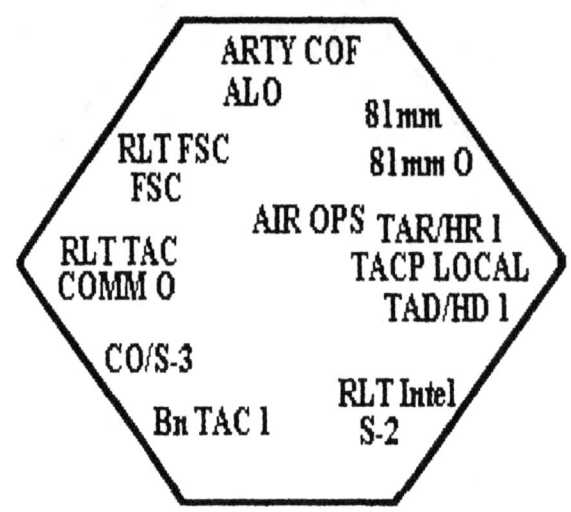

TENT CONFIGURATION FOR TACTICAL/ALPHA COMMAND

TRANSITION: At this time are there any questions?

PRACTICE (CONC)

    a. Students will practice what was taught in upcoming training evolutions.

PROVIDE HELP (CONC)

    a. The instructors will assist the students when necessary.

OPPORTUNITY FOR QUESTIONS (3 Min)

1. **QUESTIONS FROM THE CLASS**

2. **QUESTIONS TO THE CLASS**

    Q. What are two considerations in using radio communications?

    A. (1) Battery power.
       (2) Material failures.

    Q. What are two considerations in using wire communications?

    A. (1) Battery power.
       (2) Material failures.

    Q. What are the problems that a messenger may encounter in a cold weather environment?

    A. (1) Personal survival.

(2) Transportation over the snow.
(3) Wild animals.
(4) The enemy.

SUMMARY (2 Min)

a. During this period of instruction we have covered the problems encountered when using radio, wire, messenger, visual, audio, maintenance, and safety procedures in a cold weather/mountainous environment. We have also discussed the techniques that should be used to combat these problems.

b. Those of you with IRF's please fill them out at this time and turn them into the instructor. We will now take a short break.

UNITED STATES MARINE CORPS
Mountain Warfare Training Center
Bridgeport, California 93517-5001

WML
WMO
WSVX
09/31/00

LESSON PLAN

WINTER TRACKING

INTRODUCTION (5 Min)

1. **GAIN ATTENTION**. Tracking is a lost skill. Although many individuals believe it is impossible, there are many different skills that will enable a Marine to identify his prey, determine how far behind he is, and follow the track. The skills learned will not only assist a Marine in increasing game take, but will enable the Marine to increase his survival on a battlefield against the enemy.

2. **PURPOSE**. The purpose of this period of instruction is to introduce the student to the basic fundamentals of tracking by discussing basic terminology, track identification, and tracking methods. This period of instruction relates to Traps and Snares.

3. **INTRODUCE LEARNING OBJECTIVES**

    a. TERMINAL LEARNING OBJECTIVE. In a winter mountainous environment, conduct tracking, in accordance with the references.

    b. ENABLING LEARNING OBJECTIVES

       (1) Without the aid of references, describe in writing the types of gaits, in accordance with the references.

       (2) Without the aid of references, describe in writing the tracks of the major animal families, in accordance with the references.

       (3) Without the aid of references and given an animal hide, identify the animal, in accordance with the references.

(4) Without the aid of references, list in writing the factors that determines track age, in accordance with the references.

4. **METHOD / MEDIA**. The material in this class will be presented by lecture. You will practice what you have learned in upcoming field training exercises. Those of you with IRF's please fill them out at the end of the this lesson and return them to the instructor.

5. **EVALUATION**. You will be tested later in this course by written and performance evaluations.

TRANSITION: Before will begin tracking, we must understand some basic terminology.

BODY (50 Min)

1. (10 Min) **BASIC TERMINOLOGY**. Prior to discussing tracking, some basic terms must be understood by all.

    a. Trails and Runs. In any area, there will be many thoroughfares or trails and runs. Some may be seasonal, while others may be used by many different species. Runs are infrequently or intermittently used thoroughfares that connect trails to specific feeding, bedding, or watering areas. If trails are like highways connecting cities and towns, runs are like streets providing access to the gas stations, supermarkets, and neighborhoods.

    b. Beds and Lays. Beds are frequently used sleeping areas commonly referred to as dens or burrows. These can be found in hollow logs, trees, rock piles, brush piles, grass, thickets, or even out in the open. A lay is an infrequently used resting or sleeping spot. It is rarely used more than once.

    c. Rubs. Some rubs are accidental and some are deliberate. Accidental rubs can be in a burrow, on a trail, or over/under a fallen tree across a trail. Deliberate rubs can be when an animal scratches a hard-to-reach spot, or when a deer scrapes its antlers against a tree to remove its velvet.

    d. Scratches. They also can be accidental or deliberate. Accidental scratches are left by animals climbing trees or on a log where it left a belly rub. Deliberate scratches can be found at the base of trees where they have reached up and raked their claws downward for any number of reasons. Scratches can also be found in the ground where cats have buried scat, squirrels have cached nuts, or animals are digging at a scent.

    e. Transference. Transference is the removal of material from one area onto another. Transference can occur when walking along a muddy stream bank and crossing a log. The mud left on the log is considered transference.

    f. Compression. Compression is the actual flattening of the soil or snow pack. It is caused by the pressing down or leveling of soil, sand, stones, twigs, or leaves by the weight of the body. Compression is more likely to be found in frozen, hard, dry, sandy conditions where there is no moisture to hold a clear and lasting imprint.

g. <u>Disturbance</u>. Disturbance is the eye-catching effect of unnatural patterns.

h. <u>Gait</u>. A gait is generally the way an animal moves. Gaits are very critical in the identification of animal tracks. Although certain gaits are more indicative of certain animals, they may (depending on the circumstances) modify or alter their gait to another style.

   (1) <u>Diagonal Walker</u>. Normal pattern for all predatory animals, which includes all dogs, cats, and hoofed animals.

   (2) <u>Pacers</u>. Normal pattern for all wide-bodied animals such as bears, raccoon, opossum, skunk, wolverine, badger, beaver, porcupine, muskrat, and marmot. Instead of moving opposite sides of the body at the same time like diagonal walkers, they find it easier to move both limbs on one side of the body at the same time.

   (3) <u>Bounders</u>. Normal pattern for most of the long-bodied, short-legged weasel family such as marten, fisher, and mink. Bounders walk by reaching out with the front feet and bringing the back feet up just behind them.

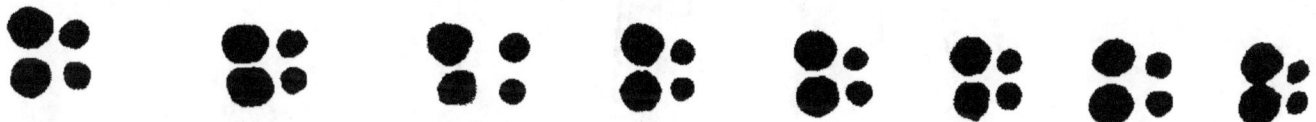

   (4) <u>Gallopers</u>. Normal pattern for rabbits, hares, and rodents (except wide-body beaver, muskrat, marmot, and porcupine). As these animals move, they push off with their back feet, hit with their front feet, and bring their back feet into position. Tree dwelling

gallopers will land with their front feet side by side, while ground dwelling gallopers will land with the front feet at a diagonal.

i. <u>Gnawing</u>. All animals will chew on vegetation; some as a food source, while predators need certain vitamins. Gnawings can be on trees (cambium layer) or on vegetation.

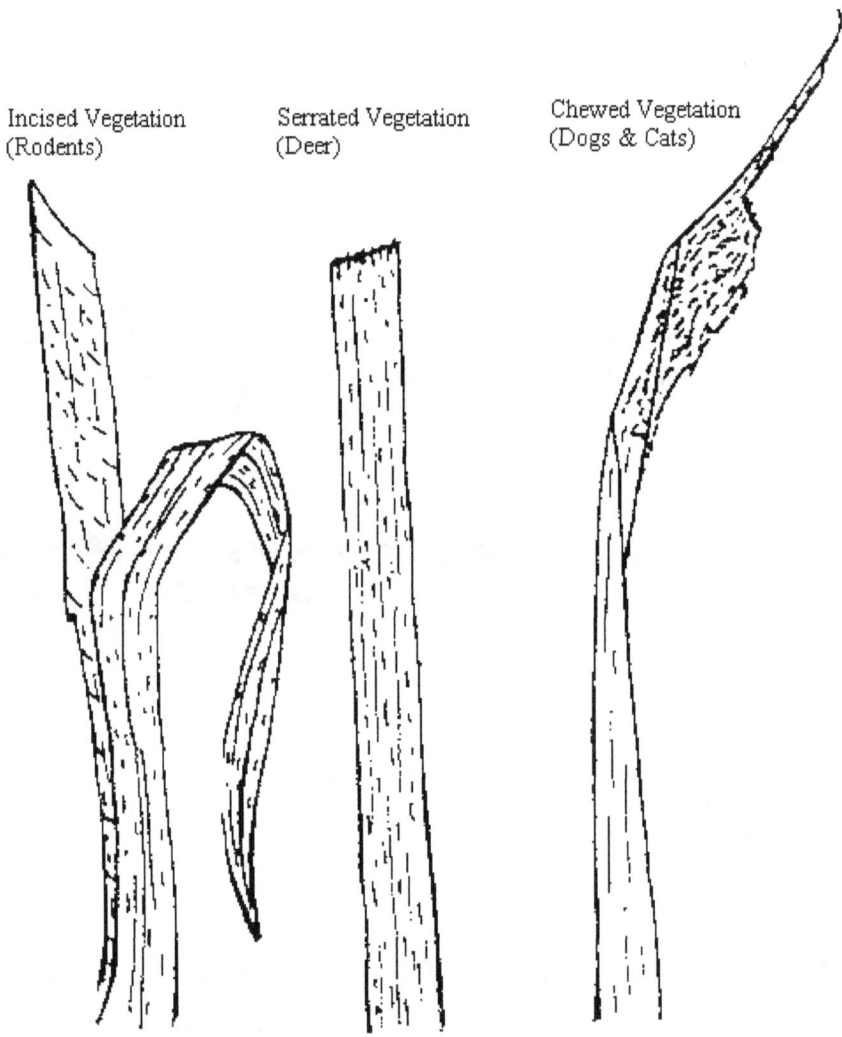

j. <u>Scat</u>. Scat is actual animal droppings.

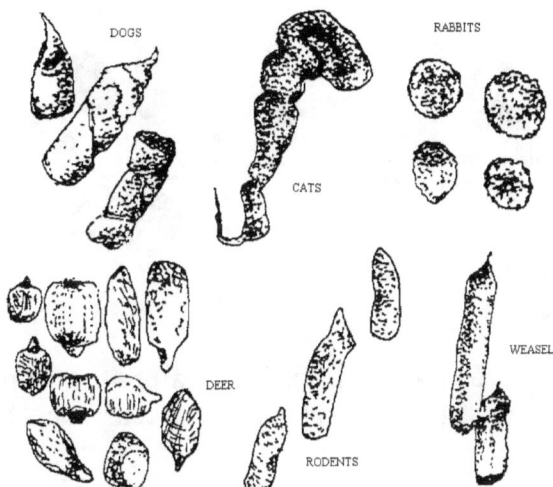

k. <u>Sign</u>. Any disturbance of the natural condition which reveals the presence or passage of animals, persons, or things. Examples of sign include stones that have been knocked out of their original position, overturned leaves showing a darker underside, sand deposited on rocks, drag and scuff marks, displaced twigs, and scuff marks on trees.

l. <u>Spoor</u>. The actual track or trail of an animal which can be identified as to size, shape, type and pattern. This word is generally interchanged with track. Spoor is broken down into two segments; aerial and ground.

TRANSITION: Are there any questions about the basic terminology? Let us now discuss reading spoor.

2. (10 Min) **READING SPOOR**. Unless a clearly visible ground spoor is readable, interpretations must be made in order to determine "what animal made this?" Prior to ever attempting to read spoor, one must be thoroughly knowledgeable about what animals are in the area. The first step is to look at the gait. This will generally narrow down the species. The next step is determined which animal family the track belongs to.

a. <u>Cat Family</u>. Bobcat, Lynx, and Mountain Lion (Cougar). 4 toe pads, no visible claw print. It moves with a sense of purpose and direct registers its paws. Its heel pad is much more defined than one from a dog.

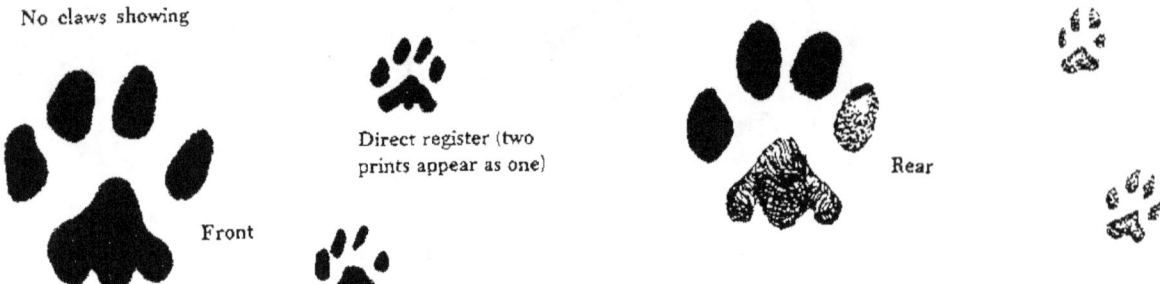

b. <u>Dog Family</u>. Fox, Coyote, and Wolves. Visible claw print, 4 toe pads. No sense of purpose, except fox, which steps like a cat and likes depressions. Heel pad is much rounder.

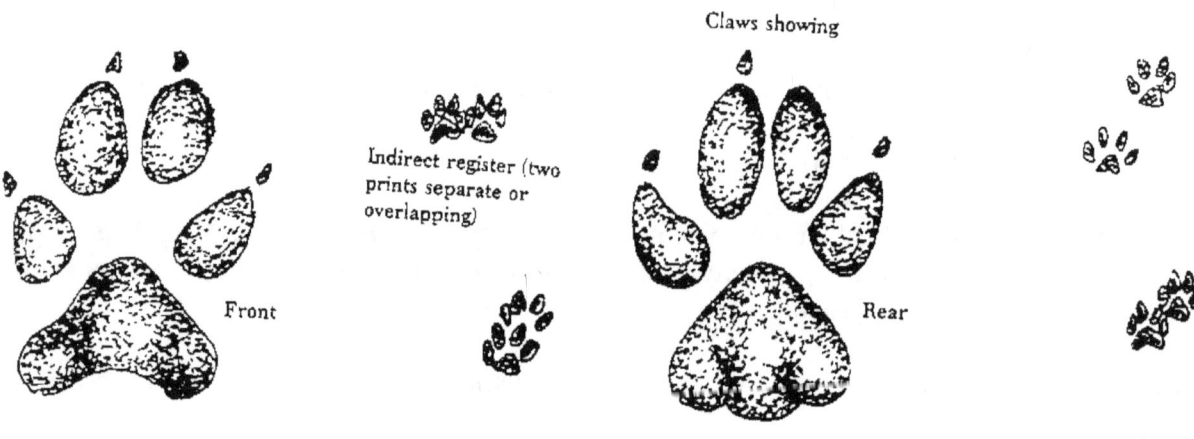

c. <u>Rabbit Family</u>. The main difference between rabbits and hares (which include the jackrabbit) is that rabbits are born almost hairless and with eyes closed, while hares are born with a thick coat of fur, open eyes, and an ability to run very soon afterwards. They have four toes with relatively enormous hind feet as compared to their front.

d. <u>Rodent Family</u>. Voles, Mice, Rats, Chipmunks, Squirrels, Woodchucks, Muskrats, & Beaver. Track size varies greatly because of the different species, but one fact remains, all have 5 toe prints on their rear feet, while having 4 toe prints on their front feet.

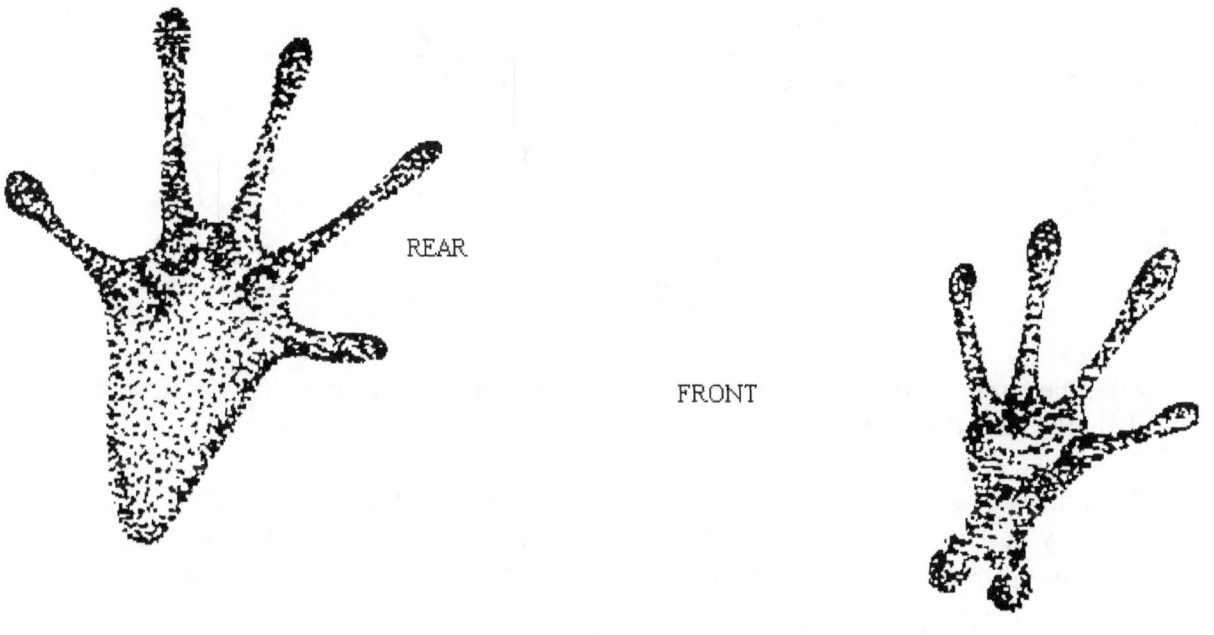

e. <u>Weasel Family</u>. Weasel, Mink, Marten, Fisher, Skunks, Otter, Badger, Porcupine, & Wolverine. All have 5 toe prints.

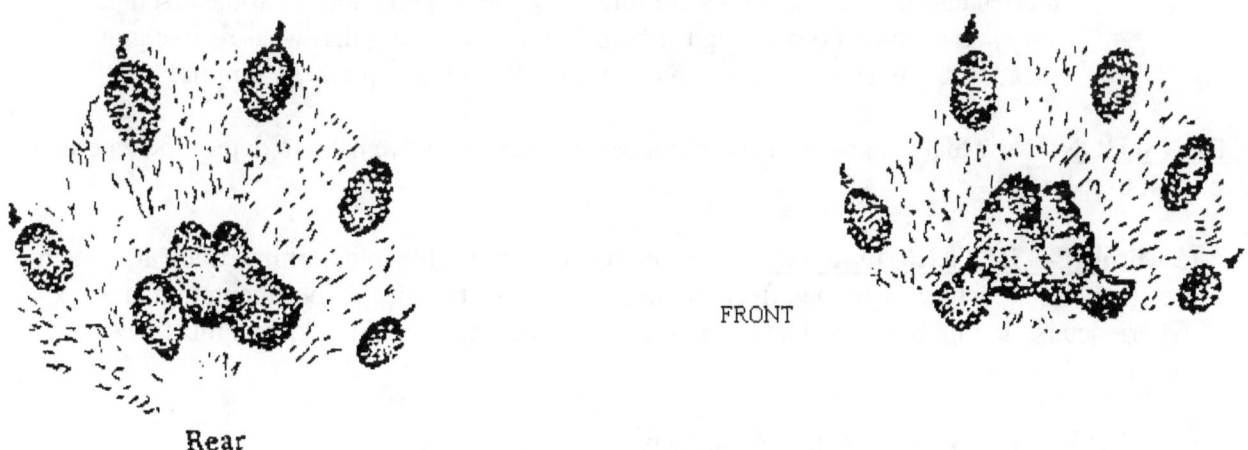

f. <u>Raccoon, Opossums, & Bear</u>. All have 5 toe prints while looking like a baby's hand print.

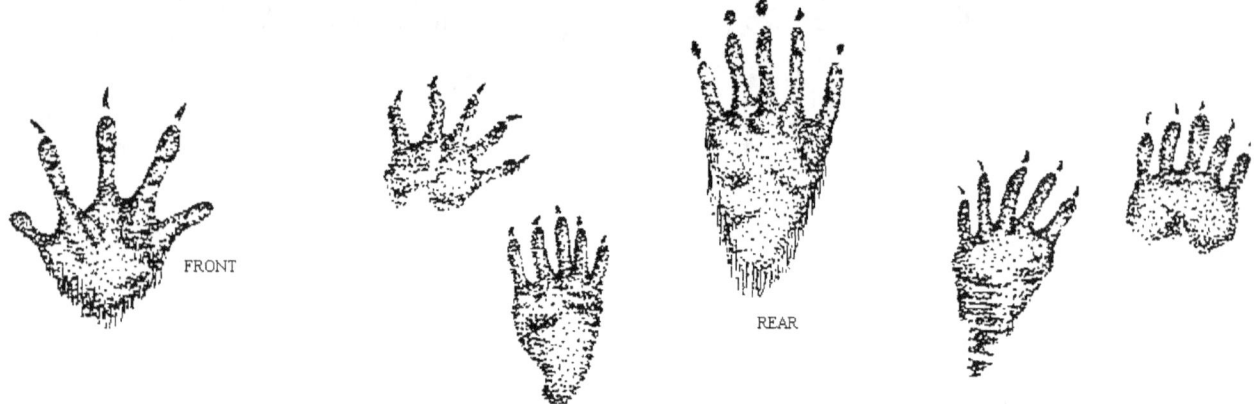

g. After the family is known, we must identify the individual species. Using various clues about the habits of animals, a determination can be made.

   (1) If the tracker is educated on the behavior and habits of animals, he can determine the individual species. This information can be used for better employment of traps and snares. The following is an example.

   (a) Walking along a creek bank, you notice a set of tracks that have five toe prints for both front and back feet with visible claw prints. The information tells you the prints belong to the weasel family. The tracks have a bounding gait pattern, which eliminates wide-bodied animals such as badger, skunk, porcupine, and wolverine. Because the tracks are approximately the size of a dime, you have eliminated marten and fisher. The tracks are following the stream bank for some distance, stopping at small holes along the bank's edge. Knowing that weasels like grassy meadows, you can determine that the track is probably made by a mink.

<u>TRANSITION</u>: Now that we know what animal we are looking at, we have to determine whether we want to follow it.

3. (5 Min) **AGE DETERMINATION**. It is very critical to be able to determine track age. Each area and climate will vary in the effects of aging tracks, so practice, experimentation, and experience is vital in that area. The following factors deteriorate all tracks and must be factored.

   a. Weather - Last snow or rain, fog, and dew.

   b. Sun.

   c. Wind.

   d. Soil content - hard, sandy, firm, or moist.

e. <u>Track Erosion</u>. All tracks will erode over a given period of time. The following time table can be used as a general guideline.

   (1) Minutes-Top edges begin to dry.

   (2) 24 hours-Top edges begin to erode.

   (3) 48 hours-"S" curve is seen.

   (4) 72 hours-Pock marks from rain or dew may be seen. Track is almost flat. Debris may fill the track.

f. <u>Aging Scat</u>. All scat dries on the inside first. Therefore, relatively wet scat on the outside could be old. The only way to determine the age is by analyzing the inside. When assessing scat, care must be taken to avoid the possibility of contracting disease.

TRANSITION: Now that we know how to determine age, let's talk about size estimation.

4. (5 Min) **SIZE ESTIMATION**. Although there is no exact method to determine actual unit size, an approximation can be made.

   a. <u>Up to a Squad</u>. It is possible to count basket marks, if on skis. The track is generally clean and straight. It is possible to identify multiple ski and/or snowshoe tracks.

   b. <u>Squad to Platoon Size</u>. If on skis, basket marks are difficult to distinguish from each other and may look like a small ditch. The track is somewhat clean and straight, but maybe a half-a-width wider than normal (i.e., 3 ski tracks or 3 snow shoes tracks wide).

   c. <u>Platoon to Company Size</u>. The track is sloppy and wide, possibly 2-3 times wider than normal. The edges of the track are destroyed at bends and curves.

TRANSITION: Are there any questions about estimating the size of a unit, let's discuss tracking.

5. (10 Min) **TRACKING**

   a. The best time to track is early in the morning or late in the afternoon due to the height of the sun to cast shadows. When reading spoor, always place it between yourself and the sun.

   b. Do not move past the last sign until you have found the next sign, this is called "sign cutting" and will be discussed later. In training, always try to find every track.

   c. Once the initial track is found, completely document and sketch it for future reference. This sketch will prevent you from following the wrong track later on. Record the following information.

d. <u>Determining Direction</u>. This generally is not a problem with animals. Man's over-the-snow equipment may confuse a tracker. All forward movement will displace snow forward, or referred to as "fluffing". This fluffing is the key. As it begins to melt, pockmarks will be left on the level snow pack.

(1) <u>Fluffing</u>

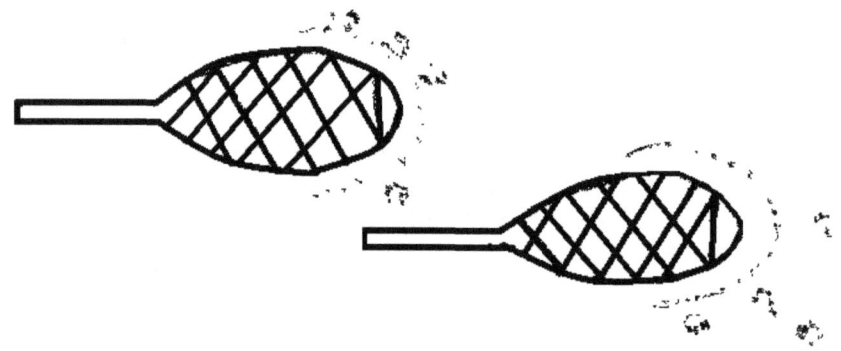

(2) <u>Backward Snow Shoes</u>. Inexperienced personnel may believe that walking backwards in snow shoes will fool someone. This type of activity is extremely exhausting and will not confuse an experienced tracker.

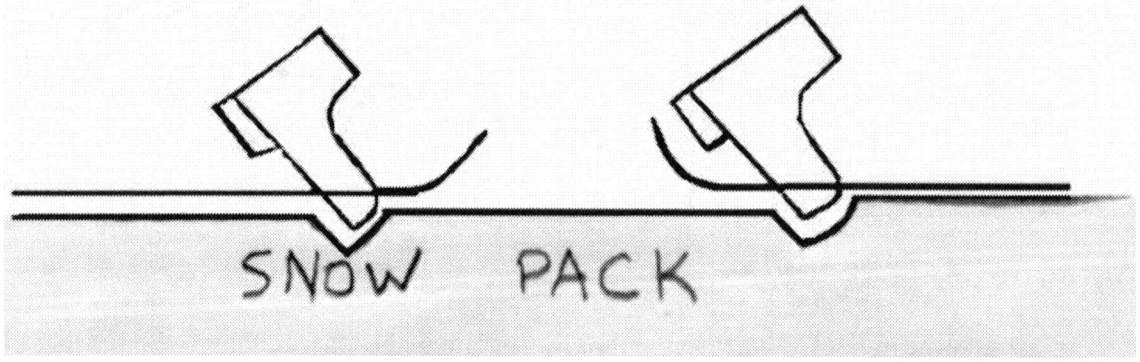

WALKING BACKWARDS IN SNOW SHOES

(3) <u>Skier</u>. As the ski pole is planted and the ski moves forward, the basket will also angle forward, causing the basket to dig into the snow, leaving an indent on the forward edge indicating direction. The point of the ski pole will also contact the snow before the pole is planted, making a line pointing away from the direction.

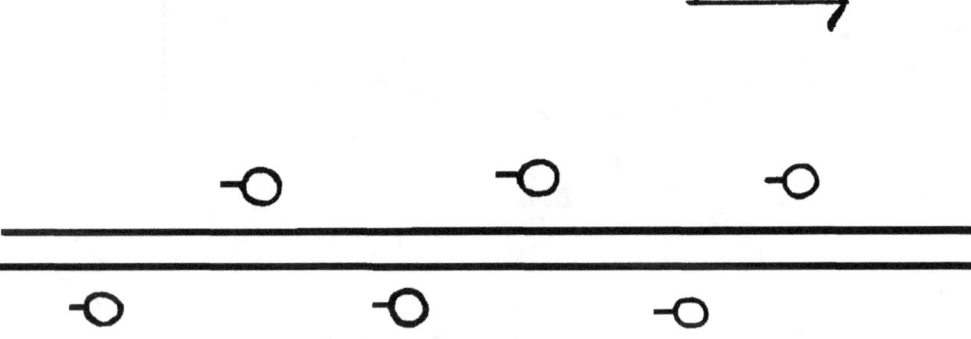

SKIING DIRECTION

(4) <u>Snowmobile and tracked vehicles</u>. Fluffing and direction determination is difficult to determine with snowmobile and tracked vehicles. The tracks will compress snow inside the track, forming plates. These plates are the keys to determining direction. At ground level, the vast majority of plates will face the direction of travel, although a small amount will face in the opposite direction.

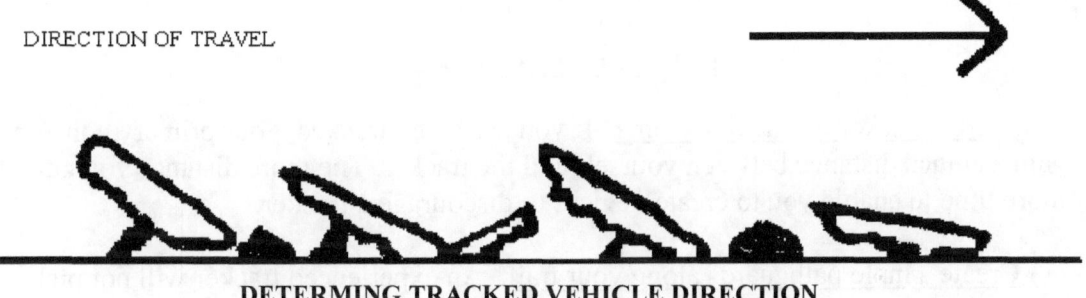

DETERMING TRACKED VEHICLE DIRECTION

e. <u>Tracking Teams</u>. If tracking teams are available, "sign cutting" can speed up the tracking process.

    (1) All tracking teams (minimum of two) must document and sketch the initial track.

    (2) The initial team continues to track until another tracking team has positively found the same track further ahead on the trail. The last track is marked for future reference.

    (3) The second team assumes the responsibility of locating each track until they have been radioed by the first team that "they have found the track". This is "sign cutting".

    (4) "Sign cutting" is accomplished by making large sweeping arcs ahead of the primary tracking team until the track is located at which time the teams change roles by "leap frogging".

    (5) If the track is lost or misidentified, the teams will move back to the last marked track.

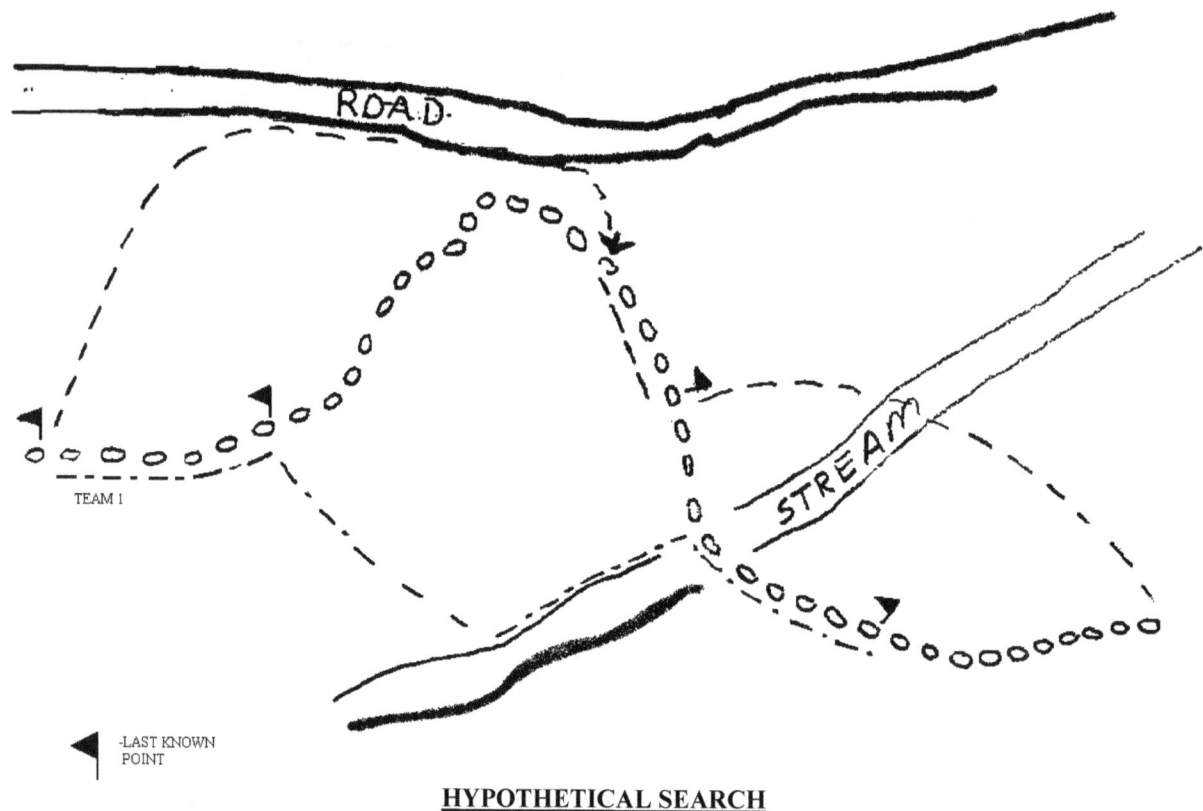

HYPOTHETICAL SEARCH

f. <u>Delaying a tracker or tracking teams</u>. If you are being tracked, your primary concern is to gain as much distance between yourself and the tracker. The more distance you gain, the more time to enable you to create devices to discourage a tracker.

(1) <u>Create simple pathguards along your trail</u>. An experienced tracker will not pick up things along the trail because of the possibility of being "booby trapped". If he notices possible traps, he will use more caution and slow his pace.

(2) <u>Use caution when moving along the trail</u>. When traveling, make it difficult for the tracker to find your tracks.

   (a) Although this is difficult in snow, staying in the tree line will reduce the possibility of being discovered by aircraft.

   (b) Do not brush up against snow covered saplings and large branches. When the snow is missing, the green foliage is like a billboard.

(3) <u>Backfill all tracks leading towards your bivouac</u>. When in a small unit, or the tactical situation dictates, backfill your jump off point with snow. This is accomplished by filling some type of bag with snow from your bivouac site and completely filling in all holes from the main track until no longer seen from the main track. Make sure that the filler snow is completely blended in with the top layer of the snow pack.

TRANSITION: Tracking is an underestimated skill. There are many things to factor in to be effective.

PRACTICE (CONC)

   a. The students will implement what has been taught concurrently throughout the course.

PROVIDE HELP (CONC)

   a. The instructors will assist the student when necessary.

OPPORTUNITY FOR QUESTIONS (3 Min)

1. QUESTIONS FROM THE CLASS

2. QUESTION TO THE CLASS

   Q. What are the four gaits?

   A. (1) Diagonal
      (2) Bounder
      (3) Pacer
      (4) Galloper

   Q. What are the six family of tracks?

   A. (1) Cat
      (2) Dog
      (3) Weasel
      (4) Rodent
      (5) Rabbits & Hares
      (6) Bears, Raccoons, & Opossum

   Q. What are the four factors that age tracks?

   A. (1) Sun
      (2) Weather
      (3) Wind
      (4) Soil

SUMMARY (2 Min)

   a. We have discussed many items very briefly in this class. Although the basic fundamentals of tracking are taught in a classroom environment, tracking is an art, which can only be mastered through practical application.

b. Those of you with IRF's please fill them out at this time and turn them in. We will now take a short break.

UNITED STATES MARINE CORPS
Mountain Warfare Training Center
Bridgeport, California 93517-5001

WML
WMO
09/12/02

LESSON PLAN

MOUNTAIN LOGISTICAL CONSIDERATIONS

INTRODUCTION

1. **GAIN ATTENTION**. The mountainous environment presents considerable challenges to combat operations. Supporting and sustaining combat operations requires more attention and planning than in less arduous environments.

2. **OVERVIEW**. The purpose of this period of instruction is to present support considerations and principles specific to combat operations in a mountainous/cold-weather environment.

3. **METHOD/MEDIA**. The material in this lesson will be presented by lecture. The student will apply the considerations and principles during training in a mountainous/cold-weather environment. Students with IRFs please fill them out at the conclusion of this period of instruction.

4. **EVALUATION**. There is no formal evaluation over this period of instruction, however the success of the student during training in a mountainous/cold-weather environment is dependent upon how well the considerations and principles are applied throughout the exercise. The success of any unit conducting combat operations in a mountainous/cold-weather environment is dependent upon how well these support considerations and principles are applied.

TRANSITION: Are there any questions over the purpose, learning objectives, how the class will be taught, or how you will be evaluated? Let's discuss the responsibilities of logistic support.

(55 Min) **BODY**

1. (5 Min) **LOGISTICS PLANNING IS GOOD LEADERSHIP.** A logistically sound operation is one conducted by a well-lead unit at all levels. Leaders properly prepare for success before execution. The leadership of a unit should be constantly asking the question, "what does my unit need and how do we get the supplies and equipment to achieve the mission tomorrow, the next week, the next month, etc." Success at the tactical level (success on the battlefield) is the primary concern. Failures in logistic planning can be extremely lethal in a mountainous/ cold-weather environment.

2. (10 Min) **LEVELS OF LOGISTICS.**

    a. Strategic. The movement of troops, equipment and sustainment from the continental United States (CONUS) or based station to anywhere on the Earth.

    b. Operational. The movement of troops, equipment and sustainment inside a theater of war in order to achieve the operation's objective.

    c. Tactical. Sustaining the combat capability of a tactical unit in order to achieve the tactical tasks assigned.

TRANSITION: Now that we have reviewed the levels of logistics, let's turn our focus to the tactical level and look at the spectrum of a Marine Unit's links to logistics.

3. (10 Min) **THE MARINE AIR-GROUND TASK FORCE (MAGTF)**

    a. Operating as a MAGTF. It should be clarified that the Marine Corps will conduct combat operations as a MAGTF. This combined force of Ground Combat, Combat Service Support, and Air Wing Marine units works together under a command element, which receives its mission from a higher command. Sustaining the MAGTF is the Marine Corps's responsibility. Support required beyond the MAGTF's organic capabilities must be identified to the higher command and coordinated through the MAGTF's Command Element (CE).

b. Size of a MAGTF. A MAGTF can be any size required to achieve the specific mission. The doctrinally standard MAGTFs are the Marine Expeditionary Force, the Marine Expeditionary Brigade, and the Marine Expeditionary Unit. The table below presents the sizes of the elements within a MAGTF.

MAGTF	GROUND COMBAT ELEMENT	COMBAT SERVICE SUPPORT ELEMENT	AIR COMBAT ELEMENT
MARINE EXPEDITIONARY FORCE (MEF)	DIVISION	FORCE SERVICE SUPPORT GROUP	WING
MARINE EXPEDITIONARY BRIGADE (MEB)	REGIMENT	BRIGADE SERVICE SUPPORT GROUP	MARINE AIR GROUP (REIN)
MARINE EXPEDITIONARY UNIT (MEU)	BATTALION	MEU SERVICE SUPPORT GROUP	SQUADRON (REIN)

c. Tactical logistic links.

    (1) MAGTF. A MAGTF requires a beach, port, or an airfield to transport the necessary sustainment for combat operations. The MAGTF Command Element coordinates logistic support and sustainment for the entire MAGTF. The CSSE is the unit with the specific mission of providing combat service support to the GCE. The CSSE coordinates with the ACE in order to integrate aviation assets into the combat service support scheme of maneuver.

    (2) DIVISION/REGIMENT. A division or regiment will require a ground main supply route (MSR) in order to employ its heavier weapon systems and sustain them in combat operations.

    (3) BATTALION/COMPANY. A battalion or company could operate without a ground MSR, but would require some form or aerial resupply over time.

TRANSITION: Now that we understand how a MAGTF is linked to its support, let's look at CSSE responsibilities and how they are affected by a mountainous/ cold-weather environment.

4. (20 MIN) **FUNCTIONS OF COMBAT SERVICE SUPPORT**.

  a. Transportation. The main logistical differences between mountain operations and operations in other terrain are a result of the problems of transporting and securing material along difficult and extended lines of support. Any proposed support structure must plan for redundancy in the ability to distribute supplies directly to units operating from predetermined supply routes. Some forms of transportation are:

(1) <u>Air Delivery</u>. Air delivery is a transportation of supplies capability. It is an asset controlled at the MEF level. Air delivery operations with any significant loads will require originating from an airfield. A detachment might support a MEB or MEU, but only if the MEF is willing to let it go. Drop zones are few in mountainous environments and accuracy is low due to steep terrain. Drops are best delivered to battalion or regimental level logistic trains in relatively secure meadows or valley floors. Supplies delivered to units on ridgelines or mountain peaks have a low rate of recovery. Door bundles thrown out of helicopters are better suited for mountain peaks and ridgelines. Air Delivery should never be the primary method of resupply because of its cost in equipment and limited capabilities. It should be planned as an emergency reserve that is employed when all other options are not possible.

(2) <u>Helicopter Support Teams (HST)</u>. Helicopters should be employed as much as possible to increase responsiveness. Helicopters have a decreased lift capability at upper altitudes and an increased capability in colder weather. However, colder temperatures cannot make up for the significant decreases resulting from altitude. HSTs must coordinate extensively concerning each helicopter's capabilities and requirements for trained HST personnel in the lift and drop LZs. Weather is also a contributing factor in whether or not helicopters can even be employed. Redundant modes of transportation must always be planned for when employing helicopters.

(3) <u>Motor Transportation</u>. Ever since the motorized vehicle has been utilized for military operations, operational maps have had big blue arrows up roads. Motor transport assets have been the largest contribution to ground combat maneuver in modern warfare. Historically, vehicles have been employed in severe weather and terrain conditions successfully and unsuccessfully. Creativity such as supplying besieged cities by driving across frozen lakes and employing anti-air weapons on vehicles against ground targets on steep terrain have answered the needs of modern warfare. Most vehicles are not designed to perform at their optimal capability in extreme cold or at high altitudes in severe terrain. Fuel consumption rates increase by 30 to 40 percent requiring more resupply, but the overall reduced employment of vehicles in mountain operations results in less fuel used compared to operations in most other environments. Factors which must be considered are:

- Plastic becomes brittle / belts snap if not warm.
- Never initiate a starter for more than 15 seconds (It burns out the starter and can drain the battery in one attempt).
- Don't jump or slave vehicles head to head (If they "jump" they could kill people or damage the vehicles).
- 50/50 anti-freeze is good from –32 to 265 Fahrenheit.
- At –0F, Straight 20 weight oil is best (It does not gel and allows engine to start with less drag).
- Below –15 deg F, constantly run the vehicle.

-Inspect chains before and during employment. Do not change tire pressure after chains have been applied. All hooks on the chains should face away from the tire. After season, chains should be wire brushed clean, dipped in crankcase oil and drip dried, then stored in a burlap or canvas bag.
-A covered vehicle is damaged less. Close all doors, windows, and hatches if possible.
-Gradually warm or defrost vehicles. Windshields will crack when shocked with change of temperature.
-Clear snow and ice from steering and tracks. Clear ice off tracks with wood stick like a broom handle or a shovel handle, ice covered steering mechanism can break.
-Slow Down (Extended stopping range, vehicles in front kick mud and snow up, vehicle going uphill has the right of way & vehicles parked on slopes need to be chocked with big anchors because they will gradually slide).
-Never sleep in a running vehicle (Carbon Monoxide buildup).
-SUSV Specifics- Never send out without a radio, in –0 to -15 F run vehicles 15 minutes per hour when using vehicle powered radios. When stopping overnight put grill covers in place).

(4) <u>Animal Packing</u>. The transportation of supplies via pack-animal can greatly support units operating for extended periods of time, away from a ground main-supply-route. MCMWTC is the only DoD unit currently training forces in the employment of pack-animals. Utilization of pack-animals will require acquisition of pack-animals within the area of operations.

b. <u>Supply</u>. Because of the burden on transportation, supplies should be prioritized and limited to essentials. The individual Marine's load should be lightened as much as possible. Certain items such as demolitions will most likely have an increased demand. Supply sources from the environment and even the enemy must be considered. Resupply must be a consideration in planning and the planning staff must know its limitations. During operational pauses, stockpiling and caching of supplies should take place to reduce future demands during combat operations. Quantities of supplies required by the individual Marine increases. Supply route considerations are:

(1) Existing roads should be rapidly analyzed for bottlenecks, deployment areas, passing places, and turnarounds for various vehicles.

(2) Routes should be classified as one- or two-way, and staggered schedules developed for the use of one-way routes.

(3) Signs should be placed for both day and night moves on difficult and dangerous routes.

(4) Whenever possible, separate routes should be designated for vehicular and dismounted movement. Additionally, separate routes should be designated for wheeled and tracked vehicles, particularly if the latter are likely to damage road surfaces.

c. Engineering. The engineering capabilities of most units is inadequate for the massive requirements that may be encountered in a mountainous/ cold-weather environment. Creative employment and augmentation must be considered. The demands may require a combined effort of all engineering units, yet the requirement to attach engineer detachments to all units also exists due to isolation and difficult terrain. Increased demolition training to non-engineer MOS Marines should be conducted.

   (1) Mobility/Counter-mobility. Mobility/counter-mobility will be the priority of the entire operation. While this may be true, the earth moving vehicles generally employed for these operations, have difficulties in severe terrain and break down often. They can become an obstacle and add to the problem. Demolitions will be utilized in much greater numbers than in most other environments.

   (2) Mines. Historically, mines have been employed in mountainous regions because of their quick ability to block entire geographical areas from foot and vehicle movement. Increased education on detection and clearing must be conducted.

d. Maintenance. Operations of equipment in mountainous terrain has proven that maintenance failures far exceed losses due to combat, and most breakdowns can be attributed to operator training. Training should be considered on all levels concerning environmental factors on vehicles and equipment. Preventive maintenance requirements must be supervised to ensure vehicle and equipment readiness.

e. Health Service.

   (1) CASEVAC. A well led unit has a sound CASEVAC plan. CASEVAC is the process of moving casualties from a combat area to a health care facility. Transportation of casualties on foot, on a pack-animal, on a sled over snow, in a vehicle or in a helicopter can all be employed in a mountainous, cold-weather environment, but all take much longer than most other environments.

   (2) MEDEVAC. Transportation of casualties from a combat zone health care facility to a health care facility located outside the combat zone will be conducted by vehicle or aircraft. It is a more administrative operation, usually conducted in less severe terrain.

f. Services. Services such as postal, administration, legal, exchange, and others are very difficult to support in a mountainous, cold-weather environment. Efforts should be taken to satisfy communications with others through a postal service, but most other services will not be able to be provided outside a base camp installation.

g. Force Protection. Force protection is not a function of CSS, but it should be an organic capability in all environments. Force protection in a mountainous environment should be additionally emphasized because of the close terrain and predictable desire of any enemy to cut off or destroy logistic support from the combat arm units. Dispersion of assets is a force protection as well as a flexibility issue.

TRANSITION: Now that we have reviewed the functions of CSS, let's talk about the needs of the individual Marine.

5. (5 MIN) **INDIVIDUAL MARINE'S NEEDS**. Self sustainment measures must be employed at the individual level.

   a. Class V Ammunition. Certain types of ammunition such as M203 grenades and demolitions will be employed much more in mountainous terrain. While the individual Marine should attempt to travel as light as possible, ammunition should never be deficient.

   b. Class I Water/Food. Ration requirements increase in severe terrain from 3,000 calories to approximately 4,500 calories per day, per Marine. Water requirements will increase as well in a mountainous/cold-weather environment and every effort should be made to allow the individual Marine to get water from the environment utilizing chemical purification, boiling, filtration or other methods.

   c. Class II Clothing/Personal Equipment. Personal equipment should be adequate to the environment and safeguarded by the individual because loss could be life threatening and replacement may take days or weeks. Constant inspection by the chain of command must be employed.

   d. Class VIII Medical Supplies. The severe nature of a mountainous/cold-weather environment greatly increase the demands on medical supplies. The standard individual issue should be increased according to the requirements most-likely caused by the environment.

TRANSITION: Now that we have looked at all the logistic considerations of combat operations in a mountainous/ cold-weather environment, let's review the principles of logistics.

6. (5 MIN) **PRINCIPLES OF LOGISTICS IN MOUNTAIN WARFARE**.

   a. Logistics planning is good leadership.

   b. A sound logistics plan is essential to mission accomplishment.

   c. Transportation of supplies requires more time.

   d. Aircraft performance is severely degraded.

   e. Health service requirements are greatly increased.

   f. Essential equipment must be safeguarded.

   g. Measures must be taken at the individual level to safeguard sustainment.

OPPORTUNITY FOR QUESTIONS (3 Min)

1. **QUESTIONS FROM THE CLASS**

2. **QUESTIONS TO THE CLASS**

    Q. What are the seven principles of mountain warfare logistics?

    Q. What are the ways supplies are transported to sustain combat operations.

SUMMARY (2 Min)

   a. During this period of instruction we have covered some of the considerations for mountain logistics. A mountainous/ cold weather environment will be one of the most challenging environments for Marines to conduct combat operations.

   b. Students with IRFs, please fill them out at this time and turn them in to the instructor. Take a short break until the next class.

UNITED STATES MARINE CORPS
Mountain Warfare Training Center
Bridgeport, California 93517-5001

WML
WMO
09/31/00

LESSON PLAN

COLD WEATHER AND MOUNTAIN HELICOPTER OPERATIONS

INTRODUCTION (3 Min)

1. **GAIN ATTENTION**. Survive, move, and fight; the three basic requirements to success on any battlefield. In cold weather or in the mountains the ability to move rapidly and with little expenditure of energy may greatly reduce the demands on the individual rifleman and significantly enhance his ability to fight.

2. **OVERVIEW**. The purpose of this period of instruction is to familiarize the student with those aspects of cold weather and mountainous terrain that effect usage of helicopters. This lesson relates to operational planning.

3. **METHOD/MEDIA**. This class will be taught by the lecture method.

4. **EVALUATION**. There will be no formal evaluation for this period of instruction.

TRANSITION: Are there any questions over the purpose how the class will be taught? Let's take a look at a general view of helicopter operations.

BODY (60 Min)

1. (5 Min) **GENERAL**. The helicopter is the single best tactical mobility asset available to Marines during cold weather operations. It can move you farther and faster than any other means of transportation. The helicopter has limitations. The greatest of which are the lack of dependability due to unpredictable weather and the extreme difficulty of performing maintenance in the cold. Additional maintenance personnel and maintenance shelters may be required. This means that unit leaders must always have an alternate movement plan to get to the destination in time to accomplish the mission.

TRANSITON: We will now discuss considerations effecting heliborne operations.

2. (5 Min) **CONSIDERATIONS EFFECTING HELIBORNE OPERATIONS**. The following items should be considered when deploying helicopters in a cold weather and mountainous environment.

   a. Reduction in Operational Tempo. Remember that everything takes longer in a cold weather environment. It takes the mechanics longer to fix, fuel, and do routine maintenance on the aircraft. The aircraft may also have more maintenance problems due to the cold weather.

   b. Vulnerability in the LZ. Delays in the LZ will make helicopters particularly vulnerable targets to both direct and indirect fire. Helicopters often create large snow signatures when conducting landings and takeoffs in snow covered terrain.

   c. Temperature and Altitude. As temperature and altitude increase, helicopter performance decreases. This effects not only payload capability but also time on station, airspeed, and maneuverability. Decreased temperature will not offset the effect of increased altitude when operating in high mountainous terrain.

   d. Weight/Bulk Load. Fewer men can be carried because he has more equipment and takes more space in the aircraft. It will take 1 1/2 normal seating space for a Marine with a full cold weather combat load.

   e. Weather. Mountainous or arctic terrain is compartmentalized and is characterized by rapid change. Weather may be good in the pickup LZ and bad in the destination LZ. Consequently, commanders must have alternate plans for insertion and extraction if possible.

   f. Rotor Wash Identification and Visibility. On landing and takeoffs, helicopters re-circulate large snow clouds that can be observed from considerable distances.

TRANSITION: Now that we've covered the environmental considerations effecting helicopter operations, are there any questions? Let's review assembly areas.

3. (5 Min) **ASSEMBLY AREAS**

   a. Assembly areas should provide security, concealment, dispersion, and a windbreak for troops.

   b. Anytime Marines wait longer then 40 minutes; they should erect warming shelters.

   c. This period may be substantially shorter in extreme cold temperatures or under severe wind-chill conditions.

TRANSITION: Are there any questions about assembly areas? Let's now discuss safety considerations.

4. (5 Min) **SAFETY CONSIDERATIONS**. Marines must understand safety considerations to reduce the risk of injury during cold weather and mountainous helicopter operations

   a. Frostbite. Frostbite is a constant danger due to the combination of wind-chill and cold temperatures. Use the buddy method to check for possible signs of frostbite.

   b. Rotor Blade Hazards. In deep snow-covered LZs, the helicopter may sink into the snow. This reduces the rotor-blades-to-surface clearance. Using the ahkio huddle for loading and unloading reduces the danger of being struck by the rotor blades. In sloping LZs, do not approach the helicopter from the upslope side as rotor-blades to surface clearance is further reduced. Extreme caution must be exercised when operating in close proximity to the tail rotor.

   c. Cargo Ramp Hazards. In deep snow the crew chief may not be able to lower the cargo ramp completely. Marines must be aware that this significantly reduces head clearance. Caution needs to be observed when operating near the cargo ramp; it is hydraulically operated and can easily crush personnel. Hydraulic fluid or ice can also cause the ramp to be slick causing a hazard for debarking Marines.

   d. Ice Shedding. Under various conditions ice may accumulate on the rotor blades of a helicopter. When it sheds it produces a myriad of flying projectiles. The safest place to be is on the ground with your face covered.

   e. Unprepared LZs. When landing in an unprepared LZ, the fuselage will float on the snow's surface. Landing points should be probed and tramped down to determine possible obstacles.

   f. Dynamic Rollover Damage. A helicopter will normally settle through the snow surface. If the ground is uneven or there are obstacles beneath the surface of the snow, it may cause the helicopter to contact the ground at an angle. In this condition the helicopter may be in danger of dynamic rollover. Dynamic rollover is a condition where the helicopter could rollover onto itself due the landing gear/skids coming in contact with the surface while power is being applied to the aircraft. Time permitting, the landing spot should be probed for any obstacles.

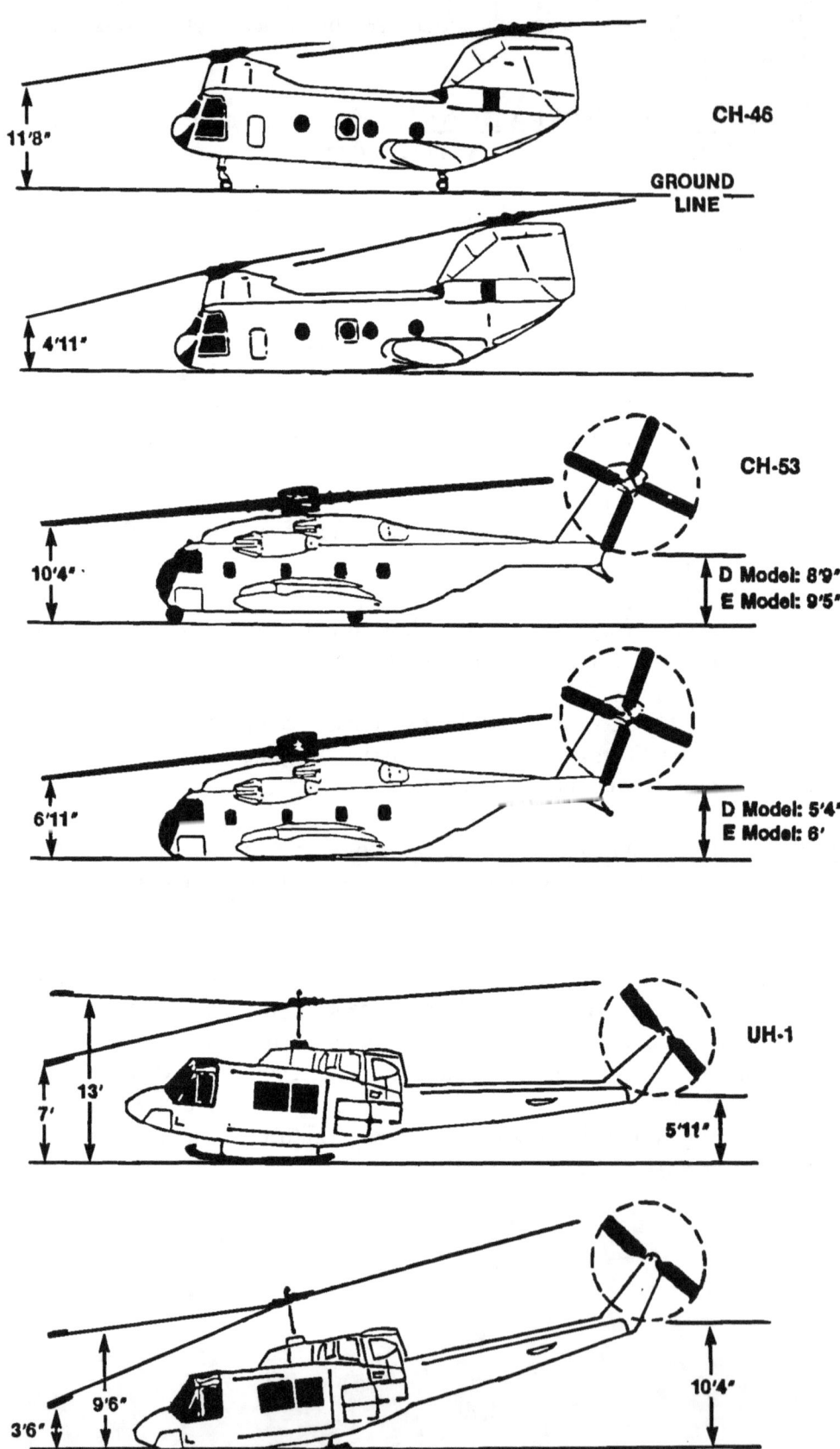

TRANSITION: Now that we talked about safety considerations, let us discuss LZ selection and preparations.

5. (5 Min) **HELICOPTER LANDING ZONE SELECTION AND PREPARATION**.

    a. <u>LZ Selection</u>. LZ size is determined by the number and type of helicopters to be employed.

        (1) <u>Size</u>. A LZ that is 50m by 50m is generally sufficient to land any helicopter. Consideration should be given though, to the altitude that operations are being conducted. As operating altitudes increase toward 10,000 ft. MSL, the size of the LZ should also increase. This is based on the performance loss that a helicopter will experience while operating at higher altitudes. At higher altitudes (close to 10,000 ft. MSL) a LZ that measures approximately 100m by 100m will provide the necessary clearance for a safe approach and departure.

        (2) <u>Approaches and Exits</u>. If the LZ in question has significant obstacles, surrounding the zone or on the approach /departure path, size considerations should be increased.

NOTE: In all situations, a face to face brief with the pilots in the operating area will give you an idea of the capability/limitations of the aircraft and what size zones they will be looking for. The Air Officer/FAC should be the leading authority on this matter.

        (3) <u>Wind Direction</u>. The wind direction determines the approach and departure directions. Helicopters normally take off and land into the wind.

        (4) <u>Ground Surface.</u> Debris/snow/ice will be kicked up when the helicopter comes in to land in the zone. The majority of the danger from this flying debris is to the Marines on the ground. If rocks/ice or other debris exists, ensure Marines use caution when the aircraft is coming into land. Time permitting, packing down the area where the Ahkio huddle will be located will minimize the effects that blowing snow and ice will have on the Marines.

        (5) <u>Ground Slope.</u> Terrain which slopes more than 8° is usually considered to steep for helicopter landing due to dynamic rollover characteristics of all helicopters. Individual helicopters have different limitations for sloped landings. A briefing with aircrew in the operating area will serve to clarify what the capabilities/limitations of the aircraft.

        (6) <u>Concealment.</u> LZs should be selected which will conceal both the helicopter and the snow signature from direct enemy observation. The signature that develops from the rotor wash can be observed up to 30 kilometers away.

        (7) <u>Obstacles</u>. The unit to be loaded should look for obstacles that may be hidden under snow. Obstacles that are hidden are potentially dangerous to the helicopter. Probing the LZ should be conducted to locate tree stumps, large rocks, etc., which could place the helicopter in a dynamic rollover situation or rupture the skin on the belly of the aircraft.

(8) <u>Snow.</u> Depth and consistency of snow will have a major impact on LZ operations. Loose snow may cause the pilots to white-out, lose reference to the horizon, and have to conduct a wave off. Hard or crusted snow may break up and become a hazard to Marines on the ground.

(9) <u>Lakes and Rivers as LZs</u>. Commanders should consider using frozen lakes and rivers as alternate LZs. Frozen lakes and rivers do make excellent LZs since they are level and have little loose snow due to the scouring winds. CH-53 and CH-46s need 15 inches of ice to conduct operations. UH-1N and AH-1W need 8 inches of ice thickness.

b. <u>LZ Preparation</u>. Marines should make every effort to walk through the LZ to determine snow depth and appropriate locations for helicopter landing points.

(1) <u>Packing the LZ</u>. Packing the LZ makes it easier for a pilot to find the landing point and for Marines to move about. This consideration is particularly important when conducting external operations. Packing does take more time and the possibility of detection by the enemy may be increased.

(a) Time, conditions, and tactical situation permitting, the first area to be packed should be the area around the Ahkio huddle. The next area will be for the landing point which should be approximately 50 meters square. This will decrease the amount of snow that will be kicked up by the rotor wash.

(b) If an LZ is in a safe area, and will be used frequently, a request for Engineer Support to pack the snow should be made. Over the snow vehicles are most effective for packing landing points quickly. Marines on snowshoes, ski, or the boot packing method can also be used but is more time intensive and exhausting.

(2) <u>Marking the LZ</u>. Marking the LZ and the landing points is critical. The white snow-covered zones can provide a difficult background for the pilots. Blowing snow can cause a white-out condition and may cause the pilots to lose reference to the horizon. A reference point must be visible at all times. Any object that will contrast with the snow and does not move will provide a reference point.

(a) <u>Air Panels</u>. Air panels contrast in color with the snow. They must be secured to ensure that they are not blown away by rotor wash.

(b) <u>Smoke Grenades</u>. These should be used to mark the LZ only and not the landing points. When used in snow covered LZs use a platform to prevent the smoke grenade from sinking in the snow.

(c) <u>Chemical lights</u>. Chemical lights provide good close-in lighting at night but are hard to see beyond ½ mile.

(d) <u>Tree Boughs</u>. Lay or stick these into the snow to provide a contrasting reference for pilots to orient on.

(e) <u>Sled teams</u>. Ahkio huddles are the primary method of marking landing points. The huddle should contrast in color to the background in the LZ. Individuals should remove overwhites, wear a protective face mask, and be sure that no bare skin is exposed to the rotor wash.

(3) The following is the minimum required information for a landing zone brief.

(a) Your call sign

(b) LZ Location

(c) LZ Marking

(d) Wind Direction and Velocity

(e) LZ Size

(f) LZ Elevation

(g) Obstacles/Snow Conditions

(h) Visibility

(i) Approach/Retirement Direction (Recommended)

TRANSITION: Now that we have covered LZ selection and preparation, are there any questions? Next, we'll discuss preparations for embarkation in cold weather snow covered LZs.

6. (5 Min) **PREPARATIONS FOR EMBARKATION**.

a. <u>Planning</u>. Helicopters will often have reduced payloads when operating at higher altitudes. In addition, high temperatures, high humidity, and high Density Altitude will degrade helicopter performance. Consequently, helicopter payloads may change significantly due to both the current and forecasted weather and LZ altitudes.

b. <u>Payload</u>. The following chart is estimates only and should be used as a guideline. Actual lift capacity will vary depending on fuel consumption, ordnance on board, time of flight, weather, etc.

HELICOPTER	SEA LEVEL	5000 FT MSL	10,000 FT MSL
UH-1N	6 Pax and gear	4 Pax and gear	2 Pax and gear

| CH-46E | 16 Pax and gear | 8 Pax and gear | 4-6 Pax and gear |
| CH-53E | 37 Pax and gear | 24 Pax and gear | 18 Pax and gear |

    c. <u>Personnel</u>. A major hazard to personnel operating around helicopters in the cold weather is the wind chill generated by the rotor wash. Exposed skin should be kept to a minimum. If a long wait is expected, warming tents should be erected.

    d. <u>Equipment</u>

        (1) The team sled should be staged as near the landing point as possible. To prevent the team sled from being moved by rotor wash, the Marines embarking on the aircraft should lay on top of the sled. This procedure is known as the Ahkio Huddle and will be discussed later.

        (2) Weapons should be in Condition 4 when embarking the aircraft. Muzzles should be pointed down on CH-46 and CH-53s. Muzzles should be pointed up or outward on UH-1Ns.

        (3) No equipment (skis, poles, radio antennas) should be allowed to protrude above the height of a man. This is to prevent any equipment from going into the rotor blades.

        (4) Packs should not be worn aboard helicopters due to the restricted movement and the requirement to fasten seat belts before departure. Packs and team sleds should be staged at the center of the aisle on assault aircraft.

<u>TRANSITION</u>: Now that we talked about preparation for embarkation, are there any questions? Let us now discuss the Ahkio huddle.

7. (10 Min) **<u>AHKIO HUDDLE PROCEDURES</u>**. The Ahkio huddle is designed to get Marines on and off a helicopter as quickly as possible with minimum exposure to wind chill. This must be rehearsed so that it can be accomplished in extreme weather and reduced visibility.

    a. <u>Ahkio Huddle</u>.

        (1) The ahkio huddle is established around the sled/tent equipment on the landing point. Packs off and skies are bound together. Marines group together on top of the equipment, face down, to keep the equipment from blowing away.

        (2) All of the tent teams equipment necessary for survival against the environment is loaded on the same aircraft as the personnel.

        (3) The helicopter will land so as to place the sled team huddle under its rotor arc at the 2 o'clock position.

    b. <u>Rehearsals</u>. Before conducting sled team huddle operations, all Marines, including pilots, aircrews and troops to be lifted must be properly trained in ahkio huddle procedures. This will:

(1) Eliminate the dangers of troops walking into the helicopter blades.

(2) Reduces the problems of wind chill.

(3) Reduces the amount of time the helicopter must remain in zone by: providing the pilots a solid reference point, reducing the distance Marines must move through the snow to the aircraft, and reduce the loading/unloading time.

TRANSITION: Are there any question concerning the Ahkio huddle? If not, let's discuss embarking and debarking of the helicopters.

8. (7 Min) **EMBARKING AND DEBARKING THE HELICOPTER**. The ahkio team leader supervises the loading of the sled and any other equipment.

   a. The Ahkio Team Leader

      (1) Loads first, moves to the front of the helicopter, and secures his gear by the left most forward seat.

      (2) Should immediately communicate with the pilot.

      (3) When in flight, observe from between the pilots to maintain orientation.

      (4) Designates Marines to load / unload equipment.

      (5) Straps in for take off and landing.

      (6) Maintains accountability of the ahkio team.

   b. The Ahkio Team

      (1) Loads the helicopter only when directed by the crew chief who will direct the team to load through either the side or rear door.

      (2) Enters the aircraft quickly and moves to pre-assigned seats.

      (3) Hand carries their snowshoes on board. Once seated, they hold their snowshoes and weapons between their legs.

      (4) Bind skis and poles together in pairs. When loading or unloading, keep them parallel to the deck at waist level. Once loaded, place skis on the deck of the aircraft beneath the feet.

   c. The Assistant Ahkio Team Leader

      (1) The assistant ahkio team leader supervises the loading and unloading.

(2) Ensure that all gear and Marines have boarded.

(3) He boards last and signals thumbs up to the crew chief.

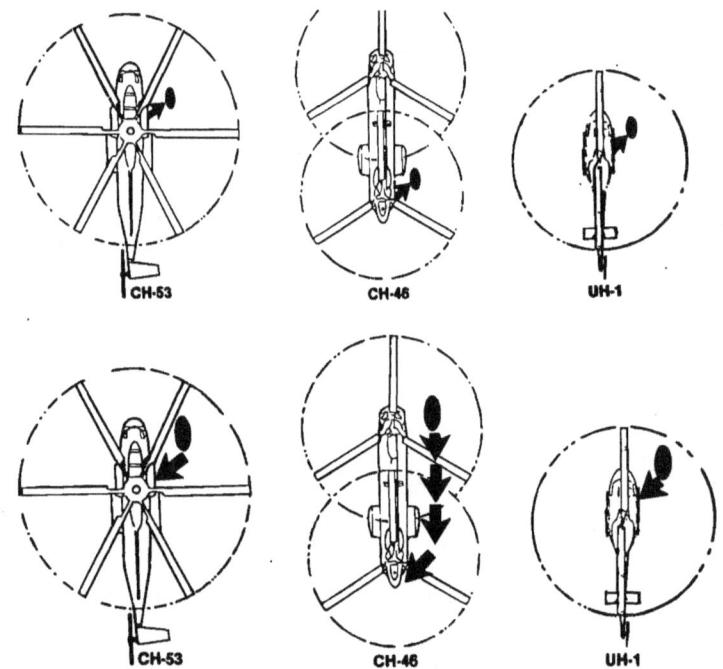

EMBARKING AND DEBARKING BY THE AHKIO HUDDLE METHOD

   a. <u>Debarking</u>. As with embarking, the object during debarking is efficiency and safety. The off load generally follows a reverse order of the on load.

      (1) <u>Sequence</u>

         (a) The assistant ahkio team leader supervises the off load.

         (b) Unload the sled and any other equipment first.

         (c) Then all remaining Marines exit in reverse order of embarking.

         (d) Assume the ahkio huddle just off the ramp or outside the door under the rotor arc.

      (2) <u>Tactical Deployment</u>. Only after the helicopter lifts do ahkio teams tactically deploy.

<u>TRANSITION</u>: Now that we've discussed embarking and debarking procedures, are there any questions? Let's talk casevacs.

9. (5 Min) **CASEVAC CONSIDERATIONS IN COLD WEATHER**. Combat casualties are complex enough for a small unit leader to handle; the cold weather will make casevac extremely critical. Some points to remember:

   a. If possible, pre-plan your casevac LZs and have a dedicated casevac helicopter on alert.

   b. Ensure your Marines are cautious when loading a patient aboard a helicopter in deep snow. Remember the reduced rotor clearance.

   c. Protect the patient from the rotor wash. Any exposed skin will be subject to frostbite.

   d. Establish a warming tent for your patients and your loading teams. Handling casevac is physically and mentally exhausting.

TRANSITION: Now that we have discussed safety and casevacs, are there any questions? If there are none for me, than I have some for you.

PRACTICE (CONC)

   a. Students will plan a helicopter operation.

PROVIDE HELP (CONC)

   a. The instructors will assist the student when necessary.

OPPORTUNITY FOR QUESTIONS (3 Min)

1. QUESTIONS FROM THE CLASS

2. QUESTIONS TO THE CLASS

   Q. What happens to helicopter lift performance as altitude increases?

   A. Helicopter lift is greatly reduced.

   Q. Should snow be packed down in the LZ prior to helicopter landing?

   A. Yes (time permitting)

SUMMARY (2 Min)

   a. We have looked at some of the considerations for using helicopters in cold weather and mountainous environments. Your battalion Air Officer should be able to provide additional information as well as arrange for helicopter support to help train your Marines to the needed level.

b. Those of you with IRF's please fill them out and turn them in. We will now take a short break.

UNITED STATES MARINE CORPS
Mountain Warfare Training Center
Bridgeport, California 93517-5001

WML
WMO
09/04/02

LESSON PLAN

COLD WEATHER CONSIDERATIONS FOR NBC DEFENSE

INTRODUCTION (5 Min)

1. **GAIN ATTENTION**. Not long ago, the Soviet forces, or Soviet-bloc forces, used chemical agents in countries such as Yemen, Kampuchea, and Afghanistan. More recently, the Iranians and Iraqis have employed chemical agents, thus causing mass casualties in single attacks. The nuclear arms race is a topic of constant conversation in governments around the world. It doesn't take a genius to figure out that any future conflicts have a very distinct possibility of being fought in an NBC environment. If that NBC environment is also in a cold weather environment, we as Marines must still be able to move, fight, and survive to accomplish the mission.

2. **PURPOSE**. The purpose of this period of instruction is to discuss those aspects of NBC warfare that are peculiar to a cold weather environment. This includes a review of basics of NBC defense, cold weather peculiarities of MOPP gear and other individual equipment, nuclear specifics in a cold weather environment, and the freeze point of common chemical agents.

3. **INTRODUCE LEARNING OBJECTIVES**

   a. TERMINAL LEARNING OBJECTIVE. In a cold weather mountainous environment and given an NBC scenario, conduct military operations, in accordance with the references.

   b. ENABLING LEARNING OBJECTIVES.

      (1) Given NBC detection equipment and an NBC scenario in a cold weather environment, conduct NBC monitor survey operations, in accordance with the references.

(2) Given full MOPP gear and an NBC scenario in a cold weather environment, operate in MOPP gear, in accordance with the references.

(3) Given NBC decon equipment and an NBC scenario in a cold weather environment, decontaminate personnel and equipment, in accordance with the references.

4. **METHOD / MEDIA**. The material in this lesson will be presented by lecture and demonstration. Those of you with IRF's please fill them out at the conclusion of this period of instruction.

5. **EVALUATION**. You will be tested later in the course by performance evaluation.

TRANSITION: This class is not designed to teach NBC defense, but to help you apply what you should already know about NBC defense to this environment. Let's start by applying our fundamentals of NBC defense to the cold.

BODY (55 Min)

1. (5 Min) General Effects.

   The general effects of NBC agents are altered in the cold weather environment. Temperatures are a constant problem. All NBC protective equipment and supplies must Be kept from freezing. Below 0 F, the radius of a nuclear blast is increased. Most chemical agents remain hazardous down to –50 F. The survival of most biological agents is significantly enhanced as the temperature drops. Marine Corps NBC operations are covered in OH/FMFM 11, MAGTF *Nuclear, Chemical, and Defensive Biological Operations* (currently under development).

   a. Severely Degraded NBC Equipment Effectiveness. Secondary frostbite occurs when current equipment is used. Materials become brittle, crack, And tear easily in extreme cold. Frozen solutions clog and damage equipment. Heat must be supplied when conducting decontamination operations. Agent detection reactions are slow or will not work even when re-warmed with re-agents. Contamination problems arise due to the cold's effects on donning and doffing procedures. There is an increased logistics requirement foe NBC defensive operations against chemical attack in extreme cold.

   b. Psychological Stress Extreme cold and high altitudes produce psychological problems, which numb the intellect and degrade personal and unit security. Current knowledge is insufficient to estimate what the added impact on military operation will be. Casualty rates cannot help but increase if psychological stress slows or stops normal reaction of individual Marines or units to a chemical attack.

   c. Mountainous Areas Breathing is more difficult at higher altitudes. It is unknown if NBC equipment will work better in the extreme cold or high altitudes than in the cold at low altitudes. At the lower barometric of pressure of high altitudes, chemical agents will evaporate or sublime more rapidly. The increase isolation may, by localized surface heating, speed the vaporization of chemical agents. Contamination may spread quicker by greater wind speeds at higher altitudes. Contamination may collect in depressions and small valleys.

TRANSITION: Now that we have an understanding of the general effects, we will look at the effects of the cold on nuclear weapons.

2. (15 min) Cold's Effects on Nuclear Weapons

    a. Blast. Below 25 F, the effectiveness radius of a nuclear blast increases up to 20 percent. Icepack or snow pack extends the distance of static overpressure, the crushing effect of the blast. Conversely, the distance of dynamic pressure may decrease due to soft, absorbent characteristics of drifts, snow cover, etc. Tundra and ice formations can break up pressure waves. The cratering effects in ice and frozen ground may be reduced, similar to those in solid rock.

        (1) Primary and reflected waves and ground shock from even small yield nuclear weapons may create earthquake-like fissures, crevasses, avalanches, and rock slides as far as 30 kilometers from ground zero.

        (2) Secondary effects include snowstorms, avalanches, quick thaws, and ice breakup on lakes and rivers, which can interfere with troop movement.

        (3) Blast will increase damage on equipment due to its inflexibility from cold soaking, which makes metals and alloys brittle.

        (4) In heavily forested areas, blown down trees will make large areas virtually impassible to vehicles and personnel.

        (5) During winter months, trees freeze and become brittle. In a nuclear blast, they can be converted into many splinter-like projectiles. Expect puncture injuries.

        (6) Personnel injury from flying debris increase because of layered clothing.

    b. Thermal. Minimum safe distances may need to be increased by as much as 50 percent due to increased density of cold air and the high reflection of snow and ice. However, cold temperatures, cover or frost, ice, and snow may reduce thermal effects on combustibles. The thermal effect may produce flash flooding in low-lying areas when ice and snow pack melt. Thawing can greatly reduce troop movement in some zones. The snow, ice, and low temperatures may reduce destruction from post-blast fires. Muskeg, tundra, and wet terrain are average reflecting surfaces, which reduce thermal radiation. Ice fog and snow cloud cover reduce thermal effects if the device is detonated as an air burst. Snow is a good reflecting surface, which increases due to winter darkness dilating the pupils. Injurious heat absorption by personnel and equipment may be reduced by reflective overwhite camouflage clothing and netting.

c. <u>Radiation</u> Weather plays an important part in radioactive fallout patterns. Snow can mask radiological hot spots from detection. Snow deposition is erratic due to rapidly changing winds. High winds extend radiation fallout patterns, but correspondingly, may reduce radiation dose rates due to dispersion of contamination. At extremely low temperatures, the increased density of the atmosphere may reduce the distance of initial radioactive fallout by as much as 25 percent.

1. Contaminated snow may still spread the fallout. Amounts of induced radioactivity in the soil are reduced and even prevented by deep snow. Poorly drained areas, such as meadows, limit natural flushing and may act as collective points or radioactive contaminants. Most of the arctic is poorly drained. New snow may lessen fallout contamination of areas; hence, personnel and equipment may safely cross it.

2. Levels of local radiation can change quickly in windy conditions, which can lead to hot spots far removed from ground zero and very low intensity areas near ground zero. Consequently, the need for radiological surveying is increased.

3. Either M-1950 or ECWCS clothing provide excellent protection from fallout. Removal of radioactive particles may be done by rigorous shaking and brushing of the outer garments. Snow caves and below ground shelters provide excellent shielding against radiation. Avoid melting snow and ice for drinking water.

d. <u>Electromagnetic Pulse (EMP)</u> Effects are expected to be the same as in temperature zones. EMP will hamper or negate radio and tactical – satellite communication for extended periods. EMP mitigating practices: e.g., burying cables and wire links, may be difficult or impossible because or permafrost. Recovery may be out of the question.

<u>TRANSITION</u>: The cold also affects biological agents as well as nuclear weapons. Here are some of the things to look for.

2. (5min) <u>Cold's Effects on Biological Agents</u>

    a. Biological attacks cannot be ruled out. Up-to date immunizations, acclimated personnel, and strictly enforced personal hygiene (often considerably more difficult in the cold) are the best ways to avoid secondary spread of any infection (see FMFM 4-50, Health Services Support of the FMF). Cold weather discourages the use of vector borne agents since they can't survive. Toxins are less susceptible to the cold and the possibility of their use by covert means must be considered. Research by the Dugway Proving Grounds indicates that the survival of most microorganisms increases significantly below 32 F. Layers of snow and reduced sunlight in northern regions lengthens the hazard period for biological agents. Organisms remain alive but dormant. They become active when exposed to warmer temperatures. The most effective means of biological warfare in cold weather is by delivery of live organisms by covert means. After a known biological attack, precautions to prevent the spread must be taken just as they would in a temperate climate.

    b. Temperature inversions frequently found over snowfields and bodies of water tend to prolong aerosolized biological clouds. With the exception of thermal inversions, aerosolization of biological is less effective.

    c. Aerosolizing live biological agents becomes more difficult at extremely cold temperatures. Only some spores form bacteria and certain viruses survive.

    d. Tents and other areas where temperatures are higher than ambient are likely areas of biological agent is likely to spread rapidly in crowded living conditions.

<u>TRANSITION</u>: As you can see there are many effects that cold has on biological agents. The cold affects chemicals in many ways as well.

4. (10min) <u>Cold's Effects on Chemical Agents</u>

  Chemical agents work so well in the cold that they must be considered a significant hazard. Exposure to any chemical agents will require masking. Aerosol dispersal is good. Most agents freeze at – 50 F. Chemical agents either thickened or frozen on clothing or equipment produce deadly off gas concentration once into heated areas.

a. <u>Blood and Choking Agents</u>. Blood and especially choking agents remain extremely hazardous and nonpersistent throughout low temperatures. Theses agents may be disseminated as a liquid, solid, or aerosol. Masks are require whenever they are used. Hazard times may be longer. Agents will remain nonpersistent. The blood agent AC is extremely hazardous even as low as –65 F.

b. <u>Blister Agents</u>. Blister agents freeze below 0 F They can be brushed from clothing and equipment. However, some mixtures such as HL remain a liquid hazard at fairly low temperatures. The standard winter blister agent is considered to be a mixture of mustard and lewisite. In areas that lack water, frozen or otherwise, this blister agent expected to persist in liquid form for up to 6 months. If water is present, then the agent decomposes to form pure mustard that freezes at 58 F. Blister agents usually are employed to cause blisters on contact with the skin, blindness, and inhalation casualties.

c. <u>Nerve Agents</u>. Nerve agents will freeze in severe cold. They present a very serious vapor hazard when brought into warm areas. They present a very serious vapor hazard when brought into warm areas. When used to contaminate key facilities; i.e., LOC or major population centers, nerve agents become an immediate long –term hazard. This hazard may require tremendous decontamination efforts or even a change in seasons to reduce below lethal levels. Persistency is controlled by three factors; an increased temperature, a smooth terrain surface, and wind speed. Nerve agents, particularly VX, are effective when absorbed through the skin or eyes but the low volatility of VX makes the vapor h   hazard negligible below 32 F. The physical behavior of persistent nerve agents is only slightly affected by decreasing temperature. As the temperature nears 32F, persistent nerve agents dissolve in water, have reduced vapor hazard, and increase in persistency. No persistent nerve agents; e.g., GB tend to become semipersisitent, lasting from 2 to 10 days.

<u>TRANSITION</u>: Now that we have an understanding of how cold affects these agents, let's talk about the methods chemical agents are employed.

6. (2.5 min) <u>Employment of Chemical Agents</u>.
The offensive capabilities (excluding aerial delivery) of units in terms of ability to deliver toxic chemicals will be limited. The usual method for con- ducting toxic chemical attacks will be to place the available concentration of fires directly on those small, well –located targets, which are most vulnerable to chemicals attack. " Time on target" fire techniques will be used by artillery and mortars to place a maximum number of rounds on the target in minimum time. Fuse settings should be varied depending on the nature of soil, depth of snow, and the type of target being attacked. Minefields placed to restrict the enemy's use of key terrain should be

composite minefields. The chemical mines should be placed to force the enemy off the road net and to use undesirable terrain. Approaches to bridges and bridge abutments can be contaminated when destroying the bridge, causing delay in reconstruction. Chemical weapons function in cold temperatures. However, a general rule to use when analyzing friendly unit vulnerabilities is to double the munitions requirements for each 20 F decrease in temperature to achieve the same result expected at the warmer temperature. Agents have been developed with a variety of regulated persistencies, which give the field commander a known factor to plan operations with a high degree of certainty in results. Weather, terrain, and logistical considerations limit the arctic areas where forces can operate effectively. The size of available targets for chemical attacks will usually be small.

TRANSITION: With an understanding of how these chemicals are employed, we can begin to talk about ways to defend ourselves.

7. (10 min) Nuclear, Biological, and Chemical Defense.

To effectively defend against the effects of NBC weapons, four fundamentals must be applied: detection, contamination avoidance, protection, and decontamination. In a mountainous cold weather environment, the first three fundamentals become extremely important, as decontamination quickly becomes a logistical nightmare.

    a. Detection. Automatic detectors must be heated. Detector paper is not always readable. Detection is vital to identify a hazard. The vapor hazard of chemical agents may be limited making the M-256 chemical agent detector kit unreliable. Persistent agents will probably freeze into solids that may or may not be identifiable. This will create a pick- up hazard that will not materialize until temperatures warm up. Re- agents in the M-256 chemical agent detector kit will freeze and provide inaccurate reading in temperatures – 25 F. These kits must be kept close to the body to prevent freezing. Cold slows the response of M-8 chemical detector paper. Extra time must be allowed for the paper to work. M-9 paper is of little value because all the substances that react to it are affected by the cold. Extreme cold and /or the physical state of a chemical agent may make the M-8 alarm system ineffective. Detection will also determine the extent of the hazard and if the direction of advance should change to avoid contamination. Every 50 to 100 feet, detection teams melt snow, heat it to 70 F, and test with M-8 paper or CAM. If an agent is suspected, water is takes\n into a heated shelter and heated until a vapor is given off above 70 F. It is then tested with the M-256A1 kit. Data is then reported and plotted according to unit SOP. Because of the restraints the mountains place on radio communications, it may be necessary to use messengers. Contaminated areas are then marked with the standard NATO contamination signs.

    Contamination Avoidance.
        1. Passive measures such as concealment or dispersion are used to avoid offering the enemy a lucrative target. Suggested passive measures include:

            a. Harden positions by improving overhead cover.

      b. Avoid detection.
      c. Provide early warning.
      d. Develop NBC environmental discipline.
      e. Seek protection from a potential chemical attack
      f. Remain mobile.
      g. Keep supplies and equipment covered.
      h. Limit exposure to NBC hazards.
      i. Prevent spread of contamination.

2. Active measures include:
    a. Find and destroy enemy munitions and/or delivery systems.
    b. Use NBC reconnaissance teams to monitor contamination of specific areas.
    c. Use the standard NATO warning and reporting system to warn others of hazards or to pass the alarm locally by the most expedient means when a hazard is detected. Mountains or EMP can disrupt radio transmissions from detonation of a nuclear device.
    d. Separate mission-essential equipment from nonessential equipment when both are contaminated. Cover equipment vital to mission accomplishment with readily available material. Restrict personnel movement in the contaminated area.
    e. Collective protection shelters. They provide a contamination free working environment for selected personnel for tactical or administrative use. They also provide a relief from continuous wear of full MOPP gear.

b. <u>Decontamination</u>. Decontamination (decon) is the process of making any person, object, or area safe by absorbing, destroying, neutralizing, making harmless, or removing chemical or biological agents, or by removing radioactive materials clinging to or around it. (Joint Pub 1-02) The extensive time and logistical support needed to perform deliberate decon requires commanders to avoid contamination if at all possible. Temperatures below 32 F limit the effectiveness of decontamination operations. Current chemical decontamination procedures that require wet water rinses are unacceptable in freezing weather. No water decontamination procedures have not been developed. Decontamination must be done in heated facilities. To decontaminate Marines must:

    a. Perform *Hasty Decon!* Reduce the contamination hazard as quickly as possible by removing or neutralizing the chemical.
    b. Continue to fight after *Hasty Decon*!

c. Understand the effects of chemical agents. Be proficient in individual decontamination procedures.

TRANSITION: As you see there are different things to consider when talking about NBC defense. Shelter control is also an important factor to consider.

7. (2.5min) Shelter Control

In cold weather operations, decon and detection must be accomplished in heated shelters. One of the most challenging problems is how to prevent contamination from entering warm areas. For example, if individuals get frozen agents on their clothing, it will be hard to detect because low temperature have slowed the effects of the agents. However, if the temperature warms or if the contaminated individual enters a heated area, the agents will give off gas. It may be necessary to set up a thawing station for each warm shelter and unmasks. Otherwise, all occupants may be subject to hazardous liquids or vapors. Additional personnel and equipment will be necessary to man these warming stations.

TRANSITION: Now we can begin talking about protecting ourselves with the equipment that we carry.

8. (15min) Individual Protection Equipment.

   a. Mask Carrier. This should be adjusted to be worn in a slide carry beneath the cold weather parka. Body heat will help keep the mask warm and flexible. But masking will be slow and difficult. Marines must be aware of this requirement when donning their cold weather clothing. They must ensure that the ECWCS parka is large enough to accommodate insulating layers and the gas mask with its carrier.

   b. Outserts. Two outserts are provided to prevent fogging of the eye lenses. Green and amber tinted outserts can be used for bright light and fogging. Avoid wiping eyepiece with glove when firing because it will smear.

   c. Donning the Mask. Hold your breath. Remove the mask from under your parka, lower the parka hood, and don the mask. Adjust the head harness only tight enough to create a good seal. Raise the parka and fasten outer garment. Put on gloves or mittens.

   d. Removing the Mask. Remove gloves or mittens before removing the mask. Loosen the outer garment and lower the hood. Brush snow or ice particles from the mask. Remove the mask and immediately dry face and the inside of the mask. Raise your parka hood and fasten outer garment. Wipe the mask thoroughly before storing to prevent ice formation. Store the mask in the carrier. Put on gloves or mittens. When possible, further dry the mask by placing it in a warm area but not in direct heat.

   e. Wearing the Chemical Protective Over garment. Contamination avoidance measures may fail. The enemy may find and attack with NBC weapons. Marines may be downwind of an

NBC attack or the mission may require them to cross-contaminated areas. Protection options must be available so Marines can survive and continue the mission. Wearing the chemical protective over-garment presents a particularly difficult operational/logistics decision that the MAGTF commander must make before entering the operational theater. Tariff sizes and the impact of MOPP conditions on the operations must be considered. Tariffs must be up-sized. Marine Corps policy is that the chemical protective over garment will be worn over the ECWCS. The policy allows the possibility of exercising MOPP II TO IV, providing the commander with the greatest mission accomplishment flexibility. If the ECWCS parka is not needed because of the additional warmth provided by the chemical protective over garment, store and seal it in a plastic bag. Wearing insulated layers of the ECWCS beneath the chemical protective over garment will necessitate that the tariffs be sized up for cold weather operations. If the ECWCS were worn outside the chemical protective over garment, it would permit the laying required for additional warmth. However, it would require that the contaminated clothing worn over garment be removed before entering a heated shelter where the agents can re-vaporize. Upon decon, the external layers of the over garment would have to be discarded. Also, once the ECWCS is donned, it's not possible to adjust the over garment. Replacement issues of cold weather clothing would be needed. Stocks of ECWCS are not adequate to support this requirement. A method of decontamination of this cold weather clothing would be necessary. The chemical protective over garment will produce internal moisture and consideration and dehydration problems, which can lead to hypothermia. In full MOPP gear, speed is reduced to 50 percent (probably greater due to build up of condensation and the potential danger of hypothermia).

f. <u>Chemical Protective Glove Set. Vapor barriers</u>; e.g., medical examination gloves, the wool inserts, and the leather shell covered by the butyl-rubber glove, make up the chemical protective glove set. This set allows the insulator (the wool insert) to remain warm and dry. If either outer glove is punctured or torn, the glove set must be replaced. In extreme cold when mittens are needed, this combination may result in frostbite.

g. <u>VB Boots</u>. The VB boots are effective and an adequate replacement for the normal over-boots. However, the natural rubber composing the VB boot will be penetrated by chemical agents, which contain mustard gas. After 50 hours exposure, they must be replaced and thoroughly decontaminated.

h. <u>Decon Kits M-258A1 and M-13</u>. The M-258A1 decon kit is good to 32F. It must be kept next to the body to prevent freezing. Because it assumes surrounding air temperature, the solution should be used quickly once it is opened to avoid possible frostbite. The M-13 consists of a bag of Fuller's earth equipment decon. It is not affected by the cold.

i. <u>NAAK M-K1</u>. The NAAK M-K1 should be protected from freezing. At temperatures –40F, remove from the mask carrier, and store in a parka pocket.

<u>TRANSITION</u>: Also important when talking about protection is the units protection. There are several things to consider.

9. (5 min) <u>Unit Protection</u>

Units will have to be self-sufficient in NBC defense. Terrain and weather in mountainous areas will dictate a requirement for a high degree of NBC defense preparedness. Cold weather conditions will impose many special considerations on defense planning. Because deployments of large forces are limited, adjacent units may not be able to provide mutual support. Logistical support may be drastically reduced and rapid maneuver difficult.

    a. <u>Bivouac Location</u>. The well dug infighting position with overhead cover remains the best means of protection for the individual. Because of terrain and weather restrictions, bivouac location for effective NBC defense requires consider- able planning. Gullies, ravines, ditches, natural depressions, fallen trees, and caves become effective only when used with normal precautions. However, low-lying areas are places where toxic clouds can settle. If tactically feasible, positions should be constructed upwind of a possible threat. Reserves slopes are good protection from nuclear attack. The heat, light, and initial radiation are absorbed by hills while the slope deflects the rest up. Predicting the actual point of an enemy nuclear attack is almost impossible. Likely targets such as MSR's and mountain passes should be avoided. Wooded areas present a mixed blessing. Biological agents persist longer as the area retains moisture and keeps out sunlight. The heavily wooded forested overhead found in the coniferous forests of Europe and the northwestern United States will reduce the possibility of liquid contamination from chemical weapons. However, chemical vapor hazard will increase.

    b. <u>Construction of Fighting Positions</u>. Once the bivouac location is chosen, the construction of fighting positions capable of providing NBC defense is paramount. Open NBC fighting positions provide protection mainly against nuclear effects. It reduces exposure to thermal effects, initial alpha and beta radiation fallout, live biological agents, and a liquid chemical agent provides missile protection. To construct effective overhead cover:

        1. Use dense covering materials.
        2. Be sure to construct cover in depth.
        3. Use strong supports
        4. Cover as much of the opening as possible.
        5. Thoroughly line or revert positions on the sides using log forms, ammo boxes, or empty MRE boxes filled with ice crete or snow-crete.

    c. <u>Work Rate</u>. Dehydration injuries are likely when Marines are required to perform difficult physical tasks. Marines will not suffer the same degree of heat buildup as in warmer climates. However, they are likely to suffer subsequent cold weather injuries. Freezing air inside of protective clothing or perspiration which cannot evaporate leads to chilling or hypothermia. Generally, Marines in MOPP IV conditions at 20 F will require twice the normal amount of time to accomplish tasks. Additional supervision of workloads is necessary to alleviate cold weather injuries at all levels.

d. <u>MOPP Gear Exchange</u>. If a unit becomes contaminated, decon will be done by MOPP gear exchange (gloves and over garment only). Two complete sets of MOPP gear and waterproof bags to seal contaminated clothing should be available for each Marine. Care must be taken with packaging the over garment. Tears and exposure to air degrade protective qualities. Extra suits must be provided when crossing water obstacles or conducting amphibious operations. Any contaminated cold weather-clothing item must be replaced.

TRANSITION: Even in the best situation, you may become contaminated with a NBC agent. You will need to perform some form of decontamination.

10 (10 min) Decon of Equipment.

Equipment and supplies needed for decon will generally be more intensive. Cold temperatures can be expected to adversely affect aqueous solutions, pumps, sponges, swabs, and brushes. Everything will be difficult and time consuming. The list of equipment and supplies can be used as general guidelines for decon needs.

a. Water. The most common ingredient in decon operations, is useless if frozen. It should not be used when temperatures are so low that it freezes on contact with equipment. In these temperatures, use undiluted DS2, but remember DS2 corrodes equipment quickly. Equipment will need to be replaced rapidly. Because of their low freezing points, solvents such as jet petroleum 4 (JP4), diesel fuel, or kerosene may be used to physically contamination. With present technology, equipment decon problems are difficult to overcome in an arctic environment.

b. Commanders will have to consider fighting dirty in cold weather areas. Fresh units can be rotated into the contaminated areas so that dirty units can be moved to decon stations. DS2 and STD freeze at approximately –15F. In temperatures below –15f, JP4 can be used to remove contamination. This does not deactivate the chemicals. It only removes them. The ground and equipment used to remove contaminates must be destroyed or removed properly. JP4 is also highly flammable. Use extreme caution! The large amount of static electricity in cold dry climates could ignite the fuel. There is a definite risk of frostbite when fuel contacts exposed flesh. Rinse water must be heated or antifreeze added to the rinse solution to prevent freezing on contact with a cold vehicle. If the vehicle cannot be rinsed, the DS2 will quickly corrode the vehicle.

c. The chemical agent monitor is limited by the cold's degradation on battery life.

d. Two nitrogen cylinders may be required to expand the contents of the M-11decon apparatus at temperatures below 32F (0 C)

e. Difficulties will develop in dispensing DS2 as temperatures near 32 F. To overcome this problem, the decon apparatus portable (DAP) must be warm enough for the DS2 to pump through the brush assembly. Commanders should consider positioning a contingency supply of the M-13 inside heated vehicles and develop a plan to rotate the outer need DAP's into heated shelters.

f. The M-17 sanator decon apparatus has problems in cold temperatures because system relies on a water- based decon method. Normal engine cold-soaking problems have been observed along with internal pumps and lines cracking from the expanding freezing water. This system must be used within a heated shelter.

g. Decon stations should be suited in built-up areas, near road junctions, intersections of forest lanes, or where they may be approached from several directions.

h. Snow can be used to cover contaminated areas. However, when the snow blows away or when vehicles or personnel break through this surface, the contamination will reappear. Snow cover provides some protection if left undisturbed, but this protection is unreliable.

i. Unfrozen earth may not be available to make STB dry mix. Use snows in place of dirt (same proportion) in shuffle pits and for other decon purposes. Burying contaminated materials in frozen ground will be difficult. Mark burned or abandoned materials with standard contamination signs

TRANSITION: Just because we are already engaged with our enemy by direct and indirect fire, and with the weather that can be murderous by itself, does not allow us to forget that the enemy might other weapons waiting to take our lives. Are there any questions?

PRACTICE (CONC)

a. Students will practice what was taught in upcoming field evolutions.

PROVIDE HELP (CONC)

a. The instructors will assist the students when necessary.

OPPORTUNITY FOR QUESTIONS (2 Min)

1. QUESTIONS FROM THE CLASS

2. QUESTIONS TO THE CLASS

Q. What is the best means of protection for an individual?

A. The fighting position with overhead cover.

SUMMARY (2 Min)

a. This period of instruction has covered NBC operations in a cold weather environment including NBC general effect, cold weather effects on NBC agents, NBC defense, shelter control, individual protective equipment, unit protection, and decontamination of personnel and equipment.

b. Those of you with IRF's please fill them out at this time and turn them in to the instructor We will now take a short break.

UNITED STATES MARINE CORPS
Mountain Warfare Training Center
Bridgeport, California 93517-5001

WML
WMO
08/15/01

LESSON PLAN

FIRE SUPPORT IN A COLD WEATHER ENVIRONMENT

INTRODUCTION (5 Min)

1. **GAIN ATTENTION**. Fire support is extremely important in any operation. However, in a cold weather or mountainous environment, there are many factors that if not considered, could lead to disaster.

2. **PURPOSE**. The purpose of this period of instruction is to familiarize the student with fire support considerations and limitations in a cold weather environment. This lesson relates to mountain operations planning.

3. **INTRODUCE LEARNING OBJECTIVES**

    a. TERMINAL LEARNING OBJECTIVE. In a cold weather mountainous environment, employ fire support, in accordance with the references.

    b. ENABLING LEARNING OBJECTIVES

    (1) Given a scenario in a cold weather mountainous environment and map, write a fire support plan in accordance with the references.

    (2) Given a scenario in a cold weather mountainous environment, write a target list in accordance with the references.

    (3) Given a scenario and radio in a cold weather mountainous environment, call for fire in accordance with the references.

    (4) Given a list of surface conditions, select the appropriate fuse in accordance with the references.

4. **METHOD / MEDIA**. The material in this period of instruction will be covered by lecture, you will be evaluated during upcoming field training evolutions. Those of you with IRF's please fill them out at the conclusion of this period of instruction.

5. **EVALUATION**. This lesson will be evaluated during the field training portion of the course by performance.

TRANSITION: If there are no questions on what is to be covered, we will begin this lecture by discussing the ammunition used and how it is effected by the cold.

BODY (45 Min)

1. (5 Min) **AMMUNITION**

    a. Storage. Storage of ammunition is important. The cold greatly effects the capabilities of all ordnance. It is best to store ordnance in the cold. You want to keep the ammunition and the weapon that will be firing it at the same temperature.

    (1) First of all, if your weapon is in a cold weather environment, you want to keep it at the same temperature as the surroundings. It is virtually impossible for you to maintain your weapon at a temperature other than the ambient temperature when conducting operations. If you were to bring your weapon from the cold into a warmer climate, it will sweat for up to one hour. In field conditions, this condensation is difficult to remove from the smaller parts of the weapon, which may result in freezing of the parts.

    (2) Secondly, if you are firing ammunition that is stored at inconsistent temperatures, the FDC will have a difficult time maintaining accurate registration data. The rounds will have varying ranges based upon how cold the propellant is. The colder the propellant, the slower the burn rate and thus the lower the velocity and range of the projectile. Proper handling and continual updating of propellant temperatures will help insure effective gunnery.

    (3) Ammunition has a high malfunction/dud rate when brought from warm to extreme cold temperatures, due to condensation and freezing. The ammunition should not be stored inside tents or shelters that are warmed above freezing which can cause condensation. Projectiles, powder canisters, fuses, or primers may freeze when exposed to the colder air. This may damage propellants, increase difficulty in handling heavy projectiles, or prevent the proper mating of fuses and projectiles. Old firing positions can be effectively used as ammunition storage points.

    (4) Additionally, the weapons will tend to malfunction as a result of suddenly brittle parts, expansion and contraction of the weapon system, slow burn, etc. The bottom line is to keep the ammunition at the same temperature as the weapon. When beginning to fire it is important to begin firing at a slower sustained rate to allow the weapon to warm up and the moving parts to free themselves if frozen.

b. When storing ordnance in the cold, keep it off of the deck. You want to have at least 4 inches of air circulating beneath the ammunition and keep it as free of ice and snow as possible. Putting the ammunition on dunnage or pallets and covering it with a canvas tarp will suffice. As a field expedient method, you could stack the ammunition on tree branches and cover it from the elements.

c. The effects of winter conditions on munitions will in all probability cause an increase in ammunition requirements due to a high dud rate, reduced visibility, and the decreased effects of ordnance in the snow. Up to 80% of the fragmentation effects of fuse quick will be absorbed by only 12 inches of snow. If fuses are improperly planned for, a 5 times increase in ammunition could occur. It may be necessary to establish pre-designated levels of on-hand quantities and restrict firing when these levels have been reached. Every effort should be made to keep the basic load on-hand to guard against interruption in re-supply operations. Additionally, history has shown that ammunition requirements will increase two to four times for night operations. This is due to more indiscriminate firing of weapons.

TRANSITION: Are there any questions about the ammunition in general? Let's talk about the effects of rounds in the snow.

2. (5 Min) **EFFECTS OF ROUNDS**

   a. During extremely cold periods when temperature changes are sudden, the ballistic characteristics of weapons and ammunition are effected. The standard temperature for firing table is 70°F. During extremely cold periods, indirect fire support rounds impacting 100 meters short for every 1000m of range is not uncommon. This is a result of the propellant burning slower thus the round having a slower muzzle velocity when exiting the weapon system.

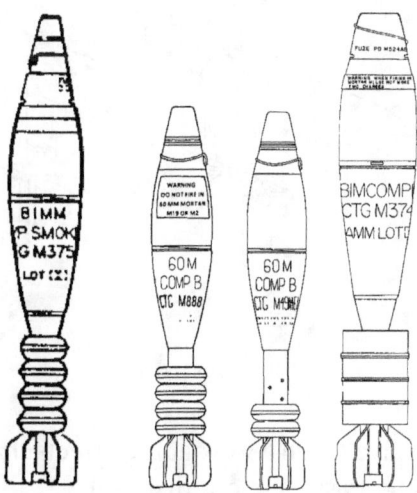

   b. On frozen ground, ice-covered ground, and/or rocky terrain, use fuse quick (point detonation), which will increase the fragmentation of high explosive (HE) rounds due to secondary projectiles. Fuse quick is ineffective in snow-covered terrain, as up to 80% of the fragmentation is absorbed. Use proximity or VT fuses in snow-covered terrain. When targeting frozen waterways (lakes or deep rivers), use a delay fuse. This will use the water as tamping in order to heave up and break up more ice (more casualties can be caused by drowning and cold weather injuries rather than fragmentation).

c. During extreme cold, the rate of fire will be slow until the weapons have warmed up, this is especially true of weapons that have a hydro-pneumatic type recoil. The rate of fire will also be affected by the fact that personnel will react slower and less deliberately in a very cold climate. Keeping personnel warm is of prime importance.

d. Smoke rounds and WP can be used very effectively for marking for air. The main consideration is ensuring that the rounds are deployed above the snow pack, so that the smoke is not absorbed by the snow. Another consideration is the color of the smoke or WP. Illumination rounds are not very effective for adjusting air, however, they are useful for FO's to use when spotting. The smoke cloud that the round makes helps the Artillery or Mortar FO's locate the spot of the round whereas a HE round is difficult to see.

e. Phosphorus shells, although producing desired smoke, contaminate the area of impact with phosphorus particles, which may remain buried under the snow. The will effect the area and surrounding areas during the spring melt off much like chemical weapons as discussed in COLD WEATHER CONSIDERATIONS FOR NBC DEFENSE.

f. FASCAM/DCPICM. Rapid temperature changes may cover or expose FASCAM/DCPICM rendering them ineffective. FASCAM/DCPICM may come to rest in the snow at angles causing a non-killing orientation of some of the mines, making them less effective. This is the same for ground mine dispensers that shoot out mines from canisters.

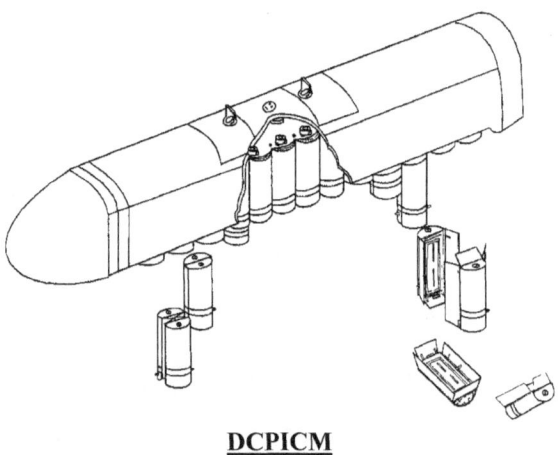

DCPICM

TRANSITION: Now that we have taken a look at the effects that the cold and snow has on rounds, let's take a look at the fire support assets at our disposal.

3. (10 Min) **FIRE SUPPORT**. Fire support assets must be as mobile as the force that they support.

(1) <u>Mortars</u>. Mortars are the most dependable of supporting arms in mountainous snow covered terrain. Refer to COLD WEATHER EFFECTS ON WEAPONS AND OPTICS for mortar considerations.

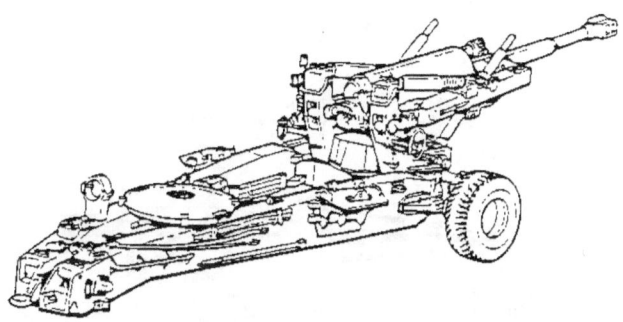

a. <u>Medium field artillery</u> gives the longer range required and can be positioned further to the rear, but it is often limited by higher terrain crest clearance and mobility. The advantage of medium weight howitzers is that it has to displace less and has longer ranges. This adds more fire power and flexibility to maneuver elements during operations. 155's are difficult to move in snow-covered terrain, as their prime movers are trucks. This limits gun line selection to flat areas next to a maintained MSR, there are few MSRs in a winter mountainous environment.   The BV-206 is not capable of towing the M198. Their ability to maneuver on and off roads is restricted due to their size and weight. Defensively, well positioned artillery batteries will be able to deny advancing enemy units use of MSR's and likely avenues of approach. The rate of fire for the M198 will be slower until the hydro-pneumatic recoil system warms up. Also, when firing at a high angle of fire, the rate of fire will be slower because of breech loading (lowering and raising the tube between each round). In snow, the M198 must be dug in all the way to the ground.

b. <u>Naval gunfire</u>, if available and within range, will be limited by its flat trajectory. Fire support stations with several alternate stations in phased sequence may be required instead of more flexible fire support areas. This will permit shifting ships, to support the operation and will allow them to be prepositioned to fire up valleys and corridors. In areas such as Norway or the Aleutian Islands, the MAGTF most likely will be amphibious with its own naval gunfire capability. The capabilities of this asset will be restricted by mountainous terrain and the extended coastal shelves found in these theatres. Targets in valleys or on reverse slopes will be very difficult to effectively engage due to the flat trajectory and high muzzle velocity of the rounds. The advantages of naval gunfire are that the weapons systems are not affected by environmental conditions and are not effected by limited rounds or support.

c. <u>Close Air Support</u>

(1) Tactical air operations provide the most mobile and often the most economical fire support available. The hazards of flying place some limitations on the use of low-flying aircraft, but the restrictive nature of the terrain and limited road nets present many opportunities for aircraft to render critical support, particularly against enemy positions on reverse slopes. Terrain may also limit attack options available to pilots and forward air controllers.

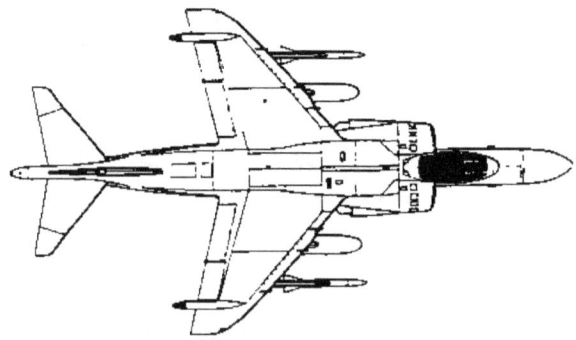

(2) Cold weather operating areas are constantly plagued by foul weather, which may at times completely negate aviation capabilities for extended periods of time. Command and control may prove to be difficult due to the multiple problems of communications in polar regions and visibility problems encountered by FAC's. CAS is less affected by climate and can play a key role in fire support in the cold. FAC's must be ski or snowshoe trained.

(3) Due to the unit's remoteness, air support may be your best option for fire support.

   (a) Provides the most mobile and economical fire support available.

   (b) Exploits critical vulnerabilities of restrictive terrain and road networks.

   (c) Renders critical support against reverse slope defenses.

   (d) Rotary wing CAS at high altitude will have reduced ordnance/time on station due to reduced lift capability.

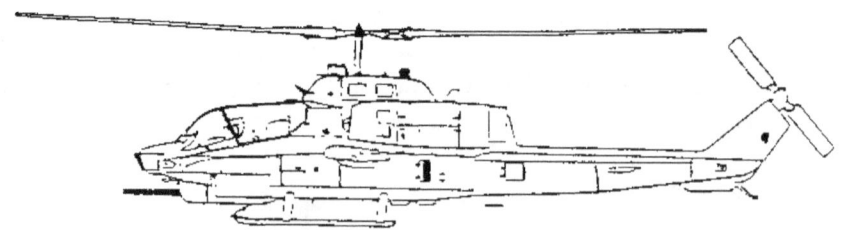

(4) Marking for targets in a winter environment is going to be similar to a temperate environment with the exception of a few nuances. MCWP 3-23.1, Close Air Support,

describes numerous ways to mark for CAS aircraft. The following section will discuss some of these techniques, and how they may be affected in a snow covered environment.

(a) Ultimately, this environment is nothing but limiting to CAS aircraft. For the most part, CAS missions will be completed in the same fashion as they would in a temperate, low lying area. The limiting factors are mainly in the ability to get the pilots eyes, which ultimately means his bombs, on the target. It is much more difficult to pick out targets, or even see the mark, in terrain that is characterized with steep valleys, rapidly rising mountains, different colored rocks, trees, grasses. This may be coupled with some other problems such as: dead space due to surrounding terrain for artillery/mortar rounds to mark, or the inability for a FAC to maneuver into a position to make good corrections.

(b) The most popular mark used in training environments today is White Phosphorus (WP). In the desert, jungle, or coastal plains, this mark provides the pilot with a distinctive reference point to adjust from. In a snow-covered environment though, WP may prove almost useless. Attempting to discern a white cloud of smoke out of a white background from high altitudes will be next to impossible. An important aspect of CAS and marking that should be kept in mind is the reduction of effects from the blast. Studies by CORAL have shown up to 80% reduction in the blast affect of a round/bomb that detonates in the snow (Depending on snow depth). This will ultimately mean that it is that much more important to have a good mark in order to have effects on target. This may also affect SEAD missions that are being conducted in conjunction with CAS.

(c) Colored Smoke: Colored smoke is available in 105mm rounds, which the Marine Corps no longer uses. However, you may see other nations use the 105mm rounds and hence the following information is included. The round is equipped with a mechanical time fuse and several base ejecting smoke canisters, the height where the round is activated can be adjusted to provide colored smoke streamers form a point over the target down to it. The round activation height should be as low as possible but still able to get the desired effect. If the round is activated too high, the dispersion of the canisters when they hit the snow will be too great to pinpoint the target. Increased smoke from the canisters can also obscure the target.

(d) White Phosphorus: As stated earlier, WP may prove useless in this environment as the primary means to mark a target. A possibility exists though, to use WP for aircraft at low altitudes. In the case of a helicopter that is providing CAS or an A-10 that has the ability to remain low and at relatively slow airspeeds, they will still be able to see the vertical development from the white smoke. As an aircraft increases in altitude this vertical development becomes increasingly difficult to discern. For night CAS, WP may still prove effective due to reflectivity off the snow.

(e) Illumination Rounds/Illum on Deck: An Illum on the deck mark will quickly melt through the snow. The round should be timed to detonate at an altitude above the ground that will allow the pilot to obtain a visual before it disappears into the snow. The difficulty will be for the FAC to make a correction from the round while it is still descending. This, coupled with the fact that the mark may move due to winds, from the time the correction is given, may make this technique difficult. These factors, of course, are dependant upon the depth of the snow. This option may be useful at night due to the amount of light that will be reflecting off/through the snow.

(f) LASER Systems: LASER systems are still useful in a snow-covered environment. One consideration is the reflectivity of snow/ice and how that may affect the LASER energy. There have been documented cases, during exercises in Norway, of missiles tracking off the reflected energy from a LASER.

(g) High Explosive (HE): HE rounds may work well in this environment depending upon the snow depth. A dark cloud of debris may be easier to discern against the white background.

(5) Perhaps this is also a good time in which to discuss anti-air defense. Considerations for the STINGER. In mountainous terrain, correctly employing these weapons will require compromises. Elevated firing stations will provide greater area of coverage, but may permit the enemy air to transit below your position with relative safety. Conversely, firing stations on low ground will restrict coverage and mask possible targets. In cold, there is increased missile warm up time and a slower engagement sequence due to cold weather clothing. The Stinger can be fired down to –40 degrees F and stored to –50F. All personnel within 50m of a launch at –25 degrees F or below must hold their breath for 20 seconds. Inhalation of the argon exhaust gas can be fatal, the dissipation time in cold is slower.

TRANSITION: We have just discussed the majority of the fire support assets in an Infantry Battalion, now lets discuss how to emplace and position these assets.

4. (5 Min) **EMPLACEMENT AND POSITIONING**

   a. Emplacements. Prior to occupation of a position, the terrain should be carefully reconnoitered and gun positions, traffic lanes, and snow parapets should be prepared. Even in deep snow, dig all the way down. Bulldozers will be needed.

   b. Positioning. If no suitable position can be located off the MSR, positioning immediately adjacent on the shoulders or in one lane can be done. Care should be taken to ensure that positions on commanding terrain provide defilade. The relative scarcity of good firing positions increases the probability of receiving enemy fires when occupying a desirable position. Good gun positions are hard to find. These are selected for flash defilade, cover, and accessibility to road nets and landing zones. Positions on commanding terrain are preferable to low ground positions because of:

(1) Less chance of being struck by rock slides or avalanches.

(2) Reduced amount of dead space in the target area.

(3) Less exposure to small arms fire from surrounding heights.

(4) Avoids flooded, marshy areas during spring snow melt.

    c. Camouflage. Camouflage discipline must be strictly enforced. Limited camouflage can be obtained by the application of paints and nets. Tracks left in the snow cannot be effectively covered except by fresh snowfall. Therefore, vehicles and troops must move only on designated trails and roads.

TRANSITION: As stated above, camouflage is extremely important in this environment. White backgrounds are the most difficult to camouflage against.

5. (3 Min) **FIRE SUPPORT PLANNING**

    a. Rugged terrain and reduced mobility place increased reliability on artillery fire support. Planners must make sure that increased consumption rate of ammunitions is taken into account. The duration of a suppression mission will take longer to cover the slower moving unit.

    b. Communication between supporting fire elements and maneuver elements and the coordination of fire with organic infantry support weapons require special attention. Retrans sites may often be necessary. An infantry attack over rising terrain is easier to support than one over descending terrain. In the final stage of the attack, organic infantry support weapons may provide the most effective fire support. Fire support must be closer to maneuvering elements than normal to be effective. Combat outposts should always have priority on supporting fires. Obstacles, barriers, and dead space should be covered by observed fires.

    c. More suppression missions, no destruction missions in snow or compartmented terrain.

    d. Cover frozen waterways, which are avenues of approach in the winter, not obstacles.

TRANSITION: Now that we talked about fire support planning, are there any questions? Let's now talk about targeting.

6. (2 Min) **TARGETING**

    a. Because of the decentralized nature of cold weather/mountainous operations, targets warranting massed fires may present themselves less often than in open terrain. Narrow defiles used as routes of supply, advance, or withdrawal by the enemy are potentially profitable targets for interdiction fires or heavy surprise concentrations. Large masses of snow or rocks above enemy positions and along main supply routes are also good targets.

In WW1, over 80,000 troops died in avalanches on the Austrian-Italian front (most initiated by artillery).

    b. When targeting frozen waterways, there are several considerations. If a barrier is desired, then the ice must breached daily, depending on temperatures, to avoid refreezing. If the enemy is likely to cross it, a perpendicular sheaf to the enemy will split his forces on each side of the ice obstacle. If the enemy is using it as an MSR, a parallel sheaf to the enemy will deep 6 as many as possible (if any enemy reach the safety of the banks, change to airburst at that time). Remember, only one round in effect to breach ice using a delay fuse. Shift and repeat the sheaf in effect based on the length of the column or the enemy's attempt to maneuver around unbroken ice.

TRANSITION: Let's now talk about fire control.

7. (5 Min) **FIRE CONTROL**

    a. Gunnery can be difficult because of:

        (1) High angles of fire that increases the time of flight for rounds to impact and the time between each round due to breech loading.

        (2) Vast reverse slope areas hidden from observation.

        (3) Increased amounts of dead space, which cannot be hit by artillery fire.

        (4) Differences in altitude between firing units and targets.

    b. The majority of all indirect fires are observed, especially close support and defensive fires. Unobserved fires are frequently unreliable since weather conditions change rapidly, and registration corrections for high-angle fire are valid for only short periods of time.

    c. Firing tables will have to be constructed during extreme cold due to slow burn of propellants. Many variables have to be inputted into the FDC computer: temperature of the air and powder, winds, barometric pressure, altitude, humidity. This would be near impossible by hand, which may be an advantage to us depending on the OpFor capability.

    d. Unreliable communication may be the biggest obstacle to overcome. Digital is preferred to voice.

8. (10 Min) **OBSERVATION AND ADJUSTING FIRE**

        (1) The capabilities of aerial observers should be exploited, particularly for the adjustment of fires in dead spaces.

(2) Those adjusting indirect fire need to understand that a shift of deflection will often also affect range, if on a slope. This spatial problem will cause more rounds in adjustment than on flat ground.

(3) During the winter months, good observation is limited due to short periods of daylight. It should be noted; however, that reflected light caused by snow covering greatly enhances night observation and can be a valuable asset. Snow cover also reduces depth perception and obscures ground features and landmarks.

(4) Ice fog can limit observation during adjustment, if the conditions exist for it.

(5) Lasers can reflect off of ice or smooth snow when designating targets.

(6) Ground bursts may be difficult to observe due to the dampening effect of the snow, preliminary adjustments may have to be made off of air bursts.

(7) In compartmented terrain, there will be an increased number of lost rounds.

(8) Sheaf elongation will occur on slopes. The steeper the slope, the more spread out the sheaf will become. This will reduce effect on target.

(9) An F. O. inexperienced in compartmented, snow-covered terrain will likely call a few ineffective missions at first. Plan for it! The use of computer simulators and walking rounds up, down, and oblique on a slope at 29 Palms every chance your FO's get can be of great help in peacetime.

TRANSITION: These are the major factors effecting the ammunition and its delivery systems, but remember that the cold will also seriously effect one other link in the system; the operator. Are there any questions?

PRACTICE (CONC)

a. The students will apply what was taught in upcoming field evolutions.

PROVIDE HELP (CONC)

a. The instructors will assist the students when necessary.

OPPORTUNITY FOR QUESTIONS (3 Min)

1. **QUESTIONS FROM THE CLASS**

2. **QUESTIONS TO THE CLASS**

   Q. Ammunitions should be stored where?

A. Outside, 4" off of the ground.

Q. What are some factors that can make gunnery difficult in this terrain?

A. (1) Large angles of <u>elevation</u> and increased time of flight for rounds to impact.
   (2) Vast <u>reverse slope</u> areas hidden from observation.
   (3) Increased amounts of <u>dead space</u> which cannot be hit by artillery fires.
   (4) Differences in <u>altitude</u> between firing units and targets.

SUMMARY (2 Min)

a. During this period of instruction we have discussed the effects that a cold weather/ mountainous environment have on fire support, the ammunition used, and its delivery systems. We have covered the difficulties in adjusting fires and discussed some ways to overcome these shortcomings.

b. Those of you with IRF's please fill them out at this time and turn them in to the instructor We will now take a short break.

UNITED STATES MARINE CORPS
Mountain Warfare Training Center
Bridgeport, California 93517-5001

WML
WMO
09/31/00

LESSON PLAN

COLD WEATHER / MOUNTAINOUS OPERATIONAL PLANNING

INTRODUCTION   (5 Min)

1. **GAIN ATTENTION**. Some of you may remember the World War II book entitled "The Jungle is Neutral". The point of this course is that knowledge can permit you to use the environment. While this is still true to a degree, the arctic is not neutral. It can be accommodated, and it can be used to your advantage, but it is still a killer. Matthias Zdarsky, an Austrian alpine soldier in writing of his World War I experiences stated, "The mountains in winter were more dangerous than the Italians". We may smile at this reference to a historically inept fighting force-but it highlights my point.

2. **PURPOSE**. The purpose of this period of instruction is to familiarize the student with cold weather/mountain operational planning. This lesson relates to offensive and defensive operations.

3. **METHOD / MEDIA**. This lesson will be presented by lecture. Those of you with IRF's please fill them out at the conclusion of this period of instruction.

4. **EVALUATION**. This lesson is informational and no testing is scheduled.

TRANSITION: It is not just the Austrians who respect winter mountain operations; the Germans do also. Adolph Schmidkunz in his book "Kampf Uber die Gletschern" said this of cold weather combat. "The white death, thirsty for blood, claimed countless victims in the mountains ... Hundreds upon hundreds were the men gripped by the white strangler ... It was no glorious death at the hands of the enemy, I have seen the corpses. 1t is a pitiful way to die, a comfortless suffocation in an evil element, an ignominious extinction for the Fatherland."

BODY (50 Min)

1. (1 Min) **OPERATIONAL PLANNING**. As an operational planner, it is not being gratuitous to say from the onset that all planning in cold weather must be conducted in the realm of the possible. This means being logistically sensitive. This is the mark of every good operator but never is this more apparent than in the arctic.

TRANSITION: Even though the terrain is different, we can still apply normal planning steps and use the same acronyms, such as METS.

2. (5 Min) **METS**

    a. <u>Mission</u>. The first planning factor is the mission. METS assumes that the mission is possible. We do not want to quench the Marine Corps "can do" attitude at this point, but it is my opinion that many of our senior officers have no experience in, or understanding of the rigors of cold weather operations. Therefore, unfortunately "Mission" may have to be analyzed not only as to "how", but also as to "if". Also, consideration must be made to include those specific missions that must be accomplished in support of our specific mission.

    b. It is believed that it is a commander who is conservative, who will be most successful in his mission. Not conservative in his philosophy or his tactics, but one who conserves his men. This is best done by sensitive leadership. Driving your men when it is required, and keeping them rested when that is possible. It means keeping men sheltered from the enemy and the elements. The first rule of the arctic is "Never let the men stand around in the cold", and the second is to make his shelter the best possible one. The importance of bivouacs must be emphasized. The combat efficiency of a mountain soldier, and sometimes his life, depends on his ability to get all the rest possible under difficult conditions.

    c. <u>Enemy</u>. Evaluating the enemy requires knowledge of his organization, equipment, doctrine, habits, present dispositions and leadership. This can only be obtained by study and good reconnaissance. How often have we seen Marines execute an operation based on the CO's and S-3's desires, rather than on the enemy's known weaknesses. We need to return to intelligence driven planning. Critical to this in cold weather is discovering the enemy's mobility capability. If his is superior, we cannot turn inside of his observation, orientation, decision, reaction cycle; because he can overcome our initiatives. In this case, a single crushing blow delivered with surprise is the only hope of success. If our over the snow capability is equal or superior, our options are considerably broadened. Reconnaissance assets must be integrated. For example: Engineer and reconnaissance patrols evaluating the route for Marines and equipment. One-third of your assets will be devoted to reconnaissance/security missions.

    d. <u>Terrain and Weather</u>. The effects of altitude on combat laden troops, the tremendous opportunities for long range observation in the mountains, and the information available for track intelligence are all important considerations. Variable mountain weather must also be taken into account, the infantry officers must begin to be as aware as our aviators have always been. Planning must consider the worse case weather possibilities. Movement at night or during reduced visibility becomes even more important than in other environments. The key to moving in the mountainous/cold weather environment is good detailed route selection. Other factors include, length of day or night, temperature lapse rate, higher winds at attitude, the avalanche dangers or leeward slopes, inaccurate maps or large contour intervals and distance underestimation's.

e. <u>Support</u>. The new letter stands for the full spectrum of support: Mobility and logistics are highlighted here.

   (1) Mobility includes not only assets available, but all or your other movement decisions. Using the rule of 3 kph for ski mobile Marines and adding one hour for every 300 meters ascent, or 800 meters descent, it is obvious why night movement becomes even more of a requirement. With slow mobility, only one enemy OP which has not been neutralized can devastate a column.

   (a) Other movement considerations include control measures such as checkpoints, phase lines, wide boundaries, etc. Your sub-units cannot move from a bivouac to a march formation instantaneously. Considerable time is required, therefore, orders are best given using "no move before..." and "warning to move" control measures. This allows subordinate commanders to plan backwards, and arrive on time at the line of departure without needlessly standing around. For larger moves in column, staggered start times are highly recommended to prevent bunching. Obviously, trail breaking parties must precede any major movement. March discipline must be enforced. Straggling must be kept to a minimum. Strive to arrive at your objective with 90% of your combat power.

   (b) Waiting procedures at halts should be standardized. Many warm, short halts are better than a few long, cold ones. Here are a few time guidelines for procedures based on the length of the halt.

   1. Greater than five minutes. Packs off, men sitting down resting.

   2. Greater than fifteen minutes. Squad stoves used to brew hot drinks down in individual fighting positions.

   3. Greater than thirty minutes. Watch rotation instituted with men off duty allowed to use their sleeping bags/mats in the trenches.

   4. Greater than one hour. Tents up.

   (c) In addition to being properly fed and rested, they must also be properly dressed for the mission. This only begins with cold weather clothing. A common camouflage policy is also essential for control. Camouflage changes must be unit wide, and reported to higher and adjacent units. Identification is difficult enough without compounding the problem with several uniform variations.

(2) <u>Logistics.</u> Rule of thumb for logistic support must be that the logistic train must be as mobile as the advancing column. There will be few roads for a Main Supply Route (MSR) in the mountains. This will hinder re-supply and evacuation. Increased engineer assets will be needed to improve and maintain MSR's and convoy planning and control is essential on single lane mountain roads.

<u>TRANSITION</u>: Any operation conducted in the Marine Corps must have a logical organization of all units involved if it is to succeed.

3. (2 Min) **TASK ORGANIZATION**. Also under troops, review task organization carefully. Because of the amount of security and reconnaissance required, a commander may want additional reconnaissance/surveillance assets. He will need additional engineer assets to meet the myriad of engineer related tasks. In task organizing units, the number one concern is to insure relative mobility of the whole unit. If a Dragon Section is assigned to a ski mobile company, it must be able to ski to the abilities of the company. Insure that adequate plans are made for mobility of all of your assets. A prime concern will be your over the snow capability of your assets. Over the snow vehicles such as the BV-202/206 are invaluable. The rule of thumb is to have these vehicles task organized with your most terrain dependent units. BV-206's can really lighten the load of your skiborne infantry. Command groups, heavy crew-served weapons and mobile anti-tank/anti-air assets will benefit tremendously from this asset. Other considerations for task organization should include the capability for each unit that is operating independently to be self-sustained with logistical mobility and combat assets necessary to accomplish their mission.

<u>TRANSITION</u>: After we have task organized our unit, we must review the concept of operation.

4. (2 Min) **CONCEPT OF OPERATION**

   a. In developing the concept of operation, the CO and S-3 must consider the environment first, then the mission. Of primary concern are the area of operation, type of operation, the tactical consideration of time, space, and finally, mission.

   b. The following information should by used as a rough guide for the concept of operations.

      (1) Keep your concept simple. You may wish to incorporate a phased operation and write it accordingly.

      (2) If multiple columns are used, designate the base unit. The base unit will cover the most difficult terrain. Keep the reserve element as the most experienced and most mobile.

      (3) Include the commander's intent to insure that all subordinates understand fully what is expected.

      (4) Give the maneuver commanders maximum leeward to accomplish their mission. Wide zones of action are preferred and the only way to operate effectively in the mountainous terrain.

(5) Additional coordinating instructions may be required to coordinate and execute the concept of operations.

(6) Multiple columns will enhance deception. Incorporate deceptions as much as possible.

(7) Plan for envelopments rather than frontal attacks. Command and control should be to the front, ready to influence the action.

(8) Various annexes such as mobility plan, track plan, and route selection will have to be written-establish a unit SOP.

TRANSITION: Another important aspect of operations planning is the fire support plan.

5. (5 Min) **FIRE SUPPORT**. In cold weather/mountainous terrain, we, as infantry leaders, are going to need to exercise closer oversight over those units who support us. Operational planning should be parallel and concurrent according to FMFM 3-1, but it is often only parallel in temperate climates as we have reasonable assurance that our counterparts are as knowledgeable in winter mountain operations. Concurrent planning must be done to ensure meeting the requirement for more and different ordnance for air support. Many of our newest munitions which are scattershot, such as ICM and Rockeye, are virtually useless in deep, soft snow because they will penetrate without detonation. In all areas of ordnance delivery, the dud rate will be higher. When this is coupled with the smothering effect of the snow, sharply higher ammunition expenditures will be needed, a requirement which will cause ripples throughout the entire logistics support structure. Therefore, a large increase in VT and time fuses must be planned. Here are some major topics of concern.

   a. When adjusting artillery in the mountains, a shift in deflection will often also affect range, and the center of impact above the target can enhance effects with rockfall and avalanching. Battery position preparation will draw off snow removal equipment and displacement must be covert, timely, and innovative to ensure full coverage of the many dead spots in mountainous terrain.

   b. Artillery registrations will provide indelible evidence on the snow of possible Final Protective Fire (FPF) locations, but these must be pre-registered to ensure accurate coverage of avenues of approach in inclement weather. The only way to provide some means of deception is to over-register, increasing once again the ammunition requirements.

   c. Air support difficulties can be summarized in four words, "unreliable due to weather".

   d. Perhaps this is also a good time in which to discuss <u>anti-air defense</u>. We presently have REDEYE and STINGER. In mountainous terrain, correctly employing these weapons will require compromises. Elevated firing stations will provide greater area of coverage, but may permit the enemy air to transit below your position with relative safety. Conversely, firing stations on low ground will restrict coverage and mask targets.

e. Naval gunfire, if available and within range, will be limited by its flat trajectory. Fire support stations with several alternate stations in phased sequence may be required, in one of the more flexible fire support areas.

TRANSITION: Even though the terrain is rugged and tiring for our men, we must stress the offensive and keep the enemy off guard.

6. (5 Min) **OFFENSIVE OPERATIONS**. In the offense, we can only reemphasize the need for operations to be driven by intelligence, and not by the desires of the CO or S-3. Good reconnaissance and security are therefore mandatory, especially in the discontinuous fronts in mountainous terrain. Also, the high ground is the key, it must be seized or controlled. Every Marine must know the following: "Get high - stay high, attack from high to low". Again, the high ground is the key; it must be seized or controlled.

   a. More time is required to execute movements because of decreased mobility, so secrecy must be maintained longer to ensure surprise. This certainly points to increased night operations and increased smoke use. Daytime operations should only be attempted to pursue a disorganized enemy.

   b. On advance to contact, several routes are usually preferred. This places enveloping forces in position early, improves security through dispersion, and if in a populated area, allows better use of limited shelter. The advance guard must be heavy in machine guns and mortars because it must be able to carry on longer, unsupported-because attacks develop more slowly.

      (1) In deliberate attacks, two quotes are appropriate. It may be advisable to approach the enemy in stages, reorganizing several times undercover, and allowing the men to rest to keep the unit fresh.

      (2) Fire support is easier on ascending terrain, but the troops suffer the danger of rock fall or avalanches.

   c. Attacks must be simple and decisive. Limited mobility requires reserve to be held closer to the assault echelon. The enemy's defense will be desperate, and counterattacks immediate in order to recover his shelters, which are essential to survival. This must be anticipated.

   d. Because defensive positions in arctic areas are seldom continuous lines, envelopment is the best form of maneuver. Sharp, quick, decisive attacks in flanks and the rear will accomplish far more than slow, predictable frontal attacks. By envelopment, we sever his lines of communications and force him to meet our attack from a needed direction. Doing so forces him out of his prepared positions, exposed to the adverse effects of the weather.

   e. In order to achieve maximum surprise, attacks in the cold weather/mountainous environment should be conducted under periods of reduced visibility. Even when the moon is not up, nights are usually brighter than in more southern regions. Night operations, even at low temperatures become more of a norm. During the winter months, orthodox night combat techniques remain unchanged except for two special factors.

   (1) Movement and control may be facilitated by the increase in visibility resulting from the reflection from the snow of moon and starlight.

   (2) On windless nights, sound travels for great distances, therefore, the need for noise discipline is paramount.

f. Use of Tanks/AAV's/Wheeled Vehicles.

   (1) Use of wheeled vehicles is largely restricted to clear MSR's. Wheeled vehicles are almost no use in deep snow.

   (2) Tanks/armored vehicles can be used in the arctic, but they are more or less restricted. Deep snow will cause mobility problems. AAV's and tanks operate well up to about 20 plus inches of snow.

   (3) Frozen lakes and streams may be crossed if reasonable precautions are exercised. Prior ice reconnaissance is a must and the following rule of thumb applies for armored vehicles: 20 inches of ice will support 20 tons; each additional inch of ice will support two additional tons.

   (4) Tactical employment of armored vehicles is modified to offset the characteristics of the area and the weather.

TRANSITION: Even though we may be Marines, known around the world for our toughness and Gung-Ho attitude, we cannot stay on the offensive forever. So let us review the defense.

7. (5 Min) **DEFENSIVE OPERATIONS**

   a. To begin again, reconnaissance and security are vital. OP's should be stacked to ensure coverage regardless of weather. Tracks should not lead to defensive positions.

   b. The difficulties of camouflage enhance the effectiveness of reverse slope defenses, and the difficulties of security in independent operations often require perimeter defenses. Strong point defenses are required because of discontinuous front lines.

   c. Numerous dead spaces in mountainous terrain deserve special attention. They are increased because flanking and grazing fires are often difficult to obtain, and frontal and plunging fire predominates. Mechanized avenues of approach can be reduced in deep snow though. Tanks cannot climb over 25 degree slopes in 1 foot of snow, over 15 degree slopes in 2 feet of snow.

   d. When firing uphill, positions must be carefully prepared to avoid cramped positions and therefore inaccurate fires.

e. Reserves are held more closely, especially near crests, where they can counterattack downhill. Some manuals state that reserves should be larger in snow covered terrain. I agree only if they have superior mobility, otherwise, you simply stand the risk of reducing front line combat power without a quick reaction capability. Defenses should be deeper to put more pressure on exhausted attackers; however, and trails should be broken to counterattack positions.

f. In delays, where demolitions are used, one major task is more effective than many small ones.

TRANSITION: All the movement and fire in the world is no good if we cannot communicate, which is greatly effected by the mountains.

8. (5 Min) **COMMUNICATION PLANNING**

   a. In a winter mountainous environment, a step backward to older technology may be a step forward in offensive planning, and especially in communications. Alternate and secure means of communications are highly desirable in the face of the DF threat. Once a unit has been defiled in deep snow, their lack of mobility makes them an easy target. Fortunately, mountainous terrain restricts the use of ground DF stations, and the terrain also creates shadows and echoes that increase azimuth errors. Air DF remains a serious threat, however.

   b. As a result, alternate communications means such as lights, wire, and messengers should be considered. Hopefully, mission-type orders will reduce nonessential communications also. If wire is used, it should be strung high to prevent coverage by new snowfall. This could preclude finding and servicing breaks. Warming tents should also be planned, if required, for wire maintenance teams. Messengers/couriers should be dispatched in pairs, and the snowmobile is worth its weight in gold delivering dispatches over previously broken trails.

   c. If radio is used, a grid rebroadcast system should be planned. Helicopters and snowmobiles will be needed to establish these retransmission sites.

TRANSITION: With the enemy, the ardous terrain, the problems with comm and re-supply, and the special equipment needed for movement in the mountains, we cannot forget the threat of NBC.

9. (5 Min) **NBC PLANNING**

   a. With tactical nuclear weapons, blast, radiation, thermal effects, fallout, and EMP all differ from temperate zones.

      (1) Blast effects are blocked or channelized by terrain for greater or less effect, but surprisingly, perhaps, on flat snow covered ground blast range is reduced because wave

front does not receive reinforcement from ground heating. Secondary effects can be significant; however, with snowstorms, avalanches, and ice breaking.

(2) Thermal effects are generally enhanced because of the reflectivity of the snow especially on arctic nights. The flash could be reduced in an air-burst; however, if the source was above a snow cloud, cold weather clothing provides more protection than in temperate climates.

(3) Initial radiation is attenuated by falling snow, protective snow layers, and rough terrain.

(4) Fallout is unpredictable because of mountain winds, and increased radiological surveying is needed to discover hot spots. Obviously, in rough terrain, aerial surveys become more important.

(5) EMP is increased in the arctic because of the aurora effect.

b. Biological operations are still feasible even in the cold. The reduced sunlight improves vector life span. Vector hibernation, subsequent thawing and vector reactivation can cause long term hazards.

c. On the chemical side, low temperatures are a mixed blessing. Mustard (HD), blister (HC) and hydrogen cyanide (blood agent) all freeze below -30°F, GB freezes at -76°F. Only VX remains liquid.

(1) As a general rule, from +30°F to -30°F, chemical weapons are most effective as a pickup hazard; and below -30°F, KR is generally more effective than non-persistent agents. As a pickup hazard, snow is 20 times as effective as grass. At cold temperatures, agents will penetrate less effectively because of lower vapor pressure, and bulky cold weather clothing. This means a greater concentration must be employed to obtain the same result as in temperate climates.

(2) Pyrotechnic dispensers are hindered by deep snow; thermal or powder dispensers work best. Automatic detectors must be heated; detector paper is not always reliable.

(3) Decontamination will be very difficult because of freezing and lack of water.

(4) When wearing suits, here are a few more thoughts:

(a) Cold weather clothing must be worn outside to permit layering for additional warmth. This ensures contaminated clothing which must be removed before entering a heated shelter where the agents can re-vaporize. WP bags to seal contaminated clothing will be needed.

(b) In full MOPP, speed is reduced by 50%.

(c) On patrols, to listen effectively, all members must hold their breath simultaneously. A hand and arm signal is needed for this.

(d) Extra suits must be available, especially when crossing water obstacles.

(e) Spring melt-off can spread contamination far beyond its original boundaries.

(f) Nerve antidotes must be kept warm. Keep it inside your suit on a string so that you can get it out. Injection is more difficult.

TRANSITION: Walking, as you can surmise just by looking around, can be very arduous; therefore, we must consider alternate ways of moving. Chief among these is the helicopter.

10. (3 Min) **HELICOPTER EMPLOYMENT**

   a. Helicopters are critical to USMC cold weather operations as they are a decisive edge. All air planning; however, must have a ground mobility alternative.

   b. HLZ's must be carefully planned to avoid disclosing their location by snow balling created on landing/takeoff. Frozen lakes make excellent HLZ's following ice reconnaissance.

   c. Loading/unloading procedures may have to be modified to reduce rotor blade wind-chill. Equipment should be pre-staged at the landing point and arctic procedures can also be used.

   d. Terrain flying must be done intelligently. Rotor wash can disclose the number of aircraft and flight path when it strips trees of fresh snow.

   e. Never depend on helicopters as your primary means of maneuver. Helicopters are solely dependent upon the weather conditions.

TRANSITION: The best commander in the world can be hamstrung by a lack of organization in his command post. Let's discuss some ways to organize our CP.

11. (2 Min) **CP CONFIGURATION**. Combat operations centers in cold weather must be capable of being man-packed. There are many possible configurations, depending upon the desires of the commander, but two BV 206's backed up against each other will produce a workable static COC configuration. Also, several ECW will work for a ski mobile command group.

TRANSITION: We try to consider all aspects of an operation prior to hitting the field, this is the reason for training at all levels, contingency plans, lessons, schools, and TEWTs. Are there any questions?

PRACTICE (CONC)

   a. Students will practice what was taught in upcoming evolutions.

PROVIDE HELP (CONC)

a. Instructors will aid students as necessary.

OPPORTUNITY FOR QUESTIONS (3 Min)

1. QUESTIONS FROM THE CLASS

2. QUESTIONS TO THE CLASS

   Q. When halted for more than fifteen minutes on the march, what should be a units actions?

   A. Squad stoves are used to brew hot drinks down in individual fighting positions.

   Q. What is of primary concern when developing the concept of an operation?

   A. Area of operation, type of operation, tactical consideration of time, space, and finally, mission.

   Q. What type of alternate communications means should be planned in mountain operations?

   A. Lights, wire, and messengers.

SUMMARY (2 Min)

a. During this period of discussion we have covered operational planning, including METS, task organization, concept of the operation, offensive and defensive operations, and NBC considerations.

b. Those of you with IRF's please fill them out at this time and turn them in to the instructor. We will now take a short break.

UNITED STATES MARINE CORPS
Mountain Warfare Training Center
Bridgeport, California 93517-5001

WML
WMO
09/31/00

LESSON PLAN

PLANNING COLD WEATHER OFFENSIVE OPERATIONS

INTRODUCTION (3 Min)

1. **GAIN ATTENTION**. History has shown that the aggressor will normally have the advantage in warfare. He who is ready to commit his men to the battle at the decisive moment is the man more prepared for battle. An indecisive leader can be the doom of his unit. To plan to defend is to plan to lose. We must think of how to seek out the enemy and destroy him, for the goal of warfare is the destruction of the enemy, not the seizure of land.

2. **PURPOSE**. The purpose of this period of instruction is to acquaint the student in the aspects of offensive combat used in cold weather operations. This will include considerations, planning, and conduct. This lesson relates to cold weather mountainous operational planning.

3. **INTRODUCE LEARNING OBJECTIVES**

    a. TERMINAL LEARNING OBJECTIVE. In a cold weather mountainous environment, plan cold weather mountain operations, in accordance with the reference(s).

    b. ENABLING LEARNING OBJECTIVE. Acting as a member of an infantry patrol and given a tactical scenario in a cold weather/mountainous environment, participate in the planning of a cold weather offensive operation, in accordance with the reference(s).

4. **METHOD / MEDIA**. The material in this lesson will be presented by lecture. You will practice what you have learned during upcoming field training exercises. Those of you with IRF's please fill them out at the conclusion of this period of instruction.

5. **EVALUATION**. This lesson will be evaluated during the field training portion of this course by performance evaluation.

TRANSITION: Are there any questions concerning what we are going to cover during this period of instruction.

BODY (40 Min)

1. (5 Min) **EMPLOYMENT CONSIDERATIONS**

    a. The doctrine for the employment of the landing force does not require alteration in order to conduct combat operations in cold weather. However, the emphasis for precise planning and training does become more demanding. Combat in the cold weather/mountainous environment is considerably more demanding than in a temperate climate.

    b. In the arctic, there is need to task organize the landing force to enable it to accomplish a swift, and decisive victory. To allow the operation to lose momentum and be unduly prolonged, may result in its failure. The combat ration should favor the attacking force, and momentum should be obtained through mobility, and the use of fire support, i.e., attack helicopters, supporting fixed wing aircraft, and artillery fires.

    c. Normally, few roads are available, and often operations will be based on only one, which may not be an all-weather road. In winter, keeping roads open will be a major operation, requiring large engineering resources. To reduce the demands for ground mobility, considerable reliance will be placed on all forms of aviation support. Ground mobility itself in winter will be based primarily on skis, snowshoes, and over the snow vehicle. Naturally, whenever main supply routes can be maintained, wheeled vehicles can be used.

    d. In essence, superiority in combat will go to the force that is the least restricted to roads, and is trained to live, and move quickly in cold and snow.

TRANSITION: With these considerations in mind, let us turn our minds to the basis of operations, planning.

2. (5 Min) **COLD WEATHER PLANNING**

    a. The planning for combat operations in an arctic region must be precise, extremely detailed, continually improved, and always flexible.

    b. Extended areas of responsibility, reduction in troop density, and battle area isolation, plus difficulties in command and control, require the use of mission type orders that give maximum latitude to subordinate commanders. Winter operations require that tactical commanders be given every opportunity to exploit local situations and seize the initiative. Decentralization is the main operational mode.

    c. Planning for any operation must emphasize the support problem. The lack of roads, or snow clearance on existing roads, plus climate severity, and other environmental difficulties requires that logistical plans be flexible and adaptable to permit adjustment of supply means. Any errors or miscalculations in planning could be disastrous. Corrective actions for improper or incomplete planning are difficult to accomplish, once the operation has been set in motion.

d. The following special factors will influence operational planning in a cold weather/mountainous environment.

   (1) <u>Low population density</u>. Settlement, supplies, quartering facilities, and lines of communication are limited.

   (2) <u>Roads and railways</u>. These may be limited, and those that exist, usually are vulnerable to enemy action. Climatic conditions may effect their use also.

   (3) <u>Lakes and waterways</u>. These may either aid or hinder an operation, depending upon climatic conditions.

   (4) <u>Mapping</u>. Occasionally, maps may be unreliable or non-existent. Aerial photographs may be required as a source of terrain information.

   (5) <u>Navigation.</u> Difficulty of land navigation is increased by the lack of landmarks, periods of reduced visibility, and by large magnetic declination in extreme northern climates.

   (6) <u>Weather</u>. This is an important factor to be considered in the estimate of the situation, and may dictate a course of action. The importance of local weather prediction capability cannot be overemphasized.

   (7) <u>Snow and ice cover</u>. Snow enhances the movement of troops suitably equipped and trained; but on the other hand, reduces mobility of troops lacking proper equipment and training.

   (8) <u>Atmospheric disturbances</u>. Extended operating distances and atmospheric disturbances can make communication difficult.

   (9) <u>Delayed personnel response</u>. The extreme environmental problems encountered by personnel, requires that delay and time lag be considered in planning at all levels. "No move before..." times should be present in all orders for planning purposes. The "hurry up and wait" situation must be avoided at all costs. If troops are made to wait around in the elements before moving out, it will reduce their combat effectiveness. Tents should be left up until the unit is actually ready to move. Decisions must be made if the unit is to be stopped for any length of time, as to whether or not to erect tents or shelters.

   (10) <u>Fire support</u>. Fire support planning is basically no different than that required for temperate regions. However, limited ground mobility of artillery weapons, and the difficulties of ammunition re-supply must be considered. The importance of tactical air support is increased greatly primarily because of the remoteness of northern areas, and lack of suitable routes of supply, and resulting unavailability of normal fire support elements.

<u>TRANSITION</u>: As we plan, we must also keep in mind some specific considerations for individual and small units.

3. (5 Min) **SPECIFIC CONSIDERATIONS**. Individual and small unit tactics will remain basically unchanged, for the arctic techniques of employment will be governed by specific situations. Ideally, the objective in cold weather operations is not to pit rifleman against rifleman, but rather to use superior mobility and surprise to place the enemy in unfavorable positions from which they must withdraw. They can then be pursued in the open and destroyed by the environment and long range firepower. The following list of considerations will help you achieve this goal:

   a. Action must be sudden, violent, and decisive. An operation that is permitted to build slowly may result in a stalemate, or offer the enemy the opportunity to seize the offensive initiative.

   b. The assaulting troops should carry a light load, but must have their survival equipment. Shovel, sleeping mats and bags, water, rations, and small amounts of extra clothing must be carried. Survival gear must always be carried by the individual, whether he is on patrol, assaulting a position, or just moving from unit to unit such as a Marine carrying a message or a battalion commander checking his companies.

   c. Success depends on surprise, speed, and gaining fire superiority.

      (1) Deploy all fire support closed to the objectives than in summer. This also applies to direct fire weapons such as machine-guns, etc. However, indirect fire weapons will require careful registration because of large variations that differ from temperate firing table data.

      (2) Snow dampens the effects of all supporting arms and gives the enemy in the defense excellent protection. The ballistic characteristics of weapons and ammunition are affected. Snow smothers the lateral fragmentation of artillery rounds, and can completely absorb mortar fire and hand grenades. Using time an VT fusing to cause air bursts can help to overcome this problem. We must use more ammunition of all kinds to obtain the same results as those obtained in a temperate-type environment. This and reduced capacity off roads may require decisions that would have some weapons left behind in order to take more ammunition.

   d. Surprise must be obtained through reconnaissance and detailed planning. This can further be broken down in several other areas of planning the execution.

      (1) By timing the attack to coincide with hours of darkness or limited visibility, i.e., a snowstorm, blizzard, ground fog, low clouds, little to no illumination, etc.

      (2) The main attack should be directed toward the flanks or rear, and whenever possible, downhill.

      (3) Often, the direction of the main effort will lead across difficult terrain, to attack from that direction, from which the enemy least expects. Such a direction is of great

advantage, even though it requires great physical effort, tenaciousness, persistence, and skillfulness, especially from the artillery, tanks, and engineer support assets.

    (4) The direction of the main effort may also lead to large communication networks or supply/logistic centers.

    (5) Because of the low capacity of roads, and because of relatively small capacity of the offensive columns, units are frequently assigned a wider zone of action, especially where definable gaps are between defensive units.

    (6) In attacking defensive positions which are located on steep slopes with many dead spaces, it serves good purpose to create special assault groups which take advantage of these dead spots, approach the defensive position, and be a quick and surprise assault, to overcome particular fortified points to the enemy.

  e. The reserve must have superior over-the-snow mobility. Whenever over the snow assets are available, we should have a bigger part of the unit in reserve at every level. If over-the-snow vehicles are not available, the reserve must be kept closer to the front than normal, and possible reduced in size.

  f. A halt is made in the attack position only long enough to prepare troops for the attack. Vehicles are dispersed and artillery is moved to prepared positions, and camouflaged. Troops remain in the attack position for the length of time necessary to prepare for the attack to prevent firing positions. A warming tent per company should be established at the attack position, or at some other location between the LOD and the objective. Additionally, if along advance to contact was conducted, the establishment of a secure bivouac at the attack position for hot wets and rest may be advantageous prior to the final assault.

  g. The attack may be conducted on foot, skis, or snowshoes; or troops may be transported by tanks, assault amphibians, or helicopters. Techniques of conducting the attack are as in normal operations. If helicopters are used, a ground mobility alternate plan should also be made. The weather or enemy antiaircraft capabilities may not all the use of helicopter support.

TRANSITION: After all of the planning is completed, we must turn our energies to ensuring that they are carried out. So let us turn our energies to the actual assault here in this classroom.

4. (5 Min) **ACTUAL ASSAULT**

  a. Coordination is extremely important. At times, the distance between enveloping forces may be so great that messages must be relayed.

  b. When the attack is conducted on skis or snowshoes, the attack formation should make use of trails broken by the lead elements of the attacking force. Every attempt is made to get as close as possible to the enemy before deploying into an assault formation and delivering assault fires. Whenever possible, an attack on skis or snowshoes should be conducted down slope. Troops do not disperse or halt until reaching the final coordination line, or until the enemy fires become effective. Final coordination lines should generally be closer to the enemy during winter operations, especially if the assault is made through snow.

    c. The decision to conduct the assault on skis, snowshoes, or foot must be made by the commander based upon existing conditions. Skis and snowshoes should be dropped when they become a hindrance to the assaulting troops. This decision will depend upon the enemy's defensive position; skis and snowshoes should be brought forward by selected personnel in each fire team or squad during the reorganization phase.

    d. In continuing the attack, special efforts are directed toward rapid displacement of close support weapons using sleds or vehicles.

    e. Very often, it is possible to attack in a traditional and ordinary way. Bypassing and isolating the enemy, or denying him supplies and breaking his resistance by heavy fires and infiltration, all become more effective due to noncontinuous defenses.

TRANSITION: As soon as the assault is completed and the enemy is either dead or dying, we should move onto reorganization before the enemy can send more victims to your meat grinder.

5. (5 Min) **REORGANIZATION**

    a. Because of the difficulty of movement and lack of shelter, initial reorganization will normally take place on the enemy's positions, using his shelter and defensive positions. However, as soon as possible, they should be vacated because of the likelihood of the enemy having his own position registered as a target.

    b. Final Reorganization

        (1) Reorganize on high ground to force the enemy to counterattack uphill.

        (2) Reserves should be brought forward to relieve the assaulting troops.

        (3) Warming tents, if needed, are moved to the closest available concealed area by each unit responsible. Individual rucksacks are retrieved from the attack position, along with skis and snowshoes.

TRANSITION: We must keep in mind that the theory behind warfighting is not the seizure of hills or towns. It is the destruction of the enemy.

6. (2 Min) **EXPLOITATION AND PURSUIT**. The exploiting force is aided by cross-country vehicles and aircraft. The pursuit force, which must have higher mobility, is mounted on skis, vehicles or helicopters. During rapid advances, open flanks will be the rule rather than the exception. Security requirements in the offense during northern operations are no different than in temperate zones. However, when attacking units have large gaps between them and flanks are vulnerable, patrol and surveillance requirements will increase.

TRANSITION: Even a break through on a small front can lead to a decisive breakthrough. Are there any questions on what we've just covered.

PRACTICE (CONC)

    a. Students will practice what was taught during upcoming field exercises.

PROVIDE HELP (CONC)

    a. The instructors will assist the students when necessary.

OPPORTUNITY FOR QUESTIONS (3 Min)

1. **QUESTIONS FROM THE CLASS**

2. **QUESTIONS TO THE CLASS**

    Q. What are some special factors to be considered during the planning of cold weather ops?

    A. (1) Low population density.
        (2) Roads and railways.
        (3) Lakes and waterways.
        (4) Mapping.
        (5) Navigation.
        (6) Weather.
        (7) Snow and ice cover.
        (8) Atmospheric disturbances.
        (9) Delayed personnel response.
        (10) Fire support.

    Q. What does success depend upon in the assault?

    A. Surprise, speed, and gaining fire superiority.

SUMMARY (2 Min)

    a. During this period of instruction we have discussed the planning of cold weather mountainous offensive operations to include the considerations imposed by the terrain and weather, the assault, reorganization, and the exploitation.

    b. Those of you with IRF's please fill them out at this time and turn them in to the instructor at this time. We will now take a short break.

UNITED STATES MARINE CORPS
Mountain Warfare Training Center
Bridgeport, California 93517-5001

WML
WMO
09/31/00

LESSON PLAN

PLANNING COLD WEATHER DEFENSIVE OPERATIONS

INTRODUCTION (5 Min)

1. **GAIN ATTENTION**. There are several examples of cold weather defensive actions throughout history. Perhaps a few examples taken from the Leavenworth Papers will help to set the stage for today's class. The following examples are taken from the campaigns fought between 1918 to 1919 between the Allies and the Soviets:

INSTRUCTORS NOTE: Use these numbers 1, 2, 4, 11, 12, 13 which are located within the "Notes From Leavenworth Paper" - Cold Weather Application. These are attached at the end of the Lesson Plan.

2. **PURPOSE**. The purpose of this period of instruction is to introduce the student to the considerations of establishing a defense in a cold weather environment to include the fundamentals and organization of defenses in the mountains, organization and occupation of defensive positions, and some historical looks at classic defenses in cold weather environments.

3. **INTRODUCE LEARNING OBJECTIVES**

   a. TERMINAL LEARNING OBJECTIVE. In a cold weather mountainous environment, plan cold weather mountain operations, in accordance with the references.

   b. ENABLING LEARNING OBJECTIVE. Acting as a member of an infantry patrol and given a tactical scenario in a cold weather/mountainous environment, participate in the planning of a cold weather defensive operation, in accordance with the references.

4. **METHOD / MEDIA**. The material in this lesson will be presented by lecture. You will practice what you have learned during upcoming field exercises.

5. **EVALUATION**. You will be tested later in the course by a performance evaluation.

TRANSITION: Planning defensive operations goes beyond the execution of a defensive operation. In a cold weather/mountainous environment, the challenge constitutes a strictly detailed operational plan.

BODY (45 Min)

1. (5 Min) **GENERAL**. The defense is the employment of all means and methods available to prevent, resist, or destroy an enemy attack. It is a posture assumed by a force for the purpose of protection against enemy attack. The mission is the paramount factor that dictates the type of defense to assume, and the position or the area to be defended. However, along with the composition of opposing forces, terrain, and security, a cold weather/mountainous environment has additional peculiarities which influence defensive operations.

    a. The present Marine Corps doctrine as outlined in FMFM 6-3 (Marine Infantry Battalion) for defensive operations remains unchanged. However, there are techniques and principles learned throughout history which warrant application to that doctrine. Throughout this lesson, the following subjects will be addressed:

    (1) Fundamentals of Defensive Operations.

    (2) Organization of the Defense.

    (3) Combat Support.

    (4) Combat Service Support.

    (5) Defensive Examples.

TRANSITION: We will now look at the first subject, which is the Fundamentals of Defensive Operations.

2. (7 Min) **FUNDAMENTALS OF DEFENSIVE OPERATIONS**. Fundamentals of defense must be somewhat modified to fit the unique characteristics of mountainous terrain. The concepts of defense in the mountains, in addition to those already stated in FMFM 6-3, is to occupy or control dominating heights and key terrain, and fight from the top down. These fundamentals can thus be exemplified as advantages and disadvantages.

    a. <u>Advantages of Occupying a Defensive Position</u>

    (1) <u>A firm, multi-level defense in depth</u>. This is possible as the diversified mountain terrain is a substantial natural obstacle for an attacker. Numerous defiles and narrow valleys permit easy movement, though only for a certain period. An attacker will concentrate his forces and offer the defender the opportunity to destroy the attacking units in these places, this will be done by massing of fires at critical times.

(2) <u>Physical reconnaissance of the terrain occupied</u>. This allows the defender to anticipate the direction of attack with relative accuracy, to include possible Landing Zones (LZ's), Drop Zones (DZ's), and Avenues of Approach (AOA's).

(3) Wide zones of defense with relatively small forces and the concentration of large forces in the probable direction of attack. This is available to the defender since the predominate areas are unsuited for offensive activity.

(4) <u>The possibility to organize anti-tank defense</u>. This is possible with a great number of natural obstacles, sufficient quantities of materials to build barricades, and natural dead spaces which avail themselves as tank traps.

(5) <u>Concealed distribution of the battalion reserve echelon</u>. This is provided by dense forest and diversified terrain.

(6) In diversified mountainous terrain. "firing sacks" or kill zones are readily available for blocking a valley with flanking and cross-fire, and providing conditions for short but sudden counterattacks. These counterattacks may only slow down the tempo of the attack, but may also unfavorably impact on the enemy's entire offensive operation.

(7) Special operations, i.e., partisan, guerrilla, advance force operations, etc., are more easily conducted in the diversified, snow-covered mountainous terrain.

b. <u>Disadvantages of Occupying a Defensive Position</u>. Mountainous terrain, however, also has negative qualities which have as adverse effect on the defense in the mountains.

(1) A good possibility for reconnaissance and saboteur groups to penetrate into the depth of the defense and to operate in our rear utilizing the terrain and dense forests, as cover and concealment.

(2) The normally stoney/frozen ground much of the time will preclude utilization of our engineer assets, and present difficulties in "digging-in". Because of the complexity of the lay of the land which we will orient over the Forward Edge of the Battalion Area (FEBA) or Forward Line of Troops (FLOT) on the dispersion of units, may be all the more difficult, especially when tactical nuclear weapons are in the arsenal of our adversary.

(3) Unsuitable network of roads will make the maneuver with reserves, and task of re-supply and medevacs, more complicated. The maintenance of roads and building of subsidiary roads will consume much of our engineering assets.

(4) The lack of airfield landing zones, as well as the diversified terrain and weather conditions, makes the use of aircraft unreliable. This disadvantage fortunately affects the attacker much more so in view of this type of combat.

(5) Changing weather, to include frequent storms, prolonged rains, cold nights, and snowstorms, will have an adverse effect on the physical state of soldiers. It will therefore be necessary to take all measures, so that the effects of unfavorable climate will influence the soldier and combat techniques, as little as possible.

(6) The inability to maintain physical contact with adjacent units, especially when occupying an outpost will affect the moral of the Marines. In relatively flat terrain, units are able to maintain contact with adjacent units. However, in mountainous terrain, this is not always the case and units are frequently cut off.

TRANSITION: All these unfavorable conditions of the mountains which affect the organization of defense, and successful conduct of defensive combat, can be partially eliminated by purposeful preparation of the defense, and applying those peculiar measures to strengthen a defense in mountainous terrain.

3. (5 Min) **ORGANIZATION OF THE DEFENSE**. Before we can adequately organize the ground which will make up the defense, some peculiarities of defense in the mountains should be mentioned.

   a. Peculiarities of Defense in the Mountains

      (1) Attacks will, as a rule, be undertaken in particularly accessible directions in which normal combat techniques, and motorized/mechanical units may be used. This due to the high mountain crests and deep valleys, steep elevations and slopes, narrow and easily blocked defiles, surfaces covered by rock or deep snow, and frequently, almost impassable forests. Of course, it follows from this that we will defend also in the direction where the enemy chooses to attack. Areas where an enemy attack is less probable, will be only fortified.

      (2) A discontinuous front is characteristic of the defense in the mountains. The success of holding passes, and important Main Supply Routes (MSR's), and key terrain, will depend on the defense of the areas adjacent to them. Some areas will be defended with a large force, while other, less important areas are defended with only a part of the forces, and the areas which are passable only with difficulties are secured with a system of mutual observation, patrols, barricades, and fire.

      (3) A partial decentralization of command and independence of action. The degree of centralization and independence depends on the organization of defense, which of course follows to a decisive extent from the configuration of terrain.

      (4) Combat in encirclement is a frequent occurrence in the defense in the mountains, and is therefore another characteristic feature of the defense. This is an adverse result of a discontinuous front, frequent large gaps between units defended only with weak forces, or protected with a system of observation and obstacles, which may have been disrupted during the chaos of the ensuing battle. Therefore, there is the necessity that all units organize a strong perimeter (circular) defense.

(5) Counterattacks with surprise are possible with comparatively small units in mountainous terrain.

TRANSITION: Defensive combat on mountains is very complicated and has a great number of peculiarities. Fortunately, defensive operations conducted in the mountains are very similar to those in the flats, with the application of common sense tactical techniques.

4. (5 Min) **ORGANIZATION OF THE GROUND**

   a. As per doctrine, the outlay of the defense must include the security area, the forward defense area and the reserve area. These areas do not change, however, our thought process of actually establishing these areas must change. As we are all aware, there are two basic types of defense: area and mobile defense. Since the mobile defense is normally at division and higher echelons, we will concern ourselves with principally the area defense, which can furthermore be broken down into the perimeter and linear defense.

   b. By adopting the mountainous terrain, we can explore the defense along and across mountain crests. We will concern ourselves principally with a defense along mountain crests to secure a pass or an MSR.

      (1) Defense along the crest. A very important task of the commander is to determine the course of the main line of defense. The FEBA/FLOT may run on either the:

         (a) Military crest

         (b) Reverse slope.

   c. If the FEBA/FLOT is built on the military crest, it has these advantages:

      (1) The defender can observe the entire security echelon and beyond.

      (2) All means of fire including long, medium and short range fires can be employed to include direct and indirect fire weapons. Attrition through ever increasing concentric volumes of fire.

      (3) A defense in "height", as well as in depth can be established, thus allowing the defender the means of firing at the same time, even over our heads with direct fire weapons. During periods of partial observation, i.e., low/medium clouds, the battle area remains under observation.

      (4) The reserve echelon can remain in defilade at all times, allowing freedom of movement to support at critical points along the FEBA/FLOT.

   d. Disadvantages if FEBA/FLOT is built on the military crest:

(1) The attacker can observe the majority of the defensive system and support his movement to contract throughout the attack.

(2) The communications with the rear echelon is more difficult, and the reserves are employed with more difficulty.

e. If the FEBA/FLOT is established on the reverse slope, there are also these advantages:

(1) This permits easy deployment of the reserve with surprise.

(2) The FEBA/FLOT is not under observation by the enemy, thus the element of surprise is retained by the defender through massed surprise fires.

(3) Less concern for observation by the enemy makes the use of necessary engineer assets more favorable.

f. Disadvantages if FEBA/FLOT is established on the reverse slope:

(1) The enemy which captures the crest, now has the advantage of attacking downhill.

(2) It will not be possible for the defender to establish as dense a fire, as when the FEBA/FLOT was on the military crest.

TRANSITION: Whichever type of defense is selected, it will be dependent on the configuration of terrain, the enemy situation, and of course, the mission assigned by higher headquarters. Regardless of which defense is assumed, certain techniques will assist the commander in establishing a sound defensive perimeter.

5. (2 Min) **MOVEMENT INTO THE POSITION**. The following are considerations when moving into the defensive position:

a. <u>Routes</u>. Routes into position must use the natural concealment of overhead, such as trees and prevailing shadows. Strict adherence to track discipline is a must.

b. <u>Deception Plans</u>. Must be drawn up prior to conducting the movement. These must include a track in front of your position which can be covered by fire. The jump-off point must take advantage of concealment from any means. Utilization of existing ground vegetation, creek beds, or other obstacles which cast a shadow to cover the track, are examples of jump-off points. Finally, there must be a dummy position, which is a believable defensive position established. In all cases, the deception plan must be constantly improved, covered by observation and fire, and never neglected.

c. <u>Egress Routes Must Also Be Established</u>. This can be done enroute to the defensive position, or after establishment of the defense. In either case, if snow bridges are to be utilized for the advance to or egress from a position, they must not be destroyed. If nonexistent, they must be built on the egress route. If snow bridges exist in front of your position, they should be destroyed, and/or integrated into the barrier plan.

TRANSITION: Let's now see how we establish a defensive position.

6. (3 Min) **ESTABLISHING THE DEFENSIVE POSITION**

    a. The high ground or key terrain is paramount to an effective defensive area.

    b. The terrain that best suits the defender is narrow in width and greater in depth.

    c. Fighting positions should be located on adjacent heights, and in depth to permit covering the frontage and flanks (especially dead space) with flanking cross-fire, and strafing fire integrated with an intricate fire support plan to prevent a penetration and subsequent flanking or rear attack.

    d. Communication trenches must be established between primary, supplementary and alternate positions, and the bivouac area. The diversified terrain and snow cover will preclude a fast reaction time, unless these trenches are established.

    e. Bivouac areas must be established in such a manner as not to be affected by fire, if the forward positions come under fire. If necessary, expedient/improvised shelters should be constructed in, or as part of the firing position. The actual distance from the tent area to the fighting position will vary with terrain; however, a guidance of between 50-200 meters away is not too conservative. Whatever the distance, reaction time is of utmost concern.

    f. Observation Post's (OP's)/Listening Posts (LP's) must be established in "height" as well as in depth because of changing weather conditions.

    g. Based on the enemy threat condition, troops may have to sleep fully clothed and in shelters which are built as a part of the field defense, if a defensive position is to be ready to meet an enemy threat at all times. Keep in mind that when the enemy threat is imminent, staying and sleeping in tents is extremely dangerous.

TRANSITION: The great numbers of dead space into which direct fire weapons can fire, is reason enough for an intricate fire support and barrier plan.

7. (10 Min) **COMBAT SUPPORT**. Combat support roles and tasks are not changed by the mountainous environment; however, even routine combat support missions are more difficult because of the rugged terrain. It is of greatest advantage to distribute the means of firing in the style of a chess game along the front, into the depth, and into the heights, and in this way, achieve the possibility of frontal, flanking, oblique, and cross-firing.

    a. Some general guidance for the employment of crew-served weapons are as follows:

    (1) Machine guns, as a rule, are located on ascending terrain, they would have a good position and a large range. A large number are left on the reverse slope to repulse a possible envelopment, or attack from the flanks.

(2) Alternate and supplementary firing positions are mandatory even in directions where the enemy is not anticipated.

(3) Other direct fire weapons and tanks must be located in "tower" formations.

(4) Artillery and mortars will be spread out on the reverse slope.

(5) Anti-armor weapons are positioned in such a way that the terrain most accessible to armor is under their most effective fire.

(6) The steeper and higher the slope, the more means of firing can be distributed into the height.

b. Artillery. There are as many advantages as there are disadvantages for the employment of artillery. Rather than listing each advantage and disadvantage, the peculiarities of artillery in a mountainous environment will be discussed.

(1) Self-propelled artillery, much like tanks, will be limited to positions in the vicinity of main roads.

(2) Helicopter support will become an important prime mover for field artillery, and the re-supply of ammunition.

(3) Artillery Batteries will frequently become Operation Control (OPCON) to independent units because of communications problems.

(4) Frequent displacements may be necessary because of higher angles of fire and reduced ranges.

(5) Artillery must be as mobile as the force that it supports.

(6) Variable Time (VT) fuzes will become more difficult because of differences in altitude between firing units, and the targets.

(7) Forward observers must be employed in height, much as direct fire weapons and OP's/LP's are to preclude ineffectiveness because of weather.

c. Air. This asset can be wrapped up into one statement. "Unreliable due to weather." However, when available, they must be utilized.

(1) Helicopters are affected by weather as well as altitudes. As density altitude increases, aircraft performance decreases.

(2) Fixed-wing, all weather aircraft will continue to be effective.

(3) Forward Air Controllers (FACs) must be employed in height to preclude ineffectiveness because of weather.

(4) Snow will dampen the effect of point detonating ordnance.

(5) Diversified terrain may preclude low passes by aircraft providing Close Air Support (CAS).

d. <u>Engineers</u>. Engineer combat support is also increased in mountainous terrain because of the lack of adequate cover, and the requirement for construction of field fortifications and obstacles. The engineer's jobs may include any number of tasks, but perhaps the two most important tasks that engineers are expected to accomplish are: implementing of the barrier plan, and the establishment and maintenance of an effective road network.

(1) <u>Barrier plan</u>. The barrier plan must include:

(a) Wire obstacles in front of positions and in dead spaces, supported above the snow.

(b) Mine fields in front of positions and in dead spaces, and buffed underneath to prevent sinking further into the snow.

(c) Ice breaching/bridging conducted as necessary for obstacle creation or crossing.

(d) Anti-tank barricades constructed where tanks can move, and not waste energy in areas that tanks cannot.

(e) Dummy position is an integral part of the barrier plan, and can be constructed utilizing engineer assets.

NOTE: As a rule of thumb, the steeper the terrain, the closer to the defensive position, can the forward barrier plan be established.

(2) <u>Road network</u>

(a) Is restricted only by the capability of the engineer unit.

(b) Existing roads must be maintained to the width necessary to accommodate traffic.

(c) Subsidiary roads where subsidiary roads must be established, engineer assets will be necessary.

(d) Snow bridges are relatively easy to construct, and save an insurmountable amount of time for troop movement. Snow bridges should be established along all expected avenues of regress, and troop reinforcement.

<u>TRANSITION</u>: Now, let's take look at combat service support elements for infantry units.

8. (3 Min) **COMBAT SERVICE SUPPORT**. The re-supply lines, lines of communications, and medical evacuation routes will task the combat service support element, and if improperly prepared, could cause irreversible damage to any unit.

   a. MSR's must be maintained with snow removal equipment, if available.

   b. Warming stations must be made available for medevacs along with relief stations.

   c. Over the snow mobility must be available for the Logistical/Medical train.

   d. Communications to the rear must be maintained.

TRANSITION: The common saying "Prior proper planning prevents piss-poor performance" is always applicable. Are there any questions?

PRACTICE (CONC)

   a. Students will practice what was taught in upcoming field evolutions.

PROVIDE HELP (CONC)

   a. The instructors will assist the students when necessary.

OPPORTUNITY FOR QUESTIONS (3 Min)

1. QUESTIONS FROM THE CLASS

2. QUESTIONS TO THE CLASS

   Q. What are the advantages of occupying a defensive position?

   A. (1) A firm multi-level, defense in depth.
      (2) Physical reconnaissance of the terrain occupied.
      (3) Wide zones of defense with relatively small forces and the concentration of large forces.
      (4) The possibility to organize anti-tank defense.
      (5) Concealed distribution of the battalion reserve echelon.
      (6) In diversified mountainous terrain, "firing sacks" or kill zones are readily available.
      (7) Special operations, i.e., partisan, guerrillas, advance force operations, etc., are more easily conducted.

   Q. What are the peculiarities of a defense?

   A. (1) Attacks will be undertaken in particularly accessible directions in which normal combat techniques and motorized/mechanical units may be used.
      (2) A discontinuous front.

(3) A partial decentralization of command and independence of actions.
(4) Combat in encirclement.

SUMMARY (2 Min)

a. This period of instruction has covered the elements essential to successful operational planning in the defense.

b. Those of you with IRF's please fill them out at this time and turn them in to the instructor We will now take a short break.

UNITED STATES MARINE CORPS
Mountain Warfare Training Center
Bridgeport, California 93517-5001

WML
WMO
09/31/00

LESSON PLAN

COLD WEATHER CONSIDERATIONS FOR THE 12 PATROL STEPS

INTRODUCTION (5 Min)

1. **GAIN ATTENTION**. Movement in a cold weather environment is in itself difficult; however, when combining the inexperience of Marines on skis or snowshoes, changes in the snow conditions, tactical situation and mountainous terrain, patrolling becomes a monumental task. To patrol small or large units across varied terrain requires more than just words of encouragement or demands by higher authority depicted by broad, long arrows on a operation map. It requires positive leadership on the small unit level, and a keen sense of technical skill in detailed patrol planning.

2. **PURPOSE**. The purpose of this period of instruction is to familiarize the student with the requisite knowledge to plan and execute proper patrol planning in a cold weather environment. To include planning for additional time required for movement due to additional equipment and terrain considerations.

3. **INTRODUCE LEARNING OBJECTIVES**

    a. TERMINAL LEARNING OBJECTIVE. In a cold weather mountainous environment, plan and conduct cold weather mountain patrolling operations, in accordance with the references.

    b. ENABLING LEARNING OBJECTIVES

    (1) In a cold weather mountainous environment, organize a patrol for tactical movement, in accordance with the references.

    (2) In a cold weather mountainous environment, plan a route, in accordance with the references.

(3) In a cold weather mountainous environment, establish an FPP. In accordance with the references.

4. **METHOD / MEDIA**. The material in this lesson will be presented by lecture and demonstration. You will practice what you have learned during upcoming field training exercises. Those of you with IRF's please fill them out at the conclusion of this period of instruction.

5. **EVALUATION**. You will be tested later in the course by performance evaluation.

TRANSITION. During planning, the patrol leader uses the patrol steps; a series of mental and physical processes to ensure that all required events are planned for and all patrol members know their duties. Let's take a look at the Patrol Steps.

BODY (70 Min)

1. (5 Min) **STUDY THE MISSION**. The patrol leader, while receiving the mission takes notes and ask questions. He then carefully studies the mission, the situation, and the terrain to determine the essential tasks to be accomplished in executing the mission. These essential tasks become missions of the patrol elements and the teams for which organization, personnel, and equipment must be considered. He uses the acronym METT to estimate the situation.

    a. M - MISSION. Specific mission requirements also known as essential elements of information (EEI's) and when the information is needed. This is usually expressed as an NLT.

        (1) Recon of the battle area and particular targets, while establishing a presence forward of the FEBA/FLOT.

        (2) Harassing the enemy lines, depriving them of shelter and rest.

        (3) Deep penetration by ski and/or helicopter to destroy logistic supply lines.

        (4) Installation of observation posts and radio relay/retrans sites.

        (5) Picketing of high ground on the unit's flanks during the advance.

        (6) TRAP mission; recovery of downed aircraft/troops and equipment.

    b. E - ENEMY. Known and suspected enemy situation as it may affect your patrol.

        (1) What type of mobility is the enemy utilizing, and what is their ability level?

        (2) The units discipline.

        (3) The units morale.

(4) What was their last known action?

(5) The units re-supply capabilities.

c. **T - TERRAIN AND WEATHER.** These must be considered as to their effect on the patrol. These elements can have a direct effect on the size of the patrol, the equipment carried, patrol movement (routes and speed), and the insert/extract method to be employed. You should also consider what effect these elements will have on the enemies operations and visibility.

d. **T - TROOPS AND FIRE SUPPORT.** Under troops you have to consider personnel availability and any special skills required for the mission. Under fire support, you must consider the type of support (air, naval gun, or artillery), the range of the guns or aircraft, and the rules for engagement.

TRANSITION: METT is used throughout the planning phase of a patrol and in the execution phase. Due to change in terrain, weather, and the enemy situation, you are always estimating the situation. Now let's discuss planning the use of time.

2. (2 Min) **PLAN USE OF TIME**. In preparing the schedule, the patrol leader works backwards from the time of departure of friendly lines to present.

   a. The time schedule is then written by putting this information in a chronological order. The time schedule should cover all phases of the patrol, including preparations, inspections and rehearsals. You also need to consider EENT, BMNT and the time for insert and extract.

   b. The time schedule then becomes part of the warning order. It should be written as WHEN (time), WHAT, WHERE, and WHO.

TRANSITION: Now that we talked about timelines, let's discuss the terrain and situation.

3. (10 Min) **STUDY THE TERRAIN AND SITUATION**

   a. TERRAIN. The terrain in the vicinity of the objective influences the manner in which the patrol will be organized and conducted. The terrain to and from the objective also influences the size, organization and equipment of the patrol. A thorough map study is done by the patrol leader using the acronym KOCOA.

   (1) K - KEY TERRAIN is any terrain feature that gives an advantage to the possessor.

   (a) At high elevations, cold weather, even snow can be encountered during all times of the year. Mobility will become extremely difficult due to these factors, and very taxing to the endurance of the unit. The wide variety of terrain encountered will affect speed of movement, concealment, and security.

(b) Above the Tree line. This exposed terrain makes movement and security more difficult. Try to use micro terrain and shadows as much as possible to make observation from the enemy more difficult. Movement during the hours of darkness should be done whenever possible.

(c) Below the Tree line. Movement and bivouac areas can be concealed by vegetation.

(d) Contour. The lay of the land in the mountains will have a great deal to do with the way in which we negotiate that piece of real estate. To move directly up or down a slope without regard to gradient could be disastrous. Unnecessary cold/heat casualties could result, not to mention the fact that your Marines will not be effective to fight once on or at the objective.

(e) Natural Lines of Drift. As their name indicates, these are terrain features that tend to draw a unit into them, due to the ease of movement that they provide. All things, to include a unit of Marines carrying heavy loads who may not be the most experienced skiers, tend to take the path of least resistance whenever given the opportunity. But to allow yourself to be drawn into this situation could be leading you straight into an enemy ambush. A good leader in this environment, will set up an ambush near one of these natural lines of drift, because he knows how easily a less experienced leader can allow the terrain to dictate his route.

(2) O - OBSERVATION AND FIELDS OF FIRE. This is what you can see and shoot. Don't forget to keep in mind that this works for the enemy also.

(3) C - COVER AND CONCEALMENT. Cover stops bullets, concealment hides you from view. Planning concealed routes leaves yourself some room to maneuver.

(4) O - OBSTACLES. These can be man-made or natural. Obstacles can affect movement, observation, and communications.

(a) Narrow Depression. These include such things as creek beds, fallen logs, or similar obstacles. The preferable technique is to cross the obstacle on a snow bridge. If a snow bridge is not available, a few options exist. One option is to side step into the depression, and then side step out. Another is to build a snow bridge by shoveling snow into an appropriate area. This is recommended if a snow bridge does not exist and a large force must cross the obstacle.

(5) A - AVENUES OF APPROACH. These allow the easiest access to an area or point. Avenues of approach usually conform to your movement routes and the enemies.

(a) Frozen Waterways/Lakes. Any frozen body of water makes an ideal avenue of approach, as well as a large obstacle. Another concern is that the tracks are very difficult to conceal unless there are several overhanging trees, or the areas crossed are shaded the majority of the time. Finally, the banks around these areas are frequently very steep and may be difficult to ascend.

b. SITUATION. Several things need to be considered under the situation: friendly, enemy, and the constantly changing weather.

   (1) Friendly Situation (HAS).

      (a) <u>H</u> - HIGHER. Need to be aware of higher commands mission.

      (b) <u>A</u> - ADJACENT UNITS. Need to be aware of the location of adjacent units, and their missions, planned actions and patrol routes.

      (c) <u>S</u> - SUPPORTING UNITS. Need to be aware of who can support your patrol and how.

   (2) Enemy Situation (SALUTE, DRAW-D).

      (a) <u>S</u> - SIZE. How many enemy were observed?

      (b) <u>A</u> - ACTIVITY. What are they doing?

      (c) <u>L</u> - LOCATION. Where are they?

      (d) <u>U</u> - UNIT. This may be derived from unit markings, type of uniform worn, or through prisoner of war interrogation.

      (e) <u>T</u> - TIME. When did you see them?

      (f) <u>E</u> - EQUIPMENT. What weapons or equipment did they have at their disposal?

      (g) <u>D</u> - DEFEND. Will they defend their position?

      (h) <u>R</u> - REINFORCE. Will they reinforce their positions?

      (i) <u>A</u> - ATTACK. Will they counter-attack us?

      (j) <u>W</u> - WITHDRAW. Will they withdraw from their positions?

      (k) <u>D</u> - DELAY. Will they attempt to delay our attack?

   (3) Weather (Present and Predicted).

      (a) It is very important for the leader to gather as much information as possible about incoming weather patterns. The weather is perhaps the equalizer between actual combat and cold weather training operations. The more severe is the weather, the more possibility of cold weather injuries, the more pronounced is personnel delayed response, and of course, the slower aspects of mobility. The weather is the most important physical factor which will influence our route selection. Contributing factors to the weather are temperatures, visibility, precipitation, and wind direction; all of which will be discussed in greater detail in the planning and execution phase.

- (b) Avalanches. The avalanche hazard in the area of operations must not be taken lightly. Moving across such a critical slope without proper prior reconnaissance could be adverse to the attainment of a unit's goal. Because avalanche initiation is a feasible means by which our adversary can destroy our forces, movement on or near critical slopes must be carefully planned.

- (c) Snow Conditions. Changing snow conditions can greatly enhance or deteriorate a skier's ability to negotiate a particular route, or segment of that route. The constant and sometimes unpredictable changes in the snow conditions, make it very difficult for combat laden Marines to utilize a particular route; especially in areas which are windswept or crusted because of temperature changes. The anticipation of merging to and from varying snow conditions, cause Marines to become rigid, thereby losing any ability to absorb the shock of that transfer.

- (d) Cold Temperature and High Winds. These affect your Marines, their weapons, and equipment. Individual response time is slowed, and your time schedule must reflect this.

- (e) Snow Cover. Affects the rate and mode of movement. Terrain can be affected if avalanche conditions exits. Snow depth and consistency may change considerably during the patrol and may greatly affect movement.

- (f) Visibility will become reduced during storms.

- (g) In the high latitudes, you will experience about four hours of daylight from November to February. Mid-December has only a few hours of twilight per day.

TRANSITION: Now let's discuss the organization of the patrol.

4. (3 Min) **ORGANIZE THE PATROL**. The patrol leader organizes the patrol into elements and teams as needed to accomplish the essential tasks at the objective. He then adapts this organization to that needed for movement to and from the objective. Organization is a two step process:

   a. General - the major elements needed; headquarters, security, and recon.

   b. Special - the elements are organized into teams to accomplish specific tasks; left flank security, recon team, trail breaking team, etc.

(1) Trailbreaking Party/Team. The trailbreaking party will normally consist of approximately 1/4 of the force making the march. Example: For a company, one platoon would break trail.

TRANSITION: Now that we talked about the organization of the patrol, let's discuss the selection of men, weapons, and equipment.

5. (5 Min) **SELECT MEN, WEAPONS, AND EQUIPMENT**. The Patrol Leader takes only those men and gear required to accomplish the mission. Unit integrity is maintained when practical.

   a. Men. The size of the patrol will depend on the mission, and the members will be employed in arctic buddy teams.

      (1) Cold weather patrols are generally larger due to the gear requirements, i.e. survival gear, casevac gear, and substantial firepower assets must be spread loaded amongst the members of the patrol. Keep the gear list limited to only that which is essential to the mission, anything in addition to this will hamper your mobility.

      (2) Group's Ability. As we all remember, being aware of the group's ability plays a major factor in the cold weather and mountainous operations. In ski or snowshoe training, Marines must be continually challenged with falls and recovery emphasis; however, to move a unit across diversified terrain without regards to degree of slope, snow conditions, temperature, etc., all of which influence ability and be detrimental to both morale and mission accomplishment.

      (3) Personnel must posses the physical capabilities required to accomplish the mission under extremely arduous conditions. If in doubt of an individual, leave him in the patrol base.

   b. Weapons. Take only those weapons necessary to accomplish the mission. Ensure weapons are properly maintained for cold weather.

   c. Equipment. There are five general areas for which equipment is chosen:

      (1) In the Objective Area. This is the equipment with which the patrol accomplishes its mission. It includes such items as ammunition, demolitions, binoculars, night vision devices, trip flares, etc.

      (2) En Route. This is the equipment which enables or assists the patrol in reaching the objective. It includes such items as maps, compasses, ski wax, skins, coyotes, 165' dynamic ice rope, etc.

      (3) Control. This equipment is used in assisting the patrol leader in controlling the patrol while moving and during actions at the objective. It includes such items as whistles, pyrotechnics, radios, flashlights, etc.

(4) Routine Equipment. This is equipment normally common to all members of the patrol; weapons, uniforms, 782 gear.

    (a) Camouflage.

        1. Unit designation prescribed for use with over-whites.

        2. Ensure they are clean.

        3. Clean white tape should be firmly applied to all 782 gear and weapons.

(5) Water and Food. Rations for time away from the bivouac.

TRANSITION: Any questions? Let's discuss the warning order.

6. (5 Min) **ISSUE THE WARNING ORDER**. The patrol leader provides patrol members maximum preparation time possible by issuing a warning order as soon as his tentative plan is made. The warning order should include the following:

a. Situation. A brief statement of the enemy and friendly situation.

b. Mission. The patrol leader reads the mission exactly as he received it.

c. General Instructions:

    (1) General and Special Organization. General tasks are assigned to units and teams. Specific details of tasks are given in the patrol order.

    (2) Uniform and Equipment Common to All.

    (3) Weapons, Ammunition, and Equipment.

    (4) Chain of Command.

    (5) Time Schedule.

d. Specific Instructions

    (1) To Subordinate Leaders. The patrol leader gives out all the information concerning the drawing of ammunition, equipment, water, etc. identifies the personnel he wants to accompany him on his reconnaissance, coordinations, etc.

    (2) To Special Purpose Teams or Key Individual. The patrol leader should address requirements of designated personnel or teams, such as having trailbreakers, pacers, and navigators make thorough map study and a terrain model.

TRANSITION: Any questions on the warning order? Now let's discuss coordination.

7. (5 Min) **COORDINATE**. Coordination should begin as early as possible, and be continuous throughout planning, preparation, and execution of the patrol.

    a. Movement in Friendly Areas. The patrol leader finds out the location of other friendly units or patrols so his patrol will not be restricted or endangered in its movement.

    b. Departure and Reentry of Friendly Lines/Area. The patrol leader checks with the small-unit leaders occupying the areas through which the patrol will depart and estimation of return. He ensures that they know about his patrol, times of its departure and return, and whether or not guides from their units will be required to lead the patrol through any friendly obstacles, such as mines or wire.

    c. Fire Support. During his briefing with the platoon commander, the patrol leader finds out what fire support is available to him during the patrol. These assets are going to be limited by the same conditions that your patrol is experiencing. Their reaction time may be slower, as well as they may have difficulties operating their equipment in the cold. Ammunition may also be less effective against certain targets.

        (1) Air Support may be grounded due to weather.

        (2) Artillery may not be able to displace in time and may have a slower response time. Also terrain may mask portions of your area of operations.

        (3) Mortars, although an effective weapon in this terrain, are limited by their firepower and range.

        (4) Rockets. Method of target engagement may be limited in extreme cold (-25°F & below) due to ice fog.

        (5) Small Arms are affected in extreme cold due to reduced rate and range.

        (6) Grenades/Smoke must be pre-rigged prior to leaving the patrol base.

    d. Logistic Support. The patrol leader must arrange for the delivery or pick up of ammunition, special equipment, demolitions, water etc. Unusual weather and terrain conditions make problems of supply, medical evacuations, transportation, and services more difficult and more time consuming. More time must be allowed for moving supplies and troops because of the environment.

    e. Information Checklist. The battalion S-2 may be able to provide the patrol leader with some valuable information about the enemy, including some information gained from prior patrol reports, aerial photographs, reconnaissance operations, etc.

TRANSITION: Now that we discussed coordination, let's talk about making the reconnaissance.

8. (3 Min) **MAKE THE RECONNAISSANCE**. Whenever possible, the patrol leader makes a physical reconnaissance of the routes he wants to follow and of the objective area. Often, because of the enemy situation, he is not able to do so. Some other sources are:

   a. The S-2 of your unit.

   b. Other units patrols, or agencies that have been in your area of operation.

   c. Recent security / reconnaissance patrols.

   d. EPW's and locals are an excellent source of information.

   e. Maps, charts, and aerial photographs. Use as many methods as possible to gain the most comprehensive intelligence.

TRANSITION: Any questions? Let's now discuss the detailed planning.

9. (15 Min) **COMPLETE THE DETAILED PLAN**. The patrol leader is now ready to plan his patrol in detail.

   a. Specific Duties of Elements, Teams, and Individuals. The warning order assigned tasks to elements, teams, and key individuals. The patrol leader now assigns specific duties to each.

   b. Route and Alternate Route. The patrol leader selects the patrol routes based on his map study, aerial photographs, his own reconnaissance, and/or consultation with others who have been over the terrain. He chooses a route that affords cover and concealment. Not only commanders and navigators, but all Marines should know the route and briefing should be held before the march. Patrol routes are pointed out to the patrol members by the following:

      (1) Indicating the routes on a map, overlay, or terrain model.

      (2) Designating objectives and checkpoints.

      (3) Route Cards.

         (a) Time-Distance Formula (TDF). Each route should show a complete TDF: however, commanders must be patient since the actual execution across that route may include hidden difficulties. This is especially true of routes crossing streams, roads, or cross compartments.

            1. 3 kph + 1 hour for every 300 meters ascent; and/or + 1 hour for every 800 meters descent.

            2. The TDF is made for troops on foot in the summertime and troops on skis in the wintertime. If on snowshoes, multiply the total time by 1.33. If on foot in deep snow, multiply the total time by 2.0.

c. Conduct of the Patrol. The patrol leader's plan must address all the following:

   (1) Patrol Formation and Order of Movement.

      (a) Movement Formations. These will vary and change with tactical demands. Normally, during approach marches, the best formation to adopt is column or wedge formation. On firmly packed snow you will probably travel in parallel columns since this is faster and gives the commander better security.

         1. Intervals. The intervals between Marines will vary according to speed, terrain, formation, size, and equipment of units, etc.

         2. Security. Security instructions issued for a patrol will be rigidly adhered to. They normally include:

            - Measures to cover the trail.

            - Need for silence on the march is paramount.

            - Camouflage clothing policy.

            - Discipline. Track discipline, deception, etc.

      (b) Track Discipline. Tracks in snow are visible from a great distance, particularly in soft snow conditions.

         1. Line of March. Line of march selected should, whenever possible, be concealed from the air as well as the ground level. The route should follow natural contours of terrain features. Never make straight tracks, especially through open areas.

         2. Existing Tracks. The use of existing tracks, providing the risk of ambush is accounted for, is encouraged.

         3. Deception Plan. In areas where tracks have already been made, a simple deception plan can often be devised using a jump-off point, or by breaking track if that is the only alternative.

         4. No Marines Move Off of the Track Plan. It is vital that unless absolutely necessary, no Marines move off the track plan. For as the number of tracks increase, so does the likelihood of detection by the enemy.

         5. Camouflage the Number of Men. Towing brushwood along the track will camouflage the number of men that have passed by obliterating telltale signs. It

will not, of course, camouflage the track itself and is physically tiring, it may also slow down your movement.

(c) Overwatch Techniques.

1. Traveling. When speed is necessary in a traveling formation and enemy contact is not likely, one fire team can be positioned approximately 20 meters apart. The unit/squad leader will be with the lead team so as to provide control and perform navigation.

2. Squad Traveling Overwatch. This method of movement is used when there is a chance for enemy contact but it is not expected. The rear/tail team falls back to approximately 50 meters behind the lead team. From this position the rear/tail team can release their sleds and maneuver into position to support, by fire, the lead team. The unit/squad leader will normally be located just in front of the rear/team in order to provide control of the unit upon enemy contact. If attached weapons are with a unit, they will be located with the unit/squad leader.

3. Platoon Traveling Overwatch. The same principles for squad overwatch apply to platoon overwatch. However, the platoon formation must be modified to provide sufficient firepower to the front and rear, in which the platoon leader can control his unit and establish security. The weapons unit will be devided - one group to the front rifle squad, the second group to the rear rifle squad. This arrangement will make up the front and rear elements. Thus, the platoon in a traveling overwatch formation will have 20 meters between each element and 10 meters between personnel.

4. Squad Bounding Overwatch. This method of travel is used when contact with the enemy is expected. The squad will be limited in movement in what is called a "one-leg bound". The lead team moves on to an overwatch position which will provide security and fire support for the traveling team. The trail team is then moved forward by the squad leader to the lead team's location. The squad leader can then give further instructions to the lead fire team regarding where to maneuver next. The squad leader will remain with the traveling team so as to maneuver his squad upon enemy contact. If there are attachments they will remain with the squad leader so he can deploy them where necessary.

5. Platoon Bounding Overwatch. Just like the squad, the "one-leg bound" is also used by a platoon-size unit when contact with the enemy is expected. In a platoon-size unit, the lead element is made up of a fire team from the lead squad and half of the attached weapons element. They will bound forward to an overwatch position as directed by the platoon commander. The remainder of the platoon minus, will provide cover to the lead team. All elements will arrive at the next position before the lead group moves to its next position.

(2) Departure from and reentry to friendly lines or areas.

(a) Insert and Extract Methods.

    1. Helo

    2. BV / Skijor

(3) Rally Points and Actions at the Halt.

  (a) Halts.

    1. First halt on the move should be made about 15 minutes after departure to give the Marines the opportunity to adjust their clothing and equipment.

    2. Subsequent halts should be taken frequently but should be short, merely long enough to allow a short rest and a change over of duties but not long enough to get chilled. 2-3 minutes for every 15-20 minutes of movement may be used as a guide.

    3. Halts should coincide with navigation checks if possible, to reduce needless waiting about in the cold. All Marines should be kept informed of progress, distance yet to be march, how long it will take, plus any changes to the rout.

    4. Do not sit or lie in the snow during halts, but use snowshoes, isopor mats or packs, etc., as an insulation from the snow.

    5. Avoid stepping off of the track during halts. Two men from each squad must be posted as security off the track. The remainder should alternate to the sides, i.e., odd numbered men to the right, even numbered men to the left. This must be maintained despite the prevailing wind and weather, or the slope angle.

    6. Weapons alternated and outboard.

    7. Frostbite check using the buddy system.

    8. Hot meals/wets. During the march, or especially just prior to the assault, the hasty establishment of warming tents can be of great valve. These stops can be used to make hot meals/wets as well as waxing skis while skiborne. These should be secure areas to warm and rest units.

(4) Final Preparation Position (FPP) and Actions at that Position.

  (a) Establishment of the FPP

    1. Leaders Recon.

2. Movement into the FPP. Deceptive actions are mandatory in a Snow-covered environment. Make use of jump-off points, deceptive tracks, and dummy positions.

   - Jump-off point is made on slopes or in dense woods where it is possible to hide the real track and where enemy pursuit will have such a high speed of travel that it is difficult for him to discover where the real track is, as well as any booby traps. The deceptive track should be continued as far away from the jump off point as time permits.

   - Button Hook. Is the preferred method of setting in a track plan to a patrol base. This technique allows for the track to be observed and covered by fire during occupation of the position.

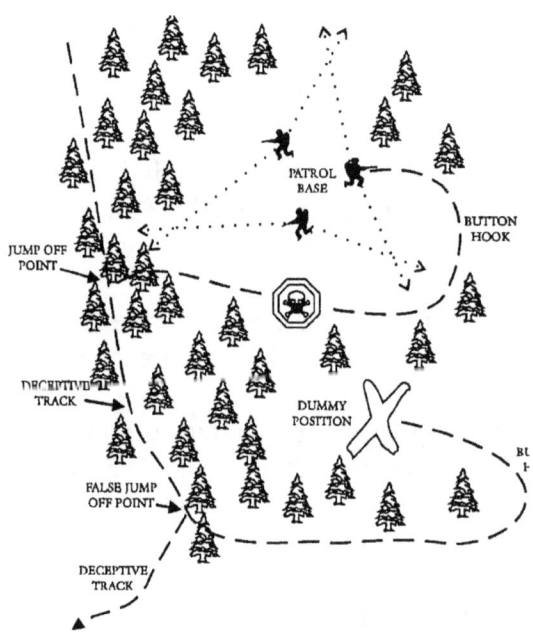

   - Deceptive Tracks. Used to mislead the enemy past your jump-off position. They should be located in an area where you can observe them and cover them with small arms fire from the patrol base.

   - Dummy Position. Make a position located at the end of the deceptive track as a decoy, to draw the enemy to this position, realism should be emphasized as much as possible. Erect tentage, antennas, etc. It should be continuously observed and located within small arms fire from your position. This was covered in greater detail during the class on Camouflage, Cover, and Concealment.

(b) Considerations in the FPP/ORP

1. On reaching the FPP, all men should put on warm gear.

2. Select an FPP out of the wind.

3. Mark the over whites of your patrol with armbands, tape, etc. Most armies over whites look similar.

4. Ensure all men have a isopor mat to lie on. One sleeping bag per buddy team may be used.

5. Ready all mission essential gear and ensure weapons are checked and functional.

6. The extremities will be the first part of the body to get cold. Especially the toes and hands which need to be kept moving at all times to maintain circulation.

7. Ensure each man has a full thermos. Don't use heat sources in the FPP to make hot-wets. The body's core temperature will start to drop. To counteract this, the buddy teams will consume one of their thermoses over a period of time. When they start to feel cold again, they will consume the other thermos.

8. For security in the FPP always employ Marines in buddy teams on 50% alert.

9. Using warming tents at the FPP may be required. If a Marine is getting near the stage where he is spending more time trying to stay warm/alert, he may be experiencing the effects of hypothermia/frostbite. That Marine, depending on the tactical situation, will have to move back to the warming tent. Unfortunately, if one Marine is in that state others could be soon to follow.

(5) Movement To and From the Objective. Infantry combat has two chief requirements to move and to fight. The difficulties caused by moving through deep snow, and the requirement for Marines to pull their combat gear in sleds is tremendous. The general considerations for a tactical movement are as follow:

(a) Terrain Negotiation. Once we have determined where we are going, the physical route must be selected by the lead element. Rather than constantly going up and down small draws and fingers, Marines on skis or snowshoes should contour around these obstacles. This reduces the amount of elevation gained and lost to a minimum, and also affords the opportunity to gain or lose necessary elevation efficiently. By traversing early on we eliminate the necessary for sudden gains or loses in elevation, which will ultimately add several minutes to hours on a previously short move. This technique will also encourage utilizing the actual concealment and cover provided by constantly moving in defilade, preventing silhouetting ourselves on the crest, and enhances track discipline.

(b) Kick Turns vs. Wide Turns. When descending steep terrain, the tendency is to traverse, then kick turn, then traverse, and kick turn through the completion of the slope. This technique is fine, provided we are not pulling sleds. Since performing kick turns while pulling sleds is very difficult, we should attempt to make wide, sweeping turns on benches or uphill from large trees or pockets of vegetation. This prevents gaining too much speed uncontrollably heading downhill. Another technique is to turn around a tree. Finally, when conducting long traverses, try to remain just inside the treeline whenever possible. By doing so, you remain in generally softer snow and usually will not have the problems of negotiating crust or windswept snow.

(c) Climbing. Swedish experiments prove that a 5% climb increases energy requirements by 30% and a 10% climb by 40%. Where possible, avoid uphill routes, and changes in elevation gained and lost by traversing left and/or right and upward and/or downward.

(6) Actions at Danger Areas. Units should follow their own SOP when crossing danger areas. Special considerations need to be observed when crossing frozen ponds or lakes, because the possibility of an enemy ambush exists, as taught in ICE RECONNAISSANCE AND BREACHING.

(7) Actions on Enemy Contact. The following are considerations:

(a) Break Contact. The conduct of the drill remains the same, however, the firing positions must be taken into consideration. The initial reaction of the personnel being engaged, i.e. point, flank, rear, will be to drop down and return fire in order to suppress engaging enemy. But, due to snow consistency, Marines may drop into the prone position and be submerged into a few inches or even a foot of snow creating a problem in the accurate return of fire. Also getting back up out of the prone position can be tiring and slow causing the Marines to be in the killing zone longer. Weapons position and how they are carried will become a critical factor to the reaction taken to break contact.

(b) Hasty Ambushes. The ultimate goal of the drill is the same as in a temperate climate. However, much more precaution must be taken in the emplacement and positioning of the Marines into the ambush site due to the mobility being used when applied to the situation of the ambush.

1. If the ambush situation is that an enemy force is tracking your patrol i.e. upon returning to a patrol base, during movement. In order to establish an effective ambush it will be necessary to "button hook" into the ambush site. This will set up an effective means of deceiving a tracking enemy into the kill zone and at the same time, maintain the units track discipline. The button hook should encompass terrain that favors the ambushing unit in security and observation of the oncoming enemy.

2. If the ambush situation is that an enemy is to the flacks or front, it may be advisable to go foot borne into the ambush site to ensure noise discipline and

avoidance of obstacles on skis/snowshoes. If the goal of the ambush is harassment, this would probably not be a good course of action due because it will not allow rapid withdrawal from the ambush site.

        3. The ambush site should facilitate an efficient avenue of withdrawal should a superior enemy force be encountered. This route should allow for rapid downhill movement in order to put as much distance between the ambush force and the enemy.

    (c) Counter-Ambush. The drills for near and far ambush remain the same. The following points should be considered.

        1. Firing positions, along with placing weapons systems into action, should be rehearsed prior to patrolling operations.

        2. It is important to note the dampening effect of snow on explosives and screening devices, so proper preparation of the ordinance prior to the patrol should be accomplished.

    (d) Encountering an Enemy Force while Skijoring.

        1. The best course of action overall is to brief the drivers that if caught in a kill zone they should accelerate through the zone while the skijoring Marines will present themselves in the lowest silhouette as possible.

        2. If the vehicles are attacked and render inoperable, then the skijoring Marines should engage in counter-ambush techniques to suppress the enemy.

(8) Actions at the Objective. This will vary with each mission.

d. Arms and Ammunition. The patrol leader checks to see if the arms and ammunition specified in the warning order have been obtained.

e. Uniform and Equipment. The patrol leader checks to see if all required equipment was available and was drawn.

f. Wounded and Prisoners.

(1) Wounded. The procedures for handling wounded may vary, depending on the seriousness of the wound and if it occurs en route to the objective, at the objective, or on the return to the friendly lines. Weather, snow condition and arduous terrain will also enhance the difficulty of a casevac.

(2) Prisoners. Prisoners will normally travel with the patrol. Special considerations have to include the means of mobility of the patrol with the prisoners; i.e. snowshoes, skis, foot etc.

g. Signals. The patrol leader's plan addresses the type of signals to be used during the patrol hand and arm signals, radio, pyrotechnic, audio. Normally he should designate a primary and alternate signal for each event requiring signals.

h. Communication with Higher Headquarters. The patrol leader includes all essential details of communication i.e. call signs, frequencies, reporting times, code words, and security requirements.

i. Challenge and Password. In addition to the parent unit's challenge and password, the patrol leader designates a challenge and password to be used within the patrol, outside of friendly lines/areas.

j. Location of Leaders. Leaders position themselves for best control and normally retain direct control of crew-served weapons.

TRANSITION: Now that we've talked about detailed planning, let's discuss issuing the patrol order.

10. (2 Min) **ISSUE THE PATROL ORDER**. When the patrol leader has completed his plan, he assembles the members of the patrol and issues the order. He issues the order in a clear, concise manner, following the standard five-paragraph order format. He should:

    a. Ensure all patrol members are present.

    b. Receive a status report from his unit/team leaders on the tasks assigned to them in the warning order.

    c. Precede the issuance of the order with an orientation.

    d. Issue the entire order utilizing a terrain model and before taking any questions.
    e. Conclude the question/answer session with a time check and announce the time of the next event.

TRANSITION: Any questions? We will now talk about supervision.

11. (3 Min) **SUPERVISE, INSPECT, REHEARSE, AND REINSPECT**. Inspections and rehearsals are vital to proper preparation. They are conducted even when the patrol leader and patrol members are experienced in patrolling.

    a. Supervision is continuous.

    b. Inspection.

        (1) The patrol leader should determine the state of readiness of the men, both mentally and physically.

(2) Clothing and equipment will be inspected in sufficient time to allow for any necessary adjustments or reloading. The equipment list is limited to only mission essential gear because anything in addition could hamper your mobility. Particular attention should be paid to:

   (a) Fitting of packs.

   (b) Sled Loading.

   (c) Ski/bindings/wax or snowshoes

   (d) Weapons/equipment readily available.

(3) Rehearsals are to ensure the operational proficiency of the patrol. Throughout rehearsals patrol members become thoroughly familiar with their actions during the patrol. Everything possible should be rehearsed.

TRANSITION: Now that we discussed supervision, let's execute the mission.

12. (2 Min) **EXECUTE THE MISSION**

TRANSITION: The twelve patrol steps is a very long process on paper, however, when these steps are performed regularly become second nature. Are there any questions up to this point?

PRACTICE (CONC)

   a. Students will plan using the CW considerations for the 12 patrol steps.

PROVIDE HELP (CONC)

   a. The instructors will assist the students when necessary.

OPPORTUNITY FOR QUESTIONS (3 Min)

1. QUESTIONS FROM THE CLASS

2. QUESTIONS TO THE CLASS

   Q. What are the first six of the twelve patrol steps?

   A. 1. Study the mission
      2. Plan the use of time
      3. Study the terrain and situation
      4. Organize the patrol

5. Select men, weapons, and equipment
6. Issue the warning order

Q. What is the last step of the twelve patrol steps?

A. Execute the mission

SUMMARY (2 Min)

a. This period of instruction has covered the cold weather considerations for the twelve patrol steps. We have given a format which shows considerations for each step, which in the future will allow for expedient planning on your part or your Marines part.

b. Those of you with IRF's please fill them out at this time and turn them in to the instructor. We will now take a short break.

UNITED STATES MARINE CORPS
Mountain Warfare Training Center
Bridgeport, California 93517-5001

MSVX
WMLC
WMO
05/24/04

LESSON PLAN

REQUIREMENTS FOR SURVIVAL

INTRODUCTION (5 Min)

1. **GAIN ATTENTION**. Although you are receiving this class as a course requirement from a formal school at MCMWTC, it can apply to you and/or any of your loved ones in many situations. If you, or they, ever backpack, cross country ski, fly in an aircraft, or simply drive or ride in a vehicle that passes through remote or isolated areas, then this class pertains to you.

2. **PURPOSE**. The purpose of this period of instruction is to introduce the student to the requirements for survival by discussing those aspects of survival that are absolutely essential for survival. This period of instruction relates to Mountain Safety.

3. **INTRODUCE LEARNING OBJECTIVES**

    a. TERMINAL LEARNING OBJECTIVE. In a cold weather mountainous environment, apply the requirements for survival, in accordance with the references.

    b. ENABLING LEARNING OBJECTIVES

       (1) Without the aid of references and given the acronym "SURVIVAL", describe in writing the acronym "SURVIVAL", in accordance with the references.

       (2) Without the aid of references, list in writing the survival stressors, in accordance with the references.

       (3) Without the aid of references, list in writing the priorities of work in a survival situation, in accordance with the references.

4. **METHOD/MEDIA**. The material in this class will be presented by lecture. You will practice what you have learned in upcoming field training exercises. Those of you with IRF's please fill them out at the end of this lesson and return them to the instructor.

5. **EVALUATION**. You will be tested later in this course by written and performance evaluations.

TRANSITION: The term "SURVIVE" brings to mind a very rough and demanding situation. In such a situation, Marines must develop a certain "mind-set" to survive.

BODY (50 Min)

1. **REQUIREMENTS FOR SURVIVAL**

    a. This mental "mind-set" is important in many ways. We usually call it the "will to survive" although you might call it "attitude" just as well. This basically means that, if you do not have the right attitude, you may still not survive.

    b. A guideline that can assist you is the acronym " SURVIVAL".

        (1) **S**ize up.

            (a) Size up the situation.

                1. Conceal yourself from the enemy.

                2. Use your senses to hear, smell, and see to determine and consider what is developing on the battlefield before you make your survival plan.

            (b) Size up your surroundings.

                1. Determine the rhythm or pattern of the area.

                2. Note animal and bird noises and their movement.

                3. Note enemy traffic and civilian movement.

            (c) Size up your physical condition.

                1. Check your wounds and give yourself first aid.

                2. Take care to prevent further bodily harm.

                3. Evaluate condition of self and unit prior to developing survival plan.

            (d) Size up your equipment.

1. Consider how available equipment may affect survival senses tailor accordingly.

(2) **U**ndue haste makes waste.

   (a) Plan your moves so that you can move out quickly without endangering yourself if the enemy is near.

(3) **R**emember where you are.

   (a) If you have a map, spot your location and relate it to the surrounding terrain.

   (b) Pay close attention to where you are and where you are going. **Constantly orient yourself**.

   (c) Try to determine, at a minimum, how your location relates to the following:

   1. The location of enemy units and controlled areas.

   2. The location of friendly units and controlled areas.

   3. The location of local water sources.

   4. Areas that will provide good cover and concealment.

(4) **V**anquish fear and panic.

   (a) Realistic and challenging training builds self-confidence and confidence for a unit's leadership.

   (b) The feeling of fear and panic will be present. The survivor must control these feelings.

(5) **I**mprovise and Improve.

   (a) Use tools designed for one purpose for other applications.

   (b) Use objects around you for different needs. (i.e. use a rock for a hammer)

(6) **V**alue living.

(a) Place a high value on living.

(b) Refuse to give into the problem and obstacles that face you.

(c) Draw strength from individuals that rise to the occasion.

(7) **A**ct like the natives.

(a) Observe the people in the area to determine their daily eating, sleeping, and drinking routines.

(b) Observe animal life in the area to help you find sources of food and water.

NOTES: Remember that animal reactions can reveal your presence to the enemy. Animals cannot serve as an absolute guide to what you can eat and drink.

(8) **L**ive by your wits, **but for now**, learn basic skills.

(a) Practice basic survival skills during all training programs and exercises.

TRANSITION: Now that we have covered the acronym "SURVIVAL", let's move on to stress

2. **STRESS**. Stress has many positive benefits. Stress provides us with challenges: it gives us chances to learn about our values and strengths. Too much stress leads to distress. While many of these signs may not be self-identified, it remains critical that all survivors remain attentive to each other's signs of distress. Listed are a few common signs of distress found when faced with too much stress:

   a. Difficulty in making decisions (**do not confuse this sign for a symptom of hypothermia**).

   b. Angry outbursts.

   c. Forgetfulness.

   d. Low energy level.

   e. Constant worrying.

   f. Propensity for mistakes.

   g. Thoughts about death or suicide.

   h. Trouble getting along with others.

        i. Withdrawing from others.

        j. Hiding from responsibilities.

        k. Carelessness.

TRANSITION: Now that we understand stress, let's discuss survival stressors.

3. **SURVIVAL STRESSORS**. Any event can lead to stress. Often, stressful events occur simultaneously. These events are not stress, but they produce it and are called "stressors". In response to a stressor, the body prepares either to "fight or flee". Stressors add up. Anticipating stressors and developing strategies to cope with them are the two ingredients in the effective management of stress. It is essential that the survivor be aware of the types of stressors they will encounter.

    a. Injury, Illness, or Death. Injury, illness, and death are real possibilities a survivor has to face. Perhaps nothing is more stressful than being alone in an unfamiliar environment where you could die from hostile action, an accident, or from eating something lethal.

    b. Uncertainty and Lack of Control. The only guarantee in a survival situation is that nothing is guaranteed. This uncertainty and lack of control also add to the stress of being ill, injured, or killed.

    c. Environment. A survivor will have to contend with the stressors of weather, terrain, and the variety of creatures inhabiting an area. Heat, cold, rain, winds, snow, mountains, insects, and animals are just a few of the challenges awaiting the Marine working to survive.

    d. Hunger and Thirst. Without food and water a person will weaken and eventually die. Getting and preserving food and water takes on increasing importance as the length of time in a survival setting increases. With the increased likelihood of diarrhea, replenishing electrolytes becomes critical. For a Marine used to having his provisions issued, foraging can be a significant source of stress.

    e. Fatigue. It is essential that survivors employ all available means to preserve mental and physical strength. While food, water, and other energy builders may be in short supply, maximizing sleep to avoid deprivation is a very controllable factor. Further, sleep deprivation directly correlates with increased fear.

    f. Isolation. Being in contact with others provides a greater sense of security and a feeling someone is available to help if problems occur.

TRANSITION: Now that we understand stressors, let's discuss natural reactions.

4. **NATURAL REACTIONS**. Man has been able to survive many shifts in his environment throughout the centuries. His ability to adapt physically and mentally to a changing world

kept him alive. The average person will have some psychological reactions in a survival situation. These are some of the major internal reactions you might experience with the survival stressors.

   a. <u>Fear</u>. Fear is our emotional response to dangerous circumstances that we believe have the potential to cause death, injury, or illness. Fear can have a positive function if it encourages us to be cautious in situations where recklessness could result in injury

   b. <u>Anxiety</u>. Anxiety can be an uneasy, apprehensive feeling we get when faced with dangerous situations. A survivor reduces his anxiety by performing those tasks that will ensure his coming through the ordeal alive.

   c. <u>Anger and Frustration</u>. Frustration arises when a person is continually thwarted in his attempts to reach a goal.. One outgrowth of frustration is anger. Getting lost, damaged or forgotten equipment, the weather, inhospitable terrain, enemy patrols, and physical limitations are just a few sources of frustration and anger. Frustration and anger encourage impulsive reactions, irrational behavior, poorly thought-out decisions, and, in some instances, an "I quit" attitude.

   d. <u>Depression</u>. Depression is closely linked with frustration and anger when faced with the privations of survival. A destructive cycle between anger and frustration continues until the person becomes worn down-physically, emotionally, and mentally. When a person reaches this point, he starts to give up, and his focus shifts from "What can I do" to "There is nothing I can do." If you allow yourself to sink into a depressed state, then it can sap all your energy and, more important, your will to survive.

   e. <u>Loneliness and Boredom</u>. Man is a social animal and enjoys the company of others. Loneliness and boredom can be another source of depression. Marines must find ways to keep their minds productively occupied.

   f. <u>Guilt</u>. The circumstances leading to your survival setting are sometimes dramatic and tragic. It may be the result of an accident or military mission where there was a loss of life. Perhaps you were the only, or one of a few, survivors. While naturally relieved to be alive, you simultaneously may be mourning the deaths of others who were less fortunate. Do not let guilt feelings prevent you from living.

TRANSITION: Now that we have covered stress, let's move on to the priorities of work in a survival situation.

5. **PRIORITIES OF WORK IN A SURVIVAL SITUATION.** Each survival situation will have unique aspects that alter the order in which tasks need to be accomplished. A general guideline is to think in blocks of time.

   a. <u>First 24 hours.</u> The first 24 hours are critical in a survival situation. You must make an initial estimate of the situation. Enemy, weather, terrain, time of day, and available resources will determine which tasks need to be accomplished first. They should be the following:

   (1) Shelter.

   (2) Fire.

   (3) Water.

   (4) Signaling.

b. <u>Second 24 hours.</u> After the first 24 hours have passed, you will now know you can survive. This time period needs to be spent on expanding your knowledge of the area. By completing the following tasks, you will be able to gain valuable knowledge.

   (1) <u>Tools and weapons</u>. By traveling a short distance from your shelter to locate the necessary resources, you will notice edible food sources and game trails.

   (2) <u>Traps and snares</u>. Moving further away from your shelter to employ traps and snares, you will be able to locate your shelter area from various vantage points. This will enable you to identify likely avenues of approach into your shelter area.

   (3) <u>Pathguards</u>. Knowing the likely avenues of approaches, you can effectively place noise and casusalty producing pathguards to ensure the security of your shelter area.

c. <u>Remainder of your survival situation.</u> This time is spent on continuously improving your survival situation until your rescue.

TRANSITION: Now that we understand the concept behind individual survival, let's talk about group survival.

6. **GROUP SURVIVAL.** In group survival, the group's survival depends largely on its ability to organize activity. An emergency situation does not bring people together for a common goal; rather, the more difficult and disordered the situation, the greater are the disorganized group's problems.

   a. <u>Groups Morale</u>. High morale must come from internal cohesiveness and not merely through external pressure. The moods and attitudes can become wildly contagious. Conscious, well-planned organization and leadership on the basis of delegated or shared responsibility often can prevent panic. High group morale has many advantages.

      (1) Individual feels strengthened and protected since he realizes that his survival depends on others whom he trusts.

(2) The group can meet failure with greater persistency.

(3) The group can formulate goals to help each other face the future.

b. <u>Factors that Influence Group Survival</u>. There are numerous factors that will influence whether a group can successfully survive.

(1) Organization of Manpower - Organized action is important to keep all members of the group informed; this way the members of the group will know what to do and when to do it, both under ordinary circumstances and in emergencies.

(2) Selective Use of Personnel - In well-organized groups, the person often does the job that most closely fits his personal qualifications.

(3) Acceptance of Suggestion and Criticisms - The senior man must accept responsibility for the final decision, but must be able to take suggestion and criticisms from others.

(4) Consideration of Time - On-the-spot decisions that must be acted upon immediately usually determines survival success.

(5) Check Equipment - Failure to check equipment can result in failure to survive.

(6) Survival Knowledge and Skills - Confidence in one's ability to survive is increased by acquiring survival knowledge and skills.

TRANSITION: Keeping a proper attitude is very important in surviving. Without the will, there is no way.

PRACTICE (CONC)

a. The students will implement what has been taught concurrently throughout the course.

PROVIDE HELP (CONC)

a. The instructors will assist the students when necessary.

OPPORTUNITY FOR QUESTIONS (2 Min)

1. QUESTIONS FROM THE CLASS

2. QUESTIONS TO THE CLASS

a. What does the "S" in the acronym SURVIVAL stand for?

Answer: Size up.

b. What priorities of work are accomplished in the first 24 hours of a survival situation?

   Answer:

   (1) Shelter
   (2) Fire
   (3) Water
   (4) Signaling

SUMMARY (3 Min)

a. We have discussed many items very briefly in this class. Many of these items will be covered in greater detail in classes you will receive later in the course of instruction you are undergoing.

b. Those of you with IRF's please fill them out at this time and turn them in. We will now take a short break.

UNITED STATES MARINE CORPS
Mountain Warfare Training Center
Bridgeport, California 93517-5001

MSVX
WMO
WML
03/16/00

LESSON PLAN

EXPEDIENT SHELTERS AND FIRES

INTRODUCTION (5 Min)

1. **GAIN ATTENTION**. Imagine yourself in this situation. While being inserted by helicopter, it suddenly loses power and crashes. Being the only survivor of the crash all of your equipment was destroyed except what you have on your body. Due to the weather your chances of being found within hours are slim. A cold front is now moving in on your position which may bring snow. What will be the first survival consideration? The first consideration in this survival situation is going to be shelter and fire. The human body can survive an amazing length of time without food or water. Without protection from the elements, particularly in a harsh environment, you will only survive for a short time. You will die from exposure to the elements long before you would ever die from lack of food or water. It is essential that every Marine has sufficient knowledge to construct some type of shelter and fire.

2. **PURPOSE**. The purpose of this period of instruction is to introduce the student to building expedient shelters and fires. This will be accomplished by breaking the class into two parts dealing with shelters and fires separately. We will discuss good shelter, hazards, methods for preparing and starting fires. This lesson relates to Requirements for Survival and Survival Medicine.

3. **INTRODUCE LEARNING OBJECTIVES**

    a. TERMINAL LEARNING OBJECTIVE.

      (1) In a cold weather mountainous environment, construct expedient shelters, in accordance with the references.

(2) In a cold weather mountainous environment, construct fires, in accordance with the references.

b. ENABLING LEARNING OBJECTIVES

(1) Without the aid of references, list in writing the characteristics of a safe expedient shelter, in accordance with the references.

(2) Without the aid of references, list in writing the hazards to avoid when using natural shelters, in accordance with the references.

(3) Without the aid of references, list in writing man-made snow survival shelters, in accordance with the references.

(4) Without the aid of references, list in writing the tactical fire lay, in accordance with the references.

(5) Without the aid of references, start a fire using a primitive method, in accordance with the references.

4. **METHOD/MEDIA**. The material in this lesson will be presented by lecture and demonstration. You will practice what you have learned during upcoming field training exercises. Those of you with IRF's please fill them out at the end of this period of instruction.

5. **EVALUATION**. You will be tested later in the course by written and performance evaluations on this period of instruction.

TRANSITION: There are several criteria to be considered when building a survival shelter.

BODY (4 hrs)

1. **BASIC CHARACTERISTICS FOR SHELTER**. Any type of shelter, whether it is a permanent building, tentage, or an expedient shelter must meet six basic characteristics to be safe and effective. The characteristics are:

    a. Protection From the Elements. The shelter must provide protection from rain, snow, wind, sun, etc.

    b. Heat Retention. It must have some type of insulation to retain heat; thus preventing the waste of fuel.

c. <u>Ventilation</u>. Ventilation must be constructed, especially if burning fuel for heat. This prevents the accumulation of carbon monoxide. Ventilation is also needed for carbon dioxide given off when breathing.

d. <u>Drying Facility</u>. A drying facility must be constructed to dry wet clothes.

e. <u>Free from Natural Hazards</u>. Shelters should not be built in areas of avalanche hazards, under rock fall or "standing dead" trees have the potential to fall on your shelter.

f. <u>Stable</u>. Shelters must be constructed to withstand the pressures exerted by severe weather.

TRANSITION: There are two categories of expedient shelters, natural and man-made. The type you choose will be dictated by the terrain, weather, availability of materials and tools, time available and tactical situation. We will first discuss natural shelters.

2. **NATURAL SHELTERS**. Natural shelters are usually the preferred types because they take less time and materials construct. The following may be made into natural shelters with some modification.

    a. <u>Caves or Rock Overhangs</u>. Can be modified by laying walls of rocks, logs or branches across the open sides.

    b. <u>Hollow Logs</u>. Can be cleaned or dug out, then enhanced with ponchos, tarps or parachutes hung across the openings.

    c. <u>Hazards of Natural Shelters</u>.

        (1) <u>Animals</u>. Natural shelters may already be inhabited (i.e. bears, coyotes, lions, rats, snakes, etc.). Other concerns from animals may be disease from scat or decaying carcasses.

        (2) <u>Lack of Ventilation</u>. Natural shelters may not have adequate ventilation. Fires may be built inside for heating or cooking but may be uncomfortable or even dangerous because of the smoke build up.

        (3) <u>Gas Pockets</u>. Many caves in a mountainous region may have natural gas pockets in them.

        (4) <u>Instability</u>. Natural shelters may appear stable, but in reality may be a trap waiting to collapse.

TRANSITION: Because of the gear Marines take to the field with them and their natural knowledge of "field living", man-made shelters are more likely to be used.

3. **MAN-MADE SHELTERS**. Many configurations of man-made shelters may be used. Over-looked man-made structures found in urban or rural environments may also provide shelter (i.e. houses, sheds, or barns). Limited by imagination and materials available, the following man-made shelters can be used in any situation.

   a. Snow Wall.

   b. Snow Cave.

   c. Tree-pit Snow Shelter.

   d. Snow Trench.

   e. A-frame Shelter.

   f. Fallen Tree Bivouac.

   g. Snow Coffin

4. **CONSTRUCTION OF MAN-MADE SHELTERS**.

   a. <u>Considerations</u>.

      (1) Group size.

      (2) A low silhouette and reduced living area will improve heat retention

      (3) Avoid exposed hilltops, valley floors, moist ground, and avalanche paths.

      (4) Create a thermal shelter by applying snow, if available, to roof and sides of shelter.

      (5) Locate in vicinity of fire wood, water, and signaling, if necessary.

      (6) How much time and effort is needed to build the shelter?

      (7) Can the shelter adequately protect you from the elements?

      (8) When in a tactical environment, you must consider the following:

         (a) Provide concealment from enemy observation.

         (b) Plan escape routes.

b. <u>Snow Wall</u>. The snow wall is an extremely expedient shelter for one or two men. This shelter is constructed when the elements will not afford time to construct a better shelter.

   (1) Basic principles for construction.

   (a) Determine wind direction.

   (b) Construct a wall of compacted snow in the shape of a horseshoe to shield you from the wind. The wall should be at least 3 feet high and as long as your body.

   (c) A poncho or tarp can be attached to the top of the wall with the other end secured to the ground for added protection. Skis, poles, branches, and equipment can be used for added stability.

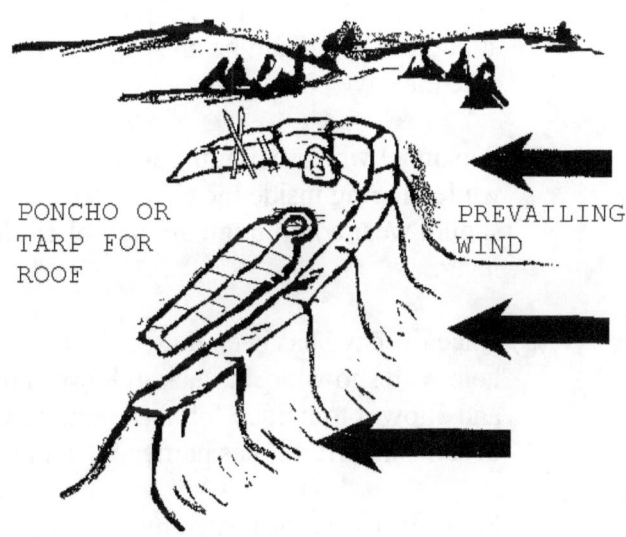

<u>SNOW WALL</u>

c. <u>Snow Cave</u>. A snow cave is used to shelter 1-16 men for extended periods of time. There must be a well-compacted snow base of at least 6 feet to construct.

   (1) Basic principles for construction.

   (a) Dig down into the snow until the desired tunnel entrance has been reached. Place all excavated snow on top of the shelter for added strength.

(b) Cut an entrance opening into the snow approximately 3 feet by 3 feet.

(c) Continue to dig out cave while removing excess snow out of the entrance. Shape the roof into a dome. If a bluish color appears through the snow in the roof, stop, there is not enough snow to support the roof.

(d) Create a cooking/working self and a sleeping bench inside the shelter.

(e) A ventilation hole should be dug through the roof at a 45-degree angle above the entrance. A ski pole or branch is left in the hole to mark the hole and allow clearing should the ventilation hole become clogged. A pine bough branch can be placed into the outside of the roof above the hole to aid in keeping the hole clear during falling snow. During the day and at night there should be an Arctic century posted for safety in case the cave collapses.

(f) Personnel who are digging will become wet from perspiration while digging inside the cave. Personnel that are digging should wear a minimum amount of clothing with a protective layer.

(g) Once the cave has been dug, completely fill in the entrance hole with snow block. Pack in loose snow between the cracks and allow it to harden for approximately 2-3 hours, weather dependant. After it has hardened, cut out a small entrance hole.

(h) Snow caves can be heated by a candle, which will raise the inside temperature, approximately 2 degrees. If a candle is left burning while individuals sleep, a fire watch must remain posted to reduce the danger of asphyxiation. Burning stoves to heat a cave will cause snow to melt and should be avoided.

(i) Packs, poncho, or snow blocks can be used to block the entrance to the cave.

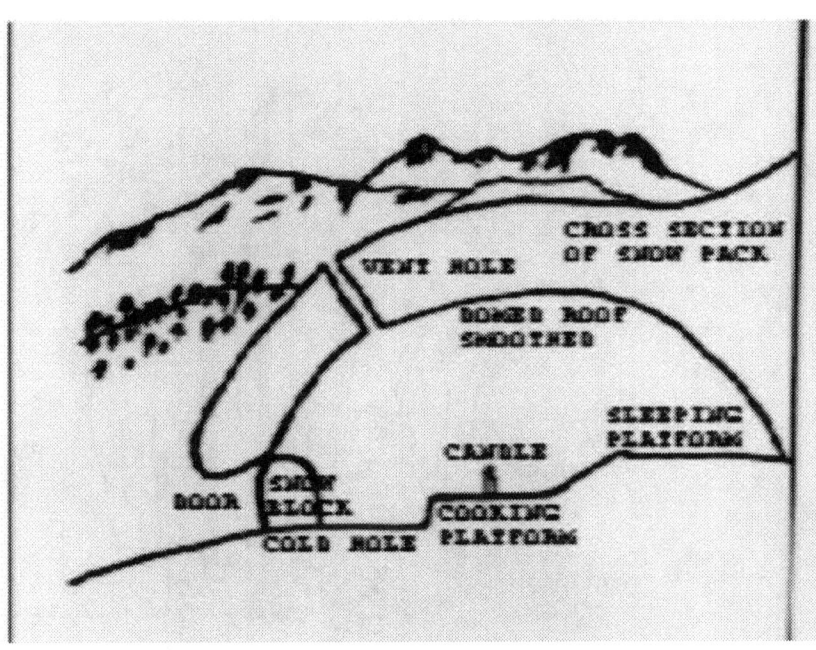

SNOW CAVE

d. <u>Tree-pit Snow Shelter</u>. A tree-pit snow shelter is designed for 1-3 men for short periods of time. It provides excellent overhead cover and concealment and should be used for LP/Ops.

   (1) Basic principles for construction.

      (a) Locate a tree with bushy branches that provides overhead cover.

      (b) Dig out the snow around the tree trunk until you reach the depth and diameter desired, or until you reach the ground.

      (c) Find and cut other evergreen boughs. Place them over the top of the pit for additional concealment. Do not utilize bough from the tree you are under.

      (d) Place evergreen boughs in the bottom of the pit for insulation

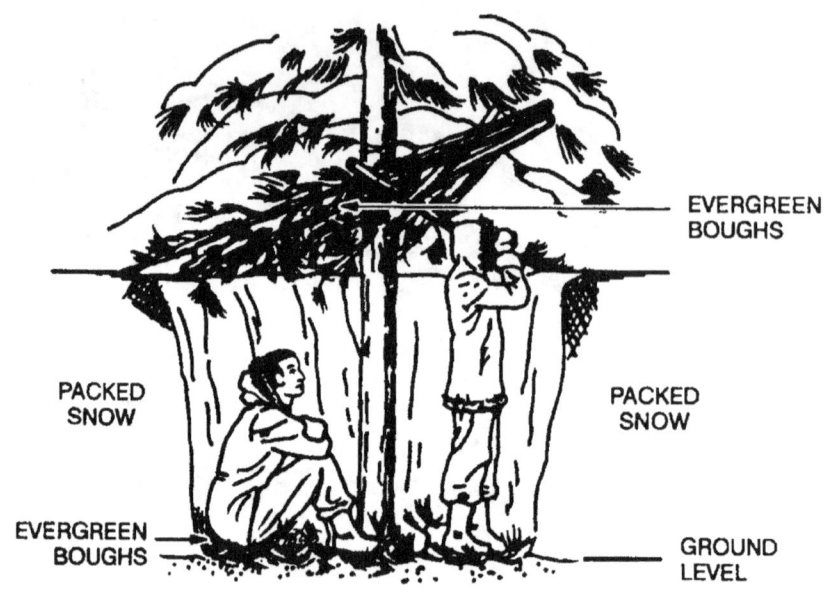

TREE-PIT SHELTER

e. <u>Fallen Tree Bivouac</u>. The fallen tree bivouac is an excellent shelter because most of the work has already been done.

  (1) Ensure the tree is stable prior to constructing.

  (2) Branches on the underside are cut away making a hollow underneath.

  (3) Place additional insulating material to the top and sides of the tree.

  (4) A small fire is built outside of the shelter.

FALLEN TREE BIVOUAC

f. A-Frame Shelter. An A-Frame shelter is constructed for 1-3 individuals. After the framework is constructed, pine bough/tentage is interwoven onto the frame and snow is packed onto the outside for insulation.

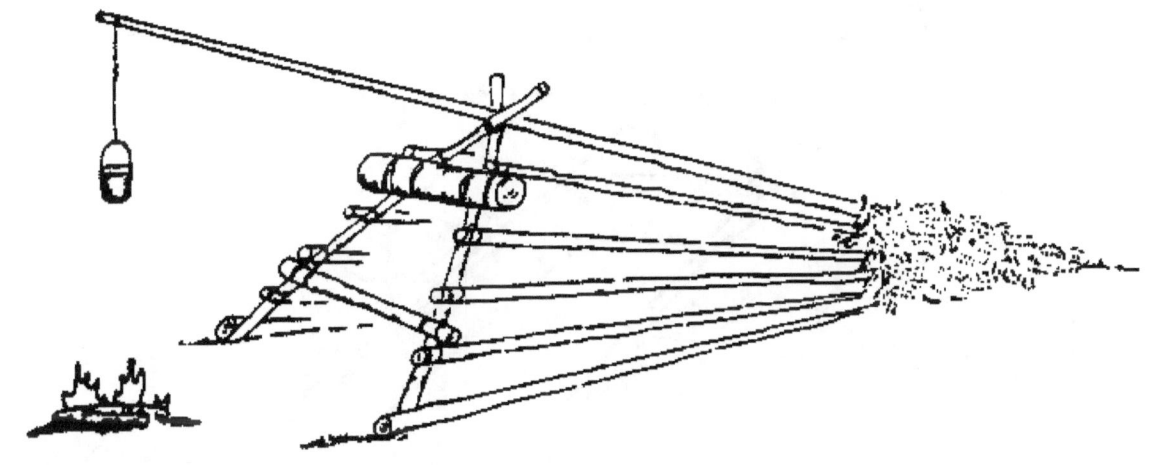

A-FRAME

g. Snow Trench. A snow trench is a short-term shelter used on extremely hard pack snow and when trees or building materials are not available, (i.e., alpine and glacier environments). Blocks of snow or ice are cut and placed to build this shelter.

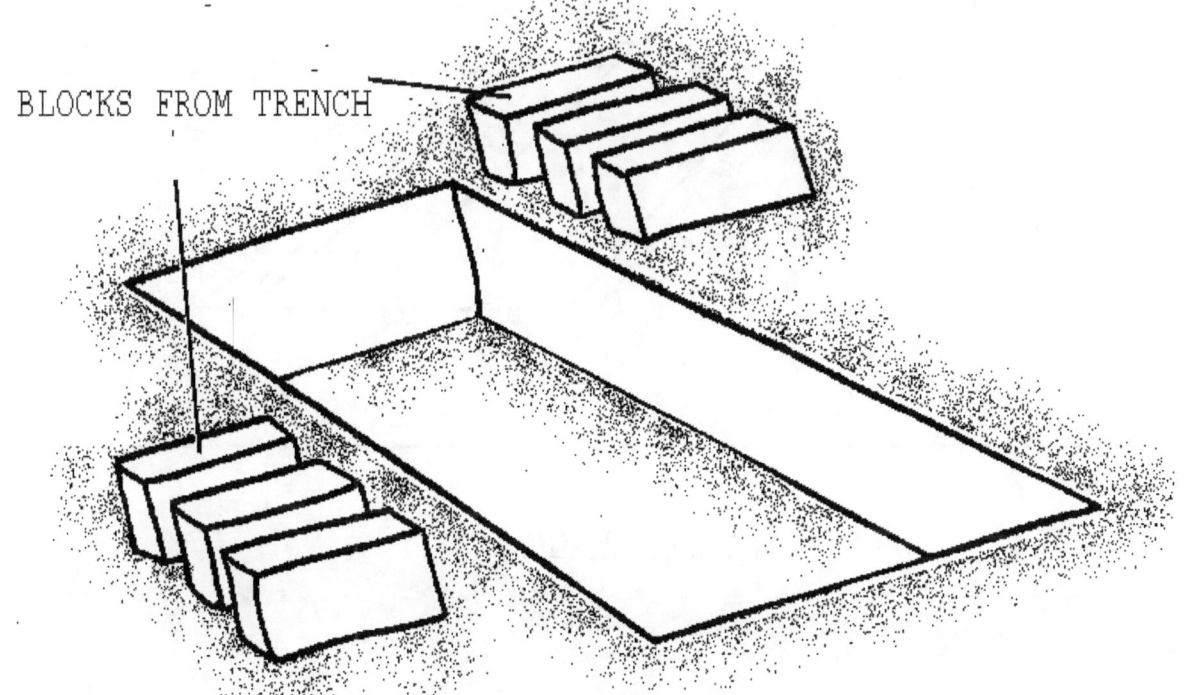

SNOW TRENCH (FIRST STEP)

A TRIANGULAR KEY BLOCK IS PLACED VERTICALLY AT THE FOOT END OF THE TRENCH. THIS WILL SERVE AS THE END SUPPORT OF THE ROOF.

THE TRENCH SHOULD BE ORIENTED SO THAT THE WIND BLOWS FROM THE FOOT SIDE AS THIS WOULD MINIMIZE SNOW EROSION. A DAILY INSPECTION SHOULD BE MADE TO AVOID A COLLAPSE OF THE ROOF DURING A WHITEOUT.

CUT A NOTCH ALONG THE INSIDE WALL OF THE TRENCH. THE SIZE AND WIDTH OF THIS NOTCH DEPENDS UPON THE CONDITION OF YOUR SNOW. WEAK SNOW WOULD REQUIRE A LARGER NOTCH.

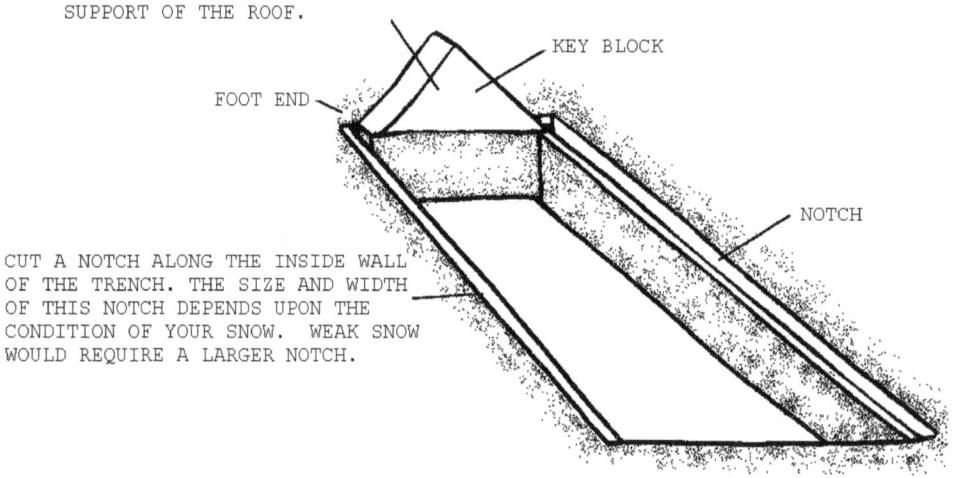

THE ROOF BLOCKS ARE ANGLE TRIMMED SO THAT THE TOPS MEET AT A POINT. THE FIRST ROOF BLOCK IS A HALF BLOCK IN WIDTH SO THAT THE JOINT LINES OF THE OTHER BLOCKS DO NOT MEET AS THIS WOULD PRODUCE A WEAKNESS IN THE ROOF

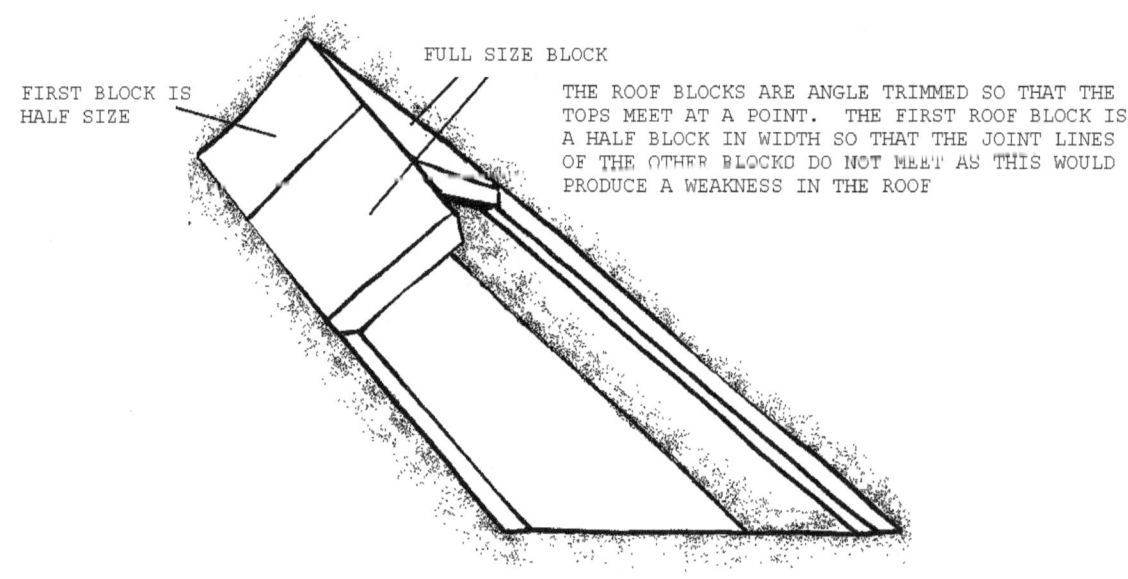

SNOW TRENCH

h. <u>Snow Coffin</u>. A snow coffin is built for 1-4 men for extended periods of time It is a variation of the snow trench and A-frame, which requires at least 4 feet of compacted snow.

   (1) Basic principles for construction.

      (a) Dig a trench into the snow approximately 3 feet wide, 8-12 fee long, and 4 feet deep.

      (b) Dig a cold hole into the floor of the trench and sleeping platforms (coffins) off the sides of the trench.

      (c) Cover the top of the trench for added protection with either an A-frame or poncho/tarp.

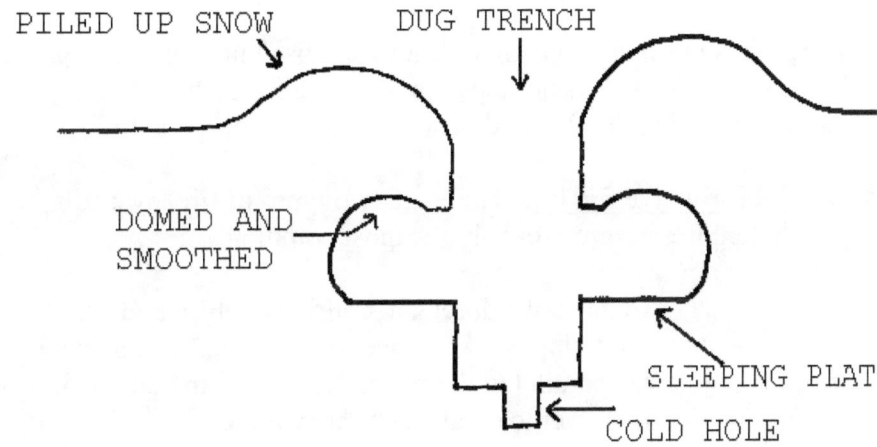

<u>SNOW COFFIN</u>

i. This table can be used as a general guideline to determine which shelter to construct.

SNOW PACK	SNOW DEPTH	EST. HRS. TO CONSTRUCT	RECOMMENDED SHELTER
LOOSE	< 2 FEET	2	A-FRAME
COMPACTED	4-6 FEET	2-3	SNOW COFFIN
COMPACTED	> 6 FEET	3	SNOW CAVE
ICED	N/A	2-3	SNOW TRENCH
N/A	N/A	1-2	FALLEN TREE
N/A	> 4 FEET	1-2	TREE-PIT
N/A	> 2 FEET	30 MIN	SNOW WALL

TRANSITION: Now that we've discussed shelters, let's look at the methods we can use to warm ourselves.

5. **FIRES**. Fires fall into two main categories: those built for cooking and those built for warmth and signaling. The basic steps are the same for both: preparing the fire lay, gathering fuel, building the fire, and properly extinguishing the fire.

   a. Preparing the fire lay. There are two types of fire lays: fire pit and Dakota hole. Fire pits are probably the most common.

      (1) Create a windbreak to confine the heat and prevent the wind from scattering sparks. Place rocks or logs used in constructing the fire lay parallel to the wind. The prevailing downwind end should be narrower to create a chimney effect.

      (2) Avoid using wet rocks. Heat acting on the dampness in sandstone, shale, and stones from streams may cause them to explode.

      (3) Dakota Hole. The Dakota Hole is a tactical fire lay. Although no fire is 100% tactical, this fire lay will accomplish certain things:

         (a) Reduces the signature of the fire by placing it below ground.

         (b) Provides more of a concentrated heat source to boil and cook, thus preserving fuel and lessening the amount of burning time.

         (c) By creating a large air draft, the fire will burn with less smoke than the fire pit.

         (d) It is easier to light in high winds.

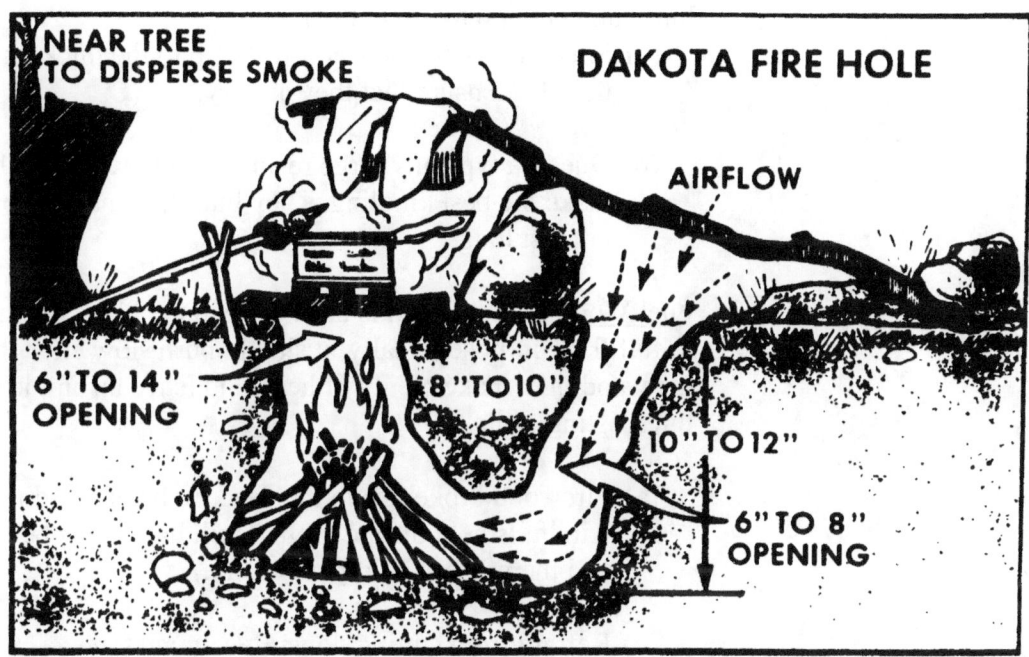

DAKATA HOLE

b. <u>Gather Fuel</u>. Many Marines take shortcuts when gathering firewood. Taking a few extra minutes can mean the difference between ease and frustration when building a fire.

    (1) <u>Tinder</u>. Tinder is the initial fuel. It should be fine and dry. Gather a double handful of tinder for the fire to be built and an extra double handful to be stored in a dry place for the following morning. Dew can moisten tinder enough to make lighting the fire difficult. Some examples are:

        (a) Shredded cedar/juniper bark, pine needles.

        (b) Dry grass.

        (c) Slivers shaved from a dry stick.

        (d) Hornet's nest.

        (e) Natural fibers from equipment supplemented with pine pitch (i.e., cotton battle dressing).

        (f) Cotton balls and petroleum jelly or Char-cloth.

Note: Sticks used for tinder should be dry and not larger than the diameter of a toothpick.

(2) <u>Kindling</u>. This is the material that is ignited by the tinder that will burn long enough to ignite the fuel.
    (a) Small sticks/twigs pencil-thick up to the thickness of the thumb. Ensure that they are dry.

    (b) Due to a typically large resin content, evergreen limbs often make the best kindling. They burn hot and fast, but typically do not last long.

(3) <u>Fuel Wood</u>. Fuel Wood is used to keep the blaze going long enough to fulfill its purpose. Ideally, it should burn slow enough to conserve the wood pile, make plenty of heat, and leave an ample supply of long-lasting coals.

    (a) Firewood broken from the dead limbs of standing trees or windfalls held off the ground will have absorbed less moisture and therefore should burn easily.

    (b) Refrain from cutting down live, green trees.

    (c) Softwoods (evergreens and conifers) will burn hot and fast with lots of smoke and spark, leaving little in the way of coals. Hardwoods (broad leaf trees) will burn slower with less smoke and leave a good bed of coals.

    (d) Learn the woods indigenous to the area. Birch, dogwood, and maple are excellent fuels. Osage orange, ironwood, and manzanita, though difficult to break up, make terrific coals. Aspen and cottonwood burn clean but leave little coals.

    (e) Stack your wood supply close enough to be handy, but far enough from the flames to be safe. Protect your supply from additional precipitation.

    (f) If you happen to go down in an aircraft that has not burned, a mixture of gas and oil may be used. Use caution when igniting this mixture.

c. <u>Building the Fire</u>. The type of fire built will be dependent upon its intended use; either cooking or heating and signaling.

    (1) <u>Cooking Fires</u>. The following listed fires are best used for cooking:

        (a) <u>Teepee Fire</u>. The teepee fire is used to produce a concentrated heat source, primarily for cooking. Once a good supply of coals can be seen, collapse the teepee and push embers into a compact bed.

TEEPEE FIRE

(2) <u>Heating and Signaling Fires</u>.

    (a) <u>Pyramid Fire</u>. Pyramid fires are used to produce large amounts of light and heat. They will dry out wet wood or clothing.

PYRAMID FIRE

d. <u>Starting Fires</u>. Starting a fire is done by a source of ignition and falls into two categories; modern igniters and primitive methods.

  (1) <u>Modern Methods</u>. Modern igniters use modern devices we normally think of to start a fire. Reliance upon these methods may result in failure during a survival situation. These items may fail when required to serve their purpose.

    (a) Matches and Lighters. Ensure you waterproof these items.

(b) Convex Lens. Binocular, camera, telescopic sights, or magnifying lens are used on bright, sunny days to ignite tender.

(c) Flint and Steel. Sometimes known as metal matches or "Mag Block". Scrape your knife or carbon steel against the flint to produce a spark onto the tinder. Some types of flint & steel designs will have a block of magnesium attached to the device which can be shaved onto the tinder prior to igniting. Other designs may have magnesium mixed into the flint to produce a higher quality of spark.

(2) <u>Primitive Methods</u>. Primitive fire methods are those developed by early man. There are numerous techniques that fall into this category. The only method that will be taught at MCMWTC is the Bow & Drill.

(3) <u>Bow & Drill</u>. The technique of starting a fire with a bow & drill is a true field expedient fire starting method which requires a piece of cord and knife from your survival kit to construct. The components of the bow & drill are bow, drill, socket, fire board, ember patch, and birds nest.

(a) <u>Bow</u>. The bow is a resilient, green stick about 3/4 of an inch in diameter and 30-36 inches in length. The bow string can be any type of cord, however, 550 cord works best. Tie the string from one end of the bow to the other, without any slack.

(b) <u>Drill</u>. The drill should be a <u>straight</u>, seasoned hardwood stick about 1/2 to 3/4 of an inch in diameter and 8 to 12 inches in length. The top end is tapered to a blunt point to reduce friction generated in the socket. The bottom end is slightly rounded to fit snugly into the depression on the fire board.

(c) <u>Socket</u>. The socket is an easily grasped stone or piece of hardwood or bone with a slight depression on one side. Use it to hold the drill in place and to apply downward pressure.

(d) <u>Fire board</u>. The fire board is a seasoned softwood board which should ideally be 3/4 of an inch thick, 2-4 inches wide, and 8-10 inches long. Cut a depression 3/4 of an inch from the edge on one side of the fire board. Cut a U-shape notch from the edge of the fire board into the depression. This notch is designed to collect and form an ember which will be used to ignite the tinder.

(e) <u>Ember Patch</u>. The ember patch is made from any type of suitable material (i.e., leather, aluminum foil, bark). It is used to catch and transfer the ember from the fire board to the birds nest. Ideally, it should be 4 inches by 4 inches in size.

(f) <u>Birds Nest</u>. The birds nest is a double handful of tinder which will be made into the shape of a nest. Tinder must be dry and finely shredded material (i.e., outer bark from juniper/cedar/sage brush or inner bark from cottonwood/aspen or dry grass/moss). Lay your tinder out in two equal rows about 4 inches wide and 8-12 inches long. Loosely roll the first row into a ball and knead the tinder to further break down the fibers. Place this ball perpendicular onto the second row of tinder and wrap. Knead the tinder until all fibers of the ball are interwoven. Insert the drill half way into the ball to form a partial cylinder. This is where the ember will be placed.

(4) <u>Producing a fire using the bow & drill</u>.

    (a) Place the ember patch under the U-shaped notch.

    (b) Assume the kneeling position, with the left foot on the fire board near the depression.

    (c) Load the bow with the drill. Ensure the drill is between the wood of the bow and bow string. Place the drill into the depression on the fire board. Place the socket on the tapered end of the drill.

    (d) Use the left hand to hold the socket while applying downward pressure.

    (e) Use the right hand to grasp the bow. With a smooth sawing motion, move the bow back and forth to twirl the drill.

    (f) Once you have established a smooth motion, smoke will appear. Once smoke appears, apply more downward pressure and saw the bow faster.

    (g) When a thick layer of smoke has accumulated around the depression, stop all movement. Remove the bow, drill, and socket from the fire board, without moving the fire board. Carefully remove your left foot off the fire board.

(h) Gently tap the fire board to ensure all of the ember has fallen out of the U-shaped notch and is lying on the ember patch. Remove the fire board.

(i) Slowly fan the black powder to solidify it into a glowing ember. Grasping the ember patch, carefully drop the ember into the cylinder of the birds nest.

(j) Grasp the birds nest with the cylinder facing towards you and parallel to the ground. Gently blow air into the cylinder. As smoke from the nest becomes thicker, continue to blow air into the cylinder until fire appears.

(5) Trouble Shooting the Bow & Drill

(a) Drill will not stay in depression- Apply more downward pressure and/or increase width/depth of depression.

(b) Drill will not twirl- Lessen the amount of downward pressure and/or tighten bow string.

(c) Socket smoking- Lessen the amount of downward pressure. Wood too soft when compared to hardness of drill. Add some lubrication: animal fat, oil, or grease.

(d) No smoke- Drill and fire board are the same wood. Wood may not be seasoned. Check drill to ensure that it is straight. Keep left hand locked against left shin while sawing.

(e) Smoke but no ember- U-shaped notch not cut into center of the depression.

(f) Bow string runs up and down drill- Use a locked right arm when sawing. Check drill to ensure that it is straight. Ensure bow string runs over the top of the left boot.

(g) Birds nest will not ignite- Tinder not dry. Nest woven too tight. Tinder not kneaded enough. Blowing too hard (ember will fracture).

e. <u>Extinguishing the Fire</u>. The fire must be properly extinguished. This is accomplished by using the drown, stir, and feel method.

(1) <u>Drown</u> the fire by pouring at water in the fire lay.

(2) <u>Stir</u> the ember bed to ensure that the fire is completely out.

(3) Check the bed of your fire by <u>feeling</u> for any hot spots.

(4) If any hot spots are found, start the process all over again.

TRANSITION: Fires are very important to our well-being. They keep us warm, cook our foods, signal for help, give us a sense of accomplishment, force us to work, and really warm your spirits when you are in a bad situation.

PRACTICE (10.5 hrs)

    a. Students will construct survival shelters and build expedient fires.

PROVIDE HELP
    (CONC)

    a. The instructors will assist the students when necessary.

OPPORTUNITY FOR QUESTIONS (2 Min)

1. <u>QUESTIONS FROM THE CLASS</u>

2. <u>QUESTIONS TO THE CLASS</u>

    a. What are the six characteristics for a safe shelter?

    Answer:
        (1) Heat retention.
        (2) Protection from the elements.
        (3) Ventilation.
        (4) Drying facility.
        (5) Free from hazards.
        (6) Stable.

    b. What are the types of man made shelters?

    Answer:
        (1) Poncho Shelter.
        (2) Sapling Shelter.
        (3) Lean-To.
        (4) Double Lean-To.
        (5) A-Frame.
        (6) Fallen Tree Bivouac.

SUMMARY (3 Min)

    a. During this period of instruction we have discussed the six basic characteristics for a safe expedient shelter, natural shelters, man-made shelters, and fires to include such things as site selection and fire starting methods.

Those of you with IRF's please fill them out at this time and turn them in to the instructor. We will take a short break.

UNITED STATES MARINE CORPS
Mountain Warfare Training Center
Bridgeport, California 93517-5001

MSVX
WML
WMO
5/24/04

LESSON PLAN

SURVIVAL NAVIGATION

INTRODUCTION   (5 Min)

1. **GAIN ATTENTION**. In almost every MOS there is a chance that a Marine may become separated from his unit. To survive, he will need some type of method to find his way back. Survival navigation could very well be the method he needs.

2. **PURPOSE**. The purpose of this period of instruction is to introduce the student to the principles and techniques of survival navigation. This lesson relates to Requirements for Survival.

3. **INTRODUCE LEARNING OBJECTIVES**

    a. <u>TERMINAL LEARNING OBJECTIVE</u>. In a cold weather mountainous environment, navigate in a survival situation, in accordance with the references.

    b. <u>ENABLING LEARNING OBJECTIVES</u>

    (1) Without the aid of references, list in writing the considerations for travel, in accordance with the references.

    (2) Without the aid of references, describe in writing the seasonal relationship of the sun and its movement during the equinox and solstice, in accordance with the references.

    (3) Without the aid of references, and given a circular navigational chart and operating latitude, determine the bearing of the sun at sunrise and sunset, in accordance with the references.

    (4) Without the aid of references, construct a pocket navigator, in accordance with the references.

(5) Without the aid of references, describe in writing the two methods for locating the North Star, in accordance with the references.

4. **METHOD/MEDIA**: The material in this lesson will be presented by lecture and demonstration. You will practice what you have learned during upcoming field training exercises. Those of you with IRF's please fill them out at the end of this period of instruction and turn them in to the instructor.

5. **EVALUATION**: You will be tested later in the course by written and performance evaluations.

TRANSITION: The decision to stay or travel needs to be well thought out. Let's discuss some considerations to making that decision.

BODY (75 Min)

1. **CONSIDERATIONS FOR STAYING OR TRAVELLING**.

    a. Stay with the aircraft or vehicle if possible. More than likely somebody knows where it was going. It is also a ready-made shelter.

    b. Leave only when:

       (1) Certain of present location; have known destination and the ability to get there.

       (2) Water, food, shelter, and/or help can be reached.

       (3) Convinced that rescue is not coming.

    c. If the decision is to travel, the following must also be considered:

       (1) Which direction to travel and why.

       (2) What plan is to be followed.

       (3) What equipment should be taken.

       (4) How to mark the trail.

       (5) Predicted weather.

    d. If the tactical situation permits leave the following information at the departure point:

       (1) Departure time.

       (2) Destination.

(3) Route of travel/direction.

(4) Personal condition.

(5) Available supplies.

TRANSITION: Once the decision to move is made, the first step is to orient yourself to your surroundings. Mother Nature has provided us a reliable method of using the sun and stars to help.

2. **DAYTIME SURVIVAL NAVIGATION**

    a. <u>Sun Movement</u>. It is generally taken for granted that the sun rises in the east and sets in the west. This rule of thumb, however, is quite misleading. In fact, depending on an observer's latitude and the season, the sun could rise and set up to 50 degrees off of true east and west.

    b. The following diagram and terms are essential to understanding how the sun and stars can help to determine direction:

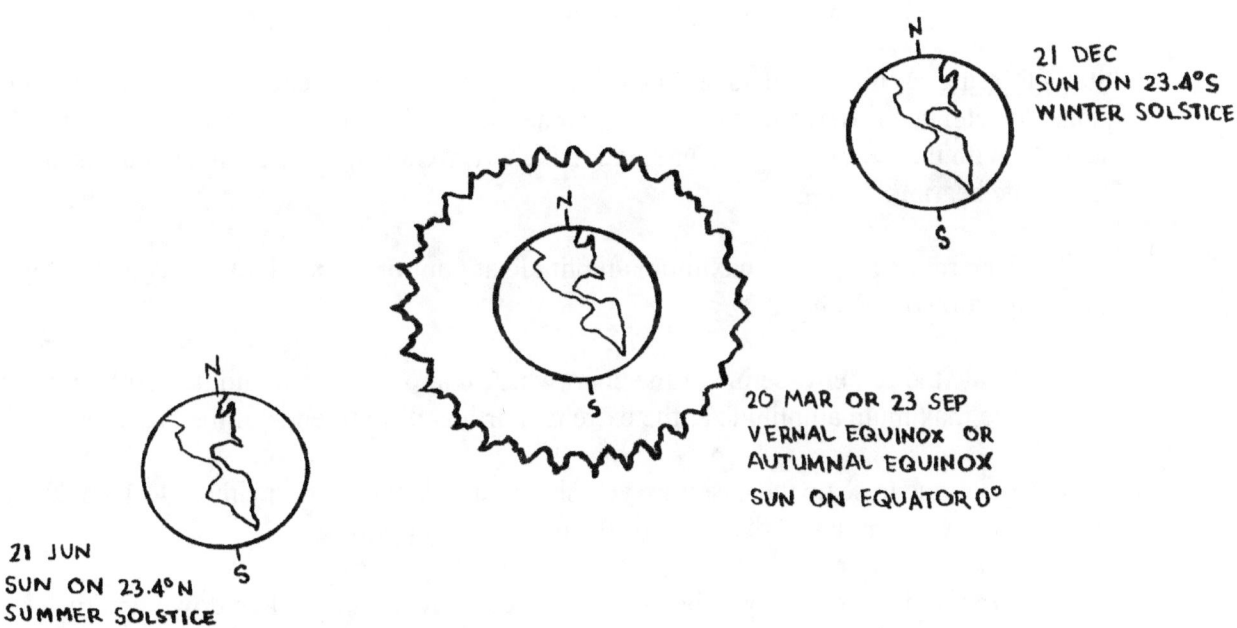

Position of the Sun at Equinox and Solstice

    (1) Summer/Winter Solstice: (21 June/21 December) Two times during the year when the sun has no apparent northward or southward motion.

    (2) Vernal/Autumnal Equinox: (20 March/23 September) Two times during the year when the sun crosses the celestial equator and the length of day and night are approximately equal.

c. <u>Sun's Movement</u>. As reflected in the diagram above, the sun continuously moves in a cycle from solstice to equinox; throughout each day, however, the sun travels a uniform arc in the sky from sunrise to sunset. Exactly half-way along its daily journey, the sun will be directly south of an observer (or north if the observer is in the Southern Hemisphere). This rule may not apply to observers in the tropics (between 23.5 degrees north and south latitude) or in the polar regions (60 degrees latitude). It is at this point that shadows will appear their shortest. The time at which this occurs is referred to as "local apparent noon."

d. <u>Local Apparent Noon</u>. Whenever using any type of shadow casting device to determine direction, "local apparent noon" (or the sun's highest point during the day) must be known. Local apparent noon can be determined by the following methods.

   (1) Knowing sunrise and sunset from mission orders, i.e., sunrise 0630 and sunset 1930. Take the total amount of daylight (13 hours), divide by 2 (6 hours 30 minutes), and add to sunrise (0630 plus 6 hours 30 minutes). Based on this example, local apparent noon would be 1300.

   (2) Using the string method. The string method is used to find two equidistant marks before and after estimated local apparent noon. The center point between these two marks represents local apparent noon.

e. <u>Sun's Bearing</u>. With an understanding of the sun's daily movement, as well as its seasonal paths, a technique is derived that will determine the true bearing of the sun at sunrise and sunset. With the aid of a circular navigational chart, we can accurately navigate based on the sun's true bearing:

   (1) Determine the sun's maximum amplitude at your operating latitude using the top portion of the chart.

   (2) Scale the center baseline of the chart where 0 appears as the middle number; write in the maximum amplitude at the extreme north and south ends of the baseline.

   (3) Continue to scale the baseline; you should divide the baseline into 6 to 10 tick marks that represent equal divisions of the maximum amplitude.

   (4) From today's date along the circumference, draw a straight line down until it intersects the baseline.

   (5) The number this line intersects is today's solar amplitude. If the number is left of 0, it is a "north" amplitude; if the number is right of 0, it is a "south" amplitude. Use the formula at the bottom of the chart to determine the sun's bearing at sunrise or sun set.

LATITUDE (N or S)	5	10	15	20	25	30	35	40	45	50	55	60
MAXIMUM AMPLITUDE	24°	24°	24°	25°	26°	27°	29°	31°	34°	38°	44°	53°

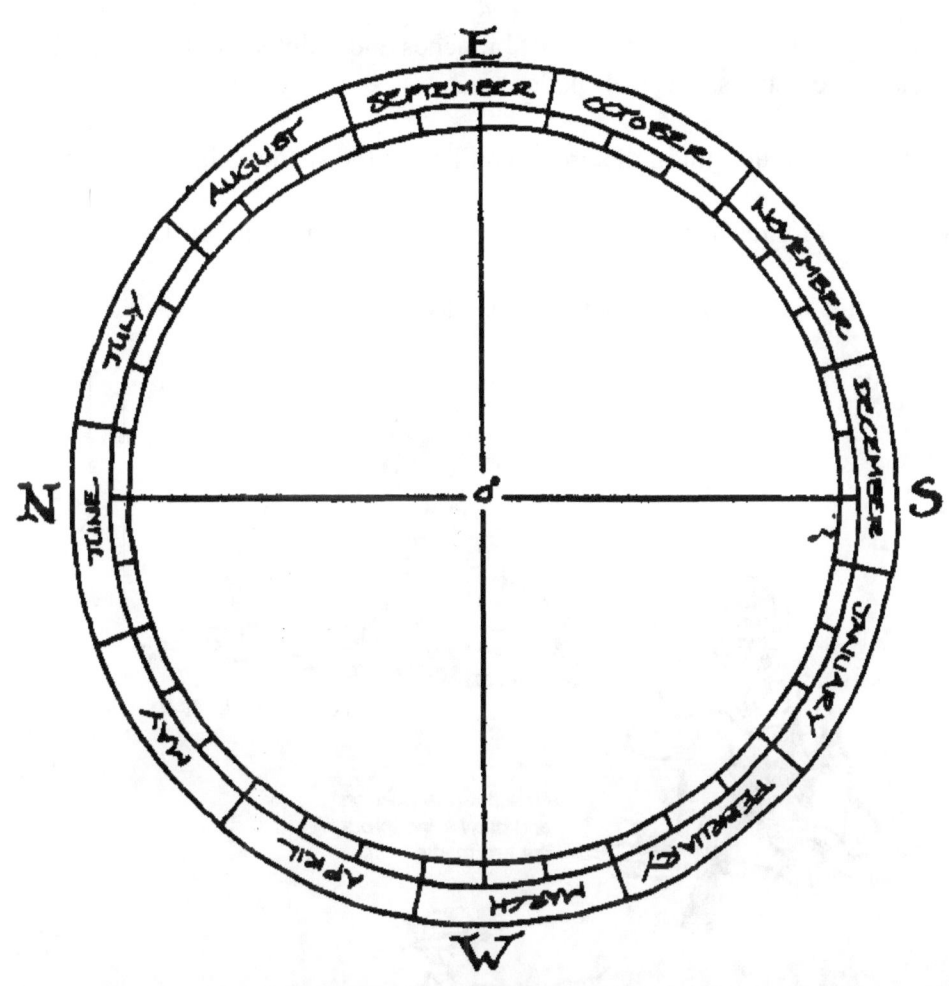

NORTH AMPLITUDES	SOUTH AMPLITUDES
SUNRISE BEARING = 90° − AMPLITUDE	SUNRISE BEARING = 90° + AMPLITUDE
SUNSET BEARING = 270° + AMPLITUDE	SUNSET BEARING = 270° − AMPLITUDE

Circular Navigational Chart

f. <u>Shadow Stick Construction</u>. This technique will achieve a cardinal direction within 10 degrees of accuracy if done within two hours of local apparent noon. Once again, this technique may be impractical near the polar regions as shadows tend to be very long; similarly, in the tropics shadows are generally very small.

    (1) Get a straight, 3-6 foot stick free of branches and pointed at the ends and 3-5 small markers: i.e., sticks, rocks, or nails.

    (2) Place the stick upright in the ground and mark the shadow tip with a marker.

    (3) Wait 10-15 minutes and mark shadow tip again with a marker.

    (4) Repeat this until all of the markers are used.

SHADOW STICK METHOD

(5) The markers will form a West—East line.

(6) Put your left foot on the first marker and your right foot on the last marker, you will then be facing north.

TRANSITION: Another method of navigating can be worked out before hand and will make for a quick and easy orientation.

3. **POCKET NAVIGATOR**. The only material required is a small piece of paper or other flat-surface material upon which to draw the trace of shadow tips and a 1 to 2 inch pin, nail, twig, wooden matchstick, or other such device to serve as a shadow-casting rod.

   a. Set this tiny rod upright on your flat piece of material so that the sun will cause it to cast a shadow. Mark the position where the base of the rod sits so it can be returned to the same spot for later readings. Secure the material so that it will not move and mark the position of the material with string, pebbles, or twigs, so that if you have to move the paper you can put it back exactly as it was. Now, mark the tip of the rod's shadow.

   b. As the sun moves, the shadow-tip moves. Make repeated shadow-tip markings every 15 minutes. As you make the marks of the shadow tip, ensure that you write down the times of the points.

   c. At the end of the day, connect the shadow-tip markings. The result will normally be a curved line. The closer to the vernal or autumnal equinoxes (March 21 and September 23) the less pronounced the curvature will be. If it is not convenient or the tactical situation does not permit to take a full day's shadow-tip markings, your observation can be continued on subsequent days by orienting the pocket navigator on the ground so that the shadow-tip is aligned with a previously plotted point.

   d. The markings made at the sun's highest point during the day, or solar noon, is the north—south line. The direction of north should be indicated with an arrow on the navigator as soon as it is determined. This north-south line is drawn from the base of the rod to the mark made at solar noon. This line is the shortest line that can be drawn from the base of the pin to the shadow-tip curve.

   e. To use your pocket navigator, hold it so that with the shadow-tip is aligned with a plotted point at the specified point. i.e.; if it is now 0900 the shadow-tip must be aligned with that point. This will ensure that your pocket navigator is level. The drawn arrow is now oriented to true north, from which you can orient yourself to any desired direction of travel.

   f. The pocket navigator will work all day and will not be out of date for approximately one week.

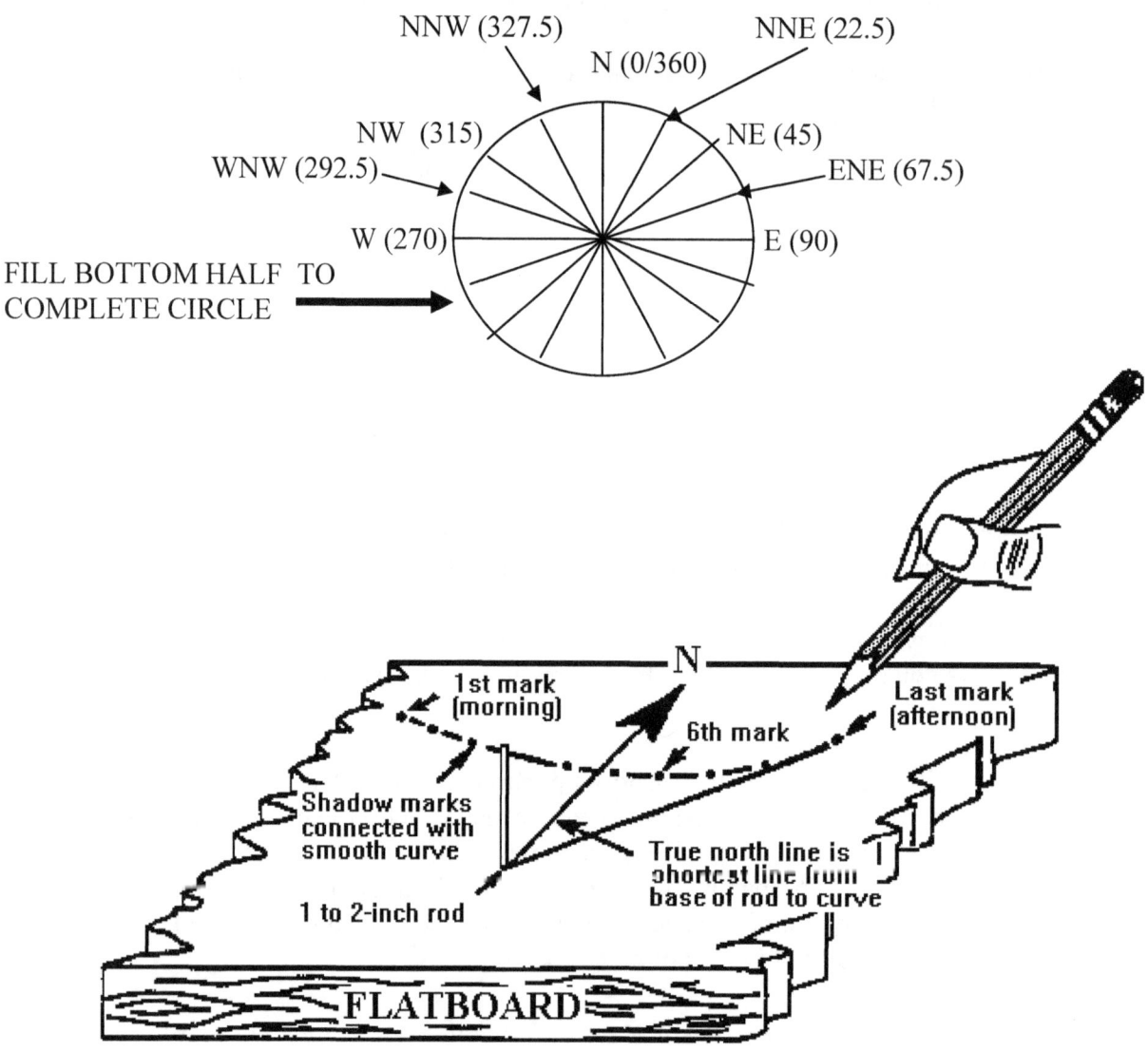

POCKET NAVIGATOR

TRANSITION: Now that we have covered some of the methods used for navigating during the daytime, let's talk about navigating at night.

4. **NIGHTTIME SURVIVAL NAVIGATION**

    a. <u>Mark North</u>. To aid you in navigating at night, it is beneficial to watch where the sun goes down. If you're going to start moving after dark mark the northerly direction.

    b. <u>Locating the North Star</u>. There are two methods used in locating the North Star.

(1) <u>Using the Big Dipper</u> (*Ursa Major*). The best indictors are the two "dippers". The North Star is the last star in the handle of the little dipper, which is not the easiest constellation to find. However, the Big Dipper is one of the most prominent constellations in the Northern Hemisphere. The two lowest stars of the Big Dipper's cup act as pointers to the North Star. If you line up these two stars, they make a straight line that runs directly to the North Star. The distance to the North Star along this line is 5 times that between the two pointer stars.

(2) <u>Using Cassiopeia</u> (*Big M or W*). Draw a line straight out from the center star, approximately half the distance to the Big Dipper. The North Star will be located there.

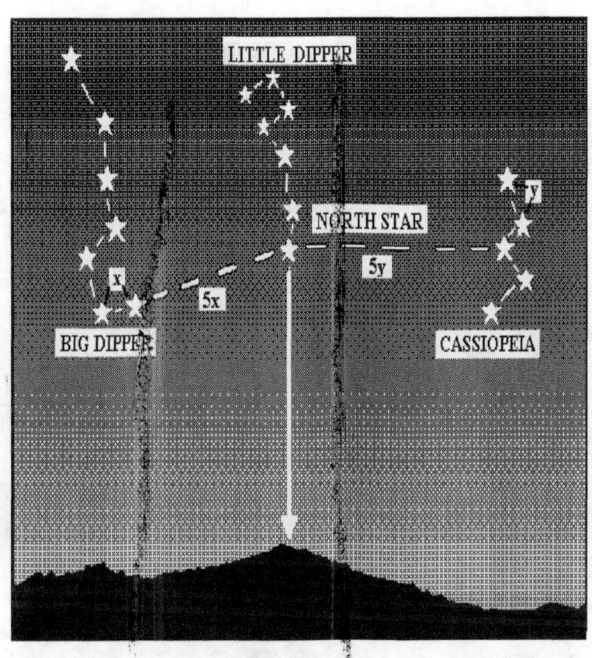

LOCATING THE NORTH STAR

NOTE: Because the Big Dipper and Cassiopeia rotate around the North Star, they will not always appear in the same position in the sky. In the higher latitudes, the North Star is less effective for the purpose of orienting because it appears higher in the sky. At the center of the Arctic circle, it would be directly overhead, and all directions lead South.

c. <u>Southern Cross</u>. In the Southern Hemisphere, Polaris is not visible. There, the Southern Cross is the most distinctive constellation. An imaginary line through the long axis of the Southern Cross, or True Cross, points towards a dark spot devoid of stars approximately three degrees offset from the South Pole. The True Cross should not be confused with the larger cross nearby know as the False Cross, which is less bright, more widely spaced, and has five stars. The True Cross can be confirmed by two closely spaced, very bright stars that trail behind the cross piece. These two stars are often easier to pick out than the cross itself. Look for them. Two of the stars in the True Cross are among the brightest stars in the heavens; they are the stars on the southern and eastern arms. The stars on the northern and western arms are not as conspicuous, but are bright.

Note: The imaginary point depicted in the picture is the dark spot devoid of stars.

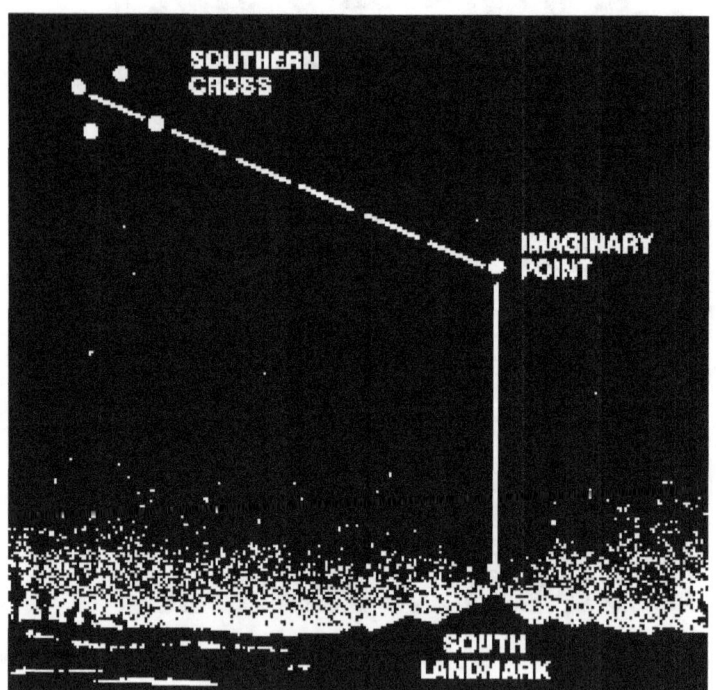

SOUTHERN CROSS

    d. <u>Moon Navigator</u>. Like the sun, the moon rises in the east and sets in the west. Use the same method of the shadow stick as you did during the day.

TRANSITION: Some day you could find yourself without the aid of celestial bodies which will require alternative navigational tools.

5. **<u>IMPROVISED COMPASSES</u>**. There are three improvised techniques to construct a compass.

    a. <u>Synthetic technique</u>. The required items are a piece of synthetic material, (i.e., parachute cloth), and a small piece of iron or steel that is long, thin, and light. Aluminum or yellow metals won't work (only things that rust will do). A pin or needle is perfect, but a straightened paper clip, piece of steel baling wire, or barbed wire could also work.

- (1) Stroke the needle repeatedly in one direction against the synthetic material. Ensure that you lift the material a few inches up into the air at the end of each stroke, returning to the beginning of the needle before descending for another stroke in the same direction. Do this approximately 30 strokes. This will magnetize the needle.

- (2) Float the metal on still water using balled up paper, wood chip, or leaf. Gather some water in a non-magnetic container or a scooped out recess in the ground, such as a puddle. Do not use a "tin can" which is made of steel. (An aluminum can would be fine.) Place the float on the water, then the metal on it. It will slowly turn to orient itself.

b. Magnet technique. You will achieve the same results by using a magnet. Follow the same steps as you did with the synthetic material. The magnets you are most likely to have available to you are those in a speaker or headphones of a radio.

c. Magnetization through a battery. A power source of 2 volts or more from a battery can be used with a short length of insulated wire to magnetize metal. Coil the wire around a needle. If the wire is non-insulated, wrap the needle with paper or cardboard. Attach the ends to the battery terminals for 5 minutes.

d. Associated problems with improvised compasses. The following are common problems with all improvised compasses.

- (1) Soft steel tends to lose its magnetism fairly quickly, so you will have to demagnetize your needle occasionally, though you should not have to do this more than two or three times a day.

- (2) Test your compass by disturbing it after it settles. Do this several times. If it returns to the same alignment, you're OK. It will be lined up north and south, **though you will have to determine by other means which end is north. Use the sun, stars, or any other natural signs in the area.**

- (3) Remember, this will give magnetic north. In extreme northern lattitudes, the declination angle can be extreme.

6. **SURVIVAL NAVIGATION TECHNIQUES**

    a. Navigator.

    - (1) Employ a navigation method.

    - (2) Find the cardinal direction.

(3) Pick a steering mark in the desired direction of travel.

b. <u>Maintain a Log</u>. The possibility may arise when you will not have a map of the area. A log will decrease the chance of walking in circles.

(1) Construction.

(a) Use any material available to you i.e., paper, clothing, MRE box, etc.

(b) Draw a field sketch annotating North, prominent terrain features, and friendly/enemy position.

(2) Maintenance.

(a) Annotate distance traveled, elevation gained and lost, and cardinal directions

(b) Maintain and update field sketch as movement progresses.

(c) Ensure readability of your field sketch. (i.e.; don't clutter the sketch so much that it can't be read.)

c. <u>During Movement Constantly Refer To</u>.

(1) Log.

(2) Steering marks.

d. <u>Actions If You Become Lost</u>.

(1) Immediate action

(a) Orient your sketch. This will probably make your mistake obvious.

(2) Corrective action

(a) Backtrack using steering marks until you have determined the location of your error.

(b) Re-orient your sketch.

(c) Select direction of travel and continue to march.

TRANSITION: Knowing where you are is just as important as knowing where you want to be, one leads naturally to another. Are there any questions?

PRACTICE (CONC)

    a. Students will practice survival navigation.

PROVIDE HELP (CONC)

    a. The instructors will assist the students when necessary.

OPPORTUNITY FOR QUESTIONS (2 Min)

1. QUESTIONS FROM THE CLASS
1. QUESTIONS TO THE CLASS

    a. What are the two daytime survival navigation techniques?

       Answer:
          (1) The shadow stick method.
          (2) The pocket navigator method.

    b. What are the two methods in locating the North Star?

       Answer:
          (1) Big Dipper (Ursa Major)
          (2) Cassiopeia (Big M or W)

    c. What is the immediate action if you become lost using navigating techniques?

       Answer: Orient the sketch.

SUMMARY (2 Min)

    a. We have covered survival navigation to include day and night techniques, methods for stationary and moving situations, and some general techniques and tips to prevent you from going astray.

    b. Those of you with IRF's please fill them out at this time and turn them in to the instructor. We will now take a short break.

UNITED STATES MARINE CORPS
Mountain Warfare Training Center
Bridgeport, California 93517-5001

MSVX
WML
WMO
03/01/00

LESSON PLAN

SURVIVAL SIGNALING AND RECOVERY

INTRODUCTION (5 Min)

1. **GAIN ATTENTION**. In a survival situation, your basic concern is to establish communications with other friendly units. In a survival/evasion situation, your basic problem is to establish communications with only the right people. Communication is generally interpreted as "giving and receiving information". The signals you use as a survivor or evader must make it easier for the rescue crew to find you. The types of signals you and the rescue crew will use depends on the ground situation. Selecting the correct signaling method will assist in the rescue.

2. **PURPOSE**. The purpose of this period of instruction is to introduce the student to signaling, specifically those devices, methods, and specific means of communicating with rescuers that are used in a survival situation. This lesson relates to Survival Fires.

3. **INTRODUCE LEARNING OBJECTIVES**

    a. <u>TERMINAL LEARNING OBJECTIVE</u>. In a cold weather mountainous environment, conduct recovery, in accordance with the references.

    b. <u>ENABLING LEARNING OBJECTIVES</u>

    (1) Without the aid of references, describe in writing the audio international distress Signal, in accordance with the references.

    (2) Without the aid of references, describe in writing the visual international distress signal, in accordance with the references.

(3) Without the aid of references, construct an improvised visual signaling device, in accordance with the references.

(4) Without the aid of reference, utilize a hoist recovery device, in accordance with the references.

4. **METHOD/MEDIA**. The material in this lesson will be presented by lecture and demonstration. You will practice what you have learned during upcoming field training exercises. Those of you with IRF's please fill them out at the end of this period of instruction

5. **EVALUATION**. You will be evaluated on this material by written and practical examinations in future field evolutions.

TRANSITION: Now let's look at some signaling devices that you may have at your disposal.

BODY (50 Min)

1. **SIGNALING DEVICES**. The equipment listed below are items that may be on your body or items inside an aircraft. Generally, these items are used as signaling devices while on the move. They must be accessible for use at any moment's notice. Additionally, in a summer mountainous environment, Marines may experience areas that are snow covered and must be familiar with the effects that snow will have on specific signaling devices.

   a. Pyrotechnics. Pyrotechnics include star clusters and smoke grenades. When using smoke grenades in snow pack, a platform must be built. Without a platform, the smoke grenade will sink into the snow pack and the snow will absorb all smoke. A rocket parachute flare or a hand flare have been sighted as far away as 35 miles, with an average of 10 miles. Pyrotechnic flares are effective at night, but during daylight their detectability ranges are reduced by 90 percent.

   b. M-186 Pen Flare. The M-186 Pen Flare is a signaling device carried in the vest of all crew chiefs and pilots. Remember to cock the gun prior to screwing in the flare.

   c. Strobe Light. A strobe light is generally carried in the flight vests of all crew chiefs and pilots. It can be used at night for signaling. Care must be taken because a pilot using goggles may not be able to distinguish a flashing strobe from hostile fire.

   d. Flashlight. By using flashlights, a Morse code message can be sent. An SOS distress call consists of sending three dots, three dashes, and three dots. Keep repeating this signal.

   e. Whistle. The whistle is used in conjunction with the audio international distress signal. It is used to communicate with forces on the ground.

f. AN/PRC-90 & AN/PRC-112. The AN/ PRC 90 survival radio is a part of the aviator's survival vest. The AN/PRC-112 will eventually replace the AN/PRC-90. Both radios can transmit either tone (beacon) or voice. Frequency for both is **282.8 for voice**, and **243.0 for beacon**. Both of these frequencies are on the UHF Band.

g. Day/Night Flare. The day/night flare is a good peacetime survival signal. The flare is for night signaling while the smoke is for day. The older version flare is identified by a red cap with three nubbins while the new generation has three rings around the body for identification during darkness. The flare burns for approximately 20 second while the smoke burns for approximately 60 seconds.

NOTE: Once one end is used up, douse in water to cool and save the other end for future use.

h. Signal Mirror. A mirror or any shiny object can be used as a signaling device. It can be used as many times as needed. Mirror signals have been detected as far away as 45 miles and from as high as 16,000', although the average detection distance is 5 miles. It can be concentrated in one area, making it secure from enemy observation. A mirror is the best signaling device for a survivor; however, it is only as effective as its user. Learn how to use one now, before you find yourself in a survival situation.

   (1) Military signal mirrors have instructions on the back showing how to use it. It should be kept covered to prevent accidental flashing that may be seen by the enemy.

   (2) Any shiny metallic object can be substituted for a signal mirror.

   (3) Haze, ground fog, or a mirage may make it hard for a pilot to spot signals from a flashing object. So, if possible, get to the highest point in your area when flashing. If you can't determine the aircraft's location, flash your signal in the direction of the aircraft noise.

AIMING THE SIGNAL MIRROR

TRANSITION: There are basically two ways to communicate, audio and visual.

2. **METHODS OF COMMUNICATION**

    a. Audio. Signaling by means of sound may be good, but it does have some disadvantages.

        (1) It has limited range unless you use a device that will significantly project the sound.

        (2) It may be hard to pinpoint one's location due to echoes or wind.

        (3) International Distress Signal. The survivor will make six blasts in one minute, returned by three blasts in one minute by the rescuer.

    b. Visual. Visual signals are generally better than audio signals. They will pinpoint your location and can been seen at a greater distances under good weather conditions

        (1) The visual international distress symbol is recognized by a series of three evenly spaced improvised signaling devices.

TRANSITION: Now, let's cover improvised signaling devices.

3. **IMPROVISED SIGNALING DEVICES**. Improvised signaling devices are generally static in nature. They must be placed in a position to be seen by rescuers. They are made from any resources available, whether natural or man-made.

   a. Smoke Generator. The smoke generator is an excellent improvised visual signaling device. It gives the survivor the flexibility to signal in either day or night conditions. This type of signal has been sighted as far away as 12 miles, with an average distance of 8 miles. Smoke signals are most effective in calm wind conditions or open terrain, but effectiveness is reduced with wind speeds above 10 knots. Build them as soon as time and the situation permits, and protect them until needed.

      (1) Construct your fire in a natural clearing or along the edge of streams (or make a clearing). Signal fires under dense foliage will not be seen from the air.

      (2) Find two logs, 6 - 10 inches in diameter, and approximately five feet long. Place the two logs parallel to each other with 3 - 4 feet spacing.

      (3) Gather enough sticks, approximately two inches in diameter and four feet long, to lay across the first two logs. This serves as a platform for the fire.

      (4) Gather enough completely-dry branches to build a pyramid fire. The pyramid fire should be 4 feet by 4 feet by 2 feet high.

      (5) Place your tinder under the platform.

      (6) Gather enough pine bough to lay on top of the pyramid fire. This serves to protect the fire and the tinder.

      (7) To light, remove the pine bough and ignite the tinder. If available, construct a torch to speed up the lighting process, especially for multiple fires.

SMOKE GENERATOR

(9) To create a smoke effect during the day light hours, place the pine bough on the ignited fire.

(10) Placing a smoke grenade or colored flare under the platform will change the color of the smoke generated. Remember, you want the fire to draw in the colored smoke which will create a smoke color that contrasts with the back ground will increase the chances of success.

TRANSITION: Next we'll discuss the arrangement or alteration of natural materials.

    a. <u>Arrangement or alteration of natural materials</u>. Such things as twigs or branches, can be tramped into letters or symbols in the snow and filled in with contrasting materials. To attract more attention, ground signals should be arranged in big geometric patterns.

        (1) <u>International symbols</u>. The following symbols are internationally known.

Number	Message	Code symbol
1	REQUIRE ASSISTANCE	V
2	REQUIRE MEDICAL ASSISTANCE	X
3	NO OR NEGATIVE	N
4	YES OR AFFIRMATIVE	Y
5	PROCEED IN THIS DIRECTION	↑

INTERNATIONAL SYMBOLS

(1) <u>Shadows</u>. If no other means are available, you may have to construct mounds that will use the sun to cast shadows. These mounds should be constructed in one of the International Distress Patterns. To be effective, these shadow signals must be oriented to the sun to produce the best shadows. In areas close to the equator, a North—South line gives a shadow anytime except noon. Areas further north or south of the equator require the use of East—West line or some point of the compass in between to give the best result.

(2) <u>Size</u>. The letters should be large as possible for a pilot or crew to spot. Use the diagram below to incorporate the size to ratio for all letter symbols.

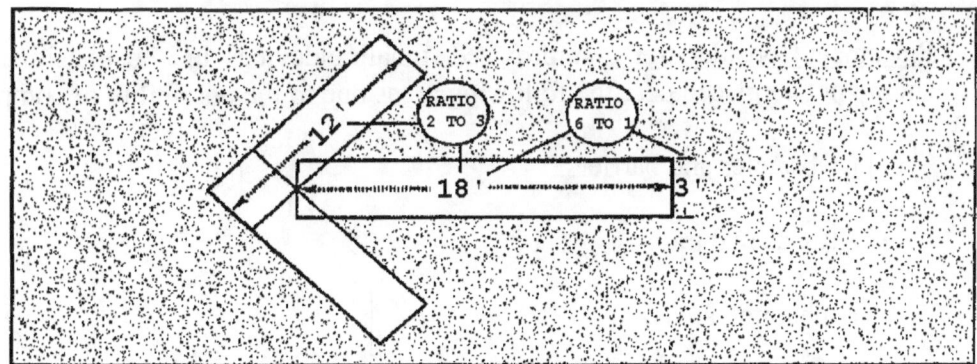

SIZE AND RATIO

(3) <u>Contrast</u>. When constructing letter symbols, contrast the letter from the surrounding vegetation and terrain. Ideally, bring material from another location to build the letter. This could be clothing, air panels, space blanket, etc.

(a) On snow, pile bough or use sea dye from a LPP (Life preserver, personal). Fluorescent sea dye markers have been sighted as far away as 10 miles, although the average detection distance is 3 miles.

TRANSITION: Now that we know how to signal an aircraft, let us find out how an aircraft will signal us.

4. **AIR TO GROUND COMMUNICATIONS**. Air to ground communications can be accomplished by standard aircraft acknowledgments.

   a. Aircraft will indicate that ground signals have been seen and understood by:

      (1) Rocking wings from side to side. This can be done during the day or in bright moonlight.

   b. Aircraft will indicate that ground signals have been seen but not understood by:

      (1) Making a complete, clockwise circle during the day or in bright moonlight.

TRANSITION: Now that we have an understanding of signaling devices, remember, you are still in a survival situation until recovered.

5. **RECOVERY**. Marines trapped behind enemy lines in future conflicts may not experience quick recovery. Marines may have to move to a place that minimizes risk to the recovery force. No matter what signaling device a Marine uses, he must take responsibility for minimizing the recovery force's safety.

   a. Placement Considerations. Improvised signaling devices, in a hostile situation, should not be placed near the following areas due to the possibility of compromise:

      (1) Obstacles and barriers.

      (2) Roads and trails.

      (3) Inhabited areas.

      (4) Waterways and bridges.

      (5) Natural lines of drift.

      (6) Man-made structures.

      (7) All civilian and military personnel.

   b. Tactical Consideration. The following tactical considerations should be adhered to prior to employing an improvised signaling device.

(1) Use the signals in a manner that will not jeopardize the safety of the recovery force or you.

(2) Locate a position which affords observation of the signaling device and facilitates concealed avenues of escape (if detected by enemy forces). Position should be located relatively close to extract site in order to minimize "time spent on ground" by the recovery force.

(3) Maintain continuous security through visual scanning and listening while signaling devices are employed. If weapon systems are available, signaling devices should be covered by fire and/or observation.

(4) If enemy movement is detected in the area, <u>attempt</u> to recover the signaling device, if possible.

(5) Employ improvised signaling devices only during the prescribed times, if briefed in the mission order.

c. <u>Recovery Devices</u>. In mountainous terrain, a helicopter landing may be impossible due to ground slope, snow pack, or vegetation. The survivor must be familiar with recovery devices that may be aboard the aircraft.

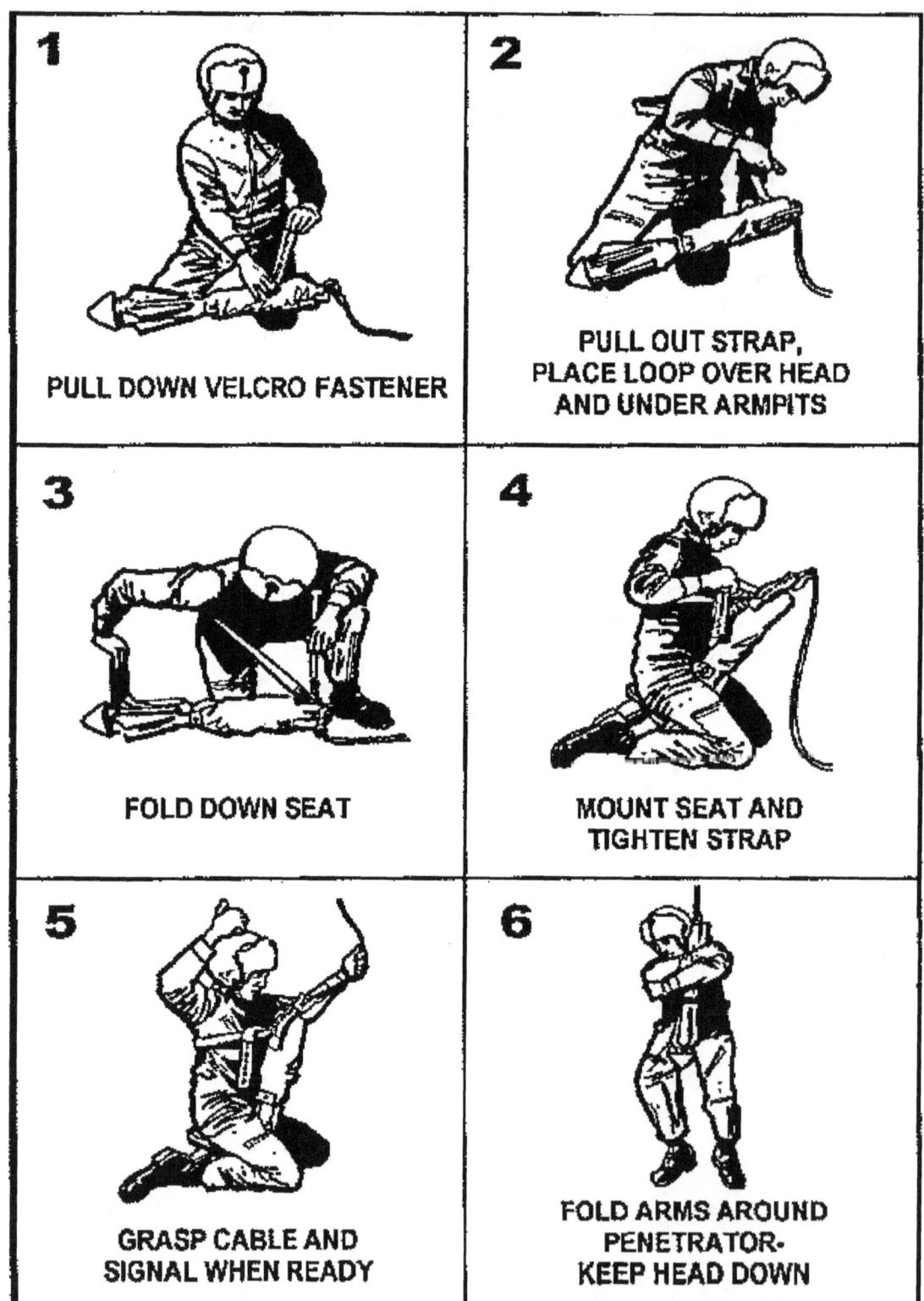

JUNGLE PENETRATOR

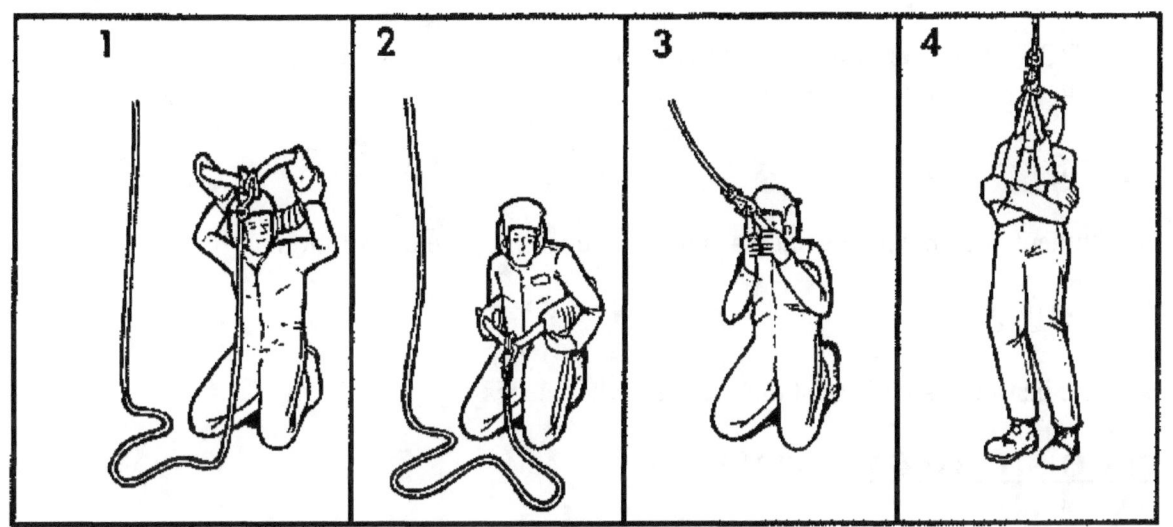

SLING HOIST

d. <u>Recovery by other than aircraft</u>. Recovery by means other than aircraft may occur. Unit SOP's should include signaling and link-up with forces at the following locations:

   (1) Border Crossings. The evader who crosses into a neutral country is subject to detention by that country for the duration of the war.

   (2) FEBA/FLOT.

      (a) <u>Static</u>. Recovery along a static FEBA is always difficult. Under these conditions, enemy and friendly forces can be expected to be densely arrayed and well camouflaged, with good fields of fire. Attempts to penetrate the FEBA should be avoided.

      (b) <u>Advancing</u>. Individuals isolated in front of advancing friendly units should immediately take cover and wait for the friendly units to overrun their position.

      (c) <u>Retreating</u>. Individuals between opposing forces should immediately take cover and wait for enemy units to pass over their position. After most enemy units have moved on, evaders should try to link up with other isolated friendly elements and return to friendly forces.

   (3) Link-up with friendly patrols. Unit authentication numbers and/or locally developed codes may assist the evader to safely make contact in or around the FEBA and when approached by friendly forces.

TRANSITION: If you are lost and incapable of moving, signaling may be the only way that you are going to get saved. Are there any questions?

PRACTICE   (CONC)

   a. Students will practice what was taught in upcoming field evolutions.

PROVIDE HELP   (CONC)

   a. The instructors will assist the students when necessary.

OPPORTUNITY FOR QUESTIONS   (2 Min)

1. QUESTIONS FROM THE CLASS

2. QUESTIONS TO THE CLASS

   a. Describe the International Distress Signal using audio?

   Answer: The survivor makes six blasts in one minute, returned by three blasts in one minute by the rescuer.

SUMMARY   (1 Min)

   a. During this period of instruction we have discussed survival signaling to include devices, techniques, procedures, and recovery used to signal rescuers in a survival situation.

   b. Those of you with IRF's please fill them out at this time and turn them in to the instructor. We will now take a short break.

UNITED STATES MARINE CORPS
Mountain Warfare Training Center
Bridgeport, California 93517-5001

                              MSVX
                              WML
                              WMO
                              09/31/00

LESSON PLAN

WATER PROCUREMENT

INTRODUCTION   (5 Min)

1. **GAIN ATTENTION**. More than three-quarters of the human body is composed of liquids. Heat, cold, stress, and exertion can cause a loss or expenditure of body fluids that must be replaced, if you are to function effectively in a survival situation.

2. **PURPOSE**. The purpose of this period of instruction is to introduce the student to the basics of water procurement. This will be accomplished by discussing locating, gathering, and disinfecting water. This lesson relates to all other lessons on survival here at MWTC.

3. **INTRODUCE LEARNING OBJECTIVES**

    a. TERMINAL LEARNING OBJECTIVE. In a summer mountainous environment and given water procurement materials, obtain potable water, in accordance with the references.

    b. ENABLING LEARNING OBJECTIVES

        (1) Without the aid of references, list in writing the types of incidental water, in accordance with the references.

        (2) Without the aid of references, list in writing the hazardous fluids to avoid substituting for potable water, in accordance with the references.

        (3) Without the aid of references, list in writing the methods for disinfecting water, in accordance with the references.

(4) Without the aid of references and given a military bottle of water purification tablets, state in writing its self-life, in accordance with the references.

(5) Without the aid of references, and given the water temperature and chemical concentration, state in writing the contact time, in accordance with the references.

(6) Without the aid of references, construct a solar still, in accordance with the references.

4. **METHOD / MEDIA**. The material in this lesson will be presented by lecture and demonstration. You will practice what you have learned during upcoming field exercises. Those of you with IRF's please fill them out at the end of this period of instruction.

5. **EVALUATION**. You will be tested later in the course by written and performance evaluations.

TRANSITION: Now that we know what is expected of us, let's talk about what determines water intake.

BODY (45 Min)

1. (5 Min) **WATER INTAKE**

   a. Thirst is not a strong enough sensation to determine how much water you need.

   b. The best plan is to drink, utilizing the OVER DRINK method. Drink plenty of water anytime it is available and particularly when eating.

   c. Dehydration is a major threat. A loss of only 5 % of your body fluids causes thirst, irritability, nausea, and weakness; a 10% loss causes dizziness, headache, inability to walk, and a tingling sensation in limbs; a 15% loss causes dim vision, painful urination, swollen tongue, deafness, and a feeling of numbness in the skin; also a loss of more than 15% body fluids could result in death.

   d. Your water requirements will be increased if:

   (1) You have a fever.

   (2) You are experiencing fear or anxiety.

   (3) You evaporate more body fluid than necessary. (i.e., not using the proper shelter to your advantage)

   (4) You have improper clothing.

   (5) You ration water.
   (6) You overwork.

TRANSITION: Now that we have a general idea of what determines water intake, let's move on to techniques of locating water.

2. (5 Min) **INCIDENTAL WATER**

   a. During movement, you may have to ration water until you reach a reliable water source. Incidental water may sometimes provide opportunities to acquire water. Although not a reliable or replenished source, it may serve to stretch your water supply or keep you going in an emergency. The following are sources for incidental water:

   (1) <u>Dew</u>. In areas with moderate to heavy dew, dew can be collected by tying rags or tuffs of fine grass around your ankles. While walking through dewy grass before sunrise, the rags or grass will saturate and can be rung out into a container. The rags or grass can be replaced and the process is repeated.

   (2) <u>Rainfall</u>. Rainwater collected directly in clean container or in plants that contain no harmful toxins is generally safe to drink without disinfecting. The survivor should always be prepared to collect rainfall at a moments notice. An inverted poncho works well to collect rainfall.

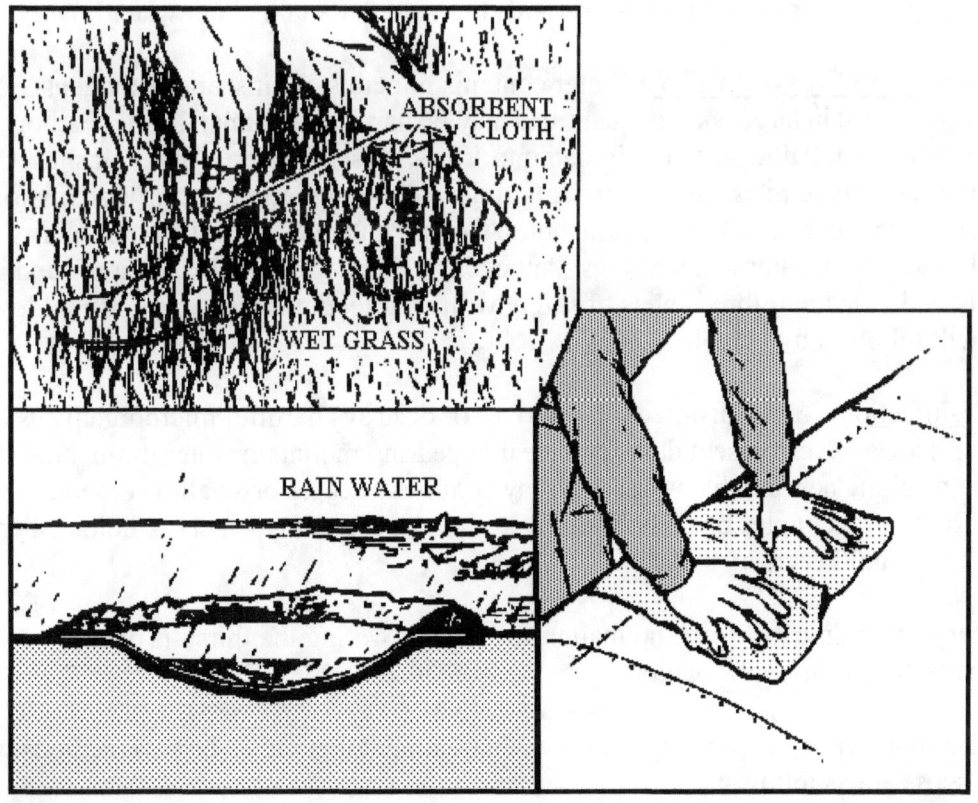

TRANSITION: Now that we know where to obtain water, let's discuss some hazards of substituting other fluids for potable water.

3. (5 Min) **HAZARDOUS FLUIDS**

a. Survivors have occasionally attempted to augment their water supply with other fluids, such as alcoholic beverages, urine, blood, or seawater. While it is true that each of these fluids has a high water content, the impurities they contain may require the body to expend more fluid to purify them. Some hazardous fluids are:

   (1) <u>Sea water</u>. Sea water in more than minimal quantities is actually toxic. The concentration of sodium and magnesium salts is so high that fluid must be drawn from the body to eliminate the salts and eventually the kidneys cease to function.

   (2) <u>Alcohol</u>. Alcohol dehydrates the body and clouds judgment. Super-cooled liquid, if ingested, can cause immediate frostbite of the throat, and potential death.

   (3) <u>Blood</u>. Blood, besides being salty, is a food. Drinking it will require the body to expend additional fluid to digest it.

   (4) <u>Urine</u>. Drinking urine is not only foolish, but also dangerous. Urine is nothing more than the body's waste. Drinking it only places this waste back into the body, which requires more fluid to process it again.

TRANSITION: Now that we know what fluids to avoid, let's talk about water quality.

4. (10 Min) **WATER QUALITY**. Water contains minerals, toxins, and pathogens. Some of these, consumed in large enough quantities may be harmful to human health. Pathogens are our primary concern. Pathogens are divided into Virus, Cysts, Bacteria, and Parasites. Certain pathogens are more resistant to chemicals and small enough to move through microscopic holes in equipment (i.e., T-shirt, parachute). Certain pathogens also have the ability to survive in extremely cold water temperatures. Pathogens generally do not live in snow and ice. Water quality is divided into three levels of safety with disinfection as the most desired level, then purified, followed by potable.

   a. <u>Disinfection</u>. Water disinfection removes or destroys harmful microorganisms. Giardia cysts are an ever-present danger in clear appearing mountain water throughout the world. By drinking non-potable water you may contract diseases or swallow organisms that could harm you. Examples of such diseases or organisms are: Dysentery, Cholera, Typhoid, Flukes, and Leeches.

   b. Remember, impure water, no matter how overpowering the thirst, is one of the worst hazards in a survival situation.

   c. The first step in disinfecting is to select a treatment method. The two methods we will discuss are as follows:

      (1) <u>Heat</u>. The Manual of Naval Preventive Medicine (P-5010) states that you must bring the water to a rolling boil before it is considered safe for human consumption. This is the most preferred method.

(a) Bringing water to the boiling point will kill 99.9% of all Giardia cysts. The Giardia cyst dies at 60°C and Cryptosporidium dies at 65°C. Water will boil at 14,000' at 86°C and at 10,000' at 90°C. With this in mind you should note that altitude does not make a difference unless you are extremely high.

(2) <u>Chemicals</u>. There are numerous types of chemicals that can disinfect water. Below are a few of the most common. In a survival situation, you will use whatever you have available.

(a) Iodine Tablets.

(b) Chlorine Bleach.

(c) Iodine Solution.

(d) Betadine Solution.

(e) Military water purification tablets. These tablets are standard issue for all DOD agencies. These tablets have a shelf-life of four years from the date of manufacture, unless opened. Once the seal is broken, they have a shelf-life of one year, not to exceed the initial expiration date of four years.

49703
Month / Year / Batch Number

(3) <u>Water Disinfection Techniques and Halogen Doses</u>

Iodination techniques Added to 1 liter or quart of water	Amount for 4 ppm	Amount for 8 ppm
Iodine tablets Tetraglcine hydroperiodide EDWGT Potable Aqua Globaline	½ tablet	1 tablet
2% iodine solution (tincture)	0.2 ml 5 gtts	0.4 ml 10 gtts
10% povidone-iodine solution*	0.35 ml 8 gtts	0.70 ml 16 gtts

Chlorination techniques	Amount for 5 ppm	Amount for 10 ppm
Household bleach 5% Sodium hypochlorie	ml 2 gtts	ml 4 gtts
AquaClear Sodium dichloroisocyanurate		1 tablet

AquaCure, AquaPure, Chlor-floc Chlorine plus flocculating agent		8 ppm 1 tablet

*Providone-iodine solutions release free iodine in levels adequate for disinfection, but scant data is available.

Measure with dropper (1 drop=0.05 ml) or tuberculin syringe

Ppm-part per million   gtts-drops   ml-milliliter

Concentration of Halogen	Contact time in minutes at various water temperatures		
	5°C / 40°F	15°C / 60°F	30°C / 85°F
2 ppm	240	180	60
4 ppm	180	50	45
8 ppm	60	30	15

NOTE: These contact times have been extended from the usual recommendations to account for recent data that prolonged contact time is needed in very cold water to kill *Giardia* cysts.

NOTE: Chemicals may not destroy Cryptosporidium.

   d. <u>Purification</u>. Water purification is the removal of organic and inorganic chemicals and particulate matter, including radioactive particles. While purification can eliminate offensive color, taste, and odor, it may not remove or kill microorganisms.

   (1) <u>Filtration</u>. Filtration purifying is a process by which commercial manufacturers build water filters. The water filter is a three tier system. The first layer, or grass layer, removes the larger impurities. The second layer, or sand layer, removes the smaller impurities. The final layer, or charcoal layer (not the ash but charcoal from a fire), bonds and holds the toxins. All layers are placed on some type of straining device and the charcoal layer should be at least 5-6 inches thick. Layers should be changed frequently and straining material should be boiled. Remember, this is not a disinfecting method, cysts can possibly move through this system.

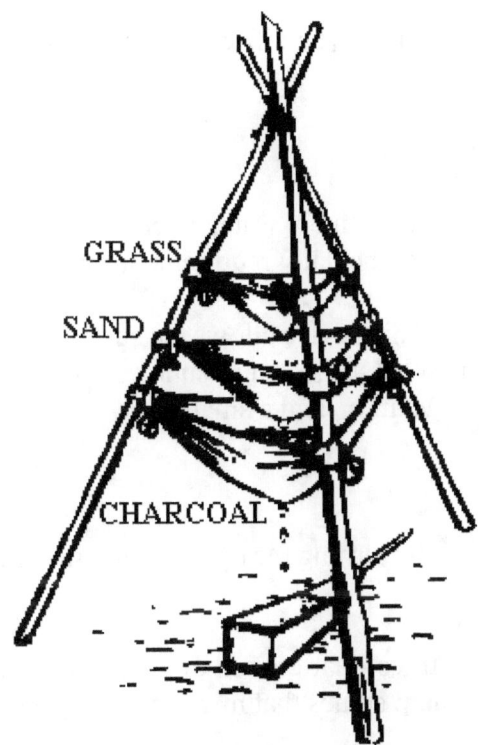

WATER FILTER

- (2) <u>Commercial Water Filters</u>. Commercial water filters are generally available in most retail stores and may be with you. Understanding what the filter can do is the first step in safeguarding against future illnesses.

    (a) A filter that has a .3 micron opening or larger will not stop Cryptosporidium.

    (b) A filter system that does not release a chemical (i.e., iodine) may not kill all pathogens.

    (c) A filter that has been overused may be clogged. Usage may result in excessive pumping pressure that can move harmful pathogens through the opening.

e. <u>Potable</u>. Potable indicates only that a water source, on average over a period of time, contains a "minimal microbial hazard," so the statistical likelihood of illness is acceptable.

   (1) <u>Sedimentation</u>. Sedimentation is the separation of suspended particles large enough to settle rapidly by gravity. The time required depends on the size of the particle. Generally, 1 hour is adequate if the water is allowed to sit without agitation. After sediment has formed on the bottom of the container, the clear water is decanted or filtered from the top. Microorganisms, especially cysts, eventually settle, but this takes longer and the organisms are easily disturbed during pouring or filtering. Sedimentation <u>should not</u> be considered a means of disinfection and should be used only as a last resort or in an extreme tactical situation.

TRANSITION: Since locating water may be difficult, let's look at two stills that may supplement your existing water supply.

5. (10 Min) **SOLAR STILLS**

   a. Solar stills are designed to supplement water reserves. Contrary to belief, they will not provide enough water to meet the daily requirement for water.

   b. Above-Ground Solar Still. This device allows the survivor to make water from vegetation. To make the aboveground solar still, locate a sunny slope on which to place the still, a clear plastic bag, green leafy vegetation, and a small rock.

   (1) Construction

   (a) Fill the bag with air by turning the opening into the breeze or by "scooping" air into the bag.

   (b) Fill the bag half to three-quarters full of green leafy vegetation. Be sure to remove all hard sticks or sharp spines that might puncture the bag.

   > **CAUTION**
   > Do not use poisonous vegetation. It will provide poisonous liquid.

   (c) Place a small rock or similar item in the bag.

   (d) Close the bag and tie the mouth securely as close to the end of the bag as possible to keep the maximum amount of air space. If you have a small piece of tubing, small straw, or hollow reed, insert one end in the mouth of the bag before tying it securely. Tie off or plug the tubing so that air will not escape. This tubing will allow you to drain out condensed water without untying the bag.

   (e) Place the bag, mouth downhill, on a slope in full sunlight. Position the mouth of the bag slightly higher than the low point in the bag.

   (f) Settle the bag in place so that the rock works itself into the low point in the bag.

   (g) To get the condensed water from the still, loosen the tie and tip the bag so that the collected water will drain out. Retie the mouth and reposition the still to allow further condensation.

   (h) Change vegetation in the bag after extracting most of the water from it.

   (i) Using 1 gallon zip-loc bag instead of trash bags is a more efficient means of construction.

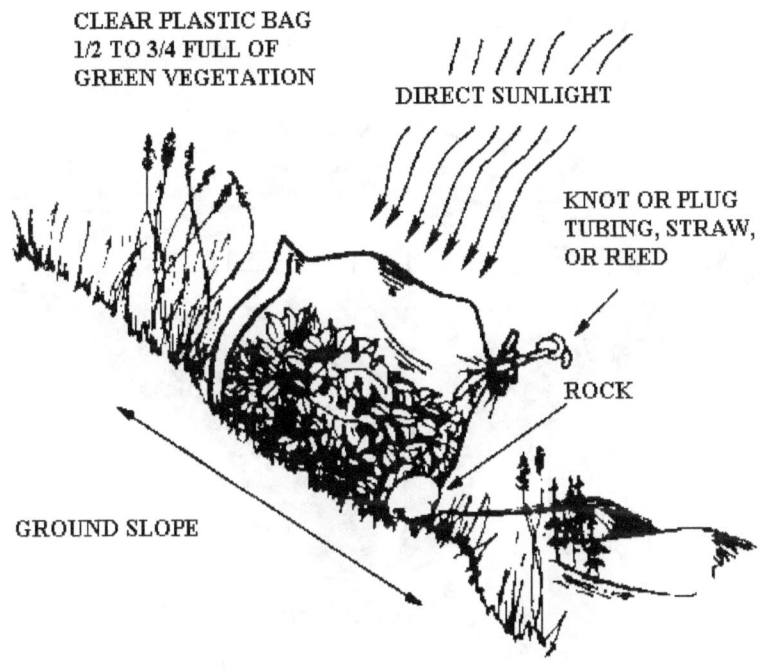

ABOVE GROUND SOLAR STILL

c. <u>Below-Ground Solar Still</u>. Materials consist of a digging stick, clear plastic sheet, container, rock, and a drinking tube.

(1) Construction

    (a) Select a site where you believe the soil will contain moisture (such as a dry streambed or a low spot where rainwater has collected). The soil should be easy to dig, and will be exposed to sunlight.

    (b) Dig a bowl-shaped hole about 1 meter across and 24 inches deep.

    (c) Dig a sump in the center of the hole. The sump depth and perimeter will depend on the size of the container you have to place in it. The bottom of the sump should allow the container to stand upright.

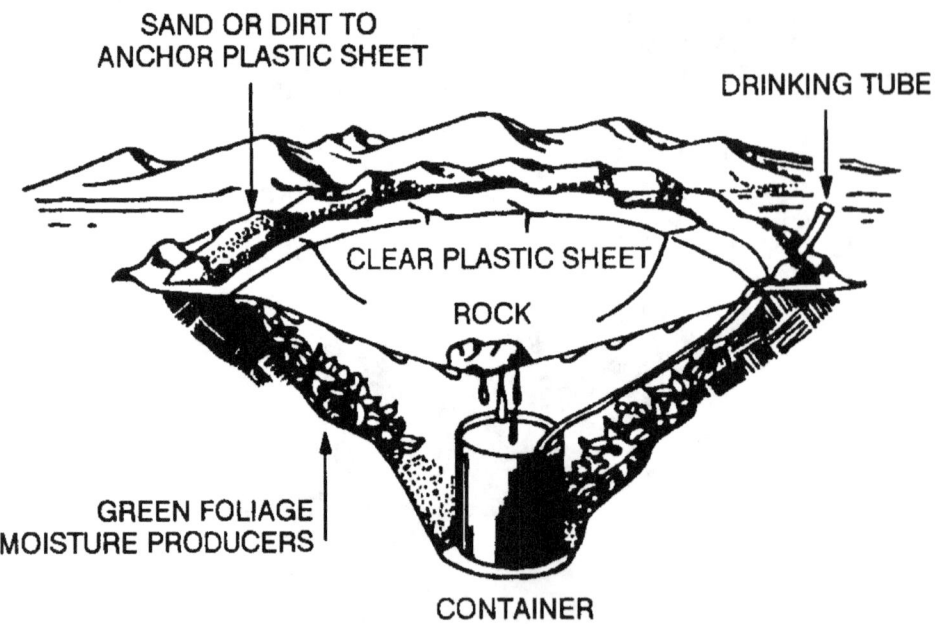

BELOW GROUND SOLAR STILL

(d) Anchor the tubing to the container's bottom by forming a loose overhand knot in the tubing. Extend the unanchored end of the tubing up, over, and beyond the lip of the hole.

(e) Place the plastic sheet over the hole, covering its edges with soil to hold in place. Place a rock in the center of the plastic sheet.

(f) Lower the plastic sheet into the hole until it is about 18 inches below ground level. Make sure the cone's apex is directly over the container. Ensure the plastic does not touch the sides of the hole because the earth will absorb the moisture.

(g) Put more soil on the edges of the plastic to hold it securely and prevent the loss of moisture.

(h) Plug the tube when not in use so that moisture will not evaporate.

(i) Plants can be placed in the hole as a moisture source. If so, dig out additional soil from the sides.

(j) If polluted water is the only moisture source, dig a small trough outside the hole about 10 inches from the still's lip. Dig the trough about 10 inches deep and 3 inches wide. Pour the polluted water in the trough. Ensure you do not spill any polluted water around the rim of the hole where the plastic touches the soil. The trough holds the polluted water and the soil filters it as the still draws it. This process works well when the only water source is salt water.

   d. Three stills will be needed to meet the individual daily water intake needs.

PRACTICE (CONC)

   a. The students will practice these skills in a survival situation.

PROVIDE HELP (CONC)

   a. The instructors will provide help as necessary.

OPPORTUNITY FOR QUESTIONS (3 Min)

1. **QUESTIONS FROM THE CLASS**

2. **QUESTIONS TO THE CLASS**

   Q. What factors increase your need for water?

   A. You have a fever.
   You are in fear of something.
   You evaporate more body fluid than necessary by not using available shelter to your advantage.
   You overwork.

   Q. What are the four hazardous fluids to replace water with?

   A. Sea water.
   Alcohol.
   Blood.
   Urine.

SUMMARY (2 Min)

   a. During this period of instruction we have covered what determines water intake, what hazardous fluids to avoid substituting for potable water, how to procure, and how to purify water.

   b. Those of you with IRF's please fill them out at this time and turn them in to the instructor. We will now take a short break.

UNITED STATES MARINE CORPS
Mountain Warfare Training Center
Bridgeport, California 93517-5001

MSVX
WMO
WML
09/31/00

LESSON PLAN

EXPEDIENT SNOWSHOES

INTRODUCTION (5 Min)

1. **GAIN ATTENTION**. Survival in a snow-covered environment will depend on the ability to move. Without over-the-snow mobility, movement can be slow, tiring, and even dangerous. The survivor must have a working knowledge in the construction of expedient snowshoes.

2. **OVERVIEW**. The purpose of this period of instruction is to introduce the student to expedient snowshoes; including their nomenclature and construction. This lesson relates to Field Expedient Tools, Weapons, and Equipment.

3. **TERMINAL LEARNING OBJECTIVE**. In a cold weather mountainous environment, execute snowshoe movement, in accordance with the references.

4. **ENABLING LEARNING OBJECTIVES**

    a. Without the aid of references, state in writing the purpose of expedient snowshoes, in accordance with the references.

    b. Without the aid of references, list in writing the characteristics for good expedient snowshoes, in accordance with the references.

    c. Without the aid of references, construct a pair of expedient snowshoes, in accordance with the references.

5. **METHOD / MEDIA**. The material in this lesson will be presented by lecture and demonstration. You will practice what you have learned during upcoming field training exercises.

6. **EVALUATION**. You will be tested later in the course by performance evaluation.

TRANSITION: Let's discuss the advantages and disadvantages of expedient snowshoes.

BODY (35 Min)

1. (5 Min) **ADVANTAGES AND DISADVANTAGES OF EXPEDIENT SNOWSHOES**

    a. Advantages of Expedient Snowshoes

    (1) Mobility. The purpose of expedient snowshoes is to give a means of over the snow mobility by providing floatation.

    (2) Maintenance. Limited maintenance is required to keep them serviceable.

    (3) Heavy Loads. Carrying and pulling heavy loads on gentle terrain is relatively easy.

    (4) Confined Areas. Movement in confined areas and around equipment is relatively easy.

    b. Disadvantages of Expedient Snowshoes

    (1) Materials. Constructing a pair of expedient snowshoes will consume a lot of material (i.e., it takes approximately 70 feet of 550 cord to build one shoe).

    (2) Terrain. Movement over moderate to steep slopes is extremely difficult.

    (3) Vegetation. Movement through thickly forested areas or terrain with branches protruding through the snow is difficult.

    (4) Natural Material. The green staves used in construction will shrink during the drying process. This shrinkage will require all lashing to be retied.

TRANSITION: Now let's examine the nomenclature of the expedient snowshoes.

2. (5 Min) **NOMENCLATURE OF EXPEDIENT SNOWSHOES**

    a. Tip. The front portion of the snowshoe frame.

    b. Tail. The rear portion of the snowshoe frame.

    c. Crossbars. Three crossbars are lashed to the frame for reinforcement.

d. <u>Webbing</u>. Webbing is designed to provided floatation of the shoe and release snow when the shoe is lifted. It is generally made from some type of cord (i.e., 550, comm wire).

e. <u>Window</u>. This is the opening in the snowshoe, which allows the toe of the boot to pivot through. Without the window, the boot will continue to force the tip into the snow during forward movement.

f. <u>Frame</u>. Made from resilient green saplings (i.e., willow, aspen).

g. <u>Binding</u>. The binding is used to attach the boot to the shoe. It is made from an 8 to 10 foot piece of cord.

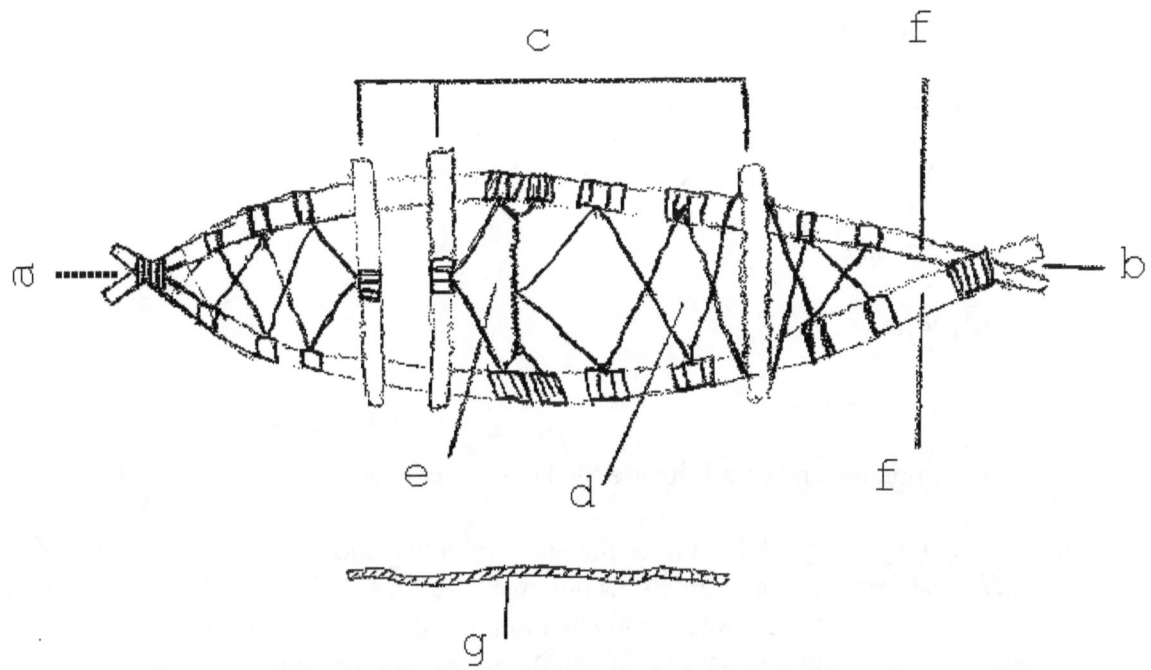

NOMENCLATURE OF EXPEDIENT SNOWSHOE

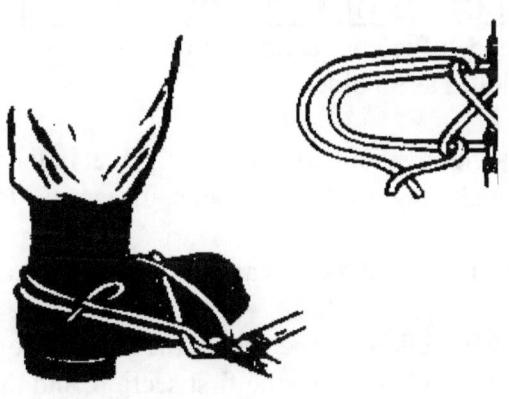

BINDING

TRANSITION: Now that we know the nomenclature, let's discuss the characteristics and materials used on construction.

3. (5 Min) **CHARACTERISTICS OF EXPEDIENT SNOWSHOES**

   a. <u>Stability</u>. Stability is the most important characteristic due to the density of the snow. Snowshoes must be able to undergo great strain and pressure. Careful consideration must be given to the selection of materials used in construction.

      (1) Materials used can vary.

         (a) Resilient green saplings.

         (b) Aluminum aircraft skin and ribbed framing.

         (c) Cordage.

         (d) Wire.

         (e) Cargo strapping.

         (f) Initiative and creativity are the key to success.

   b. <u>Provide Floatation</u>. Without over-the-snow mobility, movement becomes impossible and/or dangerous. Clothing will become wet and "post holing" will consume too much energy, which is vital in a survival situation. Increasing the surface area upon which your body rests will facilitate more efficient movement on the snow.

TRANSITION: Now that we understand the principles for expedient snowshoes, let's discuss the construction.

4. (5 Min) **CONSTRUCTION OF EXPEDIENT SNOWSHOES**. These snowshoes will work if staves are thick and sufficient quantity of cordage is available. Use these general construction techniques as a guide.

   a. Select three straight, resilient, green staves: 5 feet in length and 1 to 1.5 inches in diameter. Cut one of these staves into three sections, measuring 15 inches for each section.

   b. Join the two 5 feet staves at the tips and tails using a shear lash. This is the snowshoe frame.

   c. Attach the 15 inch sections to the frame: the first section 12-14 inches from the tip, the second section 4-6 inches below the first section, and the last section 15 inches below the second section. All sections are secured to the frame using a square lash. These sections are the crossbars.

d. Affix the latticework to the frame to form the webbing, working from the first crossbar towards the tip. Attach a second latticework from the second crossbar towards the tail.

NOTE: If cord or wire is limited, space out the latticework. Branches or bough can be interwoven to increase floatation.

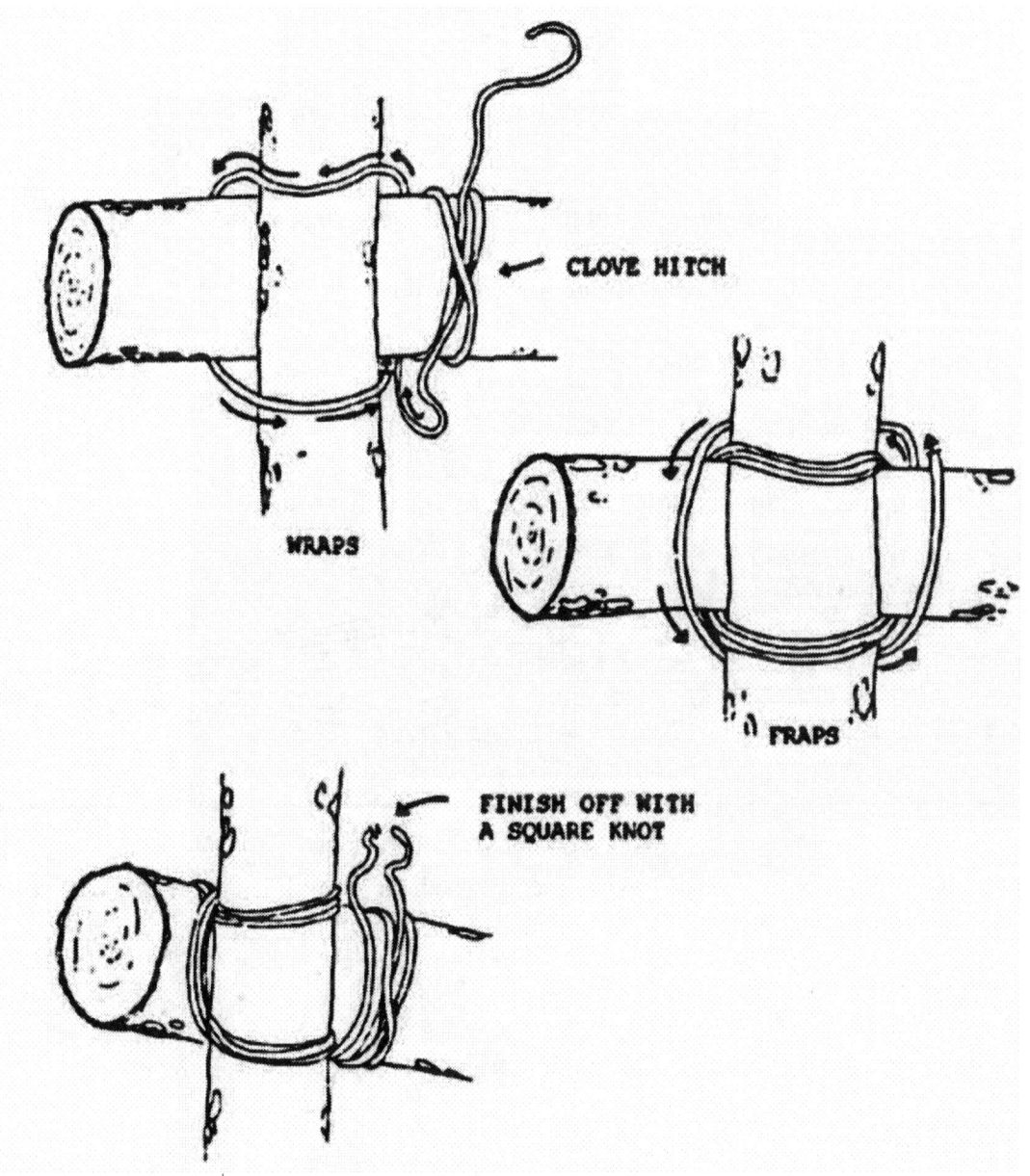

SQUARE LASH

e. Attach the boot to the snowshoe using the cord binding system.

NOTE: All pigtails are secured together by a square knot with two overhand knots.

44-5

TRANSITION: Another type of expedient snowshoe to build is the Canadian Emergency Snow Shoe.

5. (5 Min) **CANADIAN EMERGENCY SNOWSHOES**. The Canadian emergency snowshoes are an excellent method if saplings are available and cordage is limited.

   a. Select 5 poles, 6 feet long (individual's height), ¾ inch (thumb size) at the base, ¼ inch (little finger size) at the tip for one shoe.

   b. Cut 5 sticks approximately 10 inches long and ¾ inch wide for each shoe and tie them in the following steps.

   c. Lash one stick to the snowshoe's tail (across the heavy end of the 5 poles).

   d. Lash 2 sticks across the poles where the heel will rest.

   e. Lash 2 sticks across the poles where the toe will rest.

   f. Secure the tips together.

   g. Attach the boot to the snowshoe using the cord binding system.

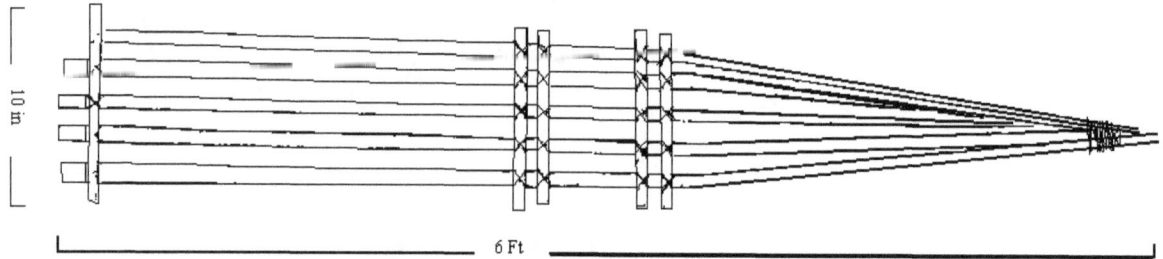

TRANSITION: In a snow covered environment, expedient snowshoes will become a necessary tool. Are there any questions?

PRACTICE (CONC)

   a. Students will practice what was taught during upcoming field training exercises.

PROVIDE HELP (CONC)

   a. The instructors will assist the students when necessary.

OPPORTUNITY FOR QUESTIONS (3 Min)

1. QUESTIONS FROM THE CLASS

2. QUESTIONS TO THE CLASS

    Q. What is the purpose of expedient snowshoes?

    A. The purpose of expedient snowshoes is to give a means of over-the-snow mobility by providing flotation.

    Q. What are the two characteristics for good expedient snowshoes?

    A. (1) Stability
       (2) Provide flotation

SUMMARY (2 Min)

   a. During this period of instruction, we have discussed expedient snowshoes, including the advantages and disadvantages, nomenclature, characteristics, and techniques.

   b. Those or you with IRF's please fill them out at this time and turn them in to the instructor We will now take a short break.

UNITED STATES MARINE CORPS
Mountain Warfare Training Center
Bridgeport, California 93517-5001

WML
SS
091702

LESSON PLAN

CROSSING OBSTACLES IN WINTER

INTRODUCTION (5 Min)

1. **GAIN ATTENTION**. Movement over mountainous terrain in a winter mountainous environment may seem difficult and even impossible at times. The trained military mountaineer can use these techniques to negotiate obstacles that would otherwise stop you in your tracks.

2. **OVERVIEW.** The purpose of this period of instruction is to introduce the student with the basics of rope work as it applies to negotiating obstacles in a deep snow pack on skis. A summer qualified mountain leader or assault climber will more easily understand the techniques taught here, but all can understand the basics. This lesson relates to all installations and climbing performed here at MWTC.

INSTRUCTOR NOTE: Have students read learning objectives.

3. **INTRODUCE LEARNING OBJECTIVES**

   a. TERMINAL LEARNING OBJECTIVE. In a summer mountainous environment, utilize rope techniques to negotiate terrain, in accordance with the references.

   b. ENABLING LEARNING OBJECTIVES (WMLC) and (SS)

   (1) In a winter mountainous environment and given a sling rope and a designated amount of time, tie the four mountaineering knots required for winter operations, in accordance with the references.

   (2) In a winter mountainous environment and given the proper equipment, set up and utilize a body belay, in accordance with the references.

4. **METHOD/MEDIA**. The material in this lesson will be presented by lecture and demonstration. You will practice what you have learned immediately following this class and during your tactical movements. Those of you with IRF's please fill them out at the conclusion of this lesson.

5. **EVALUATION**. You will be tested later in the course by performance evaluations on this period of instruction.

   TRANSITION: Are there any questions over the purpose, learning objectives, how the class will be taught, or how you will be evaluated? Now let's start with some common rope terms.

BODY (80 Min)

1. (5 Min) **TERMS USED IN ROPE WORK**

    a. <u>Bight</u>. A simple bend in the rope in which the rope does not cross itself.

    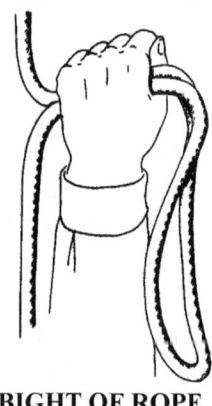

    BIGHT OF ROPE

    b. <u>Loop</u>. A simple bend in the rope in which the rope does cross itself.

    LOOP

c. <u>Half Hitch</u>. A loop that runs around an object in such a manner as to bind on itself.

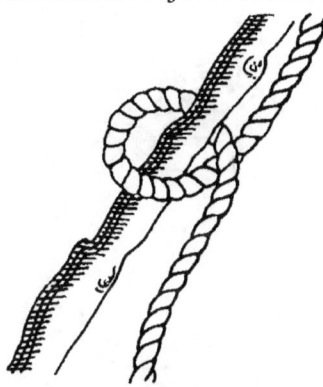

HALF HITCH

d. <u>Standing End</u>. The part of the rope that is anchored and cannot be used, also called the static end.

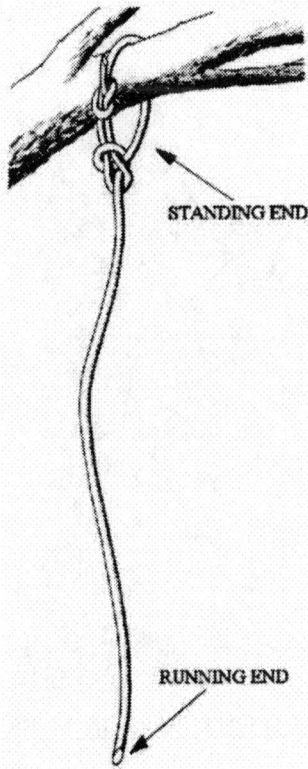

e. <u>Running End</u>. The free end of the rope which can be used.

f. <u>Lay</u>. The same as the twist of the rope. (Applies only to hawser laid ropes, such as manila.)

g. <u>Pigtail</u>. The short length left at the end of a rope after tying a knot or coiling a rope. It may or may not be tied off with a secondary knot, depending on the circumstance.

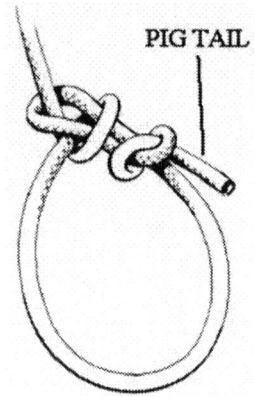

PIG TAIL

h. <u>Stacking (or Flaking)</u>. Taking off one wrap at a time from a coil, and letting it fall naturally to the ground.

FLAKED OUT ROPE ON THE GROUND

i. <u>Dressing the knot</u>. This involves the orientation of the entire knot parts so that they are properly aligned, straightened, or bundled, and so the parts of the knot look like the accompanying pictures. Neglecting this can result in an additional 50% reduction in knot strength.

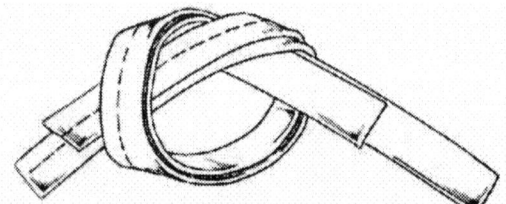 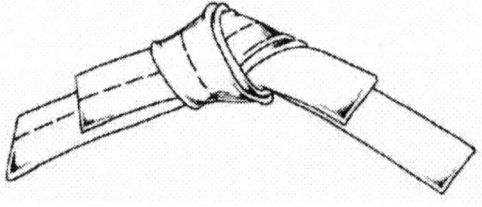

NON-DRESSED DOWN WATER TAPE KNOT   **DRESSED DOWN WATER TAPE KNOT**

j. <u>Setting the knot</u>. This involves tightening all parts of the knot so that all of the rope parts bind upon other parts of the knot so as to render it operational. A loosely tied knot can easily deform under strain and change character.

TRANSITION: Are there any questions over the terminology? Since the rope is the climbers' lifeline it deserves a great deal of care and respect. Next we will talk about the considerations for the care of the rope and how to inspect it.

2. (10 Min) **CONSIDERATIONS FOR THE CARE OF ROPE**

   a. The rope should not be stepped on or dragged on the ground unnecessarily. Small particles of dirt will get into and through the sheath causing unnecessary wear to the rope within.

   b. The rope should never come in contact with sharp edges of any type. Nylon rope is easily cut, particularly when under tension. If a rope must be used around an edge, which could cut it, then that edge must be padded or buffed using fire hose if available, or several small sticks.

   c. Never leave a rope knotted or tightly stretched longer than necessary.

   d. The rope should be inspected prior to each use for frayed or cut spots, mildew, rot or defects in construction.

   e. The rope should not be marked with paints or allowed to come in contact with oils or petroleum products for these products will weaken it.

   f. The rope should never be subjected to high heat or flame as this can significantly weaken it.

   g. When not in use, ropes should be coiled and hung on wooden pegs rather than on nails or any other metal object. They should be stored in a cool place out of the direct rays of the sun.

   h. When in areas of loose rock, the rope must be inspected frequently for cuts and abrasions.

3. (10 Min) **INSPECTION OF ROPE**

   a. All ropes have to be inspected before, during, and after all operations. Kernmantle rope is harder to inspect than a laid rope. i.e. green line. The winter mountain leader must know what to look and feel for when inspecting a rope. Any of the below listed deficiencies can warrant the retirement of a rope.

   (1) <u>Excessive Fraying</u>. Indicates broken sheath bundles or PIC breakage.

DAMAGED AND EXPOSED ROPE

   (2) <u>Exposed Core Fibers</u>. Indicates severe sheath damage. (When you can see the inner core fibers)

   (3) <u>Glossy Marks</u>. Signify heat fusion damage, also called a booger.

   (4) <u>Uniformity of Diameter / Size</u>. May indicate core damage, noted by an obvious depression (hour glass) or exposure of white core fibers protruding from the sheath (puff).

   (5) <u>Discoloration</u>. A drastic change from the ropes original color may indicate chemical change or damage.

   (6) <u>Stiffness or Soft Spots</u>. Could signify core damage.

NOTE: Dynamic ropes measuring between 10mm and 12mm are marked at each end of its pigtails with a number "1" indicating that the rope is UIAA approved for single rope lead climbing. Dynamic ropes measuring between 8mm and 9mm are marked at each end of its pigtails with a "1/2" number indicating that two of these ropes are required to conduct a lead climb.

TRANSITION: Now let's discuss the various ways of coiling a rope.

4. (10 Min) **COILING A ROPE**. There are two types of rope coils frequently used at MWTC. The Mountain Coil and the Butterfly Coil.

   a. <u>Mountain Coil</u>. This coil is useful for carrying the rope over a pack or over a climber's shoulder and neck. It can be used for short time storage. The mountain coil can be tied in the following manner:

   (1) Sit down with your leg bent at a 90-degree angle, heel on the deck. Starting at one end, the rope is looped around the leg in a clockwise fashion, going over the knee and under the boot sole until the entire rope is coiled.

(2) If coiling a 150-foot rope, use only one leg and offset the other, when coiling a 300-foot rope, use two legs and keep them together.

(3) With the starting end of the rope, form a 12-inch bight on the top of the coils.

(4) Uncoil the last loop and along the top of the coils, wrap 4-6 times towards the closed end of the bight.

(5) The end of the rope being wrapped is then placed through the closed end of the bight.

(6) The running end of the bight is then pulled snuggly to secure the coil.

(7) To prevent the coil from unraveling, the two pigtails are tied together with a square knot.

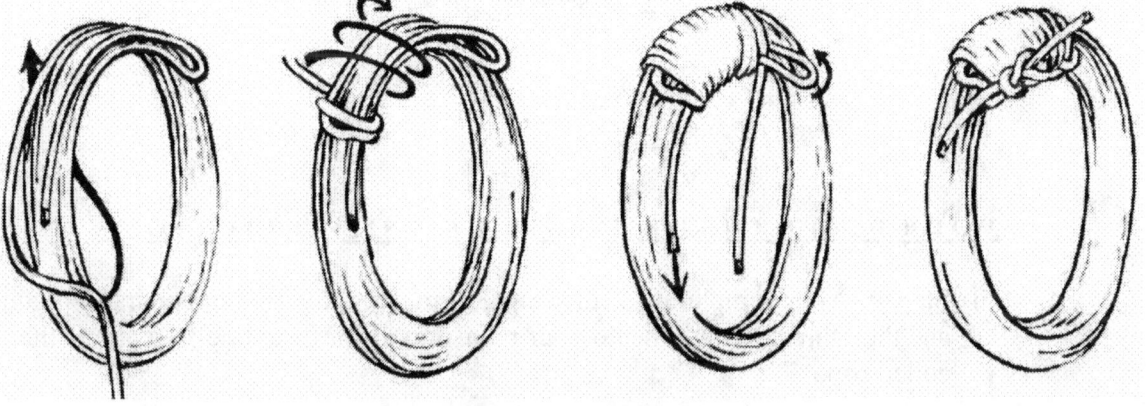

MOUNTAIN COIL

b. <u>Butterfly Coil.</u> This method is used for carrying a rope when the individual needs to have maximum use of his upper body, (i.e. while climbing), without the encumbrance of a large rope coil hanging across his chest.

(1) Coiling the Butterfly Coil

(a) <u>Step 1</u>: Find the middle of the rope, then form a three foot bight laying both ropes in the upraised palm at the two foot point.

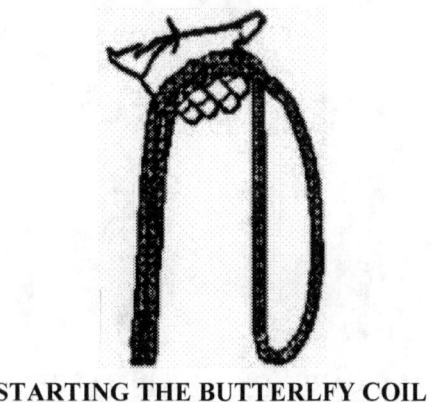

STARTING THE BUTTERLFY COIL

45-7

(b) Step 2: Form another two-foot bight with the running end. Place the rope at the two-foot bight along side on top of the original bight ensuring the running end is on the same side as the original bight.

(c) Step 3: Continue making two foot bights, laying them alternately into your palm until there are only six to eight feet remaining. At that point, begin wrapping the two pigtails horizontally four to six times at the mid way point of the ropes in a bight from bottom to top.

TWO FOOT BIGHTS ON BOTHS SIDES WITH 6 TO 8 FEET REMAINING

(d) Step 4: After completing your wraps, form a bight with the remaining pigtail and then thread it underneath your palm and upwards to one-foot above the coiled rope.

(e) Step 5: With the remaining pigtail, thread it through the one-foot bight in step four.

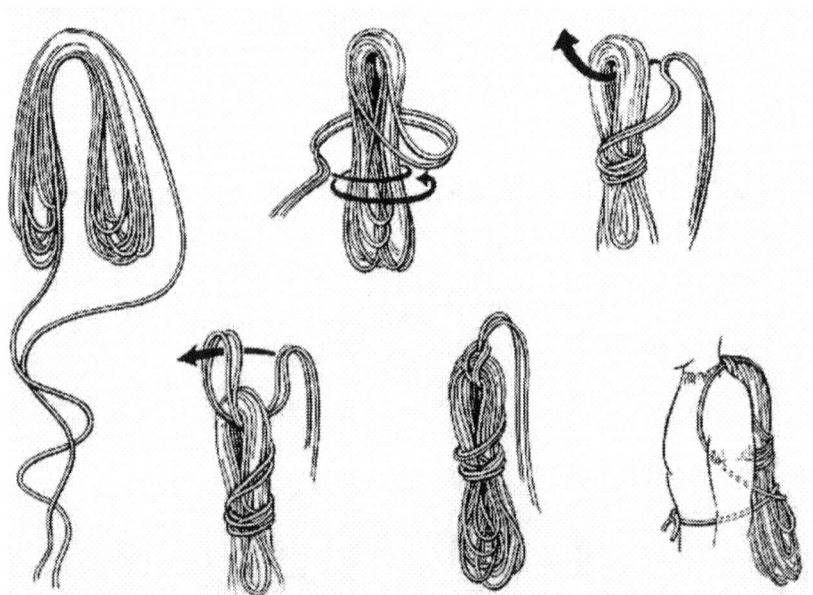

TYING AND CARRYING THE BUTTERFLY COIL

(2) Carrying the Butterfly coil. Separate the running ends, placing the coil in the center of the back of the carrier, then fun the two ends over his shoulders so as to form shoulder straps. The running ends are then brought under the arms, crossed in the back over the coil, brought around the body of the carrier and tied off with a square knot at his stomach.

TRANSITION: Are there any questions over coiling and securing a rope? Now we will discuss the required mountaineering knots, their uses, and how to tie them.

5. (10 Min) **MOUNTAINEERING KNOTS**.

   a. **Class I** – *End of the Rope Knots*

      (1) Square Knot. Used to tie ends of two ropes of equal diameter together. It should be secured by overhand knots on both sides of the square knot.

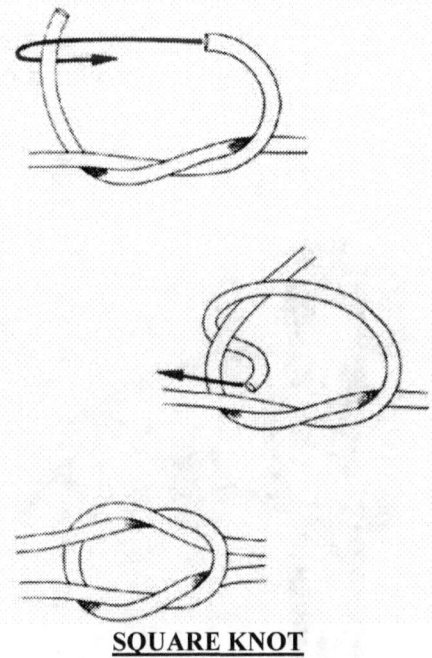

SQUARE KNOT

   b. **Class II** - *Anchor Knots*

      (1) Bowline. Used to tie a fixed loop in the end of a rope. This knot is always tied with the pigtail on the inside and secured with an overhand knot.

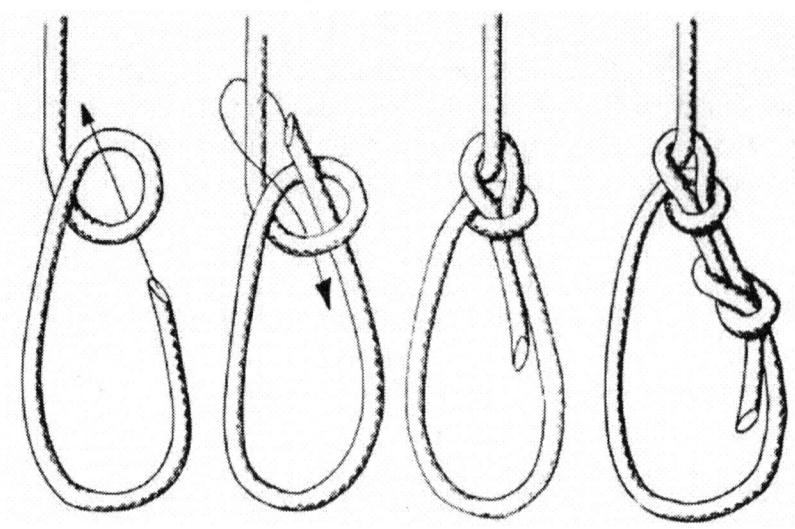

BOWLINE

(2) <u>Round Turn with Two Half Hitches or a Bowline</u>. A loop that runs around an object in such a manner as to provide 360 degree contact and may be used to distribute the load over a small diameter anchor. It will be secured with two half hitches or a bowline.

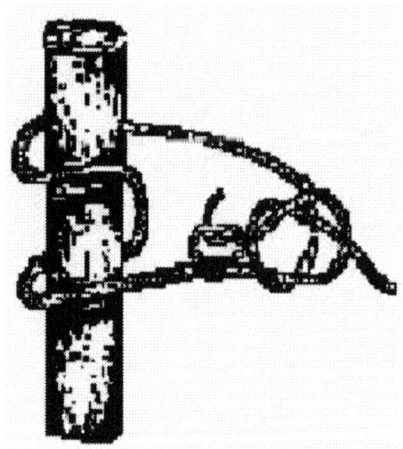

ROUND TURN AND A BOWLINE

(3) <u>Clove Hitch</u>. This knot is an adjustable hitch. It could be considered a middle-of-the-rope anchor knot at the end-of-the-rope when used in conjunction with a bowline or round turn and two half hitches.

    (a) Around-the-object Clove Hitch.

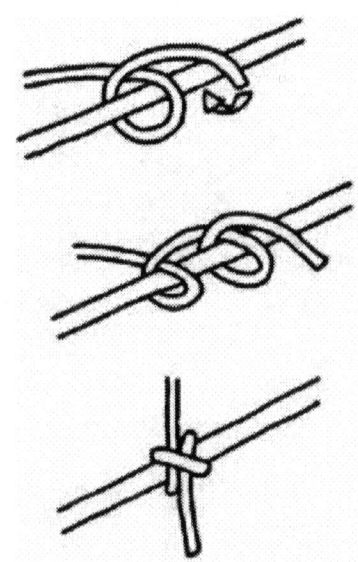

AROUND THE OBJECT CLOVE HITCH

(b) Over-the-object Clove Hitch.

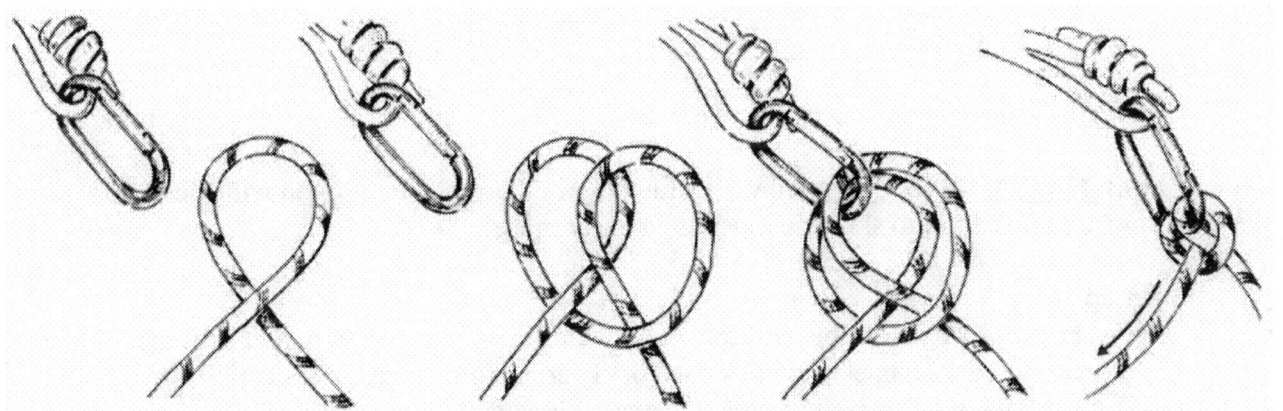

OVER THE OBJECT CLOVE HITCH

c. **Class III -** *Middle-of-the-rope*

   (1) <u>Figure-of-Eight-Loop</u>. This is a strong knot that can be readily untied after being under load.

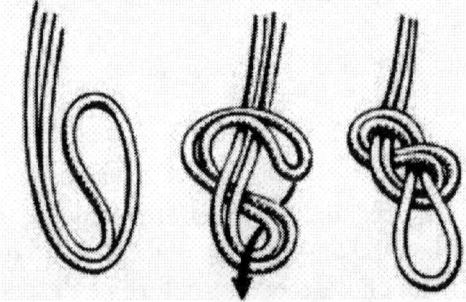

FIGURE 8 LOOP

(1) <u>Overhand Knot</u>. Can be used to secure primary knots.

45-11

OVERHAND

    d. (5 Min) <u>KNOT TESTING TIME LIMIT</u>. The following times must be met to pass the knot tying portion of this course.

KNOT	**WML, SSkier TIME LIMIT**
Square knot	30 Seconds
Round Turn and a Bowline	30 Seconds
Clove Hitch (around the object)	30 Seconds
Figure 8 Loop	30 Seconds

<u>TRANSITION</u>: Are there any questions over the knots? Next we will talk about the hasty rappel.

6. (5 Min) **<u>HASTY RAPPEL.</u>** Often while traveling in the backcountry you will encounter slopes that are impossible to ski. You must use a hasty rappel.

    a. <u>Set up</u>
        (1) The hasty rappel can be used to descend dangerous slopes that would be impossible to ski. The hasty rappel cannot be used on a vertical cliff face, only steeply sloping hills. Sleds and possibly packs would have to be lowered.

        (2) Find the half way point of the rope and place that against the backside of a sturdy (6" diameter) tree.

        (3) Bring each end of the rope around. Tie the ends together.

        (4) Backstack the rope with the knotted ends on top.

        (5) Throw rope down slope.

        (6) If a longer rappel is required and the ropes are available, two ropes may be tied together with a square knot. Then the knot will be placed just to the side of the tree. This is done to ensure a rappeller can check the knot before he descends.

    b. <u>Conduct</u>. A hasty rappel is conducted in the following manner:

(1) Sleeves will be rolled down and gloves will be put on.

(2) Face slightly sideways.

(3) Place the rappel rope across your back, grasping it with both hands, palms forward, and arms extended.

(4) The hand nearest the anchor is the guide hand. The hand farthest from the anchor is the brake hand.

(5) Lean out at a moderate angle to the slope.

(6) Descend down the hill facing half sideways, taking small steps and continually looking downhill while leading with the brake hand.

(7) Feet should not cross and the downhill foot should lead at all times.

c. <u>Braking</u>. The steps for braking during a hasty rappel are as follows:

(1) Bring the lower (brake) hand across the front of the chest to brake.

(2) At the same time, turn to face up toward the anchor point.

HASTY RAPPEL

7. (2 MIN) **NARROW DEPRESSIONS**. These include such things as creek beds, fallen logs, or similar obstacles. The preferable technique is to cross the obstacle on a snow bridge. One option is to side step into the depression, and then side step out. If you point your skis down and ski directly into it you will most likely fall. If each man in a company falls at this obstacle then it will take too long to cross it.

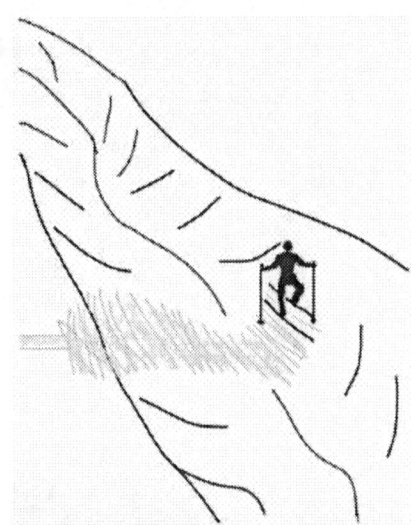

SIDE STEPPING INTO AND OUT OF AN OBSTACLE

8. (10 MIN) **BELAYING.** A belay is a system that safeguards an individual from falling. It consists of the anchor, the belayer, the rope and the person who is being belayed. Often a skier will venture onto terrain that would be dangerous if he fell. Avalanche prone slopes and water covered with ice or snow are the most common terrain requiring a belay.

   a. <u>Body belay</u>. A summer qualified mountain leader or assault climber should rig this.
      (1) Tie rope to a tree (or other sturdy anchor such as a large rock or a vehicle) with a round turn and a bowline.

      (2) Tie other end of rope to person being belayed with a round the chest bowline.

      (3) The belayer will stack the rope near his belay stance.

      (4) Belayer will run the rope from the stack under his arms, across his back to the person being belayed.

      (5) Belayer will assume a strong belay stance by positioning himself behind a tree, rock or mound. If none exist he will dig a bucket seat in the snow to sit in.

      (6) Gradually feed rope out to the person needing the belay. To brake, hold rope tightly in brake hand and place it across your chest and lean back against the pull of the rope. This method may also be used to lower equipment down a steep slope.

**BODY BELAY;
WITH THE LEFT HAND AS THE GUIDE HAND
AND THE RIGHT AS THE BRAKE HAND**

   b. <u>Expedient Belay</u>: Nearly the same as a body belay except the tree will provide the braking power in case of a fall. A tree with rough bark is better than a smooth tree.

(1) Tie rope to a tree with a round turn and a bowline.

(2) Tie other end of rope to person being belayed with a round the chest bowline.

(3) The belayer will stack the rope near the tree.

(4) The belayer will pass the rope around the backside of the tree.

(5) Gradually feed rope out to the person needing the belay. To brake, hold rope tightly against the tree.

(6) This method may also be used to lower equipment down a steep slope.

EXPEDIENT TREE BELAY

9. (3 MIN) **WALKING AND LOWERING SLEDS.** At times it is often easier, faster and safer to simply walk down steep slopes instead of attempting to ski them. If it is too awkward or dangerous to haul the sleds, then they must be lowered. To lower set up as in a body belay or expedient tree belay and tie other end to sled with round turn and a bowline. Gently lower to the bottom where other personnel will untie the rope. Pull it back to the top and repeat.

TRANSITION: Since trees may not always be available to set up belays and rappels we will discus artificial anchors.

10. (5 MIN) **DEAD MAN ANCHOR**. Any long sturdy item may be used as a dead man anchor in snow. A set of skis, several sets of ski poles, or a large log will all work.

   a. Dig a T-trench perpendicular to the intended load and about a foot deep and as long as the object being used as the anchor. As you dig, undercut the trench toward the load.

   b. Tie a round turn with a bowline around the center of the anchor then stamp it into the trench ensuring that the rope runs down the T- slot and is fully stretched with no slack.

DEAD MAN ANCHOR

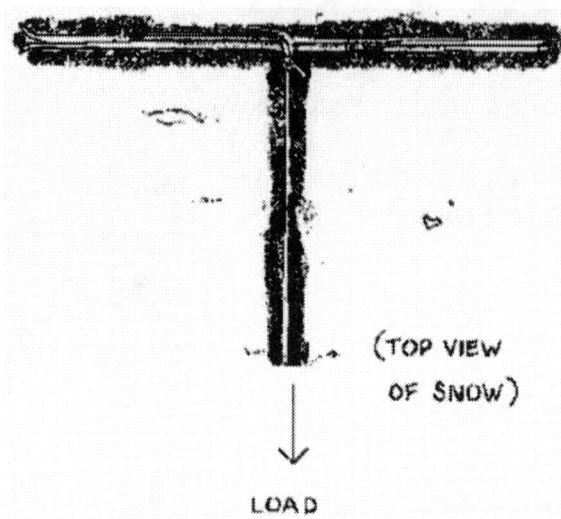

(TOP VIEW OF SNOW)

LOAD

c. Finally bury and stamp down everything except the rope, which will be used as the anchor point. You can strengthen this anchor by plunging another ski or two immediately in front of the anchor.

11. (5 MIN) **SNOW BOLLARD.** A solid bollard distributes the load around a fairly even curve without the anchor rope running over any high spots.

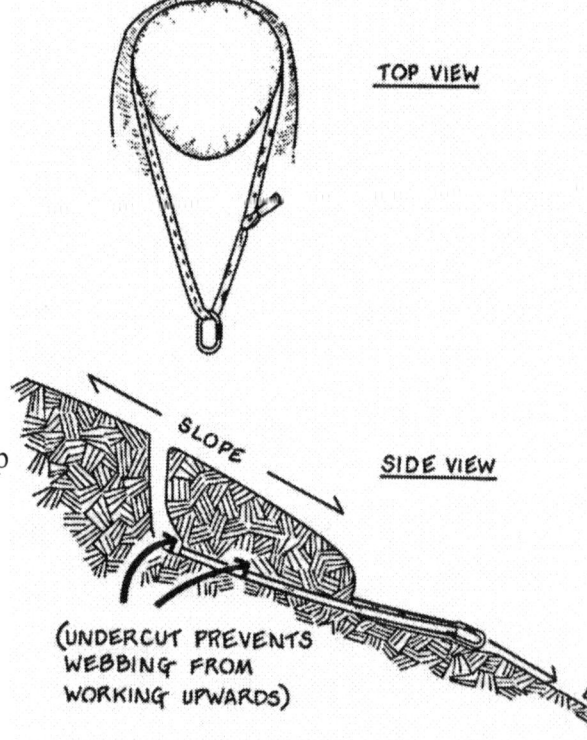

a. Construction of a Snow Bollard. A mound of snow is the basis of this type of anchor. These teardrop shaped anchors offer sturdy and reliable protection but are time consuming to construct.

(1) Look for a relatively high spot where the load will naturally pull downward, then scrape away any superficial snow that won't hold its shape.

(2) Create the mound by chopping the teardrop outline, gradually concentrating more on the three sides that will bear most of the weight.

(3) The minimum dimensions of a snow bollard are 4-10 feet wide, 12-15 feet long and 18 inches deep.

(4) As the groove gets deeper, small blows with the axe's pick or a shovel can define the undercut lip. **BOLLARD**

(5) Place the rope around the bollard to see if any protrusions might lift off and eliminate them.

(6) The rope is placed around the bollard only once and secured with a suitable anchor knot.

(7) An axe or shovel placed spike first or handle down in the back of the bollard can provide additional support.

12. (5 MIN) **CROSSING SNOW COVERED STREAM.**

   a. <u>Natural Snow Bridge</u>. Look for natural snow bridges. Usually the larger the better. Make sure that it is not severely undercut.

   (1) The first skier will be belayed as he inspects the snow bridge. First he will use a ski pole or probe pole to check snow depth and consistency. If good then the skier will step out onto the bridge, gently easing his weight onto it, one ski at a time.

   (2) One he is standing on it. Test it using the same methods as when testing a rutschblock. If solid then other skiers may cross unbelayed.

   b. <u>Artificial Snow Bridge</u>. If no natural snow bridge exists then one must be constructed.

   (1) Find an area that is shallow with slow moving water that is less than 10 feet wide. If the water is exceptionally deep, wide and/or fast moving, a snow bridge will not be possible.
   (2) Place several men on the shores with shovels. The men will throw snow into the stream until a suitable bridge has been constructed. Add tree branches to help give a base for the snow if necessary.
   (3) Belay the first man across to ensure it is safe. The shovel team must stay in place and continue to pile snow onto the bridge.

   c. <u>No Bridge</u>. If a snow bridge cannot be constructed the unit must cross the stream as taught in STREAM CROSSING or ONE ROPE BRIDGE (SUMMER MANUAL). Crossing ice covered water obstacles is covered in ICE RECONNAISSANCE AND BREACHING (WINTER MANUAL).

TRANSITION: Now that we have discussed crossing obstacles in winter environment are there any questions? If you have none for me, then I have some for you.

PRACTICE (CONC)

   a. Students will practice proper rope management.

PROVIDE HELP (CONC)

   a. The instructors will assist the students when necessary.

OPPORTUNITY FOR QUESTIONS (3 Min)

1. QUESTIONS FROM THE CLASS

2. QUESTIONS TO THE CLASS

    Q. What are the two types of rope coils and in what situation is each used?

    A. (1) Mountain Coil - used in movement from point to point when frequent use of the rope is anticipated.

    (2) Butterfly Coil - used when you have to carry the rope and require maximum freedom of movement of your arms.

SUMMARY (2 Min)

   a. During this period of instruction we have discussed the various considerations of rope management, including the mountaineering knots that you will be using.

   b. Those of you with IRF's please fill them out at this time and turn them in to the instructor. We will now take a short break and then you will be guided to your next class by one of your company instructors.

UNITED STATES MARINE CORPS
Mountain Warfare Training Center
Bridgeport, California 93517-5001

BIBLIOGRAPHY OF REFERENCES

Mountain Safety (Winter)

a. TC 90-6-1, Military Mountaineering
b. USMC Battle Skills Training Book
c. Mountaineering, The Freedom of the Hills, 6th Edition, The Mountaineers, Seattle WA 1996.
d. MCRP 3-35.1, Commander's Guide to Cold Weather Operations
e. MCRP 3-35.1A, Small Unit Leader's Guide to Cold Weather Operations
f. MCWP 3-35.2, Mountain Operations
g. The Royal Marine's Mountain and Arctic Warfare Handbook, 1972

Cold Weather Clothing and Personal Equipment

a. The Soldier's Load and the Mobility of a Nation, Col. S.L.A. Marshall, MCA Quantico, VA 1980.
b. MCRP 3-35.1, Commander's Guide to Cold Weather Equipment
c. MCRP 3-35.1A, Small Unit Leader's Guide to Cold Weather Operations
d. FM 31-70, Basic Cold Weather Manual.

Cold Weather Mountain Leadership Challenges

a. Cold Weather Operations Manual, U.S. Army Alaska, NWTC, December 1999

Mountain Health Awareness

a. Wilderness and Travel Medicine, Eric A. Weiss M.D., Adventure Medical Kit, Oakland, CA 1997.
b. Wilderness Medicine, 4th Edition, Wm. Forgey M.D., ICS Books Inc., Merillville, IN 1994.
c. Mountain Sickness, Peter H. Hackett M.D., The American Alpine Club, Golden, CO 1980.
d. J.A. Wilkerson, Medicine for Mountaineering, Third Edition.
e. MCRP 3-35.1, Commander's Guide to Cold Weather Operations.
f. MCRP 3-35.1A, Small Unit Leader's Guide to Cold Weather Operations.
g. M. J. Lentz, Mountaineering First Aid, Third Edition.

Winter Warfighting Load Requirements

a. The Soldier's Load and the Mobility of a Nation, Col. S.L.A. Marshall, MCA Quantico, VA 1980.
b. MCRP 3-35.1, Commander's Guide to Cold Weather Operations
c. MCRP 3-35.1A, Small Unit Leader's Guide to Cold Weather Operations
d. FM 31-70, Basic Cold Weather Manual.

Military Ski Equipment

a. Cold Weather Operations Manual, U.S. Army Northern Warfare Training Center, U.S. Army Alaska, December 1999
b. Swix Catalog, Swix Company, Lillehammer, Norway, 2000
c. MCRP 3-35.1B, Combat Skiing

Mountain Weather

a. FMFRP 3-29, U.S. Navy Oceanographic and Meteorological Support Syetem Manual
b. Jeppesen Sanderson, Private Pilots Manual, 1983, Jeppesen Sanderson, Inc. Englewood, CO
c. Lehr, Paul E; Burnett, R. Will; Zim, Herbert S. Weather-1975, Western Publishing Company, Inc., Racine, WI.

Mountain Casualty Evacuations (Winter)

a. Mountaineering, the Freedom of the Hills, 6th Edition, The Mountaineers, Seattle WA 1996.
b. Rope Rescue Manual, 2nd Edition, Frank and Smith, Santa Barbara, CA 1992.
c. TC 90-6-1, Military Mountaineering
d. The Royal Marine's Mountain and Arctic Warfare Handbook, 1972.
e. MCRP 3-35.1, Commander's Guide to Cold Weather Operations
f. MCRP 3-35.1A, Small Unit Leader's Guide to Cold Weather Operations

Avalanche

a. Mountaineering, the Freedom of the Hills, 6th Edition, The Mountaineers, Seattle WA 1996
b. MCRP 3-35.1, Commander's Guide to Cold Weather Operations
c. MCRP 3-35.1A, Small Unit Leader's Guide to Cold Weather Operations
d. Jill A. Fredston and Doug Fesler, Snow Sense; A Guide to Evaluating Snow Avalanche Hazard, 1988, Revised Third Edition.
e. U.S. Forest Service, Avalanche Handbook, July 1978
f. Mountain and Arctic Warfare Cadre Royal Marines, Manual of Military Mountaineering.
g. E.D. LaChapelle, The ABC's of Avalanche Safety, 1985, Second Edition
h. National Avalanche School Handbook, 1978.
i. David McClung and Peter Schaerer, Avalanche Handbook, The Mountaineers, Seattle WA 1993

Snow Stability Evaluation

a. Mountaineering, the Freedom of the Hills, 6th Edition, The Mountaineers, Seattle WA 1996
b. MCRP 3-35.1, Commander's Guide to Cold Weather Operations
c. MCRP 3-35.1A, Small Unit Leader's Guide to Cold Weather Operations
d. Jill A. Fredston and Doug Fesler, Snow Sense; A Guide to Evaluating Snow Avalanche Hazard, 1988, Revised Third Edition.
e. U.S. Forest Service, Avalanche Handbook, July 1978
f. Mountain and Arctic Warfare Cadre Royal Marines, Manual of Military Mountaineering.
g. E.D. LaChapelle, The ABC's of Avalanche Safety, 1985, Second Edition
h. National Avalanche School Handbook, 1978.
i. David McClung and Peter Schaerer, Avalanche Handbook, The Mountaineers, Seattle WA 1993

Avalanche Search Organization

a. Mountaineering, the Freedom of the Hills, 6th Edition, The Mountaineers, Seattle WA 1996
b. MCRP 3-35.1, Commander's Guide to Cold Weather Operations
c. MCRP 3-35.1A, Small Unit Leader's Guide to Cold Weather Operations
d. Jill A. Fredston and Doug Fesler, Snow Sense; A Guide to Evaluating Snow Avalanche Hazard, 1988, Revised Third Edition.
e. U.S. Forest Service, Avalanche Handbook, July 1978
f. Mountain and Arctic Warfare Cadre Royal Marines, Manual of Military Mountaineering.
g. E.D. LaChapelle, The ABC's of Avalanche safety, 1985, Second Edition
h. National Avalanche School Handbook, 1978.
i. Operations Handbook for Pieps, Ortovox and Ramer Avalanche Transceivers

Avalanche Transceivers

a. S.O.S. Transceiver Handbook

Military Snowshoes

a. MCRP 3-35.1, Commander's Guide to Cold Weather Operations
b. MCRP 3-35.1A, Small Unit Leader's Guide to Cold Weather Operations
c. FM 31-70, Basic Cold Weather Manual
d. Cold Weather Operations Manual, U.S. Army Alaska, NWTC, December 1999

Military Ski Movement

a. MCRP 3-35.1B, Over-the-Snow Mobility
b. Expeditionary Nordic Ski Instructor Manual
c. Skiing Right, Horst Abraham, Johnson Books, Boulder, Colorado, 1983

Skijoring

    a. UD 6-81-5E, A Guide to Cold Weather Operations, Book 5, Movement, Headquarters Defense Command Norway, U.S. Army Staff, 1987
    b. MCRP 3-35.1A, Small Unit Leader's Guide to Cold Weather Operations
    c. FM 31-70, Basic Cold Weather Manual
    d. MCRP 3-35.1B, Combat Skiing
    e. Cold Weather Operations Manual, U.S. Army Alaska, NWTC, December 1999

Marine Corps Cold Weather Infantry Kit

    a. The Soldier's Load and the Mobility of a Nation, Col. S.L.A. Marshall, MCA Quantico VA 1980.
    b. MCRP 3-35.1, Commander's Guide to Cold Weather Equipment
    c. MCRP 3-35.1A, Small Unit Leader's Guide to Cold Weather Operations
    d. FM 31-70, Basic Cold Weather Manual.
    e. Peak 1 Stove Instructional Manual.

Ten Man Tent & Yukon Stove

    a. Cold Weather Operations Manual, U.S. Army Alaska, NWTC, December 1999

Team Sled Movement

    a. Cold Weather Operations Manual, U.S. Army Alaska, NWTC, December 1999
    b. The Royal Marine's Mountain and Arctic Warfare Handbook, 1972.

Bivouac Routine

    a. Cold Weather Operations Manual, U.S. Army Alaska, NWTC, December 1999

Light and Noise Discipline in a Cold Weather Environment

    a. Cold Weather Operations Manual, U.S. Army Alaska, NWTC, December 1999

Camouflage, Cover and Concealment

    a. Cold Weather Operations Manual, U.S. Army Alaska, NWTC, December 1999

Defensive Positions and Field Fortifications

    a. Cold Weather Operations Manual, U.S. Army Alaska, NWTC, December 1999
    b. MCRP 3-35.1, Commander's Guide to Cold Weather Equipment
    c. MCRP 3-35.1A, Small Unit Leader's Guide to Cold Weather Operations

Route Planning in Cold Weather Operations

  a. Cold Weather Operations Manual, U.S. Army Alaska, NWTC, December 1999
  b. MCRP 3-35.1, Commander's Guide to Cold Weather Operations
  c. MCRP 3-35.1A, Small Unit Leader's Guide to Cold Weather Operations

Cold Weather Patrolling

  a. Cold Weather Operations Manual, U.S. Army Alaska, NWTC, December 1999
  b. MCRP 3-35.1, Commander's Guide to Cold Weather Equipment
  c. MCRP 3-35.1A, Small Unit Leader's Guide to Cold Weather Operations

Winter Tracking

  a. David Scott-Donelan, Tactical Tracking Operations
  b. Tom Brown, Field Guide to Nature Observation and Tracking

Ice Breaching and Reconnaissance

  a. UD 6-81-5E, A Guide to Cold Weather Operations, Headquarters Defense Command Norway, U.S. Army Staff, 1987
  b. Cold Weather Operations Manual, U.S. Army Alaska, NWTC, December 1999
  c. MCRP 3-35.1A, Small Unit Leader's Guide to Cold Weather Operations
  d. FMFM 7-21, Tactical Fundamentals for Cold Weather Warfighting
  e. MCRP 3-35.1, Commander's Guide to Cold Weather Operations
  f. The Royal Marine's Mountain and Arctic Warfare Handbook

Cold Weather Navigation

  a. Cold Weather Operations Manual, U.S. Army Alaska, NWTC, December 1999
  b. MCRP 3-35.1A, Small Unit Leader's Guide to Cold Weather Operations

Effects of Cold Weather on Infantry Weapons and Optics

  a. Cold Weather Operations Manual, U.S. Army Alaska, NWTC, December 1999
  b. MCRP 3-35.1, Commander's Guide to Cold Weather Equipment
  c. MCRP 3-35.1A, Small Unit Leader's Guide to Cold Weather Operations
  d. FM 31-72, Mountain Operations
  e. TC 90-6-1, Military Mountaineering

Communications Considerations in a Cold Weather Environment

  a. Cold Weather Operations Manual, U.S. Army Alaska, NWTC, December 1999
  b. MCRP 3-35.1, Commander's Guide to Cold Weather Equipment
  c. MCRP 3-35.1A, Small Unit Leader's Guide to Cold Weather Operations
  d. FMFRP 3-34, Field Antenna Handbook

Wave Propagation, Antenna Theory, and Field Expedient Antennas

    a. Cold Weather Operations Manual, U.S. Army Alaska, NWTC, December 1999
    b. MCRP 3-35.1, Commander's Guide to Cold Weather Equipment
    c. MCRP 3-35.1A, Small Unit Leader's Guide to Cold Weather Operations
    d. FMFRP 3-34, Field Antenna Handbook

Mountain Logistical Considerations

    a. FM 31-72, Mountain Operations
    b. MCRP 3-35.1, Commander's Guide to Cold Weather Operations
    c. MCRP 3-35.1A, Small Unit Leader's Guide to Cold Weather Operations

Cold Weather and Mountain Helicopter Operations

    a. MCRP 3-35.1, Commander's Guide to Cold Weather Operations
    b. Cold Weather Operations Manual, U.S. Army Alaska, NWTC, December 1999

Cold Weather Considerations for NBC Defense

    a. FM 3-3, Chemical and Biological Contamination Avoidance
    b. MCRP 3-35.1, Commander's Guide to Cold Weather Operations
    c. MCRP 3-35.1A, Small Unit Leader's Guide to Cold Weather Operations
    d. MCWP 3-37, MAGTF NBC Defensive Operations

Fire Support in a Cold Weather Environment

    a. MCWP 3-16, Techniques and Procedures for Fire Support Coordination
    b. MCWP 3-16.1, Marine Artillery Support
    c. MCWP 3-35.1, Commander's Guide to Cold Weather Operations
    d. MCRP 3-42.1, Fire Support in MAGTF Operations

Cold Weather and Mountain Operational Planning

    a. MCWP 3-11.1, Marine Rifle Company/Platoon (FMFM 6-4)
    b. MCWP 3-11.3, Scouting and Patrolling (FMFM 6-7 w/CH 1)
    c. Cold Weather Operations Manual, U.S. Army Alaska, NWTC, December 1999
    d. MCRP 3-35.1A, Small Unit Leader's Guide to Cold Weather Operations
    e. FM 3-31.1, Cold Weather Operations
    f. MCRP 3-35.1, Commander's Guide to Cold Weather Operations
    g. Notes from Leavenworth Papers, Number 5

Planning Cold Weather Offensive Operations

    a. MCWP 3-11.1, Marine Rifle Company/Platoon (FMFM 6-4)
    b. MCWP 3-11.3, Scouting and Patrolling (FMFM 6-7 w/CH 1)

    c. Cold Weather Operations Manual, U.S. Army Alaska, NWTC, December 1999
    d. MCRP 3-35.1A, Small Unit Leader's Guide to Cold Weather Operations
    e. FM 3-31.1, Cold Weather Operations
    f. MCRP 3-35.1, Commander's Guide to Cold Weather Operations
    g. Notes from Leavenworth Papers, Number 5

Planning Cold Weather Defensive Operations

    a. MCWP 3-11.1, Marine Rifle Company/Platoon (FMFM 6-4)
    b. MCWP 3-11.3, Scouting and Patrolling (FMFM 6-7 w/CH 1)
    c. Cold Weather Operations Manual, U.S. Army Alaska, NWTC, December 1999
    d. MCRP 3-35.1A, Small Unit Leader's Guide to Cold Weather Operations
    e. FM 3-31.1, Cold Weather Operations
    f. MCRP 3-35.1, Commander's Guide to Cold Weather Operations
    g. Notes from Leavenworth Papers, Number 5

Cold Weather Considerations for the 12 Patrol Steps

    a. Cold Weather Operations Manual, U.S. Army Alaska, NWTC, December 1999

Requirements for Survival

    a. MCRP 3-02F, Survival

Expedient Shelters and Fires

    a. MCRP 3-02F, Survival
    b. MCRP 3-02H, Survival, Evasion and Recovery

Survival Navigation

    a. MCRP 3-02F, Survival

Survival Signalling and Recovery

    a. MCRP 3-02F, Survival
    b. MCRP 3-02H, Survival, Evasion and Recovery

Survival Traps and Snares

    a. MCRP 3-02F, Survival

Water Procurement

    a. MCRP 3-02F, Survival

Survival Fishing

    a. MCRP 3-02F, Survival

Survival Uses for Game

    a. MCRP 3-02F, Survival

Field Expedient Tools, Weapons, and Equipment

    a. MCRP 3-02F, Survival

Expedient Snowshoes

    a. MCRP 3-02F, Survival

www.ingramcontent.com/pod-product-compliance
Lightning Source LLC
Chambersburg PA
CBHW081412230426
43668CB00016B/2212